Applied Multivariate Methods
for Data Analysts

Dallas E. Johnson
Kansas State University

Duxbury Press
An Imprint of Brooks/Cole Publishing Company

I(T)P® An International Thomson Publishing Company

Pacific Grove • Albany • Belmont • Bonn • Boston • Cincinnati • Detroit • Johannesburg • London
Madrid • Melbourne • Mexico City • New York • Paris • Singapore • Tokyo • Toronto • Washington

Sponsoring Editor: *Carolyn Crockett*
Marketing Team: *Marcy Perman, Margaret Parks*
Editorial Assistant: *Kimberly Raburn*
Production Editor: *Nancy Velthaus*
Manuscript Editor: *Lorretta Palagi*

Permissions Editor: *May Clark*
Cover Design: *Roy R. Neuhaus*
Typesetting: *Spectrum Publisher Services*
Cover Printing: *Courier*
Printing and Binding: *Courier*

For more information, contact:

BROOKS/COLE PUBLISHING
COMPANY
511 Forest Lodge Road
Pacific Grove, CA 93950
USA

International Thomson Publishing Europe
Berkshire House 168–173
High Holborn
London WC1V 7AA
England

Thomas Nelson Australia
102 Dodds Street
South Melbourne, 3205
Victoria, Australia

Nelson Canada
1120 Birchmount Road
Scarborough, Ontario
Canada M1K 5G4

International Thomson Editores
Seneca 53
Col. Polanco
11560 México, D. F., México

International Thomson Publishing GmbH
Königswinterer Strasse 418
53227 Bonn
Germany

International Thomson Publishing Asia
221 Henderson Road
#05–10 Henderson Building
Singapore 0315

International Thomson Publishing Japan
Hirakawacho Kyowa Building, 3F
2-2-1 Hirakawacho
Chiyoda-ku, Tokyo 102
Japan

Printed in the United States of America.

10 9 8 7 6

Library of Congress Cataloging-in-Publication Data

Johnson, Dallas E., [date]
    Applied multivariate methods for data analysts / Dallas E.
Johnson.
        p.   cm.
    Includes bibliographical references and index.
    ISBN 0-534-23796-7
    1. Multivariate analysis.   I. Title.
QA278.J615   1998
519.5′35—dc21   97-47133
                                                        CIP

Contents

Preface

I once attended a conference at which George Box stated that "Statistics is much too important to be left entirely to statisticians." A bit later, Walt Federer stated that "Science is much too important to be left entirely to scientists." Both of these famous statisticians were correct! Never before in the history of science and statistics has there been a greater need for interactions and collaborations between scientists and statisticians. This book helps to facilitate such collaborations and interactions. I have been fortunate in that I have had substantial contact with scientists during my tenure at Kansas State University. These collaborations have greatly influenced my approach to teaching multivariate methods. I believe that multivariate methods are too important to be taught only to statisticians.

Furthermore, I have been teaching public seminars and college courses in applied multivariate analyses for the last 20 years. In these seminars and courses, students have posed many important questions that multivariate methods can help answer. As data sets grow in size, multivariate methods become ever more useful. Today's technologies make it very easy to collect large amounts of data; multivariate methods are needed to determine whether such massive amounts of data actually contain information. It has been said that while it is easy to collect data, it is much harder to collect information. Multivariate methods can help determine whether there is information in data, and they can also help to summarize that information when it exists.

To date, textbooks have emphasized only the theory of multivariate methods or only the application of the methods. Readers were given information that was either too advanced to apply or too elementary to illustrate the power of the methods. This text has broken the mold by focusing on the why, when, how, and what of multivariate analyses and answering the following questions:

Why should multivariate methods be used?

When should they be used?

How can they be used?

And what has been learned by the application of the methods?

Ideally, users of this book will have had a previous course in statistics that included multiple regression. Some familiarity with matrix algebra is desirable, but not crucial. My approach assumes familiarity with most of the statistical concepts encountered in a first course in statistics, such as means and standard deviations, correlations, p-values, hypothesis tests, and confidence limits.

While this text is loaded with examples using real data, several of the exercises are directed at data sets that students are asked to provide from their own experiences. I find that students enjoy working with their own data. So, when I teach multivariate methods, I require each student to provide a data set for class use along with a description of the data's important features and the reasons behind its being collected. These data sets are then placed in a computer directory that every student in the class can access. I then use these data sets as much as possible when assigning exercises to the class. I strongly encourage instructors who use this book to do the same.

Other unique features of this text include:

- annotated computer output, emphasizing SAS and SPSS
- extensive use of graphics to explain concepts
- data disk that contains data files from text discussion and exercises, as well as computer commands used to create the analyses described throughout the text

I owe much of the development of this text to those who have participated in my seminars and courses. From these "students," I learned about their needs, their concerns, and their abilities. In writing this text, I have tried to address their needs and concerns, while recognizing their differing abilities.

Acknowledgments

I wish to express my appreciation to all who helped me with the development of this text. I am particularly grateful to the students at Kansas State University and students who have taken public seminars through the Institute of Professional Education. These students have provided numerous valuable suggestions that have greatly improved the content of this text. I would also like to thank Ms. Carolyn Crockett and Mr. Alexander Kugushev for their valuable suggestions. I would like to thank the following reviewers for their helpful comments: Marcia Gumpertz, North Carolina State University; John E. Hewett, University of Missouri, Columbia; Linda S. Hynan, Baylor University; Dipak Jain, Northwestern University; Lincoln Moses, Stanford University; Mack C. Shelley II, Iowa State University; Eric Smith, Virginia Polytech Institute; and Richard Sundheim, St. Cloud State University. I also thank Mrs. Jane Cox for her help in creating many of the formulas in this text. Finally, I would like to thank my parents, Chet and Dorothy Johnson, for giving me the opportunity for furthering my education and my wife, Erma, for the help and support that she provided during this endeavor.

Dallas Johnson

Applied Multivariate Methods

Multivariate data occur in all branches of science. Almost all data collected by today's researchers can be classified as multivariate data. For example, a marketing researcher might be interested in identifying characteristics of individuals that would enable the researcher to determine whether a certain individual is likely to purchase a specific product. Furthermore, a wheat breeder might be interested in more than just the yields of some new varieties of wheat. The wheat breeder may also be interested in these varieties' resistance to insect damage and drought. Finally, a social scientist might be interested in studying the relationships between teenage girls' dating behaviors and their fathers' attitudes. Each of these endeavors involves multivariate data.

To begin a discussion of multivariate data analysis methods, the concept of an experimental unit must be defined. An *experimental unit* is any object or item that can be measured or evaluated in some way. Measuring and evaluating experimental units is a principal activity of most researchers. Examples of experimental units include people, animals, insects, fields, plots of land, companies, trees, wheat kernels, and countries. *Multivariate data* result whenever a researcher measures or evaluates more than one attribute or characteristic of each experimental unit. These attributes or characteristics are usually called *variables* by statisticians.

The next section gives an overview of some multivariate methods that are discussed in this text.

1.1
An Overview of Multivariate Methods

Multivariate methods are extremely useful for helping researchers make sense of large, complicated, and complex data sets that consist of a lot of variables measured on large numbers of experimental units. The importance and usefulness of multivariate methods increase as the number of variables being measured and the number of experimental units being evaluated increase.

Often, the primary objective of multivariate analyses is to summarize large amounts of data by means of relatively few parameters. The underlying theme behind many multivariate techniques is simplification.

Multivariate analyses are often concerned with finding relationships among (1) the response variables, (2) the experimental units, and (3) both response variables and experimental units. One might say that relationships exist among the response variables when several of the variables really are measuring a common entity. For example, suppose one gives tests to third-graders in reading, spelling, arithmetic, and science. Individual students may tend to get high scores, medium scores, or low scores in all four areas. If this did happen, then these tests would be related to one another. In such a case, the common thing that these tests may be measuring might be "overall intelligence."

Relationships might exist between the experimental units if some of them are similar to each other. For example, suppose breakfast cereals are evaluated for their nutritional content. One might measure the grams of fat, protein, carbohydrates, and sodium in each cereal. Cereals would be related to each other if they tended to be similar with respect to the amounts of fat, protein, carbohydrates, and sodium that are in a single serving of each cereal. One might expect sweetened cereals to be related to each other and high-fiber cereals to be related to each other. One might also expect sweetened cereals to be much different than high-fiber cereals.

Many multivariate techniques tend to be exploratory in nature rather than confirmatory. That is, many multivariate methods tend to motivate hypotheses rather than test them. Consider a situation in which a researcher may have 50 variables measured on more than 2000 experimental units. Traditional statistical methods usually require that a researcher state some hypotheses, collect some data, and then use these data to either substantiate or repudiate the hypotheses. An alternative situation that often exists is a case in which a researcher has a large amount of data available and wonders whether there might be valuable information in these data. Multivariate techniques are often useful for exploring data in an attempt to learn if there is worthwhile and valuable information contained in these data.

Variable- and Individual-Directed Techniques

One fundamental distinction between multivariate methods is that some are classified as "variable-directed" techniques, while others are classified as "individual-directed" techniques.

Variable-directed techniques are those that are primarily concerned with relationships that might exist among the response variables being measured. Some examples of this type of technique are analyses performed on correlation matrices, principal components analysis, factor analysis, regression analysis, and canonical correlation analysis.

Individual-directed techniques are those that are primarily concerned with relationships that might exist among the experimental units and/or individuals being measured. Some examples of this type of technique are discriminant analysis, cluster analysis, and multivariate analysis of variance (MANOVA).

Creating New Variables

We quite often find it useful to create new variables for each experimental unit so they can be compared to each other more easily. Many multivariate methods help researchers create new variables that have desirable properties.

Some of the multivariate techniques that create new variables are principal components analysis, factor analysis, canonical correlation analysis, canonical discriminant analysis, and canonical variates analysis.

Some brief overviews of the multivariate techniques that are considered in this book are given next.

Principal Components Analysis

When a researcher is beginning to think about analyzing a new data set, several questions about the data should be considered. Important questions include these: (1) Are there any aspects of the data that are strange or unusual? (2) Can the data be assumed to be distributed normally? (3) Are there any abnormalities in the data? (4) Are there outliers in the data? Experimental units whose measured variables seem inconsistent with measurements on other experimental units are usually called *outliers*.

By far, the most important reason for performing a *principal components analysis* (PCA) is to use it as a tool for screening multivariate data. New variables, called *principal component scores*, can be created. These new variables can be used as input for graphing and plotting programs, and an examination of the resulting graphical displays will often reveal abnormalities in the data that you are planning to analyze. For example, plots of principal component scores can help identify outliers in the data when they exist. In addition, the principal component scores can be analyzed individually to see whether distributional assumptions such as normality of the variables and independence of the experimental units hold. Such assumptions are often required for certain kinds of statistical analyses of the data to be valid.

Principal components analysis uses a mathematical procedure that transforms a set of correlated response variables into a new set of uncorrelated variables that are called *principal components*. Principal components analysis can be performed on either a sample variance–covariance matrix or a correlation matrix. The type of matrix that is best often depends on the variables being measured. Occasionally, but not often, the newly created variables are interpretable. One cannot always expect to be able to interpret the newly created variables. In fact, it is considered to be a bonus when the principal component variables can actually be interpreted. When using PCA to screen a multivariate data set, you do not need to be able to interpret the principal components because PCA is extremely useful regardless of whether the new variables can be interpreted.

Principal components analysis is usually quite helpful to researchers who want to partition experimental units into subgroups so that similar experimental units belong to the same subgroup. In this case, principal component scores can be used as input to clustering programs. This often increases the effectiveness of the clustering programs, while reducing the cost of using such programs. Furthermore, the principal component scores can and should always be used to help validate the results of clustering programs.

Factor Analysis

Factor analysis (FA) is a technique that is often used to create new variables that summarize all of the information that might be available in the original variables. For example, consider once again giving tests to third-graders in reading, spelling, arithmetic, and science, whereby individual students may tend to get high scores, medium scores, or low scores in all four areas. If this really does happen, then one might say that these test results are being explained by some underlying characteristic or factor that is common to all four tests. In this example, it might be reasonable to assume that such an underlying characteristic is "overall intelligence."

Factor analysis is also used to study relationships that might exist among the measured variables in a data set. Similar to PCA, FA is a variable-directed technique. One basic objective of FA is to determine whether the response variables exhibit patterns of relationships with each other, such that the variables can be partitioned into subsets of variables so that the variables in a subset are highly correlated with each other and so that variables in different subsets have low correlations with each other. Thus, FA is often used to study the correlational structure of the variables in a data set. One similarity between FA and PCA is that FA can also be used to create new variables that are uncorrelated with each other. Such variables are called *factor scores.*

One advantage that FA seems to have over PCA when new variables are being created is that the new variables created by FA are generally much easier to interpret than those created by PCA. If a researcher wants to create a smaller set of new variables that are interpretable and that summarize most of the information in the measured variables then FA should be given serious consideration.

Discriminant Analysis

A company specializing in credit cards would certainly like to be able to classify credit card applicants into two groups of individuals: (1) individuals who are good credit risks and (2) individuals who are bad credit risks. Individuals deemed to be good credit risks would then be offered credit cards, while those deemed to be bad risks would not. To help make this determination, the credit card company might consider several demographic characteristics that can be measured on each individual. For example, the company may consider educational level, salary, indebtedness, and past credit history as possible predictors of creditworthiness. The company would then attempt to use this information to help to decide whether an applicant for a credit card should be approved. The multivariate method that would help the company classify applicants into one of the two credit risk groups is called discriminant analysis.

Discriminant analysis (DA) is primarily used to classify individuals or experimental units into two or more uniquely defined populations. To develop a discriminant rule for classifying experimental units into one of several possible categories, the researcher must have a random sample of experimental units from each possible classification group. Then, DA provides methods that will allow researchers to build rules that can be used to classify other experimental units into one of the classification groups.

In the credit card example, a discriminant rule is constructed using demographic data from individuals known to be good credit risks and similar data from individuals known to be bad credit risks. Then new applicants for credit cards are classified into one of the two risk groups using the resulting rule.

Canonical Discriminant Analysis

Canonical discriminant analysis (CDA) is a procedure that creates new variables that contain all of the useful information for discrimination that is available in the original variables. These new variables often lead to simpler rules for actually classifying experimental units into the different classification groups.

Logistic Regression

Logistic regression is often used to model the probability that an experimental unit falls into a particular group based on information measured on the experimental unit. Such models can be used for discrimination purposes. In the credit card example described previously, one can model the probability that an individual with certain demographic characteristics will be a good credit risk. After developing this model, it can be used to predict the probability that a new applicant will fall into a certain risk group. Individuals whose predicted probability for the "good risk" group is greater than 0.5 are determined to be good credit risks.

Cluster Analysis

Suppose that an archaeologist discovers a large cache of pottery fragments and takes small drill samples from each fragment. Suppose that each drill sample can be analyzed, and the relative amounts of different chemical elements, such as zinc, magnesium, iron, and so on, can be ascertained. The archaeologist wants to separate the fragments into distinct piles so that all of the fragments within a pile come from the same piece of pottery and those in different piles come from different pieces of pottery. Clearly, this may be a difficult task because the archaeologist does not know how many piles of fragments there should be, how many fragments should be in each pile, or whether there are any fragments that actually belong in the same pile. Cluster analysis is a multivariate method that can help solve this problem.

Cluster analysis (CA) is similar to discriminant analysis in that it is used to classify individuals or experimental units into uniquely defined subgroups.

Discriminant analysis can be used when a researcher has previously obtained random samples from each of the uniquely defined subgroups. Cluster analysis deals with classification problems when it is not known beforehand from which subgroups observations originate.

Multivariate Analysis of Variance

Multivariate analysis of variance (MANOVA) is a multivariate generalization of (univariate) *analysis of variance* (ANOVA), which is a technique used to compare the means of several populations on a single measured variable. When several variables are measured on each experimental unit, you could produce an ANOVA on each measured variable using one variable at a time. For example, if 25 variables are measured, a researcher could produce 25 separate analyses, one for each variable. However, this is not wise. Unfortunately, a majority of the experiments are being analyzed using one-variable-at-a-time analyses.

Statisticians raise two main objectives to individual analyses for each measured variable. One objection is that the populations being compared may be different on some variables but not on others. Often a researcher finds it confusing as to which populations are really different and which are similar. Multivariate analysis of variance can help researchers to compare several populations by considering all of the measured variables, simultaneously.

A second objection is that there is inadequate protection against making Type I errors when performing one-variable-at-a-time analyses. Recall from your introductory statistics course that a Type I error occurs whenever a true hypothesis is rejected. The more variables that a researcher analyzes, the more likely it is that at least one of the variables will give rise to statistical significance. As the number of variables being analyzed increases, the probability of finding at least one of these analyses statistically significant (i.e., producing a p value of less than 0.05) approaches one.

Certainly, the large risk of making Type I errors should be a concern for experimenters. A researcher should want to be confident when claiming that two or more populations have different means with respect to a measured variable and that his or her claim would not be contradicted by other experimenters conducting similar analyses on similar data sets.

A MANOVA should be performed whenever two or more different populations are being compared to one another on a large number of measured response variables.

If a MANOVA shows significant differences between population means, then the researcher can be confident that real differences actually exist. In this case, it is reasonable to consider one-variable-at-a-time analyses to see where the differences actually occur.

If the MANOVA does not reveal any significant differences between population means, then the researcher must use extreme caution when interpreting one-variable-at-a-time analyses. Such analyses may be identifying nothing more than "false positives."

Canonical Variates Analysis

Canonical variates analysis (CVA) is a method that creates new variables in conjunction with multivariate analyses of variance. These new variables are useful for helping researchers determine where the major differences among the population means occur when the populations are being compared on many different variables by using all of the measured variables simultaneously. Occassionally, the canonical variates may suggest important differences that might otherwise be missed.

Canonical Correlation Analysis

A researcher wanted to compare fathers' attitudes with their daughters' dating behaviors. When several different variables have been measured on the fathers and several others measured on the daughters, canonical correlation analysis might be used to identify new variables that summarize any relationships that might exist between these two sets of family members.

Canonical correlation analysis is a generalization of multiple correlation in regression problems. It requires that the response variables be divided into two groups. The assignment of variables into these two groups must always be motivated by the nature of the response variables and never by an inspection of the data. For example, a legitimate assignment would be one in which the variables in one group are easy to obtain and inexpensive to measure, while the variables in the other group are hard to obtain or expensive to measure. Another would be measurements on fathers versus measurements on their daughters.

One basic question that canonical correlation analysis is expected to answer is whether the variables in one group can be used to predict the variables in the other group. When they can, then canonical correlation analysis attempts to summarize the relationships between the two sets of variables by creating new variables from each of the two groups of original variables.

Where to Find the Preceding Topics

Principal components analysis is discussed in Chapter 5, factor analysis is discussed in Chapter 6, discriminant analysis and canonical discriminant analysis are discussed in Chapter 7, logistic regression is discussed in Chapter 8, cluster analysis is discussed in Chapter 9, multivariate analysis of variance and canonical variates analysis are discussed in Chapter 11, and canonical correlation analysis is discussed in Chapter 12. Multivariate generalizations of univariate t-tests and confidence interval procedures learned in a first course in statistics are discussed in Chapter 10. Chapter 10 also contains a procedure for generalizing univariate quality control techniques to a set of multivariate responses.

Table 1.1 gives a cross-listing of multivariate data analysis problem types and the various multivariate techniques described previously. For example, if you wanted to explore relationships that might exist among the measured variables in a data set, you can see that factor analysis is definitely a useful multivariate technique for that purpose. The table also suggests that principal components analysis and canonical correlation analysis can sometimes help you explore relationships among the measured variables. However, DA and MANOVA would not be useful.

1.2
Two Examples

Table 1.2 contains body measurements on 33 African-American female police department applicants. The data are taken from Gunst and Mason (1980, p. 367). The variables measured in centimeters were height (HEIGHT), sitting height (SITHT), upper arm length (UARM), forearm length (FORE), hand

T A B L E 1.1

Cross-listing of multivariate methods and problem types

	Multivariate technique							
Problem type	PCA	FA	DA	CDA	CA	MANOVA	CVA	CCA
Exploring relationships among variables	Sometimes	Definitely	Never	Never	Never	Never	Rarely	Sometimes
Screening data	Definitely	Sometimes	Never	Never	Sometimes	Never	Never	Never
Creating new variables	Does	Does	Does not	Does	Does not	Does not	Does	Does
Predicting group membership	Does not	Does not	Does	Does	Does	Does not	Does not	Does not
Comparing group means	Possibly	Possibly	Rarely	Rarely	Does not	Does	Does	Does not
Comparing groups of variables	Possibly	Possibly	Never	Never	Never	Never	Never	Definitely
Verifying clusters	Definitely	Possibly	Never	Never	Definitely	Never	Never	Never
Reducing dimensionality	Definitely	Definitely	Never	Definitely	Never	Never	Definitely	Definitely
Creating meaningful variables	Unlikely	Usually	Never	Possibly	Never	Never	Possibly	Unlikely

T A B L E 1.2

Body measurements on police department applicants

ID	HEIGHT	SITHT	UARM	FORE	HAND	ULEG	LLEG	FOOT	BRACH	TIBIO
1	165.8	88.7	31.8	28.1	18.7	40.3	38.9	6.7	88.36	96.53
2	169.8	90.0	32.4	29.1	18.3	43.3	42.7	6.4	89.81	98.61
3	170.7	87.7	33.6	29.5	20.7	43.7	41.1	7.2	87.80	94.05
4	170.9	87.1	31.0	28.2	18.6	43.7	40.6	6.7	90.97	92.91
5	157.5	81.3	32.1	27.3	17.5	38.1	39.6	6.6	85.05	103.94
6	165.9	88.2	31.8	29.0	18.6	42.0	40.6	6.5	91.19	96.67
7	158.7	86.1	30.6	27.8	18.4	40.0	37.0	5.9	90.85	92.50
8	166.0	88.7	30.2	26.9	17.5	41.6	39.0	5.9	89.07	93.75
9	158.7	83.7	31.1	27.1	18.3	38.9	37.5	6.1	87.14	96.40
10	161.5	81.2	32.3	27.8	19.1	42.8	40.1	6.2	86.07	93.69
11	167.3	88.6	34.8	27.3	18.3	43.1	41.8	7.3	78.45	96.98
12	167.4	83.2	34.3	30.1	19.2	43.4	42.2	6.8	87.76	97.24
13	159.2	81.5	31.0	27.3	17.5	39.8	39.6	4.9	88.06	99.50
14	170.0	87.9	34.2	30.9	19.4	43.1	43.7	6.3	90.35	101.39
15	166.3	88.3	30.6	28.8	18.3	41.8	41.0	5.9	94.12	98.09
16	169.0	85.6	32.6	28.8	19.1	42.7	42.0	6.0	88.34	98.36
17	156.2	81.6	31.0	25.6	17.0	44.2	39.0	5.1	82.58	88.24
18	159.6	86.6	32.7	25.4	17.7	42.0	37.5	5.0	77.68	89.29
19	155.0	82.0	30.3	26.6	17.3	37.9	36.1	5.2	87.79	95.25
20	161.1	84.1	29.5	26.6	17.8	38.6	38.2	5.9	90.17	98.96
21	170.3	88.1	34.0	29.3	18.2	43.2	41.4	5.9	86.18	95.83
22	167.8	83.9	32.5	28.6	20.2	43.3	42.9	7.2	88.00	99.08
23	163.1	88.1	31.7	26.9	18.1	40.1	39.0	5.9	84.86	97.26
24	165.8	87.0	33.2	26.3	19.5	43.2	40.7	5.9	79.22	94.21
25	175.4	89.6	35.2	30.1	19.1	45.1	44.5	6.3	85.51	98.67
26	159.8	85.6	31.5	27.1	19.2	42.3	39.0	5.7	86.03	92.20
27	166.0	84.9	30.5	28.1	17.8	41.2	43.0	6.1	92.13	104.37
28	161.2	84.1	32.8	29.2	18.4	42.6	41.1	5.9	89.02	96.48
29	160.4	84.3	30.5	27.8	16.8	41.0	39.8	6.0	91.15	97.07
30	164.3	85.0	35.0	27.8	19.0	47.2	42.4	5.0	79.43	89.83
31	165.5	82.6	36.2	28.6	20.2	45.0	42.3	5.6	79.01	94.00
32	167.2	85.0	33.6	27.1	19.8	46.0	41.6	5.6	80.65	90.43
33	167.2	83.4	33.5	29.7	19.4	45.2	44.0	5.2	88.66	97.35

width (HAND), upper leg length (ULEG), lower leg length (LLEG), and foot length (FOOT). From these data two additional variables were created— the ratio of forearm length to upper arm length times 100 (BRACH) and the ratio of lower leg length to upper leg length times 100 (TIBIO). Thus, BRACH = 100·FORE/UARM and TIBIO = 100·LLEG/ULEG.

Table 1.3 contains data concerning rubber consumption and other variables from 1948 to 1963. The data are from Draper and Smith (1981, p. 410). The variables measured were total rubber consumption, tire rubber consumption, car production, gross national product, disposable personal income, and motor fuel consumption.

T A B L E 1.3
Rubber consumption data

Year	Total rubber consumption	Tire rubber consumption	Car production	Gross national product	Disposable personal income	Motor fuel consumption
1948	0.909	0.871	1.287	0.984	0.987	1.046
1949	1.252	1.220	1.281	1.078	1.064	1.081
1950	0.947	0.975	0.787	1.061	1.007	1.051
1951	1.022	1.021	0.796	1.013	1.012	1.046
1952	1.044	1.002	1.392	1.028	1.029	1.036
1953	0.905	0.890	0.893	0.969	0.993	1.020
1954	1.219	1.213	1.400	1.057	1.047	1.057
1955	0.923	0.918	0.721	1.001	1.024	1.034
1956	1.001	1.014	1.032	0.996	1.003	1.014
1957	0.916	0.914	0.685	0.972	0.993	1.013
1958	1.173	1.170	1.291	1.046	1.027	1.037
1959	0.938	0.952	1.170	1.004	1.001	1.007
1960	0.965	0.946	0.817	1.002	1.014	1.008
1961	1.106	1.096	1.231	1.049	1.032	1.024
1962	1.011	0.999	1.086	1.023	1.020	1.030
1963	1.080	1.093	1.001	1.035	1.053	1.029

The data illustrated in Tables 1.2 and 1.3 are similar in that both sets contain rows and columns of numbers. In both tables, each row represents a different experimental unit and each column represents a different measured response variable. In Table 1.2, the different applicants define the experimental units; in Table 1.3, the experimental units are defined by the different years.

Independence of Experimental Units

While the data sets in Tables 1.2 and 1.3 have similar forms, there is one very important feature that differentiates them. One condition that must be satisfied by almost all multivariate methods is that the variables measured on any given experimental unit must be independent of similar variables measured on any other experimental unit. In other words, the observed values of the measured variables on one experimental unit must not have any influence on the observed values of the measured variables on any other experimental unit.

This condition of independence among experimental units is certainly true for the data in Table 1.2 where the measurements made on one applicant do not and will not affect the measurements made on any other applicant. However, this independence condition may not be true for the data in Table 1.3. The data in Table 1.3 were collected over time, and it might seem reasonable that what happens during some years may have some effect on what happens in succeeding years. If this were true, then the data in Table 1.3 would not satisfy the condition of independence among the experimental units and, as a consequence, many of the multivariate techniques discussed in this book cannot be appropriately applied to such data.

It is important to note that multivariate statistical computing packages do not really care whether the rows of a data set are independently distributed or not. All computer programs care about is whether the data are read into the computer as rows and columns. If the data contains rows and columns, then the computing package will do whatever analysis it is asked to do. The data analyst must be wise enough to know whether the analysis being obtained from a computer program is appropriate and whether all the necessary conditions for that analysis are satisfied.

1.3
Types of Variables

Each of the measured variables appearing in Tables 1.2 and 1.3 is an example of a continuous response variable. Continuous variables are numeric and could feasibly occur anywhere within some interval; the only thing that limits a continuous variable's value is the ability to measure the variable accurately. Variables that are not continuous are called discrete variables. Discrete variables can be either numeric or nonnumeric.

Examples of nonnumeric discrete variables include gender of a person or animal, race or country of citizenship of a person, variety or type of a field crop, and species of an animal or insect. These are called *nonnumeric* because there is no way to quantify or order the values of the variables. For example, there is no way to order—from smallest to largest—races of people, varieties of wheat, or species of insects.

Discrete variables can also be numerically ordered in some instances. Examples of numeric discrete variables would include the number of times a person has been married, the number of children in a family unit, and the number of insects on a plant.

Quite often when survey data are collected, individuals are asked questions such as "How well do you like this product?" Possible answers may include these: (1) I like it very much, (2) I like it, (3) I am indifferent, (4) I dislike it, and (5) I dislike it very much. Although such responses are obviously nonnumeric, they can be quantified and/or ordered. For example, one might assign a 5 to response 1, a 4 to response 2, a 3 to response 3, and so on. This creates a set of numerically ordered discrete variables. When the response categories are believed to be equally spaced, the multivariate methods described in this book should work reasonably well.

Multivariate data can be discrete as well as continuous, although traditional multivariate methods deal with continuous data rather than discrete data. This is because most of the popular and useful multivariate methods originate from an assumption that the multivariate data come from a multivariate normal probability distribution. This would imply that the measured variables need to be continuous. Numeric discrete variables cannot be multivariate normally distributed because they are not continuous.

Fortunately, most multivariate methods are quite robust with respect to nonnormal data. This means that methods based on the multivariate normal distribution work well even when the data do not come from multivariate normal populations. Multivariate methods tend to work quite well when the values of each of the response variables can be quantitatively ordered in some way.

1.4
Data Matrices and Vectors

In this section we define notational conventions that are used throughout the book.

Mathematical notation is used in this book for several important reasons: (1) accuracy, (2) efficiency, (3) simplicity, and (4) interpretability. Although this book attempts to keep mathematical notation to a minimum, it does not ignore it altogether. If a researcher is planning to use multivariate methods and accurately interpret computer output, he or she must become familiar with mathematical notation similar to that used in this book.

Variable Notation

In this book, p will always be used to represent the number of numeric response variables being measured and N will always represent the number of experimental units on which variables are being measured. The variable x_{rj} is used to identify the value of the jth response variable on the rth experimental unit for $r = 1, 2, \ldots, N$ and $j = 1, 2, \ldots, p$.

Data Matrix

The x_{rj}'s can be arranged in a matrix, called the *data matrix*, so that x_{rj} is the element in the rth row and the jth column of the matrix. The data matrix is an $N \times p$ matrix and is denoted by $\mathbf{X}$. Thus

$$\mathbf{X} = \begin{bmatrix} x_{11} & x_{12} & \cdots & x_{1p} \\ x_{21} & x_{22} & \cdots & x_{2p} \\ \vdots & \vdots & \cdots & \vdots \\ x_{N1} & x_{N2} & \cdots & x_{Np} \end{bmatrix}$$

is a data matrix.

To help clarify these definitions, note that in Table 1.2, $N = 33$, $p = 10$, $x_{23} = 32.4$, and $x_{94} = 27.1$.

Caution   Some authors reverse the order of subscripts in the above definition. Always be sure to check an author's definition of a data matrix whenever referring to another book or journal article.

In this book rows always correspond to experimental units and columns always correspond to measured response variables. This is consistent with the way most statistical computing packages expect data to be entered for analysis purposes.

Data Vectors

The rows in a data matrix are called *row vectors*. The data that occurs in the rth row of $\mathbf{X}$ is denoted by $\mathbf{x}'_r$. Thus,

$$\mathbf{x}'_r = [x_{r1}\, x_{r2} \cdots x_{rp}]$$

When the data in the rth row of $\mathbf{X}$ are written in a column vector it is denoted by $\mathbf{x}_r$. In this case

$$\mathbf{x}_r = \begin{bmatrix} x_{r1} \\ x_{r2} \\ \vdots \\ x_{rp} \end{bmatrix}$$

Note   The row vector $\mathbf{x}'_r$ is called the *transpose* of the column vector $\mathbf{x}_r$. Also the data matrix is given by

$$\mathbf{X} = \begin{bmatrix} \mathbf{x}'_1 \\ \mathbf{x}'_2 \\ \vdots \\ \mathbf{x}'_N \end{bmatrix}$$

From Table 1.2, the data vector from the 32nd police department applicant is (in row format)

$$\mathbf{x}'_{32} = [167.2\ \ 85.0\ \ 33.6\ \ 27.1\ \ 19.8\ \ 46.0\ \ 41.6\ \ 5.6\ \ 80.65\ \ 90.43]$$

or (in column format)

$$\mathbf{x}_{32} = \begin{bmatrix} 167.2 \\ 85.0 \\ 33.6 \\ 27.1 \\ 19.8 \\ 46.0 \\ 41.6 \\ 5.6 \\ 80.65 \\ 90.43 \end{bmatrix}$$

Data Subscripts

The choice of r and j as the two subscripts in x_{rj} is deliberate. In this book the letters $i, j, k, \ldots$ are used as subscripts for response variables, while $r, s, t, \ldots$ are used as subscripts for experimental units. This convention helps users of multivariate methods interpret statistical results.

For example, ρ_{ij} would denote a relationship between the ith and jth response variables (perhaps the correlation between them) since both subscripts occur in the $(i, j, k, \ldots)$ range of the alphabet, while d_{rs} would denote a relationship between the rth and sth experimental units (perhaps the distance between them) since both subscripts occur in the $(r, s, t, \ldots)$ range of the alphabet. On the other hand, f_{rj} might denote a jth new variable (perhaps created by means of a factor analysis) on an rth experimental unit since r and j occur in the two different ranges of the alphabet.

1.5
The Multivariate Normal Distribution

Most traditional multivariate methods depend on data vectors being random samples from multivariate normal distributions. As previously mentioned, multivariate techniques based on the normal distribution are robust and work quite well when data vectors are not multivariate normally distributed, provided that data vectors still have independent probability distributions.

It is somewhat important to understand what is required for a vector of random variables, such as,

$$\mathbf{x} = \begin{bmatrix} x_1 \\ x_2 \\ \vdots \\ x_p \end{bmatrix}$$

to be multivariate normally distributed.

Some Definitions

D E F I N I T I O N 1.1

A vector of random variables,

$$\mathbf{x} = \begin{bmatrix} x_1 \\ x_2 \\ \vdots \\ x_p \end{bmatrix}$$

is said to have a multivariate normal distribution if

$$\mathbf{a'x} = [a_1 \, a_2 \ldots a_p] \begin{bmatrix} x_1 \\ x_2 \\ \vdots \\ x_p \end{bmatrix}$$

$$= a_1 x_1 + a_2 x_2 + \cdots + a_p x_p$$

$$= \sum_{i=1}^{p} a_i x_i$$

has a univariate normal distribution for every possible set of selected values for the elements in the vector $\mathbf{a}$.

A multivariate normal distribution can be defined in many other ways, and the preceding definition may be unique to this book. One advantage that this definition has over most other definitions is that you do not have to be a mathematical statistician to understand the definition because most readers have had previous experience in statistics, including previous knowledge about the univariate normal distribution.

One consequence of the definition just given is that every element of the vector **x** must have a univariate normal distribution. That is, a frequency distribution of the values of each and every response variable, x_i, must follow a bell-shaped curve. One might expect that if each x_i were normally distributed, then the vector **x** would have a multivariate normal distribution. While this may be close to being true for real-life data sets, it cannot be proved mathematically. In fact, mathematical counterexamples exist. See Hogg and Craig (1978, p. 114, Ex. 3.55) for such an example.

Summarizing Multivariate Distributions

In univariate situations, the probability distribution of a random variable x is often summarized by its first two moments, namely, its mean and its variance (or, equivalently, its mean and standard deviation).

The mean of a random variable x is usually denoted by μ and is defined by $\mu = E(x)$ where $E(\cdot)$ denotes expected value. Expectation is an averaging process, and the expected value of x can be conceptualized as taking the average value of x in the population being sampled. The reason for using expectation rather than average in this definition is because we are usually sampling from populations that, at least theoretically, are infinite in size and computing an arithmetic average would be impossible.

The variance of a random variable x is usually denoted by σ^2 and is defined by $\sigma^2 = E[(x - \mu)^2]$. Thus, the variance of x is, conceptually, the average value of $(x - \mu)^2$ in the population being sampled. The square root of the variance of x is called the *standard deviation* of x and is usually denoted by σ. Thus, $\sigma = \sqrt{\sigma^2}$.

To summarize multivariate distributions, we need the mean and variance of each of the p variables in **x**. Additionally, we need either the correlations between all pairs of variables in **x** or the covariances between all pairs of variables. If we are given the variances and covariances, then the correlations can be determined. Likewise, if we have variances and correlations, then one can determine the covariances.

Mean Vectors and Variance–Covariance Matrices

DEFINITION 1.2

The mean of a vector of random variables **x** is denoted by μ and the covariance matrix of **x** is denoted by Σ. These are defined by

$$\boldsymbol{\mu} = E(\mathbf{x}) = \begin{bmatrix} E(x_1) \\ E(x_2) \\ \vdots \\ E(x_p) \end{bmatrix} = \begin{bmatrix} \mu_1 \\ \mu_2 \\ \vdots \\ \mu_p \end{bmatrix} \qquad \text{and}$$

$$\boldsymbol{\Sigma} = \text{Cov}(\mathbf{x}) = E[(\mathbf{x} - \boldsymbol{\mu})(\mathbf{x} - \boldsymbol{\mu})']$$

$$= \begin{bmatrix} \sigma_{11} & \sigma_{12} & \cdots & \sigma_{1p} \\ \sigma_{21} & \sigma_{22} & \cdots & \sigma_{2p} \\ \vdots & \vdots & & \vdots \\ \sigma_{p1} & \sigma_{p2} & \cdots & \sigma_{pp} \end{bmatrix} \qquad \text{where}$$

$$\sigma_{ii} = \text{Var}(x_i) = E[(x_i - \mu_i)^2], \qquad \text{for } i = 1, 2, \ldots, p, \qquad \text{and}$$

$$\sigma_{ij} = \text{Cov}(x_i, x_j)$$
$$= E[(x_i - \mu_i)(x_j - \mu_j)], \qquad \text{for } i \neq j = 1, 2, \ldots, p$$

Correlations and Correlation Matrices

D E F I N I T I O N 1.3

The correlation coefficient between x_i and x_j is denoted by ρ_{ij} and is defined by

$$\rho_{ij} = \frac{\sigma_{ij}}{\sqrt{\sigma_{ii}\sigma_{jj}}}$$

Caution The correlation coefficient provides a measure of the *linear* association between two variables. This is very important to understand because the word *correlation* seems to be misused by the general public. It seems that many people say that things are *correlated* when they believe that they are *related*, and that things are *uncorrelated* when they believe that they are *not related*.

Being uncorrelated is equivalent to being independent when two variables have a joint normal distribution, but this is not true in general. Figure 1.1 shows a scatter plot of a random sample from a two-variate (also called *bivariate*) population for which the response variables are denoted by x_1 and x_2. In Figure 1.1 there appears to be a relationship between x_1 and x_2. In fact, x_2 was selected so that it is approximately equal to $x_1^2 - 2x_1 + 3$. However, the correlation between x_2 and x_1 is nearly zero. Consequently, one can see that when pairs of variables are uncorrelated, one cannot safely conclude that the two variables are unrelated.

Note You should never calculate a correlation coefficient between two variables without also plotting the two variables against one another. If it is important to determine the correlation between two variables, then it is equally

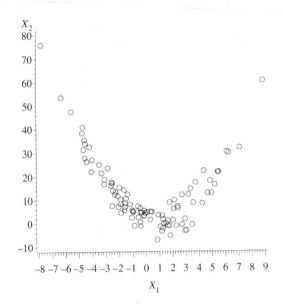

FIG. 1.1

Scatter plot of
two variables that
are uncorrelated
but related

important to construct a scatter plot of the two variables. The scatter plot will usually reveal whether the two variables are actually related to one another; and when they are related to one another, the scatter plot will often show how they might be related to one another.

In general, if two variables are independently distributed, then $\rho_{ij} = 0$ and $\sigma_{ij} = 0$. As illustrated in Figure 1.1, the converse of this statement is not always true. However, the converse is true whenever the two variables have a joint bivariate normal distribution.

We can show that $-1 \leq \rho_{ij} \leq 1$ for every $i \neq j$.

D E F I N I T I O N   1.4

The correlation matrix for a random vector **x** is denoted by **P** and is defined by

$$\mathbf{P} = \begin{bmatrix} 1 & \rho_{12} & \cdots & \rho_{1p} \\ \rho_{21} & 1 & \cdots & \rho_{2p} \\ \vdots & \vdots & \ddots & \vdots \\ \rho_{p1} & \rho_{p2} & \cdots & 1 \end{bmatrix}$$

The Multivariate Normal Probability Density Function

Suppose $\mathbf{x}$ is a p-variate random vector that has a multivariate normal distribution with mean vector $\boldsymbol{\mu}$ and variance–covariance matrix $\boldsymbol{\Sigma}$. This is often denoted by the notation:

$$\mathbf{x} \sim \mathrm{N}_p(\boldsymbol{\mu}, \boldsymbol{\Sigma})$$

The rest of this subsection can easily be ignored by most readers of this text. The theoretical description of a multivariate normal distribution is being given primarily for those who might find it interesting and useful.

In addition to being symmetric matrices, both $\boldsymbol{\Sigma}$ and $\mathbf{P}$ are nonnegative matrices and they have the same rank. (See Appendix A.1 for definitions of a nonnegative matrix and the rank of a matrix.) The rank of $\boldsymbol{\Sigma}$ is also called the rank of the probability distribution of $\mathbf{x}$. If the rank of $\boldsymbol{\Sigma}$ is equal to p, a probability density function for $\mathbf{x}$ exists. In this case the probability density function of $\mathbf{x}$ is given by

$$f_x(\mathbf{x}; \boldsymbol{\mu}, \boldsymbol{\Sigma}) = \frac{1}{(2\pi)^{p/2}|\boldsymbol{\Sigma}|^{1/2}} \exp[-\tfrac{1}{2}(\mathbf{x} - \boldsymbol{\mu})'\boldsymbol{\Sigma}^{-1}(\mathbf{x} - \boldsymbol{\mu})] \qquad \text{for } \mathbf{x} \in E_p$$

$$(1.1)$$

where $\mathbf{E}_p$ represents the p-dimensional vector space of real numbers. That is, $E_p = \{\mathbf{x}: -\infty < x_i < \infty, \text{ for } i = 1, 2, \ldots, p\}$.

Bivariate Normal Distributions

Graphs of the probability density functions of several bivariate normal distributions are shown in Figure 1.2. In these graphs the standard deviation of x_2 is twice that of x_1 and the standard deviation of x_1 is taken to be equal to 1. The values of the correlation between x_1 and x_2 are shown below each graph. Note that if we were to place a plane parallel to the (x_1, x_2) plane (i.e., the base of the figure) the intersection of the plane and the graph of the probability density function would form an ellipse. If we were to project this intersection down to the base, then we would have an ellipse in the (x_1, x_2) plane. The equation of this ellipse is given by $(\mathbf{x} - \boldsymbol{\mu})'\boldsymbol{\Sigma}^{-1}(\mathbf{x} - \boldsymbol{\mu}) = c$ where c measures the distance between the base and the intersecting plane. In Chapter 4 we learn how to plot such an ellipse on a coordinate system.

When we look at a two-dimensional scatter plot of a random sample from a bivariate normal distribution, we expect to see the data fall in an elliptically shaped region. Figure 1.3 shows scatter plots of random samples from each of the probability distributions shown in Figure 1.2. From the plots in Figures 1.2 and 1.3, we can see that if two bivariate normally distributed random variables are related (i.e., correlated), then they must be related in a linear fashion.

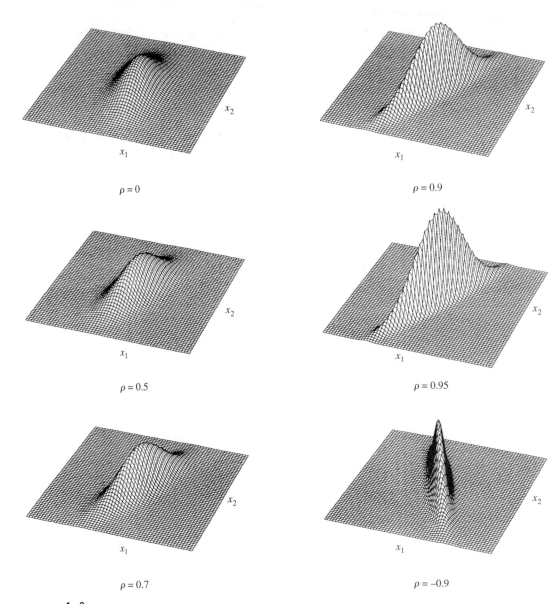

F I G. 1.2
Bivariate normal distributions

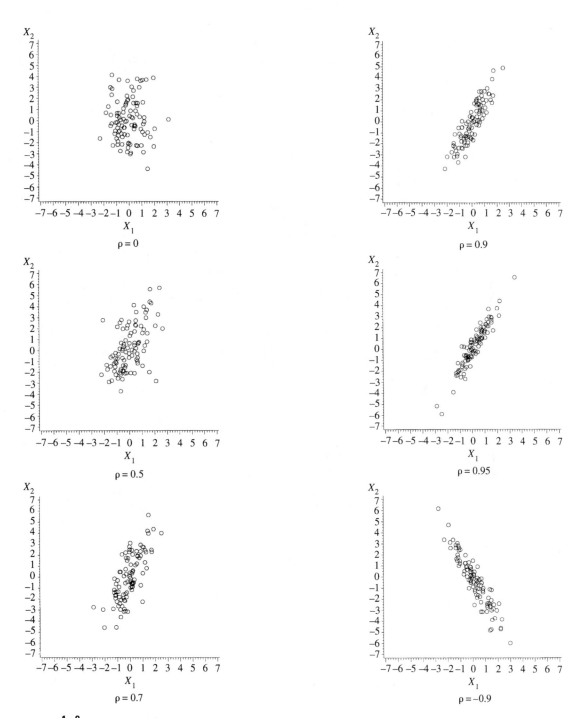

FIG. 1.3
Scatter plots of bivariate normal samples

When $p > 2$, we say that the probability density function of a p-variate multivariate normal distribution of rank p is constant on the ellipsoid defined by $(\mathbf{x} - \boldsymbol{\mu})'\boldsymbol{\Sigma}^{-1}(\mathbf{x} - \boldsymbol{\mu}) = c$. When $c = p + 2$, the resulting ellipsoid is called an *ellipsoid of concentration.*

If $p = 3$, then a scatter plot of a random sample of observations from a three-variate normal distribution would fall in a three-dimensional ellipsoid.

1.6
Statistical Computing

Many of the multivariate methods introduced in this book were developed many years ago. However, when they were being developed, they were not actually being used on very many real and complicated data sets. This was due to the fact that most multivariate methodologies required a great number of tedious numerical calculations. For example, if a researcher measures 50 variables on each experimental unit, then many multivariate procedures require that researcher to compute an inverse of a 50×50 matrix used to estimate $\boldsymbol{\Sigma}$ and/or $\mathbf{P}$.

Fortunately for potential users of multivariate methods, tedious numerical calculations are one thing computers do best. Multivariate methods are becoming popular with large numbers of researchers because of the availability of excellent statistical software and high-speed computers on which to run the software.

Cautions About Computer Usage

Readers should remember that computers compute with whatever data they are given. Computers do not care whether the data matrix has rows corresponding to experimental units or rows corresponding to variables. All the computer really cares about is that the data are in a matrix. Having the data matrix in the right form is the responsibility of the user.

Most readers have seen the acronym GIGO, which stands for "garbage in, garbage out." Nothing could be closer to the truth. Note, though, that while multivariate methods cannot perform magic with bad data, they might be able to help you determine if data are bad.

Missing Values

Researchers must also know what their computing software does with respect to missing values in a data matrix. **No existing statistical software deals with missing values adequately.** This is in spite of the fact that many software vendors advertise that their software can "handle missing values." But what does this term mean? It seems to mean that the computing program keeps on running and produces output—it does not always mean that it produces useful and meaningful output!

Replacing Missing Values by Zeros

Thirty years ago, Fortran programs often changed blank (missing?) values to zeros. So, for example, if a person's height was not entered into a data file, then the program took that person's height to be zero. How many times have you seen a 240-pound man who was 0 inches tall? This replacement method cannot be recommended, and we hope none of the present-day software programs still do this. Be aware, however, that old programs that use this unfortunate replacement method may still be in use. Users must determine what their software of choice does about missing values!

Replacing Missing Values by Averages

Some software programs allow the user to replace missing values on a given response variable by the average of that variable across all the other individuals in the data set. At first glance, this might seem very reasonable, but is not recommended. To see why, consider this scenario: Suppose a researcher is looking at a random sample of white males from some defined population. Suppose that two variables of interest are (1) the distance between each man's middle fingertips when both arms are stretched out as far as possible and (2) each man's height. Suppose also that the mean height of males in this sample is 72 inches and that the mean distance between middle fingertips is 71 inches. Now suppose there is a man in the sample whose height is 84 inches but whose distance between fingertips is not known. This technique for replacing missing values would assign a distance of 71 inches as the distance between this man's fingertips. This does not seem to be very realistic! Such an assignment would, no doubt, create an outlier in the data to be analyzed. Analysts prefer samples without outliers, and good data analysts certainly would not want to use missing value replacement techniques that create outliers.

Removing Rows of the Data Matrix

Most multivariate computing programs simply remove rows of data from the data matrix when a row contains a missing value. This is probably the most reasonable option, and while it can be recommended, it raises concerns for researchers who are collecting and analyzing data. To illustrate, suppose the value of the fourth response variable is missing for the third experimental unit. This option would remove all of the data for the third experimental unit from an analysis. Further, suppose a researcher has collected data on 2000 individuals, but has complete data on only 200 of them. Then 1800 of the individuals would be removed from the analysis, and all conclusions would be made only from the 200 individuals for which there are complete data. Although this certainly appears to be very unsatisfactory, it is still probably the safest of all possible alternatives. Researchers are well advised to consider changing their sampling strategies if their samples tend to include a lot of experimental units with missing data.

Sampling Strategies

Generally, a researcher has a finite number of resources that can be used for collecting data. How should these resources be allocated? Researchers would be wise to allocate their resources by acquiring complete data on a fewer number of experimental units rather than obtaining incomplete data on very large numbers of experimental units. Why spend resources acquiring incomplete data on many experimental units if those experimental units are going to be discarded from the analysis anyway?

Suppose a researcher wants to collect data through the use of a survey. One option might be to mail 2000 surveys to a random sample of individuals from the population of interest. A second option might be to hire individuals to conduct telephone interviews or to conduct personal interviews on a random sample of 250 individuals from the same population. A well-trained interviewer can usually get complete data from each person sampled, especially when both the interviewer and responder realize that none of the responder's answers may be usable unless the responder provides answers to all of the questions in the survey. Mail surveys are not only poor ways to actually collect a random sample from the population of interest, they are notorious for producing incomplete data from the respondents. This book recommends collecting survey data via the second option.

Data Entry Errors and Data Verification

Extreme care must be taken to ensure that no data entry errors are made as data are entered into computer files for later analyses. The easiest, cheapest, and safest way to ensure that data are entered correctly probably is to have the data "verified" by a reentry of the data by a different data entry operator. That is, have two different individuals enter the data into two different files. Most computer programs have routines that allow files to be compared with one another. If the two files are identical, then a researcher can be confident that the data being analyzed are the data that were provided to the data entry operators. If the two files are not identical, then discrepancies can be located, and the file that is in error can be corrected. A statistical analysis is only as good as the data being analyzed. Hence, all data entry errors must be identified and corrected.

Caution   Do not allow the accessibility of computing software hurry you into applying sophisticated multivariate techniques without first looking at your data very carefully. It has been said that one should never trust a large data set to be correct. Errors are unavoidable and steps must be taken to deal with them. A few unfortunate errors may cause a statistical analysis to be worthless or, even worse, misleading.

1.7
Multivariate Outliers

Outliers are generally defined as sample data points that appear to be inconsistent with a majority of the data.

Locating Outliers

Several multivariate routines exist that can help researchers locate outliers in a data set when outliers exist. As mentioned earlier, principal components analysis is useful for screening data—this includes screening data for outliers. Some graphical procedures may also help. Graphical procedures that are helpful for identifying outliers include Andrews' plots, Chernoff faces, and sun-ray or star plots. These procedures are discussed in Chapter 3. Many interactive computer exploratory data analysis packages now exist that are useful for locating and identifying outliers in multivariate data sets.

Dealing with Outliers

Locating outliers is reasonably easy using methods mentioned in the previous section. When outliers are found, a researcher should determine, if possible, whether recording errors or data entry errors were made. If such errors occur and they can be corrected, then the data should be corrected. If recording errors or data entry errors were made, but they cannot be corrected, then such data points should be removed from any multivariate analyses to be performed.

Knowing how to deal with outliers when there are no apparent recording or data entry errors is a difficult problem having no uniformly acceptable solution. No statistician or statistical technique can accurately tell an experimenter what to do with outliers in a data file. Researchers must use their own expert opinions about the populations being sampled as well as that of other experts, when possible, to make subjective, but well-informed, decisions as to what to do with outliers in data being analyzed. While robust procedures exist for dealing with one-variable-at-a-time analyses, very few robust methods exist for multivariate responses.

Researchers must decide for themselves whether they believe a particular observation is leading them toward "truth" or away from "truth." If they believe the observation is leading them toward truth, then the observation should be included in statistical analyses to be performed. If they believe the observation is leading them away from truth, then the observation should be eliminated from all statistical analyses. In this latter case, researchers should report that the observation was removed and report why it was removed. In this way, other researchers having different opinions can, if desired, perform alternative analyses that might lead to different conclusions.

Outliers May Be Influential

Sometimes it is worthwhile to analyze data in two different ways—one with all the outliers remaining in the data set and another with the outliers removed. If answers to the important questions do not change, then a researcher is not forced to make decisions about what to do with possible outliers because the outliers have no effect on the conclusions being made. More often than not, however, answers will change. This is because outliers are often the most influential points in a data set. That is, a few extreme outliers that are not removed from a statistical analysis may render the rest of the data useless for reaching truthful conclusions.

1.8
Multivariate Summary Statistics

Let $\mathbf{x}_1, \mathbf{x}_2, \ldots, \mathbf{x}_N$ be a random sample from a multivariate distribution that has a mean vector, $\boldsymbol{\mu}$, a covariance matrix, $\boldsymbol{\Sigma}$, and a correlation matrix, $\mathbf{P}$. Unbiased estimates of $\boldsymbol{\mu}$ and $\boldsymbol{\Sigma}$ are given by

$$\hat{\boldsymbol{\mu}} = \frac{1}{N}\left(\sum_{r=1}^{N} \mathbf{x}_r\right) = \frac{\mathbf{x}_1 + \mathbf{x}_2 + \cdots + \mathbf{x}_N}{N}$$

and

$$\hat{\boldsymbol{\Sigma}} = \frac{1}{N-1}\left[\sum_{r=1}^{N} (\mathbf{x}_r - \hat{\boldsymbol{\mu}})(\mathbf{x}_r - \hat{\boldsymbol{\mu}})'\right]$$

respectively. Note that

$$\hat{\boldsymbol{\mu}} = \begin{bmatrix} \overline{x}_1 \\ \overline{x}_2 \\ \vdots \\ \overline{x}_p \end{bmatrix} = \begin{bmatrix} \hat{\mu}_1 \\ \hat{\mu}_2 \\ \vdots \\ \hat{\mu}_p \end{bmatrix}$$

and

$$
\hat{\Sigma} = \begin{bmatrix}
\hat{\sigma}_{11} & \hat{\sigma}_{12} & \cdots & \hat{\sigma}_{1p} \\
\hat{\sigma}_{21} & \hat{\sigma}_{22} & \cdots & \hat{\sigma}_{2p} \\
\vdots & \vdots & \ddots & \vdots \\
\hat{\sigma}_{p1} & \hat{\sigma}_{p2} & \cdots & \hat{\sigma}_{pp}
\end{bmatrix}
$$

where

$$
\hat{\sigma}_{ii} = \widehat{\mathrm{Var}}(x_i) = \frac{1}{N-1} \sum_{r=1}^{N} (x_{ri} - \bar{x}_i)^2
$$

and

$$
\hat{\sigma}_{ij} = \widehat{\mathrm{Cov}}(x_i, x_j) = \frac{1}{N-1} \sum_{r=1}^{N} (x_{ri} - \bar{x}_i)(x_{rj} - \bar{x}_j)
$$

The estimators of the correlation coefficients ρ_{ij} are usually taken as $r_{ij} = \hat{\sigma}_{ij}/\sqrt{\hat{\sigma}_{ii}\hat{\sigma}_{jj}}$ even though r_{ij} is not an unbiased estimate of ρ_{ij} for any i and j. While r_{ij} is not an unbiased estimate of ρ_{ij}, it does have the property that $-1 \le r_{ij} \le 1$ for every $i \ne j$. It is possible to multiply r_{ij} by a constant to make it unbiased, but after doing so you might find that the estimated correlation could be greater than 1 or smaller than -1, which would not make sense. This book does not use these unbiased estimates of correlation coefficients.

The sample correlation matrix is denoted by **R** and is defined by

$$
\mathbf{R} = \begin{bmatrix}
1 & r_{12} & r_{13} & \cdots & r_{1p} \\
r_{21} & 1 & r_{23} & \cdots & r_{2p} \\
r_{31} & r_{32} & 1 & \cdots & r_{3p} \\
\vdots & \vdots & \vdots & \ddots & \vdots \\
r_{p1} & r_{p2} & r_{p3} & \cdots & 1
\end{bmatrix}
$$

1.9
Standardized Data and/or Z Scores

Sometimes data are easier to understand and compare when the response variables are standardized so that they are measured in comparable units. This is usually done by eliminating the units of measurement altogether. Define

$$
Z_{rj} = \frac{x_{rj} - \hat{\mu}_j}{\sqrt{\hat{\sigma}_{jj}}}, \qquad \text{for } r = 1, 2, \ldots, N, \qquad j = 1, 2, \ldots, p
$$

The variable Z_{rj} is called the *Z score* for the jth response variable on the rth experimental unit, and

$$
\mathbf{Z} = \begin{bmatrix}
z_{11} & z_{12} & \cdots & z_{1p} \\
z_{21} & z_{22} & \cdots & z_{2p} \\
\vdots & \vdots & \ddots & \vdots \\
z_{N1} & z_{N2} & \cdots & z_{Np}
\end{bmatrix}
$$

is called the matrix of Z scores.

Suppose one experimenter took measurements on a sample of experimental units in inches while another took similar measurements in centimeters. If these two experimenters then standardized their data, both would have exactly the same set of Z scores.

Standardization of data is recommended when the measured variables are in completely different units. For example, suppose a researcher has measurements on height and weight of individuals. These measurements, by necessity, are in completely different units. It is often much easier to compare individuals with respect to these two variables if each variable is standardized.

Caution   Most multivariate computing routines, as an initial step, standardize all of the response variables. In fact, this is usually the default; that is, standardization of the data is performed without any intervention by an user. In many cases this is desirable, but in some instances the data should not be standardized. Unfortunately, some statistical computing packages do not allow you to analyze the data in the untransformed units, also called the "raw" data values. More is said about this in Chapter 5, where examples are given.

Exercises

1   Find a multivariate data set that represents a random sample of data from a single population and enter the data into a computer file that can be shared with your classmates. This data set should have at least 5 response variables and at least 30 experimental units. Write a one-page report describing the population being sampled by your data, the variables in the data set, and some interesting questions that the data might be used to answer.

2   Are the measured variables in your data set in Exercise 1 in comparable units? Do you think the response variables should be standardized prior to performing multivariate analyses? Why or why not?

3   Consider the basic multivariate methods introduced in this chapter (namely, principal components analysis, factor analysis, discriminant analysis, cluster analysis, multivariate analysis of variance, and canonical correlation analysis). Write a short paragraph explaining why each of these multivariate techniques might or might not be appropriate techniques for answering the interesting questions identified in Exercise 1.

4   Are the experimental units in your data set from Exercise 1 likely to satisfy conditions of independence that are required for many multivariate analyses? Explain your answer.

5   Consider the data in Tables 1.2 and 1.3. Do you believe that the variables in these data sets are in comparable units? Why or why not?

6 Consider the following data matrix:

$$\mathbf{x} = \begin{bmatrix} 2 & 4 & 3 \\ 2 & 3 & 4 \\ 3 & 5 & 4 \\ 1 & 2 & 6 \\ 2 & 6 & 8 \end{bmatrix}$$

Answer each of the following questions by doing all of the computations by hand. Show your work.

a What are the values of p, N, x_{32}, and $\mathbf{x}_3$? What is $\mathbf{x}_3'$?

b Find the mean vector, $\hat{\boldsymbol{\mu}}$, the sample variance-covariance matrix, $\hat{\boldsymbol{\Sigma}}$, and the sample correlation matrix, **R**.

c Find **Z**, the matrix of Z scores for these data.

d What are the values of z_{32} and $\mathbf{z}_3$?

e Construct a scatter plot that plots the variable x_3 against the variable x_2.

f Construct a scatter plot that plots the variable z_3 against the variable z_2.

7 The file BIG8.DAT on the floppy disk that came with your book contains some statistics from 1994 football teams from the Big 8 Conference. This data set is printed below:

1994 BIG 8 FOOTBALL STATISTICS

SCHOOL	GAMES	RO_YDS	RD_YDS	PO_YDS	PD_RAT	TO_YDS	TD_YDS	SO	SD	TOM	WINS
1 COLORADO	11	291.5	114.2	203.8	125.2	495.3	343.7	36.2	19.2	0.55	10.0
2 IOWA STATE	11	178.0	272.8	137.1	137.1	315.1	460.7	17.5	33.0	−0.64	0.0
3 KANSAS	11	247.1	171.2	140.9	135.8	363.3	400.7	28.5	22.0	0.73	6.0
4 KANSAS STATE	11	125.6	167.5	237.6	94.3	363.3	312.5	27.7	14.2	1.18	9.0
5 MISSOURI	12	107.9	235.3	202.5	138.5	310.4	414.9	17.3	27.1	0.17	3.5
6 NEBRASKA	12	340.0	79.3	137.8	96.7	477.8	258.8	36.3	12.1	0.08	12.0
7 OKLAHOMA	11	182.2	148.5	173.9	107.5	356.1	295.7	19.8	21.6	−0.18	6.0
8 OKLAHOMA STATE	11	204.6	192.5	133.5	130.3	338.1	385.9	16.4	23.3	−0.45	3.5

The variables in this data set consist of the name of each school (SCHOOL), number of games played (GAMES), average yards rushing offense (RO_YDS), average yards rushing defense (RD_YDS), average yards passing offense (PO_YDS), passing defense efficiency rating (PD_RAT), average yards total offense per game (TO_YDS), average yards total defense per game (TD_YDS), scoring offense (SO), scoring defense (SD), turnover margin per game (TOM), and number of wins (WINS). Note that this data set is not a random sample of data from any population. Nevertheless, answer each of the following questions.

- **a** What are the values of p and N for these data?
- **b** Consider only the variables SO, SD, and TOM. Find the mean vector, $\hat{\mu}$, the sample variance-covariance matrix, $\hat{\Sigma}$, and the sample correlation matrix, **R**, for these three variables.
- **c** Find **Z**, the matrix of Z scores for the three variables given in part b.
- **d** Construct a scatter plot that plots the scoring offense against the scoring defense.
- **e** Which of these variables, if any, are in comparable measurement units?

8 The file labeled HOTEL.DAT on the enclosed floppy disk contains some information about Choice Hotels International in the Washington, DC, area. The variables in this data set are the location of the hotel (LOCATION), the number of rooms in the hotel (ROOMS), the minimum two-person 1995 summer rate (MIN), the maximum two-person 1995 summer rate (MAX), whether the hotel has a dining room (DINING), whether the hotel has a cocktail lounge (LOUNGE), whether the hotel serves a free continental breakfast (BKFST), whether the hotel has a pool (POOL), and the type of hotel (TYPE). The values of TYPE are CI for Comfort Inn, QI for Quality Inn, and CH for Clarion Hotel. The data in this file are given below:

CHOICE HOTELS INTERNATIONAL

OBS	LOCATION	ROOMS	MIN	MAX	DINING	LOUNGE	BKFST	POOL	TYPE
1	Capital Hill	341	69	139	yes	yes	no	yes	QI
2	Downtown	135	79	169	yes	yes	no	no	QI
3	Downtown	197	94	164	yes	yes	no	no	CI
4	Bowie, MD	110	59	135	yes	yes	no	yes	CI
5	Clinton, MD	94	58	70	no	no	yes	no	CI
6	College Park, MD	154	54	109	no	no	yes	yes	QI
7	Gaithersburg, MD	127	59	79	no	no	yes	yes	CI
8	Landover Hills, MD	84	44	64	no	no	yes	no	CI
9	Laurel, MD	118	59	150	no	no	yes	yes	CI
10	Rockville, MD	162	59	74	yes	yes	no	yes	CH
11	Silver Spring, MD	254	59	125	yes	yes	no	yes	QI
12	Alexandria, VA	207	60	99	yes	yes	no	yes	QI
13	Alexandria, VA	148	49	86	yes	yes	yes	yes	CI
14	Alexandria, VA	188	65	85	yes	no	yes	yes	CI
15	Alexandria, VA	92	49	72	no	no	yes	yes	CI
16	Arlington, VA	141	59	91	yes	yes	no	yes	QI
17	Arlington, VA	398	39	149	yes	yes	no	yes	QI
18	Arlington, VA	126	64	95	yes	yes	no	no	CI
19	Dulles, VA	140	59	85	yes	no	yes	yes	CI

20	Dulles, VA	103	59	105	no	no	yes	no	CI
21	Fairfax, VA	212	49	95	yes	yes	yes	yes	CI
22	Falls Church, VA	121	62	72	yes	no	no	yes	QI
23	Falls Church, VA	109	50	75	yes	no	yes	yes	QI
24	Vienna, VA	250	45	79	no	no	yes	yes	CI
25	Woodbridge, VA	95	59	65	no	no	yes	no	CI

a What are the values of p and N for these data?

b Which variables, if any, in this data set represent continuous variables? Which variables are discrete or categorical variables? Which variables, if any, are discrete, but quantitatively ordered?

c Construct a scatter plot that illustrates any possible relationship that might exist between minimum room cost and the number of rooms in the hotel. Does there appear to be any relationship between these two variables? Explain your answer.

d Construct a scatter plot that illustrates any possible relationship that might exist between maximum room cost and the number of rooms in the hotel. Does there appear to be a relationship between these two variables? Explain your answer.

e Based on your plot in part d, do any hotels appear to be outliers?

f Construct a scatter plot that illustrates any possible relationship that might exist between minimum room cost and maximum room cost. Does there appear to be any relationship between these two variables? Explain your answer.

g Based on your plot in part f, do any hotels appear to be outliers?

h Consider only the variables ROOMS, MIN, and MAX. Find the mean vector, $\hat{\boldsymbol{\mu}}$, the sample variance-covariance matrix, $\hat{\boldsymbol{\Sigma}}$, and the sample correlation matrix, $\mathbf{R}$, for these three variables.

i Find $\mathbf{Z}$, the matrix of Z scores for the three variables given in part h.

j Repeat part h for the CI-type hotels and for the QI-type hotels. Do the mean vectors and variance–covariance matrices for these two groups of hotels appear to be similar to one another? Explain your answer (no statistical justification of your answer is required).

k Repeat part h for hotels with and without pools. Do the mean vectors and variance–covariance matrices for these two groups of hotels appear to be similar to one another? Explain your answer (no statistical justification of your answer is required).

9 The files CEREAL.DAT and CEREAL.SAS on the enclosed disk contain nutritional data for 17 brands of breakfast cereal. Each entry uses two data lines; the first line contains the brand name followed by the recommended serving size in grams. The second line contains entries for calories, calories from fat, total fat in grams, saturated fat in grams, cholesterol in milligrams,

sodium in milligrams, potassium in milligrams, total carbohydrates in grams, dietary fiber in grams, sugars in grams, other carbohydrates in grams, and protein in grams. These are followed by entries giving the percentages of daily requirements for an adult; these entries correspond to vitamin A, vitamin C, calcium, iron, vitamin D, thiamin, riboflavin, niacin, vitamin B_6, folate, vitamin B_{12}, phosphorous, magnesium, zinc, and copper. The data are given below:

```
Grape-nuts        58
200 10 1 0 0 350 160 47 5 7 35 6 15   0 2 45 10 25 25 25 25 25 25 15 15   8 10
Raisin Bran       55
170 10 1 0 0 310 400 43 7 18 18 4 15   0 4 25 10 25 25 25 25 25 25 20 20 25 10
Golden Grahams       30
120 10 1 0 0 280 55 25   1 11 13 1 15 25 0 25   0 25 25 25 25 25 . 4 2 25 .
Bran Flakes        30
100 5 .5 0 0 220 190 24 5 6 13 3 15   0 0 45 10 25 25 25 25 25 25 15 15 10 10
All.Bran          28.4
50 0  0 0 0 140 330 22 14 0  8 4 15 25 2 25 10 25 25 25 25 25 25 25 30 25 15
Rice Krispies      30
110 0  0 0 0 320 35 26   1  3 22 2 15 25 0 10 10 25 25 25 25 25 . 4 2 4 2
Multi-Grain Cheerios 30
110 10 1 0 0 240 100 24   3  6 15 3 25 25 4 45 10 25 25 25 25 25 . 10 6 4 2
Shredded Wheat       49
170 5 .5 0 0  0 200 41 5  0 36 5  0 0 2  8 0 8 2 15 . . . 20 15 8 8
Wheaties           30
110 10 1 0 0 220 110 24   3 4  17 3 25 25 4 45  10 25 25 25 25 25  . 10 8 4 4
Frosted Flakes     32
120 0 0 0 0 220  .29 1 13 15 2 15 25 0 30 10 25 25 25 25 25 . . . . .
Special K          21
80 0 0 0 0 170  . 15   1 2 12 4 10 15 0 30  8 20 20 20 20 15 . 4 4 15 4
Crispix            18
70 0  0 0 0 140  .15 0 2 13 1  8 10 0 4 6 10 10 10 10 10 . . . 4 .
Froot Loops        27
100 5  1 0 0 135  .23 1 13 9 1 10 20 0 20 10 20 20 20 20 20 . . . 20 .
Corn Pops          28
110 0  0 0 0 90  .26 1 12 13 1 10 20 0 8 10 20 20 20 20 20 . . . 8 .
Corn Flakes        21
80 0 0 0 0 230  . 18 1 1 16 2 10 15 0 30 8 15 15 15 15 15 . . . . . .
Frosted Mini.Wheats 35
120 5 .5 0 0  0 .29 4 8 15 3  0  0 0 50 . 15 15 15 15 15 15 10 8 6 4
Fruit Granola      55
210 25 2.5 0 0 200 . 44 3 18 23 4  0 0 2 6 . 8 2 4 . . . 15 8 4 4
```

a What are the values of p and N for these data?

b Which variable, if any, identifies the experimental units in this data set?

c Which variables, if any, in this data set are continuous variables? Which variables, if any, are discrete or categorical variables?

d   Construct a scatter plot that illustrates any possible relationship that might exist between calories from fat and total fat. Does there appear to be any relationship between these two variables? Based on this plot, do there appear to be any outliers in this data set? Explain your answers.

e   Construct a scatter plot that illustrates any possible relationship that might exist between calories and total fat. Does there appear to be any relationship between these two variables? Based on this plot, do there appear to be any outliers in this data set? Explain your answers.

f   Consider only the variables sodium, potassium, and total carbohydrates. Find the mean vector, $\hat{\mu}$, the sample variance-covariance matrix, $\hat{\Sigma}$, and the sample correlation matrix, $\mathbf{R}$, for these three variables.

Sample Correlations

2

Suppose you were told that the sample correlation between two response variables was equal to 0.90. Would you find that interesting and useful? What if the sample correlation was equal to 0.30? How might your thinking change if the correlation of 0.90 was not significantly different from zero and the sample correlation of 0.30 was significantly different from zero? This chapter describes some basic and simple ways to use sample correlations as well as how sample correlations might be used to make inferences about the population being sampled.

Formulas for computing estimates of population correlation coefficients based on random samples were given in Section 1.8. The sample correlation, r_{ij}, between the ith and jth response variables measures the strength of the *linear* relationship between these two variables. Note the emphasis on the *linear* relationship. A sample correlation that is close to zero indicates that no linear relationship exists between the two variables being measured, but it does not necessarily imply that no relation exists, except in the case where the two variables are known to have a joint bivariate normal distribution.

Whenever it is important to consider the correlation between pairs of variables, it is equally important to examine scatter plots of each pair of variables. It is also important to construct confidence intervals for the true values of population correlation coefficients. Scatter plots are important because they provide visual information as to whether other relationships between variable pairs might exist that are not linear. Confidence intervals are important because they give reliable information as to the actual numerical size of a population correlation coefficient. Many statistical routines exist that allow researchers to create a single plot that shows pairwise scatter plots between many variables simultaneously. Figure 3.14 in the next chapter provides an example of such a plot.

2.1
Statistical Tests and Confidence Intervals

Many computer programs test whether the ρ_{ij}'s are equal to zero or not, that is, they test whether the r_{ij}'s are significantly different from zero. This is probably being done because it easy to do for normally distributed data. It can be shown that if x_i and x_j have a bivariate normal distribution, then

$$t_{ij} = \frac{r_{ij}\sqrt{N-2}}{\sqrt{1 - r_{ij}^2}} \tag{2.1}$$

is distributed as Student's t distribution with $N - 2$ degrees of freedom for every $i \neq j$ (see Graybill, 1976, p. 398). While such tests are easy to perform and are done automatically by many statistical computing programs, the tests are not usually useful and helpful. Often the significance of such tests is determined more by the size of the sample rather than by the magnitude of the correlation coefficient. What we really need to know is when are the correlations large enough to be useful?

Are the Correlations Large Enough to Be Useful?

Although researchers can test for the significance of sample correlations, what they need to know, in real-life situations, is which correlations, if any, are large enough to be of some practical importance. How large is *large* often depends on the kind of data being analyzed. When data are collected within controlled environments, such as laboratories, correlations greater than 0.9 are not unusual. However, when data are collected from populations in which the researcher has very little control, correlations greater than 0.7 may be hard to obtain.

A very rough guide for many experimental situations is that correlations whose absolute value is greater than 0.6 may be considered large enough to be of some practical importance. In such cases, the linear relationship between the two variables is likely strong enough that one of the variables could possibly be used as a surrogate for the other variable. For data collected from people, we may occasionally consider correlations whose absolute values are greater than 0.5 as large enough to identify important linear relationships between pairs of variables.

Sample correlations should generally not be calculated at all if the sample size is less than 12 because the number of observations is just too small to give reliable results, and it is much too easy for naive researchers to misinterpret measured correlations from small samples.

It is also important to remember that there are $p(p-1)/2$ different pairwise correlations that can be calculated when p response variables are being measured. For example, if $p = 20$, then 190 different pairwise correlations can be computed. Experimenters should be extremely cautious if only a few of these are significantly different from zero. To understand why, note that when all variables are independent we would expect 5 to 10% of the pairwise correlations to be significantly different from zero just by chance alone. Thus, when 20 variables are being measured and correlated with one another, a researcher would expect at least 10 to 20 of these pairwise correlations to be significantly different from zero just by chance alone. If a researcher sees only a few correlations that are significantly different from zero, she or he should not get overly excited about possible linear relationships between those pairs of variables.

Confidence Intervals by the Chart Method

David (1954) showed how to compute a confidence interval for a population correlation coefficient based on a sample correlation coefficient when the two relevant variables have a joint bivariate normal distribution. See Graybill (1976, p. 400) for a description of David's procedure. Confidence intervals can be approximated by using charts that were developed for that purpose. Such charts were originally given by David (1938) and a chart for 95% confi-

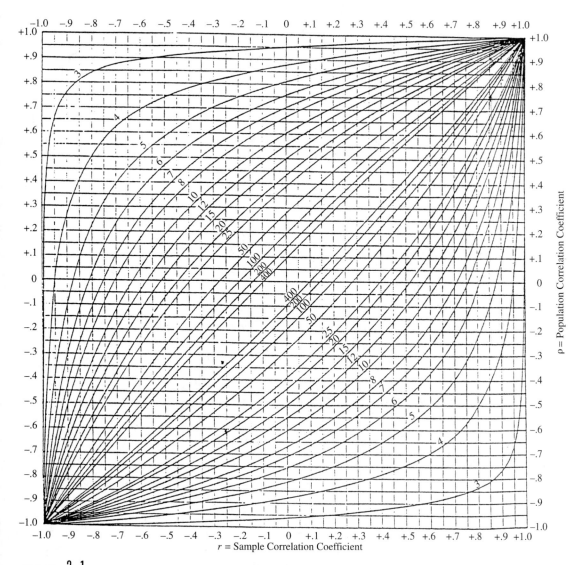

F I G. 2.1

Confidence curves for simple correlation coefficient; the curves are for $1 - \alpha = 0.95$ (Reprinted with permission)

dence intervals is given in Figure 2.1. Charts for other confidence levels are available in Graybill (1976), Pearson and Hartley (1972), and Beyer (1968).

To use one of these charts, first locate the observed sample correlation on the horizontal axis, and then draw or visualize a vertical line on the chart at this location. Next locate the intersections of this line with curves corresponding to the sample size N from which the sample correlation was computed. Finally, locate the ordinates (points on the vertical axis) of these two intersection points. This pair of ordinates forms the confidence interval for ρ.

For example, suppose $N = 6$ and $r = 0.8$; then by using the chart for a 95% confidence interval given in Figure 2.1, the confidence interval for ρ is approximately

$$-0.02 < \rho < 0.95$$

Note that even though the sample correlation of 0.8 seems to be quite large, it is not significantly different from zero since the confidence interval includes zero.

Using the chart in Figure 2.1, one can see that for $N = 25$ and $r = 0.7$, a 95% confidence interval for ρ is

$$0.41 < \rho < 0.85$$

In this case, the sample correlation of 0.7 is significantly different from zero since the confidence interval does not include zero.

Confidence Intervals by Fisher's Approximation

Fisher, as reported by Kendall and Stuart (1961), showed that when taking samples of size greater than 25 from a bivariate normal distribution with correlation ρ, the inverse hyperbolic tangent of the sample correlation is approximately normally distributed with mean equal to the inverse hyperbolic tangent of ρ and variance $1/(N - 3)$. That is,

$$U = \text{invtanh}(r)$$

is approximately normally distributed with mean

$$\delta = \text{invtanh}(\rho)$$

and variance $1/(N - 3)$ when $N > 25$. Since the inverse of the hyperbolic tangent (invtanh) function is available on many calculators, this result can be used to construct confidence intervals about ρ for large sample sizes. In particular, a $(1 - \alpha)100\%$ confidence interval for ρ is given by

$$\tanh[\text{invtanh}(r) - z_{\alpha/2}/\sqrt{N - 3}] < \rho < \tanh[\text{invtanh}(r) + z_{\alpha/2}/\sqrt{N - 3}] \quad (2.2)$$

To illustrate the use of this formula, suppose $N = 25$ and $r = 0.7$. Then invtanh$(0.7) = 0.8673$. Thus the left endpoint of a 95% confidence interval is approximately

$$\tanh[0.8673 - 1.96/22^{1/2}] = \tanh[0.8673 - 0.4179] = \tanh[0.4494] = 0.421$$

and the right endpoint is approximately

$$\tanh[0.8673 + 1.96/22^{1/2}] = \tanh[0.8673 + 0.4179] = \tanh[1.2852] = 0.858$$

Thus, a 95% confidence interval for ρ is

$$0.42 < \rho < 0.86$$

This result compares favorably with the confidence interval obtained from Figure 2.1. In this case, because $N = 25$ is still fairly small, the confidence interval obtained from the Figure 2.1 is likely to be slightly more accurate.

For readers who might not have a calculator with the hyperbolic tangent function on it, but do have one with the natural logarithm function on it, note that

$$\text{invtanh}(r) = (1/2) \log[(1 + r)/(1 - r)]$$

where log is the natural logarithm function and

$$\tanh(w) = (e^w - e^{-w})/(e^w + e^{-w}) = (e^{2w} - 1)/(e^{2w} + 1)$$

Confidence Intervals by Ruben's Approximation

Ruben as reported by Boomsma (1977) provided another numerical approximation to a confidence interval for ρ that is more accurate than Fisher's approximation. Unfortunately, it is slightly more complicated to calculate than Fisher's approximation. We provide details here for those who are interested in using this method.

Let N be the sample size and let r be the observed sample correlation as before. Let u be the upper $\alpha/2$ critical point of the standard normal distribution. Next, let $r^* = r/(1 - r^2)^{1/2}$, $a = 2N - 3 - u^2$, $b = r^*[(2N - 3)(2N - 5)]^{1/2}$, and $c = (2N - 5 - u^2)r^{*2} - 2u^2$. Next let y_1 and y_2 be the roots of the quadratic equation $ay^2 - 2by + c = 0$. Then the lower and upper $(1 - \alpha)100\%$ confidence limits for ρ are $y_1/(1 + y_1^2)^{1/2}$ and $y_2/(1 + y_2^2)^{1/2}$, respectively.

A SAS® program that can be used to calculate Ruben's confidence limits is given in the file labeled RHO_CI.SAS on the floppy disk provided with this book. The SAS commands for computing 95% confidence intervals for the two cases when $N = 6$ and $r = 0.8$, and when $N = 25$ and $r = 0.7$ are:

```
OPTIONS LINESIZE=75 PAGESIZE=54 NODATE PAGENO=1;

DATA CORR;

    /* N=SAMPLE SIZE, R=SAMPLE CORRELATION, CONF=CONFIDENCE
LEVEL*/

    INPUT N R CONF;
    U = -PROBIT((1-CONF)/2);  RTILDE=R/SQRT(1-R*R);
    A=2*N-3-U*U;
    B=RTILDE*SQRT((2*N-3)*(2*N-5));
    C=(2*N-5-U*U)*RTILDE*RTILDE-2*U*U;
```

```
Y1 = (2*B−SQRT(4*B*B−4*A*C))/(2*A);
Y2 = (2*B+SQRT(4*B*B−4*A*C))/(2*A);


/* LL=LOWER CONFIDENCE LIMIT, UL=UPPER CONFIDENCE LIMIT */


LL=Y1/SQRT(1+Y1*Y1); UL=Y2/SQRT(1+Y2*Y2);

CARDS;
6 .8 .95
25 .7 .95
;
PROC PRINT; VAR N R CONF LL UL;
RUN;
```

The results obtained from executing the above program are:

```
OBS   N    R    CONF      LL          UL


 1    6   0.8   0.95   −0.09504    0.97279
 2   25   0.7   0.95    0.41082    0.85357
```

Thus, the 95% confidence intervals for ρ are $(-0.095, 0.973)$ when $N = 6$ and $r = 0.8$ and $(0.411, 0.854)$ when $N = 25$ and $r = 0.7$.

Variable Groupings Based on Correlations

When several variables are measured on each of a large number of experimental units, it is often interesting to see if and how variables are interrelated. To examine these interrelationships when several pairs of variables are highly correlated with one another, we might try to partition the response variables into groups—so that the variables within a group have high correlations with each other, and those in different groups have low correlations with each other. Such a partitioning often reveals important aspects of the data that are useful to consider when deciding how to interpret the data. To illustrate, consider the following example.

E X A M P L E 2.1

Forty-eight individuals who had applied for a job with a large firm were interviewed and rated on 15 criteria. Individuals were rated on the form of their letter of application (FL), their appearance (APP), their academic ability (AA), their likability (LA), their self-confidence (SC), their lucidity (LC), their honesty (HON), their salesmanship (SMS), their experience (EXP), their drive (DRV), their ambition (AMB), their ability to grasp concepts (GSP), their potential (POT), their keenness to join (KJ), and their suitability (SUIT). Each criterion was evaluated on a scale ranging from 0 to 10, with 0 being a very low and very unsatisfactory rating, and 10 being a very high rating. Each individual's evaluation of these 15 criteria is shown in Table 2.1.

For fun, let us suppose the firm wanted to make job offers to the six best applicants. How should the firm select these six individuals?

One simple way to select six individuals to whom job offers could be made would be place all 48 names in a hat, mix the names thoroughly, and then draw 6 names out of the hat at random. This method would give everyone an equal chance of being selected. Most employers would not approve of this method because it uses none of the data gathered in the interviewing process.

A second method for selecting individuals would be to compute an average score for each individual by averaging the measurements on all 15 variables. Thus one might compute

$$AVG = (FL + APP + \cdots + SUIT)/15$$

for each individual and select those six individuals who have the highest average scores. Obviously, this method gives equal weight to each of the 15 criteria and, consequently, this method of computing a single score for each individual may not always be desirable.

A third alternative that is likely to be preferable to many employers would be to weight each of the criteria according to how important the employer feels each one might be, and then compute a weighted average for each individual. Thus, a weighted score could be calculated as

$$WTD\_AVG = w_1(FL) + w_2(APP) + \cdots + w_{15}(SUIT)$$

where $w_1 + w_2 + \cdots + w_{15} = 1$, and where w_i measures the relative importance of the ith criterion, $i = 1, 2, \ldots, 15$. Note that requiring the weights to sum to 1 is not actually necessary; this simply guarantees that the resulting weighted averages will have a value between 0 and 10 for these data.

To further examine whether the approaches just described seem reasonable, we could look for interrelationships among these variables. Consider the correlations among all pairs of these 15 variables. The correlations among the variables were obtained using the CORR procedure in SAS with the aid of the following statements.

```
PROC CORR NOPRINT OUTP=CORRS;
 VAR FL--SUIT;
  TITLE 'APPLICANTS SCALED ON 15 VARIABLES';

DATA; SET CORRS; IF _TYPE_='CORR';

PROC PRINT ROUND; VAR _NAME_ FL--SUIT;
```

T A B L E 2.1

Applicant data set

ID	FL	APP	AA	LA	SC	LC	HON	SMS	EXP	DRV	AMB	GSP	POT	KJ	SUIT
1	6	7	2	5	8	7	8	8	3	8	9	7	5	7	10
2	9	10	5	8	10	9	9	10	5	9	9	8	8	8	10
3	7	8	3	6	9	8	9	7	4	9	9	8	6	8	10
4	5	6	8	5	6	5	9	2	8	4	5	8	7	6	5
5	6	8	8	8	4	4	9	5	8	5	5	8	8	7	7
6	7	7	7	6	8	7	10	5	9	6	5	8	6	6	6
7	9	9	8	8	8	8	8	8	10	8	10	8	9	8	10
8	9	9	9	8	9	9	8	8	10	9	10	9	9	9	10
9	9	9	7	8	8	8	8	5	9	8	9	8	8	8	10
10	4	7	10	2	10	10	7	10	3	10	10	10	9	3	10
11	4	7	10	0	10	8	3	9	5	9	10	8	10	2	5
12	4	7	10	4	10	10	7	8	2	8	8	10	10	3	7
13	6	9	8	10	5	4	9	4	4	4	5	4	7	6	8
14	8	9	8	9	6	3	8	2	5	2	6	6	7	5	6
15	4	8	8	7	5	4	10	2	7	5	3	6	6	4	6
16	6	9	6	7	8	9	8	9	8	8	7	6	8	6	10
17	8	7	7	7	9	5	8	6	6	7	8	6	6	7	8
18	6	8	8	4	8	8	6	4	3	3	6	7	2	6	4
19	6	7	8	4	7	8	5	4	4	2	6	8	3	5	4
20	4	8	7	8	8	9	10	5	2	6	7	9	8	8	9
21	3	8	6	8	8	8	10	5	3	6	7	8	8	5	8
22	9	8	7	8	9	10	10	10	3	10	8	10	8	10	8
23	7	10	7	9	9	9	10	10	3	9	9	10	9	10	8
24	9	8	7	10	8	10	10	10	2	9	7	9	9	10	8

ID	FL	APP	AA	LA	SC	LC	HON	SMS	EXP	DRV	AMB	GSP	POT	KJ	SUIT
25	6	9	7	7	4	5	9	3	2	4	4	4	4	5	4
26	7	8	7	8	5	4	8	2	3	4	5	6	5	5	6
27	2	10	7	9	8	9	10	5	3	5	6	7	6	4	5
28	6	3	5	3	5	3	5	0	0	3	3	0	0	5	0
29	4	3	4	3	3	0	0	0	0	4	4	0	0	5	0
30	4	6	5	6	9	4	10	3	1	3	3	2	2	7	3
31	5	5	4	7	8	4	10	3	2	5	5	3	4	8	3
32	3	3	5	7	7	9	10	3	2	5	3	7	5	5	2
33	2	3	5	7	7	9	10	3	2	2	3	6	4	5	2
34	3	4	6	4	3	3	8	1	1	3	3	3	2	5	2
35	6	7	4	3	3	0	9	0	1	0	2	3	1	5	3
36	9	8	5	5	6	6	8	2	2	2	4	5	6	6	3
37	4	9	6	4	10	8	8	9	1	3	9	7	5	3	2
38	4	9	6	6	9	9	7	9	1	2	10	8	5	5	2
39	10	6	9	10	9	10	10	10	10	10	8	10	10	10	10
40	10	6	9	10	9	10	10	10	10	10	10	10	10	10	10
41	10	7	8	0	2	1	2	0	10	2	0	3	0	0	10
42	10	3	8	0	1	1	0	0	10	0	0	0	0	0	10
43	3	4	9	8	2	4	5	3	6	2	1	3	3	3	8
44	7	7	7	6	9	8	8	6	8	8	10	8	8	6	5
45	9	6	10	9	7	7	10	2	1	5	5	7	8	4	5
46	9	8	10	10	7	9	10	3	1	5	7	9	9	4	4
47	0	7	10	3	5	0	10	0	0	2	2	0	0	0	0
48	0	6	10	1	5	0	10	0	0	2	2	0	0	0	0

APPLICANTS SCALED ON 15 VARIABLES   1

OBS	_NAME_	FL	APP	AA	LA	SC	LC	HON
1	FL	1.00	0.24	0.04	0.31	0.09	0.23	-0.11
2	APP	0.24	1.00	0.12	0.38	0.43	0.37	0.35
3	AA	0.04	0.12	1.00	0.00	0.00	0.08	-0.03
4	LA	0.31	0.38	0.00	1.00	0.30	0.48	0.65
5	SC	0.09	0.43	0.00	0.30	1.00	0.81	0.41
6	LC	0.23	0.37	0.08	0.48	0.81	1.00	0.36
7	HON	-0.11	0.35	-0.03	0.65	0.41	0.36	1.00
8	SMS	0.27	0.49	0.05	0.36	0.80	0.82	0.24
9	EXP	0.55	0.14	0.27	0.14	0.02	0.15	-0.16
10	DRV	0.35	0.34	0.09	0.39	0.70	0.70	0.28
11	AMB	0.28	0.55	0.04	0.35	0.84	0.76	0.21
12	GSP	0.34	0.51	0.20	0.50	0.72	0.88	0.39
13	POT	0.37	0.51	0.29	0.61	0.67	0.78	0.42
14	KJ	0.47	0.28	-0.32	0.69	0.48	0.53	0.45
15	SUIT	0.59	0.38	0.14	0.33	0.25	0.42	0.00

OBS	SMS	EXP	DRV	AMB	GSP	POT	KJ	SUIT
1	0.27	0.55	0.35	0.28	0.34	0.37	0.47	0.59
2	0.49	0.14	0.34	0.55	0.51	0.51	0.28	0.38
3	0.05	0.27	0.09	0.04	0.20	0.29	-0.32	0.14
4	0.36	0.14	0.39	0.35	0.50	0.61	0.69	0.33
5	0.80	0.02	0.70	0.84	0.72	0.67	0.48	0.25
6	0.82	0.15	0.70	0.76	0.88	0.78	0.53	0.42
7	0.24	-0.16	0.28	0.21	0.39	0.42	0.45	0.00
8	1.00	0.26	0.81	0.86	0.78	0.75	0.56	0.56
9	0.26	1.00	0.34	0.20	0.30	0.35	0.21	0.69
10	0.81	0.34	1.00	0.78	0.71	0.79	0.61	0.62
11	0.86	0.20	0.78	1.00	0.78	0.77	0.55	0.43
12	0.78	0.30	0.71	0.78	1.00	0.88	0.55	0.53
13	0.75	0.35	0.79	0.77	0.88	1.00	0.54	0.57
14	0.56	0.21	0.61	0.55	0.55	0.54	1.00	0.40
15	0.56	0.69	0.62	0.43	0.53	0.57	0.40	1.00

F I G. 2.2

Page 1 of computer printout for Example 2.1

The results from the preceding SAS commands can be found on page 1 of the computer printout shown in Figure 2.2, except that all correlations that were 0.5 or greater in magnitude have been underlined.

Next partition these variables into subgroups so that variables within a subgroup are highly correlated with each other and so that those in different subgroups have low correlations with each other. In this process, it is probably easiest to begin to form the first subgroup by finding those two variables that are most highly correlated with one another. The correlation between LC and GSP is 0.88 as is the correlation between POT and GSP. Also the correlation between LC and POT is 0.78. Thus, these three variables should definitely be in the same subgroup because all three are highly correlated with one another. SMS should be included in this subgroup since it has correlations of 0.82, 0.78, and 0.75 with LC, GSP, and POT, respectively. AMB should also be included in this subgroup since it has correlations of 0.76, 0.86, 0.78, and 0.77 with the other variables in the subgroup.

Further examination of the correlation matrix reveals that the variables DRV and SC should also be included in this subgroup. Note that all of the variables in the first group have correlations with each other that are at least 0.67. None of the other variables appears to belong to this first group of variables, so now we begin looking for a second group of variables.

A second group of variables can be formed with the variables FL, EXP, and SUIT. The correlations between successive pairs of these variables are 0.55, 0.59, and 0.69, respectively. No other variables belong in this subgroup.

A third group of variables can be initiated by KJ and LA, which have a correlation of 0.69 with each other. HON has a correlation of 0.65 with LA, but only 0.45 with KJ. We could debate whether HON should be included in a subgroup with LA and KJ. Because these data come from measurements on "people," I would be inclined to include HON in this subgroup. What would you do?

Now every variable has been assigned to a subgroup except for AA and APP. The variable AA (academic ability) does not correlate with any of the other variables and, hence, should be placed into a group all by itself. APP has some correlation with many of the variables in group 1 and one might argue that APP can be assigned to group 1. However, I am going to keep APP in a group by itself since its correlations with many of the variables in group 1 are much lower than the correlations among the other variables that have previously been assigned to group 1. What would you do?

To summarize, the final groups of variables are:

Group 1: SC, LC, SMS, DRV, AMB, GSP, and POT.

Group 2: FL, EXP, and SUIT.

Group 3: LA, HON, and KJ.

Group 4: AA.

Group 5: APP.

A question that might be asked is "Does this grouping of variables into groups have any influence on how you might choose applicants to whom to make job offers?"

It is interesting to note that, initially, most readers probably believed that 15 different characteristics of each applicant were being measured by these 15 variables. But the preceding groupings would make most researchers believe that only 5 different characteristics were measured. As a result, many would believe that each applicant should be evaluated on these 5 characteristics. Then, to decide on those to whom to make job offers, a new overall score could be computed for each individual by either taking an average of these 5 new characteristics or a weighted average of these 5 new characteristics.

To evaluate each individual on each of these underlying characteristics, one could simply average the variables that occur within each group. Thus one would have

$$G_1 = (SC + LC + SMS + DRV + AMB + GSP + POT)/7,$$
$$G_2 = (FL + EXP + SUIT)/3,$$
$$G_3 = (LA + HON + KJ)/3,$$
$$G_4 = AA, \text{ and}$$
$$G_5 = APP$$

Then a score for each individual could be computed by using $WTD\_AVG = w_1G_1 + w_2G_2 + \cdots + w_5G_5$ where $w_1 + w_2 + \cdots + w_5 = 1$ and where w_i measures the relative importance of the ith underlying characteristic, $i = 1, 2, \ldots, 5$.

Another way to compute a score for each of the new variables would be to choose one of the variables in each group of variables to represent the underlying characteristic for each group of variables. One of the exercises at the end of this chapter will ask you to identify the six individuals to whom to make job offers by using new variables that represent each of the underlying characteristics.

Relationship to Factor Analysis

The preceding example illustrates a very "crude" type of factor analysis. To summarize, the correlation matrix of the original 15 variables was used to define 5 new, nearly uncorrelated, variables that, in some sense, explain the original variables that were being measured. These new variables describe some underlying characteristics of the applicants applying for a job with this firm. Each individual is then evaluated on these new variables, after which these new variables are used to decide whom to offer jobs.

In Chapter 6, factor analysis is developed in a much more formal way; although the process described in the preceding example may be somewhat crude, it has at least one possible advantage over factor analysis in that in the process, everyone is fully aware of what has actually taken place.

2.2
Summary

It is important to remember that correlation measures the strength of a linear relation between two variables. Two variables could be uncorrelated with one another and still be related to one another. It is important to look at confidence intervals for population correlations to determine accurately whether a sample correlation is large enough to be meaningful. A rule of thumb for determining when sample correlations are large enough to be useful depends on the kinds of data you are considering. Generally, correlations need to have a magnitude of 0.5 or greater in order to be of much practical importance. It is important to consider pairwise plots of pairs of variables to see if nonlinear relationships exist between them. It is often helpful to look at patterns of interrelationships among variables because these interrelationships often have an influence on how the variables might be utilized in future analyses. Furthermore, such interrelationships might suggest the creation of new variables that could be useful in future analyses of the data.

Exercises

1   The applicant data set from Example 2.1 can be found in the file labeled APPLICAN.DAT on the enclosed disk. Create pairwise scatter plots among all pairs of variables that fell into group 1 in Example 2.1.

2   Using the applicant data in the file labeled APPLICAN.DAT on the enclosed disk, evaluate each of the applicants on the variables G_1–G_5 created in Example 2.1. Next average these five new variables to create a single score for each applicant. Finally, determine the six applicants to whom job offers should be made.

3   Using the applicant data in the file labeled APPLICAN.DAT on the enclosed disk, find 95% confidence intervals for true correlations between the variable pairs (LC, SC), (SC, SMS), and (LC, SMS) using Eq. (2.2).

4   Repeat Exercise 3 using Ruben's approximation.

5   Repeat Exercise 3 using the correlation chart shown in Figure 2.1.

6   Use the data you created for Exercise 1 in Chapter 1 to answer each of the following.

   a   Using a statistical computing package of your choice, find the correlation matrix for your data.

   b   Can the variables in your data set be partitioned into subgroups so that the variables within the same subgroup have high correlations with one another, and so that the variables in different subgroups have low correlations with one another? If so, what variables are in each of the subgroups?

   c   Choose the three pairs of variables that have the highest correlations between them, and construct 95% confidence intervals for the true correlation between each pair using any method you wish. Do any of your intervals include zero? Which ones? Show your work.

d Using any computer plotting package of your choice, create pairwise scatter plots for the three pairs of variables that you selected in part c.

e Choose the three pairs of variables that have the lowest correlations between them, and construct 95% confidence intervals for the true correlation between each pair using any method you wish. Do any of your intervals include zero? Which ones? Show your work.

f Using any computer plotting package of your choice, create pairwise scatter plots for the three pairs of variables that you selected in part e.

g Consider your plots in parts d and f. Do there appear to be relationships between these pairs of variables that are nonlinear? Explain your answer.

7 The file labeled CITYTEMP.DAT on the enclosed data disk contains the high and low temperatures for a random sample of U.S. cities. The data are shown below. The temperature measurements were taken on August 17, 1995.

AUGUST 17, 1995 HIGH AND LOW TEMPERATURE

CITY	HIGH	LOW	CITY	HIGH	LOW
SEATTLE	63	62	PORTLAND	68	47
BOISE	69	37	BILLINGS	84	52
JACKSON HOLE	78	65	SALT LAKE CITY	85	55
SAN FRANCISCO	71	55	LAS VEGAS	100	73
LOS ANGELES	81	63	PHOENIX	103	82
ALBUQUERQUE	87	64	DENVER	93	60
MINNEAPOLIS	86	71	OMAHA	91	73
KANSAS CITY	91	74	TULSA	96	75
DALLAS	95	75	SAN ANTONIO	98	76
MILWAUKEE	93	72	DES MOINES	90	69
ST. LOUIS	95	78	LITTLE ROCK	99	75
NEW ORLEANS	93	75	DETROIT	89	72
CHICAGO	91	75	LOUISVILLE	94	76
NASHVILLE	98	75	ATLANTA	98	78
CLEVELAND	89	70	PHILADELPHIA	88	72
BOSTON	81	69	NEW YORK CITY	85	70
WASHINGTON, D.C.	87	74	CHARLOTTE	88	72
MIAMI	92	77			

a Create a two-dimensional scatter plot that plots high temperature against low temperature. Do there appear to be any outliers in this data set? Explain your answer.

b Assuming cities are independently distributed, find a 90% confidence interval for the true correlation between the high and low temperatures on this date for U.S. cities using both Fisher's and Ruben's methods.

Find a 95% confidence interval for the true correlation by using both methods.

  c Discuss the independence assumption in part b. Does it seem reasonable for these data? Why or why not?

8 The file labeled CITIES2.DAT on the enclosed data disk contains the high, low, and normal high temperatures for a random sample of non–U.S. cities. The data are shown below. The temperature measurements were taken on August 17, 1995.

August 17, 1995 Temperatures—World Cities

CITY	HIGH	LOW	NORMAL HIGH	CITY	HIGH	LOW	NORMAL HIGH
ABERDEEN	75	54	63	AMSTERDAM	75	52	68
ANKARA	88	55	88	ATHENS	88	75	91
AUCKLAND	50	50	58	BEIJING	84	68	86
BERLIN	70	54	72	BOGOTA	68	43	64
BONN	77	54	73	BRUSSELS	75	54	72
BUENOS AIRES	46	39	61	CAIRO	93	71	95
CALGARY	72	41	72	CARACAS	82	70	88
CASABLANCA	79	72	81	COPENHAGEN	68	55	70
DUBLIN	75	61	66	FRANKFURT	77	59	75
GENEVA	77	61	75	HONG KONG	84	79	88
JERUSALEM	84	66	88	KIEV	82	59	73
LIMA	64	59	63	LISBON	90	70	82
LONDON	86	64	70	MADRID	95	68	90
MECCA	100	86	102	MEXICO CITY	73	57	73
MONTREAL	88	66	77	MOSCOW	61	54	72
NAIROBI	68	55	70	NEW DELHI	93	81	93
NICE	81	70	81	PARIS	79	57	75
RIO de JANEIRO	100	66	75	RIYADH	111	81	108
ROME	86	64	86	SAO PAULO	88	63	66
SEOUL	93	75	88	SOFIA	81	61	79
STOCKHOLM	73	52	66	SYDNEY	68	46	63
TAIPEI	85	75	91	TOKYO	93	81	86
TORONTO	90	66	79	VIENNA	72	57	79
WARSAW	66	54	73	WINNIPEG	90	63	75

  a Assuming independence among cities, find a 95% confidence interval for the true correlation between the high and low temperatures for non–U.S. cities on this date using both Fisher's and Ruben's methods.

  b Assuming independence among cities, find a 95% confidence interval for the true correlation between the high and normal high temperatures for non–U.S. cities using both Fisher's and Ruben's methods.

c Create two-dimensional scatter plots between all pairs of these three variables. Do there appear to be any outliers in this data set? Explain your answer.

d Discuss the independence assumption in part b. Does it seem reasonable for these data? Why or why not?

9 Go to your local newspaper or *USA Today* and find temperatures for U.S. cities or world cities for yesterday's temperatures. Repeat Exercises 7 and/or 8 for these data.

10 A survey of 36 delivery drivers employed in the San Diego division of a national restaurant chain was conducted. This data is a small subset of a much larger study described in Appendix B. The data in this exercise are real, but for obvious reasons, the true name of the restaurant chain is not disclosed. In this book the restaurant chain is called PIZZAZZ. The survey questions used to collect the data in this exercise are given below.

USE THE FOLLOWING SCALE TO ANSWER QUESTIONS **1–15** (CIRCLE A NUMBER):

1	**2**	**3**	**4**	**5**	**6**	**7**
STRONGLY DISAGREE	MODERATELY DISAGREE	SLIGHTLY DISAGREE	NEITHER DISAGREE NOR AGREE	SLIGHTLY AGREE	MODERATELY AGREE	STRONGLY AGREE

1. I AM WILLING TO WORK HARDER THAN MOST PEOPLE AT PIZZAZZ.   1 2 3 4 5 6 7

2. I TALK UP PIZZAZZ TO MY FRIENDS AS A GREAT COMPANY TO WORK FOR.   1 2 3 4 5 6 7

3. I FEEL VERY LITTLE LOYALTY TO PIZZAZZ.   1 2 3 4 5 6 7

4. I WOULD ACCEPT ALMOST ANY TYPE OF JOB IN ORDER TO STAY WITH PIZZAZZ.   1 2 3 4 5 6 7

5. MY VALUES AND PIZZAZZ'S VALUES ARE SIMILAR.   1 2 3 4 5 6 7

6. I AM PROUD TO TELL OTHERS THAT I WORK FOR PIZZAZZ.   1 2 3 4 5 6 7

7. I COULD JUST AS WELL BE WORKING FOR ANOTHER COMPANY AS LONG AS THE WORK WAS SIMILAR.   1 2 3 4 5 6 7

8. PIZZAZZ REALLY INSPIRES ME TO DO MY BEST.   1 2 3 4 5 6 7

9. IT WOULD TAKE VERY LITTLE CHANGE IN MY WORK TO CAUSE ME TO LEAVE PIZZAZZ.   1 2 3 4 5 6 7

10. I AM GLAD THAT I CHOSE PIZZAZZ TO WORK FOR.   1 2 3 4 5 6 7

11. THERE'S NOT MUCH TO BE GAINED BY STAYING WITH PIZZAZZ A LONG TIME.   1 2 3 4 5 6 7

12. I OFTEN FIND IT DIFFICULT TO AGREE WITH PIZZAZZ'S ATTITUDES TOWARDS ITS EMPLOYEES.   1 2 3 4 5 6 7

13. I REALLY CARE ABOUT WHAT HAPPENS TO PIZZAZZ.   1 2 3 4 5 6 7

14. FOR ME THIS IS THE BEST COMPANY TO WORK FOR.   1 2 3 4 5 6 7

15. DECIDING TO WORK FOR PIZZAZZ WAS A MISTAKE.   1 2 3 4 5 6 7

The data obtained from this study are given below and can be found in the data file labeled COMMIT3.DAT on the enclosed data disk. The entries in this data file are employee ID and position number, followed by the circled responses to the 15 items given in the survey.

```
* * * * * * * * * * * * * * * * * * * * * * * * * * * * * * * * * * * * * * * * * * * * * * * * *
140   3   6   6   4   2   5   6   3   4   4   6   3   4   5   5   1
141   3   7   4   2   1   6   7   2   4   1   7   5   4   7   6   1
142   3   5   7   1   2   4   4   4   6   5   7   4   4   4   6   1
143   3   6   4   1   6   7   4   1   6   2   7   4   1   7   7   1
144   3   5   2   3   1   5   3   3   3   3   1   3   3   5   1   6
145   3   6   4   2   4   4   5   5   6   3   7   4   4   5   5   1
161   3   7   1   1   1   1   1   4   1   7   4   7   7   1   1   7
162   3   4   4   3   1   5   4   5   4   2   6   5   3   4   4   1
165   3   7   5   3   1   4   5   3   4   2   7   2   6   6   4   1
166   3   3   4   7   1   2   1   7   2   7   1   7   7   1   4   4
167   3   7   5   6   1   6   1   2   6   4   7   6   1   7   5   2
169   3   5   7   4   7   6   7   4   6   1   7   3   1   4   7   3
262   3   4   1   5   1   5   1   5   4   5   4   7   4   3   1   4
339   3   .   .   .   .   .   .   .   .   .   .   .   .   .   .   .
340   3   6   4   4   2   4   4   5   4   3   5   6   4   4   1   2
398   3   6   7   1   2   7   7   5   .   5   7   1   1   6   7   1
399   3   7   5   1   3   5   6   3   4   4   6   5   1   6   4   1
400   3   6   4   3   3   3   4   3   4   3   4   4   5   4   3   4
401   3   5   5   3   2   4   3   5   2   2   4   5   5   4   4   3
403   3   7   5   1   2   4   4   4   4   2   5   2   4   5   4   2
404   3   6   3   6   4   4   6   6   5   4   6   7   5   4   1   1
518   3   5   5   3   2   4   4   5   3   4   4   4   2   4   2   2
519   3   6   7   4   1   4   7   4   4   1   7   3   4   4   3   1
520   3   4   5   4   1   4   3   5   4   6   5   7   4   5   4   2
522   3   6   6   4   2   5   5   2   5   2   6   6   6   5   5   2
524   3   5   5   5   3   5   6   3   3   5   5   5   5   5   5   3
733   3   6   6   6   1   6   3   3   2   6   5   3   5   3   2   3
734   3   7   4   2   2   2   4   4   4   6   5   4   5   4   3   3
739   3   6   6   2   2   6   5   1   5   5   6   1   2   6   5   1
740   3   6   6   4   2   4   4   6   4   3   4   3   2   5   2   2
768   3   4   4   7   4   4   4   4   4   4   4   4   4   6   4   1
769   3   5   7   1   5   4   7   4   5   7   7   4   1   6   6   1
770   3   7   6   1   5   6   6   4   6   1   1   1   2   7   7   1
772   3   6   5   1   4   4   5   4   5   2   6   2   2   4   5   1
773   3   7   6   3   3   6   6   6   6   1   6   2   2   5   4   2
842   3   7   1   3   1   2   1   1   1   1   2   2   7   4   5   3
* * * * * * * * * * * * * * * * * * * * * * * * * * * * * * * * * * * * * * * * * * * * * * * * *
```

Answer each of the following questions:

a   Without doing any statistical analyses, please try to answer the following questions as best you can.

   i   Do you see any outliers in this data set? If so, which employees look to be outliers? Explain why you think these employees are outliers.

   ii Obviously, the responses are discrete, so these data cannot truly be distributed as a multivariate normal population. However, do you believe the variables are approximately multivariate normally distributed? Explain your answer.

   The answers to the two preceding questions are not easy. Most researchers are not able to look visually at a data set and provide accurate answers to these questions. The reason for asking you to consider these questions at this point in time is to try to excite you about some upcoming methods that may help provide good answers to these questions.

b   Using a statistical computing package of your choice, find the correlation matrix corresponding to the responses to these 15 items.

c   Can these responses be partitioned into subgroups so that the responses within a subgroup have high correlations with one another, and so that responses in different subgroups have low correlations with one another? If so, what responses belong to each of the subgroups?

d   Consider the first three responses. Construct 95% confidence intervals for the true correlation between each pair using any method you wish. Do any of your intervals include zero? Which ones? Show your work.

e   Using any computer plotting package of your choice, create pairwise scatter plots for the first five responses.

f   Consider your plots in part e. Do there appear to be nonlinear relationships between these pairs of variables? Explain your answer.

11 Consider all of the survey data obtained from 975 employees of a national restaurant chain described in Appendix B.

a   Using a statistical computing package of your choice, find the correlation matrix corresponding to the responses to the first 18 questions in the survey.

b   Can these responses be partitioned into subgroups so that the responses within a subgroup have high correlations with one another, and so that responses in different subgroups have low correlations with one another? If so, what responses belong to each of the subgroups?

c   Choose the three pairs of responses that have the highest correlations between them, and construct 95% confidence intervals for the true correlation between each pair using any method you wish. Do any of your intervals include zero? Which ones? Show your work.

d Using any computer plotting package of your choice, create pairwise scatter plots for the three pairs of responses that you selected in part c.

e Choose the three pairs of responses that have the lowest correlations between them, and construct 95% confidence intervals for the true correlation between each pair using any method you wish. Do any of your intervals include zero? Which ones? Show your work.

f Using a computer plotting package of your choice, create pairwise scatter plots for the three pairs of variables that you selected in part e.

g Consider your plots in parts d and f. Do there appear to be nonlinear relationships between these pairs of variables? Explain your answer.

12 Consider the survey data obtained from 304 married adults that is described in Appendix C.

a Using a statistical computing package of your choice, find the correlation matrix for the variables FACES1–FACES30 for the *adult males* in this data set.

b Can these variables be partitioned into subgroups so that the variables within a subgroup have high correlations with one another, and so that variables in different subgroups have low correlations with one another? If so, what variables belong to each of the subgroups?

c Repeat parts a and b for the adult females in this data set.

d Based on your answers to parts a–c, do there appear to be differences between males and females? You may answer this question using only your intuition. That is, no statistical justification of your answer is required at this point in time.

13 Repeat Exercise 12 using the variables FAD1–FAD60.

14 Consider the survey data obtained from 304 married adults that is described in Appendix C.

a Consider only the females in this data set. Find the sample covariance and correlation matrices for the Kansas Marital Satisfaction variables KMS1–KMS3.

b Construct 95% confidence intervals for the true correlations between all pairs of these three variables using any method you wish. Do any of your intervals include zero? Which ones? Show your work.

c Using a computer plotting package of your choice, create pairwise scatter plots for all pairs of these three variables.

d Repeat parts a–c for the males in this data set.

15 Consider the survey data obtained from 304 married adults that is described in Appendix C.

    a Consider only the males in this data set. Find the sample correlation matrix for the Kansas Family Life Satisfaction Scale variables KFLS1–KFLS4.

    b Using any computer plotting package of your choice, create pairwise scatter plots for all pairs of these four variables.

    c Repeat parts a and b for the females in this data set.

Multivariate Data Plots

3

In this chapter, several graphical procedures that have been suggested for plotting multivariate data are introduced. Visual displays of data are almost always more informative than printouts of large data sets. While there are many reasons to consider visual displays of multivariate data, two of the most important are (1) to help a researcher locate and/or identify abnormalities that might exist in the data and (2) to help a researcher check assumptions that may be required for certain statistical analyses to be valid.

As a simple example of why we plot multivariate data, consider the pairwise scatter plot shown in Figure 3.1. There is one point in this plot that appears to be an *outlier*, that is, a point that appears to be inconsistent with other points in the data set. If we were to examine only the x_1 values for these data points, the suspicious point would not look that unusual. This is because several of the data points have x_1 values larger than the point in question. Likewise, if we were to examine the x_2 values for these data points, the suspicious point would still not look unusual because several of the data points have x_2 values smaller than the point in question. The only thing that is really unusual about this particular point is not its x_1 value or its x_2 value, but the combination of these two values occurring at the same time. This combination of x_1 and x_2 seems very unusual for these data. It is nearly impossible to discover this kind of abnormality in a large data set without plotting the data in some way.

In addition to locating abnormalities in a data set and checking assumptions about the data that may be made, there are several other reasons for plotting multivariate data. Plotting techniques are often useful for helping to verify and/or validate the results of clustering programs. Some plotting techniques can help a researcher formulate possible relationships that might exist among the variables being measured or among the experimental units from which the data are being collected.

This book assumes familiarity with one-dimensional data plotting techniques and two-dimensional scatter plots. Many of the available statistical computing packages create such plots. Some of the more popular one-dimensional plots that are useful for displaying data include bar charts, histograms, stem-and-leaf plots, box plots, and normal-quantile plots (see Atkinson, 1985; Becker, Cleveland, and Wilks, 1987; Tukey, 1977).

Sometimes it is useful to apply one-dimensional plotting methods to new variables that can be created through a multivariate analysis. For example, factor analysis and principal components analysis are often used to create new variables that summarize the information available in an original set of variables. Univariate plots based on these newly created variables can be very useful for identifying outliers, checking distributional assumptions, and discovering abnormalities that may exist in a data set.

3.1
Three-Dimensional Data Plots

Three-dimensional data can be plotted by plotting each three-dimensional observation vector as a "blob" on a two-dimensional graph, letting the coordinates of the blob represent two of the variables and the size of the blob

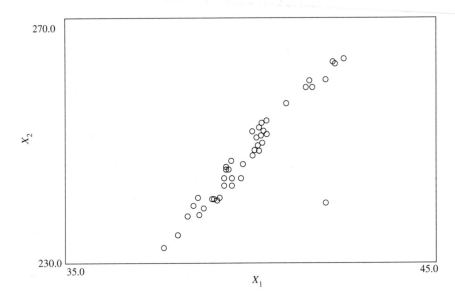

F I G. 3.1

Scatter plot show-
ing an outlier

represent the third. Some people refer to these kinds of plots as *blob plots,* others refer to them as *bubble plots.*

As an example consider the blob, or bubble, plot in Figure 3.2. The data in Figure 3.2 came from measurements taken on U.S. Navy bachelor officers' quarters. Three of the variables in this data set were the average daily occupancy (ADO) of a bachelor officer's quarter, the average number of check-ins per month (MAC), and the monthly man-hours required to operate the facility (MMH). The abscissa in Figure 3.2 corresponds to the logarithm of MAC, the ordinate corresponds to the logarithm of ADO, and the size of the bubble is proportional to the monthly man-hours required to run the facility. That is, large blobs (or bubbles) are plotted when the facility requires a large number of monthly man-hours, and small blobs are plotted when the facility requires a small number of monthly man-hours. The logarithms of MAC and ADO were used to help linearize the relationship between these two variables. The SAS commands that were used to create this plot are available in the file labeled FIG3_2.SAS on the disk provided with this book and are repeated below.

```
TITLE 'U.S. NAVY BACHELOR OFFICERS'' QUARTERS';
TITLE2 'Bubble and/or Blob Plot';
TITLE3 'Bubble Size is Proportional to Monthly Man Hours';
DATA USNAVY;
 INPUT  SITE 1-2 ADO MAC WHR CUA WNGS OBC RMS MMH;
 LOGADO=LOG(ADO);
 LOGMAC=LOG(MAC);
 LABEL ADO = 'AVERAGE DAILY OCCUPANCY'
       MAC = 'AVERAGE NUMBER OF CHECK-INS PER MO.'
```

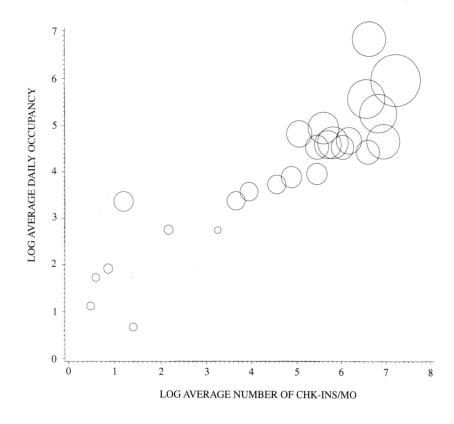

F I G. 3.2
U.S. Navy bachelor officers' quarters where bubble size is proportional to MMH (monthly man-hours)

```
        WHR  = 'WEEKLY HRS OF SERVICE DESK OPERATION'
        CUA  = 'SQ FT OF COMMON USE AREA'
        WNGS = 'NUMBER OF BUILDING WINGS'
        OBC  = 'OPERATIONAL BERTHING CAPACITY'
        RMS  = 'NUMBER OF ROOMS'
        MMH  = 'MONTHLY MAN-HOURS'
      LOGADO = 'LOG AVERAGE DAILY OCCUPANCY'
      LOGMAC = 'LOG AVERAGE NUMBER OF CHK-INS/MO.';
CARDS;
  1    2          4        4    1.26    1    6    6   180.23
  2    3          1.58    40    1.25    1    5    5   182.61
  3   16.6       23.78    40    1       1   13   13   164.38
  .    .          .        .    .       .    .    .     .
  .    .          .        .    .       .    .    .     .
  .    .          .        .    .       .    .    .     .

GOPTIONS DEVICE=PS2EGA;

PROC GPLOT;
  BUBBLE LOGADO*LOGMAC=MMH/BSIZE=15;
RUN;
```

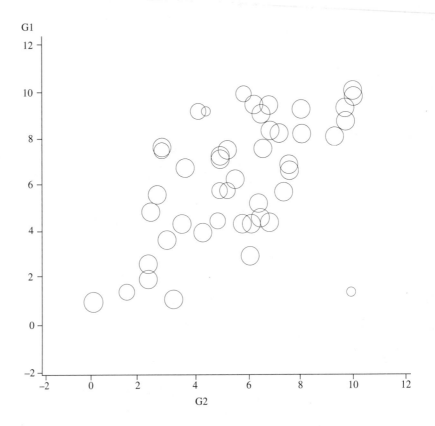

F I G. **3.3**
Bubble plot for
applicant data set
where bubble size
is proportional to
G3 (group 3)

An examination of this plot reveals that the three variables being plotted are highly correlated with one another. Not only is there a strong linear relationship between LOGADO and LOGMAC, but note how the size of the bubbles increases as both LOGADO and LOGMAC increase. This indicates a strong linear relationship between MMH and each of LOGADO and LOGMAC.

Figure 3.3 illustrates a blob plot for the applicant data described in Chapter 2. The variables plotted are G1, G2, and G3 where G1 = (SC + LC + SMS + DRV + AMB + GSP + POT)/7, G2 = (FL + EXP + SUIT)/3, and G3 = (LA + HON + KJ)/3. The size of the blob is proportional to G3. Note any possible outliers that may exist in these data.

Figure 3.4 shows a three-dimensional (3-D) plot of the data plotted in Figure 3.2. Again, note the strong linear relationships between each of the three pairs of variables. The additional SAS commands that created this plot can be found in the file labeled FIG3_4.SAS on the enclosed disk and are repeated below.

```
PROC G3D;
  SCATTER LOGADO*LOGMAC=MMH;
  TITLE2 '3-D Plot';
RUN;
```

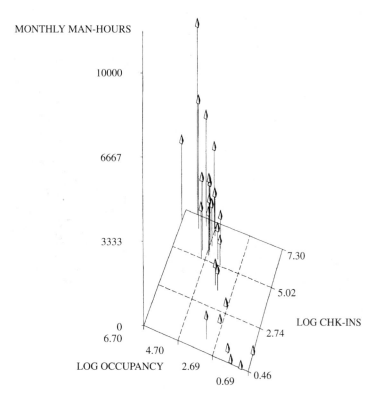

MONTHLY MAN-HOURS

F I G. 3.4
Three-dimensional
plot of U.S. Navy
bachelor officers'
quarters data

Figures 3.5 and 3.6 show 3-D plots of the new variables created in the APPLICANT data set, namely, the variables G1, G2, and G3. These plots were created by NCSS, the Number Cruncher Statistical System (1992). Note the two possible outliers that may exist in these data. Other than these two points, the rest of the data seem to fall into an ellipsoidally shaped region, which would be consistent with these variables having a trivariate normal distribution.

Many statistical graphics packages allow users to create 3-D scatter plots and then spin the cloud of data points around the vertices of the axes. These interactive plots can reveal interesting aspects of a set of data. Unfortunately, such plots cannot easily be illustrated here.

3.2
Plots of Higher Dimensional Data

The SAS GPLOT procedure has an option that allows the user to combine the features of a blob plot with those of a 3-D plot. This feature allows the user to plot bubbles or blobs whose sizes are proportional to a fourth variable in a three-dimensional plot. Figure 3.7 illustrates such a plot for variables G1 through G3 and AA from the APPLICANT data set. The size of the bubble

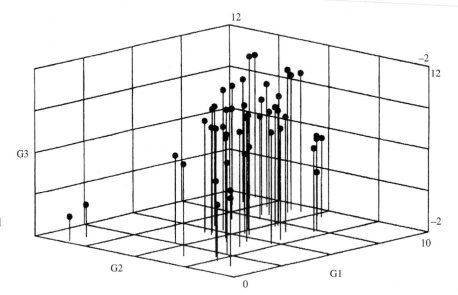

F I G. **3.5**
Three-dimensional
plot of appli-
cant data using
variables G1
through G3

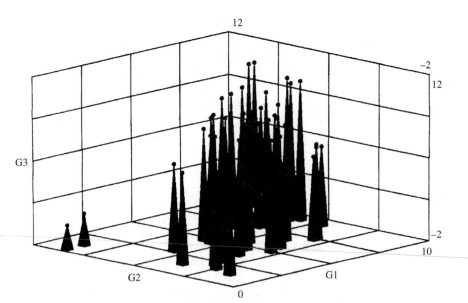

F I G. **3.6**
Three-dimensional
plot of appli-
cant data using
variables G1
through G3

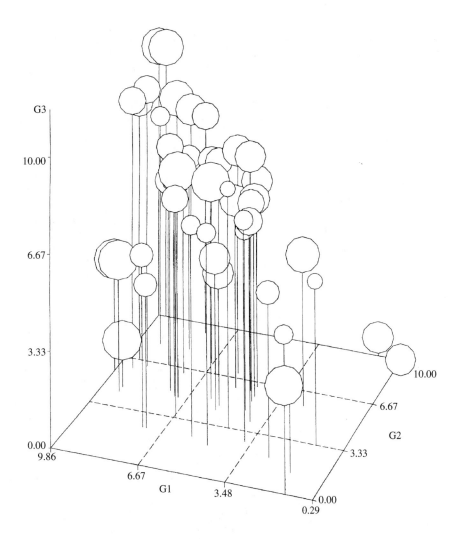

FIG. 3.7

Four-dimensional plot of applicant data where bubble size is proportional to AA

plotted for each applicant is proportional to the applicant's measure of academic ability (AA). Do the two possible outliers show up in this plot?

Chernoff Faces

Chernoff (1973) suggested using faces to represent multidimensional data. Chernoff recommended associating different facial characteristics with different variables. For example, one variable could be associated with vertical eye width, a second with horizontal eye width, a third with the size of the iris, and others could be associated with eye spacing, eye height, nose length, nose width, brow length, brow width, brow slant, ear width, ear length, ear height,

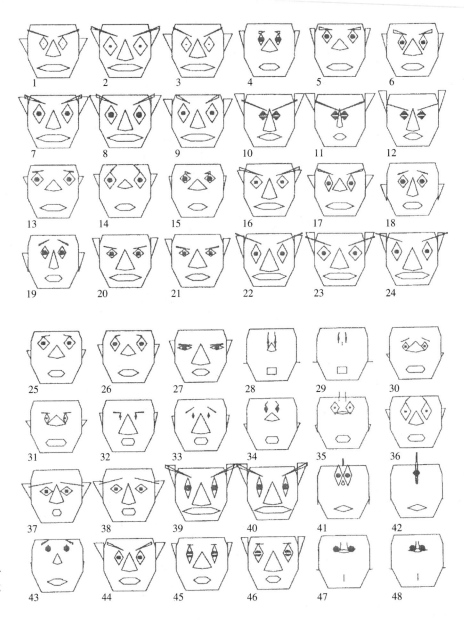

F I G. 3.8
Face plots for applicant data using all 15 responses

midmouth length, corner mouth length, mouth opening, mouth smile, etc. Figure 3.8 shows a set of Chernoff faces for each of the applicants in the APPLICANT data set. These faces were created by the NCSS graphics program.

Chernoff faces are often useful for identifying outliers in a multivariate data set. Which faces in Figure 3.8 seem to be the most unusual when compared

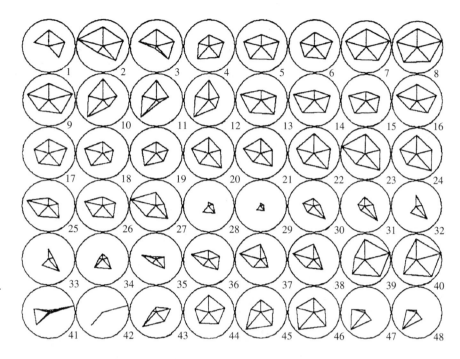

FIG. 3.9

Star plots of applicant data using variables G1 through G5

to other faces in the data set? Some of the applicants that have unusual faces include applicants 28, 29, 41, 42, 47, and 48. Do you agree? These unusual faces may identify possible outliers in the data.

Chernoff faces are also very useful for validating the results of clustering programs. Clustering programs try to partition the experimental units in a data set into subgroups, called *clusters*, so that the individuals within a cluster are similar to one another, and the individuals in different clusters are not similar to one another. If a computer program has successfully achieved this goal, then Chernoff faces for individuals within a cluster should be similar to one another, and faces for those in different clusters should be dissimilar. Notice the similarity between the Chernoff faces for applicants 7 and 8, 10 and 11, and 22, 23, and 24.

Star Plots and Sun-Ray Plots

Suppose the response variables in a data set are all positive. *Star plots*, also called *sun-ray plots*, are constructed by plotting the distance that each variable is from zero on rays or axes radiating from a central point. There is one ray for each response variable. For example, five-dimensional data vectors would require five rays or axes. For five axes, each axis would form a 72-degree angle with its adjacent axes. Figure 3.9 shows a set of star plots for the

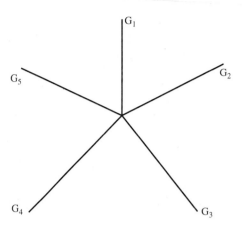

F I G. 3.10
Star plot axes

applicants in the APPLICANT data set using variables G1, G2, and G3, defined earlier, and G4 = AA and G5 = APP. In Figure 3.9, G1 is plotted along the axis pointing toward the north (i.e., the axis pointing directly upward). The other variables are plotted on the other axes in a clockwise order. See Figure 3.10 for identification of the axes.

Sun-ray and star plots are also useful for identifying multivariate outliers in a data set and for validating the results of clustering programs. Are there any applicants that appear to be outliers in Figure 3.9? How about applicants 41 and 42? How about applicants 28 and 29?

Figure 3.11 shows star plots for the individuals in the APPLICANT data set using all 15 of the original variables.

For the applicant data, the star plots have another interpretation that is kind of interesting. In Chapter 2 we tried to determine whose individuals to whom to make job offers. Generally an employer would like to offer jobs to individuals that have high values for the variables G1, G2, . . . , G5. These individuals would show up on the star plot as those individuals whose polygons have the largest areas. An examination of Figure 3.9 would seem to indicate that applicants 7 and 8 are the two best individuals in this applicant pool. Do you agree? This kind of an interpretation might also be very useful when the data consist of consumer ratings of the effectiveness of different formulations of a product such as underarm deodorants.

Usually, Chernoff faces and star plots are made subsequent to standardizing the data. In this case, the center point in the star plots corresponds to minimums for each of the variables being plotted. Standardizing the applicant data is not necessary because all variables were measured on the same 0–10 scale.

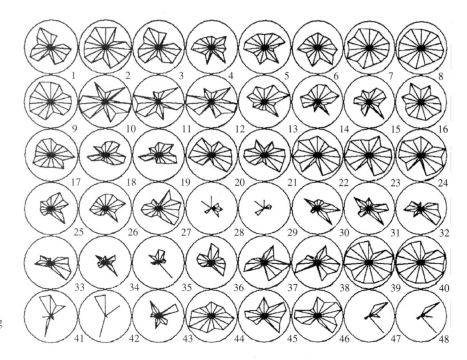

FIG. 3.11
Star plots for applicant data using all 15 responses

Andrews' Plots

Andrews (1972) suggested that the *p*-variate observation for the *r*th experimental unit, namely, $\mathbf{x}'_r = [x_{r1}, x_{r2}, \cdots, x_{rp}]$, could be represented by the function

$$f_r(t) = x_{r1}/2^{1/2} + x_{r2} \cdot \sin(t) + x_{r3} \cdot \cos(t) + x_{r4} \cdot \sin(2t) + x_{r5} \cdot \cos(2t) + \cdots$$

for $-\pi < t < \pi$. Thus the data for an individual give rise to a unique function for that individual. These functions can then be plotted as *t* goes from $-\pi$ to π. The resulting curves do not picture relationships among the variables, but are useful for finding or validating clusters that might exist in the data. They are also useful for locating outliers in the data. Any individual that produces a curve that is much different than the curves representing other individuals would correspond to an outlier.

When constructing Andrews' plots it is important that the response variables be measured in similar units. Thus, quite often the data are standardized before constructing Andrews' plots.

Interpretations of the resulting plots are affected by the order in which the variables are labeled. If some variables are thought to be more important than other variables, the most important variable should be taken to be x_1, the second most important variable should be taken to be x_2, etc.

In those cases where there are a large number of response variables,

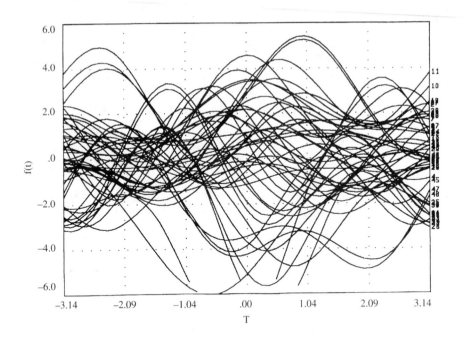

F I G. 3.12

Andrews' plots
for applicant
data using the
variables G1
through G5

Andrews' plots can be constructed subsequent to performing a principal components analysis on the data. In this case, x_1 is taken to be the first principal component score, x_2 is taken as the second principal component score, etc.

Figure 3.12 shows a set of Andrews' curves for each of the applicants in the APPLICANT data set using variables G1 through G5 as defined earlier. This plot was created by NCSS and the variables were automatically standardized prior to creating the plot. Figure 3.13 shows a set of Andrews' curves for these same applicants using the GPLOT procedure in SAS. Some of the curves that seem distinctly different than other curves have been identified by applicant numbers in Figure 3.13. For example, the curves corresponding to 28, 29, 32, 33, 41, and 42 appear to be different than most of the other curves. In Figure 3.12, the variables G1 through G5 were standardized prior to constructing the Andrews' curves. The variables were not standardized in Figure 3.13. Standardization does not seem necessary for these variables because all of the original variables are measured in similar units, and G1 through G5 were constructed so that they were also measured in similar units. The commands that created the plot in Figure 3.13 are available in the file labeled FIG3_13.SAS on the enclosed disk.

Side-by-Side Scatter Plots

Many graphical packages are available that simultaneously produce scatter plots between many pairs of variables. These plots are useful for exploring relationships that might exist among the measured variables. Figures 3.14,

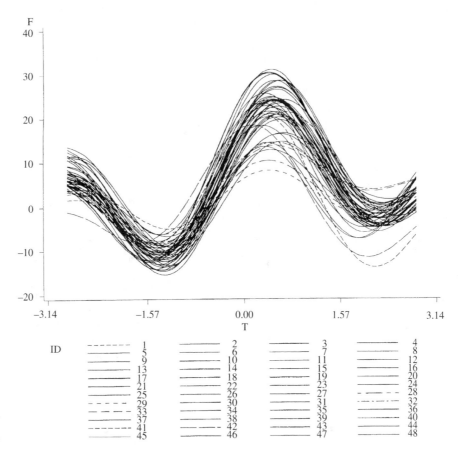

FIG. 3.13
Andrews' plots
for applicant data
using the GPLOT
procedure in SAS

3.15, and 3.16 show such plots for the subsets of variables falling into each of the first three groups of variables for the APPLICANT data set. A simple linear regression line could be fit to each pair of variables. This line could then be superimposed on each pairwise plot. Note that a positive correlation exists between all pairs of variables in each of these plots. Figure 3.17 shows a side-by-side scatter plot for the new variables G1 through G5 that were created in Chapter 2 from the APPLICANT data. Do these pairs of variables fall into elliptically shaped regions that would be consistent with the data being multivariate normally distributed?

3.3
Plotting to Check for Multivariate Normality

Methods for determining whether a set of multivariate data is distributed according to a multivariate normal distribution are limited. It is known that

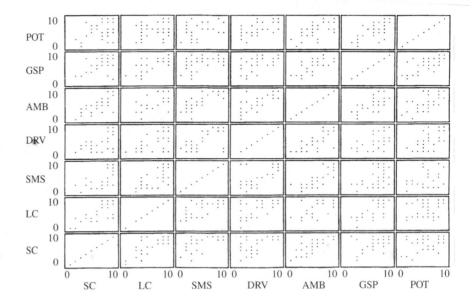

F I G. 3.14
Side-by-side correlation plots for the variables in group 1

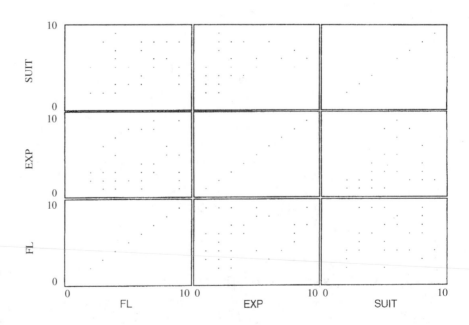

F I G. 3.15
Side-by-side correlation plots for the variables in group 2

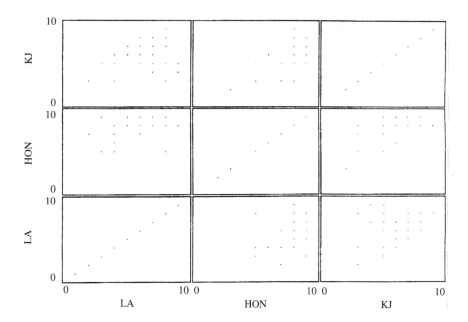

F I G. 3.16

Side-by-side cor-
relation plots for
the variables in
group 3

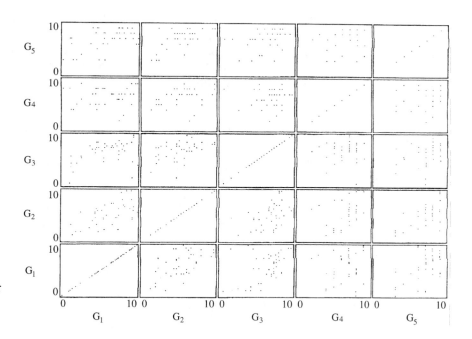

F I G. 3.17

Pairwise plots for
the variables G1
through G5

if a set of data vectors is a random sample from a multivariate normal distribution, then the principal components will have independent normal distributions. In Chapter 5, principal components of a set of data are used in univariate tests for normality.

Another method for assessing whether a set of data could be multivariate normal is described in this section. Most readers will be familiar with normal probability plots and their use in assessing whether univariate data tend to be normally distributed. For multivariate data we can develop similar types of plots.

Suppose $\mathbf{x}$ is known to have a p-variate normal distribution with mean $\boldsymbol{\mu}$ and variance–covariance matrix $\boldsymbol{\Sigma}$. Then it has been shown that $(\mathbf{x} - \boldsymbol{\mu})'\boldsymbol{\Sigma}^{-1}(\mathbf{x} - \boldsymbol{\mu})$ is distributed as a chi-square probability distribution with p degrees of freedom.

Now suppose $\mathbf{x}_1, \mathbf{x}_2, \ldots, \mathbf{x}_N$ is a random sample from $N_p(\boldsymbol{\mu}, \boldsymbol{\Sigma})$. Then $u_1 = (\mathbf{x}_1 - \boldsymbol{\mu})'\boldsymbol{\Sigma}^{-1}(\mathbf{x}_1 - \boldsymbol{\mu})$, $u_2 = (\mathbf{x}_2 - \boldsymbol{\mu})'\boldsymbol{\Sigma}^{-1}(\mathbf{x}_2 - \boldsymbol{\mu})$, $\ldots$, $u_N = (\mathbf{x}_N - \boldsymbol{\mu})'\boldsymbol{\Sigma}^{-1}(\mathbf{x}_N - \boldsymbol{\mu})$ can be considered a random sample from a chi-square distribution with p degrees of freedom. Unfortunately, $u_1, u_2, \ldots, u_N$ cannot actually be computed since $\boldsymbol{\mu}$ and $\boldsymbol{\Sigma}$ are not known. However, one could replace $\boldsymbol{\mu}$ and $\boldsymbol{\Sigma}$ with their sample estimates. This would give estimated values for the u_i's. The estimated values of the u_i's can be ordered and plotted against their theoretical counterparts for a random sample of size N from the chi-square probability distribution with p degrees of freedom. Such a plot is called a chi-square probability plot. If the plotted points tend to fall along a straight line, one concludes the data are multivariate normal. If the points do not tend to fall along a straight line, then one concludes the data are not multivariate normal.

None of the known statistical computing packages automatically creates chi-square probability plots, but it is fairly easy to create such plots using the SAS-IML procedure. The SAS commands that are required to produce a chi-square probability plot for the U.S. Navy data are given below. These commands can also be found in the file labeled EX3_1.IML on the enclosed disk.

```
OPTIONS LINESIZE=75 PAGESIZE=54 NODATE PAGENO=1;
TITLE 'U.S. NAVY BACHELOR OFFICERS'' QUARTERS';
DATA USNAVY;
 INPUT  SITE 1-2 ADO MAC WHR CUA WNGS OBC RMS MMH;
 LABEL ADO = 'AVERAGE DAILY OCCUPANCY'
       MAC = 'AVERAGE NUMBER OF CHECK-INS PER MO.'
       WHR = 'WEEKLY HRS OF SERVICE DESK OPERATION'
       CUA = 'SQ FT OF COMMON USE AREA'
       WNGS=  'NUMBER OF BUILDING WINGS'
       OBC = 'OPERATIONAL BERTHING CAPACITY'
       RMS = 'NUMBER OF ROOMS'
       MMH = 'MONTHLY MAN-HOURS' ;
```

```
CARDS;
 1    2        4       4   1.26   1   6   6  180.23
 2    3        1.58   40   1.25   1   5   5  182.61
 3   16.6     23.78   40   1      1  13  13  164.38
```
                    rest of the data go here

```
RUN;

TITLE2 'MULTIVARIATE NORMALITY PLOT';

DATA USNAVY2;
 SET USNAVY;
 DROP SITE MMH;

PROC IML; WORKSPACE=50;
   RESET NOLOG LINESIZE=75 PAGESIZE=54;
   USE USNAVY2;
   READ ALL INTO X ;
   N= NROW(X);
   P= NCOL(X);
   MEAN=( X[+,])/N; MEAN=MEAN';

PRINT "The Sample Mean is equal to"  MEAN;

   SUMSQ=X'*X-N#MEAN*MEAN';
   S=SUMSQ/(N-{1});

PRINT, "The Sample Covariance Matrix is equal to"  S;

   DIST = (X - J(N, {1})*MEAN')* INV(S)*(X -
J(N,{1})*MEAN')';

   D = VECDIAG(DIST);  CNAME={"DIST"};
 CREATE DIST FROM D[COLNAME=CNAME];
 APPEND FROM D[COLNAME=DIST];
QUIT;

PROC PRINT DATA=DIST;

DATA ; SET DIST; X=DIST;

PROC RANK OUT=RANKS;
 VAR X;
 RANKS R;

DATA PLOTDATA; SET RANKS;

   /* NOTE:  The following two numbers must be changed for
each new data set. */

NN =  25     ; * THIS IS THE NUMBER OF OBSERVATIONS IN
                    THE DATA SET;
```

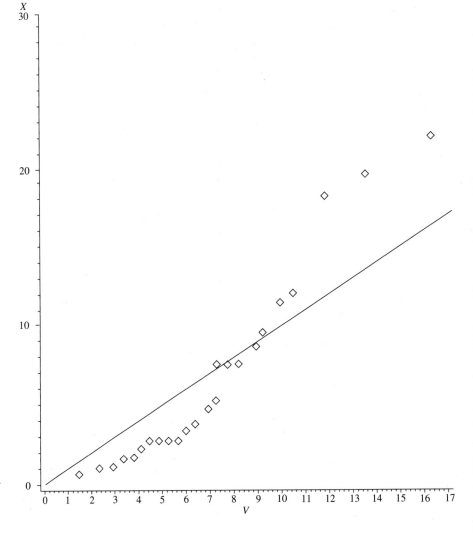

F I G. 3.18
Multivariate normal probability plot

```
P =      7    ; * THIS IS THE NUMBER OF RESPONSE VARIABLES;
RSTAR=(R-.5)/NN;
ETA=P/2;
V=GAMINV(RSTAR,ETA);
V=2*V;

PROC PRINT;
PROC SORT; BY R;

PROC PRINT; VARIABLES X R V;
FORMAT X 6.2 R 3.0 V 6.3;
```

```
PROC PLOT DATA=PLOTDATA;
  PLOT X*V='*' V*V = '+' /OVERLAY VZERO HZERO;

GOPTIONS DEVICE=PS2EGA;

  SYMBOL1 V=DIAMOND;
  SYMBOL2 V=NONE I=RL0;

PROC GPLOT DATA=PLOTDATA;
  PLOT X*V=1 V*V=2/VZERO HZERO OVERLAY;
RUN;
```

The results of these commands are shown in Figure 3.18. An examination of Figure 3.18 would seem to indicate that these data are not multivariate normal. It appears that three of the facilities are much larger than one would expect for multivariate normally distributed data. Perhaps there are three outliers in these data?

Summary

Several methods for plotting multivariate data were introduced in this chapter. These methods are most useful for verifying distributional assumptions that might be made about the data and for locating abnormalities in the data. Plots are also helpful when verifying the results of clustering programs.

Exercises

1 Choose a subset of at least five variables and at least 30 experimental units from the data you provided for Exercise 1 in Chapter 1. Use this subset of variables and experimental units to create the following plots using the raw data (i.e., the data in their original units). You may use any statistical graphics package you wish.

   a A bubble or blob plot using any three of your response variables.

   b A three-dimensional plot similar to one of the plots shown in Figures 3.4 through 3.6 using any three of your response variables.

   c A three-dimensional plot with bubbles or blobs similar to those shown in Figure 3.7 using any four of your response variables.

   d Chernoff faces using at least five of your response variables.

   e Sun-ray or star plots using at least five of your response variables.

   f Andrews' plots using at least five of your response variables.

2 Repeat Exercise 1, but standardize your data first.

3 Compare the plots in Exercise 2 with those in Exercise 1. Which set of plots seem to be the most useful? Why?

4 Consider the plots you deemed to be most useful in Exercise 3 for the data you provided for Exercise 1 in Chapter 1.

a Do there appear to be any outliers in your data? Explain your answer.

b Do there appear to be any clusters of experimental units in your data? Explain your answer.

c If SAS is available to you, use the method discussed in Section 3.3 to assess whether your data are multivariate normal or not. Do you think your data are multivariate normal? Why or why not?

d Write a short report indicating what you have learned about your data through these graphical procedures and be prepared to share information about your data with your classmates.

5 Examine the plots shown in Figures 3.5 and 3.6. Describe the location of any possible outliers in this data set. Describe the location of these same possible outliers in Figures 3.7 and 3.8.

6 Consider the Chernoff faces numbered 25 through 48 that are shown in Figure 3.8. Identify subgroups of individuals whose Chernoff faces appear to be very similar to one another.

7 Consider the star plots for the applicants shown in Figure 3.9. The text noted that applicants 7 and 8 look like the best two applicants in this applicant pool. List an additional four applicants who look very good. List the four worst applicants in this applicant pool.

8 Consider the star plots for the applicants shown in Figure 3.11. The text noted that applicants 7 and 8 look like the best two applicants in this applicant pool. List an additional four applicants who look very good. List the four worst applicants in this applicant pool. Do these answers agree with those given in Exercise 7?

9 Consider the football data from the Big 8 Conference described in Exercise 7 from Chapter 1.

a Standardize these data and create an Andrews' plot for each Big 8 team. Interpret your plot.

b Create a star or sun-ray plot for each team based on the standardized data.

c Create a Chernoff face for each Big 8 team.

d Using the plots created in parts a–c, can you make any statements about the similarities or dissimilarities of the Big 8 teams? If so, what can you say?

10 Consider the hotel data described in Exercise 8 of Chapter 1.

    a Create a three-dimensional data plot using the variables ROOMS, MIN, and MAX.

    b Create a bubble plot using the variables ROOMS, MIN, and MAX. In this plot, let the bubble size be proportional to the number of rooms in the hotel.

    c If SASGRAPH is available to you, create a three-dimensional plot so that different plot symbols are used for those hotels with and without pools.

11 Consider the world temperatures data described in Exercise 8 of Chapter 2. Create a three-dimensional data plot and a bubble plot using the high, low, and normal high temperatures. From these plots, do there appear to be any relationships between these three temperature variables? If so, what kinds of relationships? Do there appear to be any outliers in these data? If so, which countries appear to be outliers?

12 If SAS is available to you, use the method discussed in Section 3.3 to assess whether the world temperature data from Exercise 15 are multivariate normal or not. Do you think these data are multivariate normal? Why or why not?

13 Consider the survey data from 975 PIZZAZZ employees described in Appendix B and create graphs as requested below. You may use any statistical graphics package you wish.

    a A bubble or blob plot using responses to the first three questions.

    b A three-dimensional plot similar to one of the plots shown in Figures 3.4 through 3.6 using the responses to the first three questions.

    c A three-dimensional plot with bubbles or blobs similar to the one shown in Figure 3.7 using responses to questions 4–7.

    d Chernoff faces for shift supervisors using their responses to questions 8–18.

    e Sun-ray or star plots for shift supervisors using their responses to questions 8–18.

    f Andrews' plots for shift supervisors using their responses to questions 1–18.

    g Repeat parts d–f for assistant managers and managers.

14 Consider the survey data obtained from the 304 married adults described in Appendix C and create graphs as requested below. You may use any statistical graphics package you wish.

    a A bubble or blob plot using responses to the Kansas Marital Satisfaction Scale.

b   A three-dimensional plot similar to one of the plots shown in Figures 3.4 through 3.6 using responses to the Kansas Marital Satisfaction Scale.

c   A three-dimensional plot with bubbles or blobs similar to the one shown in Figure 3.7 using responses to the Kansas Family Life Satisfaction Scale.

d   Sun-ray or star plots for males using their responses to the Family Adaptability Cohesion Evaluation Scales.

e   Sun-ray or star plots for females using their responses to the Family Adaptability Cohesion Evaluation Scales.

Eigenvalues and Eigenvectors

4

This chapter requires some knowledge of matrices and vectors. Readers who need an introduction to or a review of matrix algebra should study Appendix A prior to reading this chapter. An extensive knowledge of matrix algebra is not required to make use of this text; however, those students who are comfortable with matrix ideas are likely to gain more from studying any multivariate text than those who are not. Those who plan to be experts in multivariate methods should spend whatever time and effort it takes to become extremely comfortable with matrix algebra.

Almost all multivariate methods depend on functions of the elements of a sample covariance matrix, usually denoted by $\hat{\Sigma}$, or on the elements of a sample correlation matrix, usually denoted by $\mathbf{R}$. The most important functions of the elements in a matrix for multivariate data analysis purposes are (1) its trace, (2) its determinant, and (3) its eigenvalues and eigenvectors. Although Appendix A contains most of the prerequisite matrix results required for this book, the functions just mentioned are so prevalent in multivariate methods that they are being specially introduced in this chapter.

4.1
Trace and Determinant

In each of the following definitions, let

$$\Sigma = \begin{bmatrix} \sigma_{11} & \sigma_{12} & \cdots & \sigma_{1p} \\ \sigma_{21} & \sigma_{22} & \cdots & \sigma_{2p} \\ \vdots & \vdots & \ddots & \vdots \\ \sigma_{p1} & \sigma_{p2} & \cdots & \sigma_{pp} \end{bmatrix}$$

Note that Σ is a square $p \times p$ matrix. Also let Σ be a symmetric matrix. Variance–covariance matrices and correlation matrices are always symmetric.

D E F I N I T I O N 4.1

The trace of Σ, denoted by $\text{tr}(\Sigma)$, is defined by $\text{tr}(\Sigma) = \sum_{i=1}^{p} \sigma_{ii} = \sigma_{11} + \sigma_{22} + \cdots + \sigma_{pp}$. Thus, the trace of a square matrix is equal to the sum of its diagonal elements going from the upper left-hand corner down to the lower right-hand corner.

D E F I N I T I O N 4.2

The determinant of a square matrix Σ, denoted by $|\Sigma|$, is defined by $|\Sigma| = \sum_{j=1}^{p} \sigma_{1j} \cdot \Sigma_{1j}$ where $\Sigma_{1j} = (-1)^{1+j} |\Sigma^{1j}|$ where Σ^{1j} is the matrix obtained from Σ by deleting its first row and its jth column. The determinant of a 1×1 matrix is defined as the value of the single entry in the matrix. Thus $\det(\sigma_{11}) = \sigma_{11}$.

Examples

To illustrate Definitions 4.1 and 4.2, let

$$\mathbf{\Sigma} = \begin{bmatrix} 6 & 2 \\ 2 & 3 \end{bmatrix}$$

Then

$$\text{tr}(\mathbf{\Sigma}) = 6 + 3 = 9$$

and

$$|\mathbf{\Sigma}| = 6 \cdot (-1)^{1+1} \cdot |3| + 2 \cdot (-1)^{1+2} \cdot |2| = 18 - 4 = 14$$

Special Case   The determinant of the 2×2 matrix

$$\mathbf{A} = \begin{bmatrix} a & b \\ c & d \end{bmatrix}$$

can be shown to be equal to $|\mathbf{A}| = ad - bc$.

As another illustration, let

$$\mathbf{\Sigma} = \begin{bmatrix} 6 & 2 & 1 \\ 2 & 3 & 4 \\ 1 & 4 & 8 \end{bmatrix}$$

Then

$$\text{tr}(\mathbf{\Sigma}) = 6 + 3 + 8 = 17$$

and

$$|\mathbf{\Sigma}| = 6 \cdot (-1)^{1+1} \cdot \left| \begin{bmatrix} 3 & 4 \\ 4 & 8 \end{bmatrix} \right| + 2 \cdot (-1)^{1+2} \cdot \left| \begin{bmatrix} 2 & 4 \\ 1 & 8 \end{bmatrix} \right| + 1 \cdot (-1)^{1+3} \cdot \left| \begin{bmatrix} 2 & 3 \\ 1 & 4 \end{bmatrix} \right|$$

$$= 6 \cdot (3 \cdot 8 - 4 \cdot 4) - 2 \cdot (2 \cdot 8 - 4 \cdot 1) + 1 \cdot (2 \cdot 4 - 3 \cdot 1) = 48 - 24 + 5 = 29$$

Calculating the determinants of large matrices can be very tedious. Fortunately, many excellent computer programs exist that calculate determinants of matrices very easily.

4.2
Eigenvalues

D E F I N I T I O N   4.3

The *eigenvalues* (also called *characteristic roots* or *latent roots*) of $\mathbf{\Sigma}$ are the roots of the polynomial equation defined by

$$|\mathbf{\Sigma} - \lambda\mathbf{I}| = 0 \tag{4.1}$$

Remark When one expands the determinant expression in Eq. (4.1), the resulting equation has the form

$$c_1\lambda^p + c_2\lambda^{p-1} + \cdots + c_p\lambda + c_{p+1} = 0$$

Thus, Eq. (4.1) is a polynomial equation in λ of degree p. The eigenvalues of $\mathbf{\Sigma}$ are defined as the roots of this polynomial equation. If $p = 2$, Eq. (4.1) will be a quadratic equation and, hence, it will have two roots. If $p = 3$, Eq. (4.1) will be a cubic equation and, hence, it will have three roots. In general, a pth degree equation will have p roots.

Remark If $\mathbf{\Sigma}$ is a symmetric matrix, which it will be whenever it is a variance–covariance matrix or a correlation matrix, the eigenvalues of $\mathbf{\Sigma}$ will always be real numbers.

If $\mathbf{\Sigma}$ is a symmetric matrix, its eigenvalues are real numbers and, hence, they can be ordered from largest to smallest. In this case, the eigenvalues of $\mathbf{\Sigma}$ will be denoted by

$$\lambda_1 \geq \lambda_2 \geq \cdots \geq \lambda_p$$

Thus λ_1 is the largest eigenvalue of $\mathbf{\Sigma}$, λ_2 is the second largest eigenvalue of $\mathbf{\Sigma}, \ldots$, and λ_p is the smallest eigenvalue of $\mathbf{\Sigma}$.

4.3
Eigenvectors

D E F I N I T I O N 4.4

Each eigenvalue of $\mathbf{\Sigma}$ has a corresponding nonzero vector $\mathbf{a}$ (a column of numbers) called an *eigenvector* (also called a *characteristic vector* or a *latent vector*) that satisfies the matrix equation

$$\mathbf{\Sigma a} = \lambda \mathbf{a}$$

Because $\mathbf{\Sigma}$ has p eigenvalues, it will have p eigenvectors. Let $\mathbf{a}_1$, $\mathbf{a}_2, \ldots, \mathbf{a}_p$ denote the eigenvectors of $\mathbf{\Sigma}$ corresponding to the eigenvalues λ_1, $\lambda_2, \ldots, \lambda_p$, respectively.

Remarks The eigenvectors of a matrix are not unique so they are often normalized (a mathematical term) so that $\mathbf{a}_i'\mathbf{a}_i = 1$, for $i = 1, 2, \ldots, p$. When two eigenvalues of $\mathbf{\Sigma}$ are not equal, their corresponding eigenvectors will be orthogonal to one another (i.e., $\mathbf{a}_i'\mathbf{a}_j = 0$ when $\lambda_i \neq \lambda_j$). When two eigenvalues of $\mathbf{\Sigma}$ are equal, the corresponding eigenvectors can be, and always will be, chosen to be orthogonal to one another, although in this case the eivenvectors are not uniquely determined.

Some properties of the eigenvalues of a matrix are

$$\text{tr}(\mathbf{\Sigma}) = \sum_{i=1}^{p} \lambda_i$$

$$|\mathbf{\Sigma}| = \prod_{i=1}^{p} \lambda_i = \lambda_1 \cdot \lambda_2 \cdots \lambda_p$$

Thus, the trace of a symmetric matrix is equal to the sum of its eigenvalues, and the determinant of a symmetric matrix is equal to the product of its eigenvalues.

Positive Definite and Positive Semidefinite Matrices

D E F I N I T I O N   4.5

If a matrix is symmetric and if all of its eigenvalues are positive, the matrix is called a *positive definite matrix.*

D E F I N I T I O N   4.6

If a symmetric matrix has nonnegative eigenvalues and if at least one of its eigenvalues is actually equal to zero, then the matrix is called a *positive semidefinite matrix.*

D E F I N I T I O N   4.7

If a matrix is either positive definite or positive semidefinite, the matrix is defined to be a nonnegative matrix.

Remark   Variance–covariance matrices and correlation matrices are always nonnegative matrices. That is, all of the eigenvalues of a variance–covariance matrix or a correlation matrix will be nonnegative real numbers. This is also true for sample variance–covariance matrices and sample correlation matrices.

E X A M P L E   4.1

As an example suppose

$$\mathbf{\Sigma} = \begin{bmatrix} 6 & 2 \\ 2 & 3 \end{bmatrix}$$

To compute the eigenvalues of Σ, we note that $|\Sigma - \lambda\mathbf{I}| = 0$ implies that

$$\left| \begin{bmatrix} 6 - \lambda & 2 \\ 2 & 3 - \lambda \end{bmatrix} \right| = 0$$

which implies that $(6 - \lambda)(3 - \lambda) - 4 = 0$, which implies that $\lambda^2 - 9\lambda + 14 = 0$ and that $(\lambda - 7)(\lambda - 2) = 0$, and the two roots of this equation are $\lambda = 2$ and $\lambda = 7$. Therefore, the largest eigenvalue of Σ is $\lambda_1 = 7$ and the smallest eigenvalue of Σ is $\lambda_2 = 2$. Because both eigenvalues of Σ are positive, Σ is a positive definite matrix.

To compute an eigenvector of Σ that corresponds to its largest eigenvalue, note that

$$\Sigma\mathbf{a} = \lambda_1\mathbf{a} \Rightarrow \begin{bmatrix} 6 & 2 \\ 2 & 3 \end{bmatrix}\begin{bmatrix} a_1 \\ a_2 \end{bmatrix} = 7\begin{bmatrix} a_1 \\ a_2 \end{bmatrix}$$

$$\Rightarrow \left. \begin{array}{r} 6a_1 + 2a_2 = 7a_1 \\ 2a_1 + 3a_2 = 7a_2 \end{array} \right\} \Rightarrow a_1 = 2a_2$$

So any 2×1 vector that has its first element twice its second will be an eigenvector of Σ. Thus, some eigenvectors of Σ corresponding to $\lambda_1 = 7$ are

$$\begin{bmatrix} 2 \\ 1 \end{bmatrix}, \begin{bmatrix} 6 \\ 3 \end{bmatrix}, \begin{bmatrix} -2 \\ -1 \end{bmatrix} \quad \text{and} \quad \begin{bmatrix} 500 \\ 250 \end{bmatrix}$$

Requiring that the eigenvector be normalized requires that $a_1^2 + a_2^2 = 1$ and this along with $a_1 = 2a_2$ implies that $5a_2^2 = 1$, which in turn implies that $a_2 = \pm 1/\sqrt{5}$, which implies that $a_1 = 2a_2 = \pm 2/\sqrt{5}$. Therefore,

$$\mathbf{a}_1 = \begin{bmatrix} 2/\sqrt{5} \\ 1/\sqrt{5} \end{bmatrix}$$

is a normalized eigenvector of Σ corresponding to the eigenvalue $\lambda_1 = 7$. Another normalized eigenvector of Σ corresponding to its largest root is

$$\begin{bmatrix} -2/\sqrt{5} \\ -1/\sqrt{5} \end{bmatrix}$$

In a similar fashion, we can show that

$$\mathbf{a}_2 = \begin{bmatrix} 1/\sqrt{5} \\ -2/\sqrt{5} \end{bmatrix}$$

is a normalized eigenvector of Σ corresponding to the eigenvalue $\lambda_2 = 2$. Verify that $\mathbf{a}_1$ and $\mathbf{a}_2$ are orthogonal to one another by showing that $\mathbf{a}_1'\mathbf{a}_2 = 0$.

Computing the eigenvalues and eigenvectors of a matrix is fairly tedious when $p = 2$; doing so gets even more tedious when $p > 2$. Fortunately, computer programs exist that compute these functions quite easily. Such programs are illustrated in many of the following chapters.

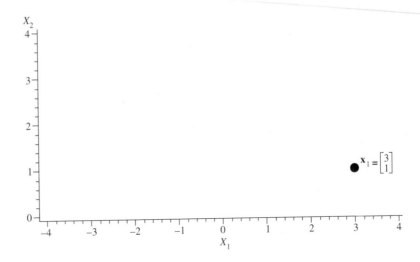

F I G. 4.1

Vector in two dimensions

4.4
Geometric Descriptions ($p = 2$)

This section provides some geometric descriptions of vectors and also shows how the eigenvalues and eigenvectors can be interpreted when they are calculated from the variance–covariance matrix for a bivariate normal distribution.

Vectors

Two-dimensional vectors can be plotted as points in a two-dimensional coordinate system with the abscissa corresponding to the first element in the vector and the ordinate corresponding to the second element in the vector. For example, to plot the vector

$$\mathbf{x}_1 = \begin{bmatrix} 3 \\ 1 \end{bmatrix}$$

on a two-dimensional graph, one would go three units to the right of the origin and one unit up. This vector is plotted in Figure 4.1.

Note that a vector is nothing more than a point in a two-dimensional space. An obvious question is "Why call it a vector rather than a point?" There is no really satisfactory answer to this question. Maybe we should credit engineers and mathematicians for this. If you draw a line from the origin to the point, and put an arrow on the end of this line as illustrated in Figure 4.2, you end up with a geometric figure that looks like a vector.

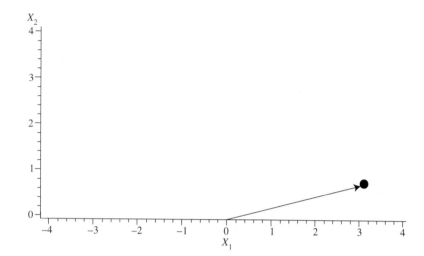

F I G. 4.2

Vector in two dimensions showing direction

In this book, we use the words *vector* and *point* interchangeably. Occasionally, vectors and/or points are used to show direction. When this is the case, the direction shown is the same as that shown in Figure 4.2.

Now let

$$\mathbf{x}_2 = \begin{bmatrix} 1 \\ -3 \end{bmatrix}$$

Since $\mathbf{x}_1'\mathbf{x}_2 = 3 \cdot 1 + 1 \cdot (-3) = 0$, the vectors $\mathbf{x}_1$ and $\mathbf{x}_2$ are orthogonal to one another. Geometrically, if both of these vectors are plotted on a two-dimensional graph and if lines are drawn from the origin to each of the points, we would see that the lines are perpendicular (orthogonal) to one another, as shown in Figure 4.3.

Bivariate Normal Distributions

Consider a bivariate normal distribution with

$$\boldsymbol{\mu} = \begin{bmatrix} 8 \\ 6 \end{bmatrix} \quad \text{and} \quad \boldsymbol{\Sigma} = \begin{bmatrix} 6 & 2 \\ 2 & 3 \end{bmatrix}$$

The ellipse shown in Figure 4.4 is proportional to the ellipsoid of concentration defined in Chapter 1, is centered at $\boldsymbol{\mu}$, and is given by the equation

$$(\mathbf{x} - \boldsymbol{\mu})'\boldsymbol{\Sigma}^{-1}(\mathbf{x} - \boldsymbol{\mu}) = 3$$

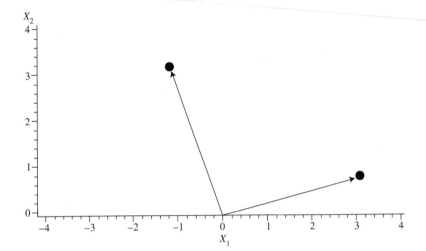

F I G. 4.3
Two orthogonal
vectors in two di-
mensions

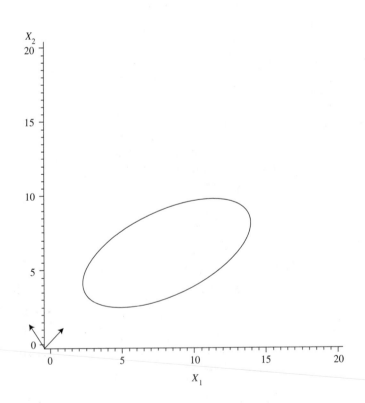

F I G. 4.4
Ellipsoid of con-
centration and
normalized ei-
genvectors

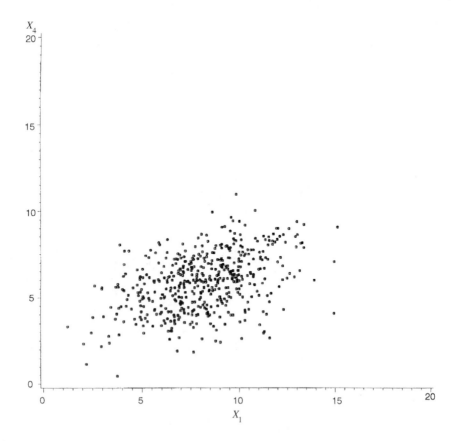

Earlier we showed that the eigenvalues of Σ are $\lambda_1 = 7$ and $\lambda_2 = 2$, and that corresponding normalized eigenvectors are

$$\mathbf{a}_1 = \begin{bmatrix} 2/\sqrt{5} \\ 1/\sqrt{5} \end{bmatrix} \quad \text{and} \quad \mathbf{a}_2 = \begin{bmatrix} 1/\sqrt{5} \\ -2/\sqrt{5} \end{bmatrix}, \text{respectively}$$

These two eigenvectors are plotted in Figure 4.4. Note that the first eigenvector points in the direction of the major axis of the ellipse; this is the direction in which the largest amount of variability occurs. Also note that the second eigenvector points in the direction of the ellipse's minor axis; this is the direction in which the smallest amount of variability occurs.

Next suppose the major axis of the ellipse is labeled y_1 and the minor axis is labeled y_2, and suppose their point of intersection is labeled $\begin{bmatrix} 0 \\ 0 \end{bmatrix}$. The equation of the ellipse in the (y_1, y_2) space is

$$\frac{y_1^2}{\lambda_1} + \frac{y_2^2}{\lambda_2} = 3$$

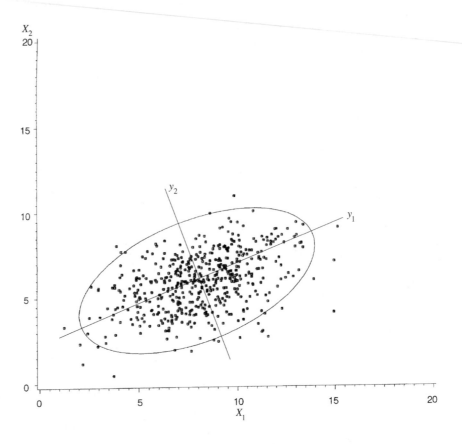

F I G. **4.6**

Bivariate normal
data with 95% tol-
erance ellipsoid

Thus, the length of the major axis of the ellipse is equal to $2\sqrt{3\lambda_1} = 9.165$ and the length of the minor axis of the ellipse is equal to $2\sqrt{3\lambda_2} = 4.899$.

Figure 4.5 shows the location of a random sample of points from the bivariate normal distribution described above. Note the elliptical shape of the plotted points. The variance of these plotted points in the x_1 direction is equal to $\sigma_{11} = 6$, the variance in the x_2 direction is equal to $\sigma_{22} = 3$, and the covariation of the points in the (x_1, x_2) system is equal to $\sigma_{12} = \sigma_{21} = 2$. The variance of the plotted points in the y_1 direction is equal to $\lambda_1 = 7$, the variance in the y_2 direction is equal to $\lambda_2 = 2$, and the covariation in the (y_1, y_2) system is equal to zero. Figure 4.6 shows an overlay of the plotted points and an ellipse that is proportional to the ellipsoid of concentration along with the y_1 and y_2 axes. Notice that most of the plotted points fall within this ellipse.

From the preceding discussion, we can see that there is information about the location of a bivariate normal distribution in its mean, μ, and information about its shape in the eigenvalues and eigenvectors of its variance–covariance matrix, Σ. Visualize how the shape of the ellipse would change as the values of the largest and smallest eigenvalues of Σ change.

If λ_2 approaches zero, the variability in the y_2 direction also approaches zero, and the plotted points would tend to fall onto a straight line. Note that $\lambda_2 = 0$ if and only if $\rho = 1$ where ρ is the correlation between x_1 and x_2; in this case the points would fall exactly on a straight line. Also note that when λ_1 and λ_2 are equal, the ellipse of concentration is a circle. In this case, the correlation between x_1 and x_2 is equal to zero. Furthermore, in this case, the eigenvectors cannot be uniquely identified because the variability in the data is the same in all directions. In other words, if we were going to place a two-dimensional axis system in this circle, we would have an infinite number of possibilities for these two axes, because the variability in all directions is the same.

4.5
Geometric Descriptions ($p = 3$)

Vectors

Three-dimensional vectors can be plotted as points in a three-dimensional coordinate system with one axis corresponding to the first element in the vector, a second axis perpendicular to the first that corresponds to the second element in the vector, and a third axis perpendicular to each of the first two that corresponds to the third element of the data vector. For example, to plot the vector

$$\mathbf{x} = \begin{bmatrix} 3 \\ 2 \\ 1 \end{bmatrix}$$

on a three-dimensional graph, we would go three units from the origin on the x_1 axis, two units from there in the direction of the x_2 axis, and one unit from there in the direction of the x_3 axis. This vector (point) is plotted in Figure 4.7.

Trivariate Normal Distributions

Next consider a trivariate normal distribution. A random sample of data from a trivariate normal distribution would tend to fall within a three-dimensional ellipsoid. Figure 4.8 shows a random sample of 400 data points from a trivariate normal distribution from several different viewing sites.

The variance–covariance matrix of a trivariate normal distribution has three eigenvalues denoted by λ_1, λ_2, and λ_3 with $\lambda_1 \geq \lambda_2 \geq \lambda_3$. When $\lambda_1 > \lambda_2 > \lambda_3$, the ellipsoid can be described as a football-shaped region in which the football has some of its air removed and two of its sides are pressed toward

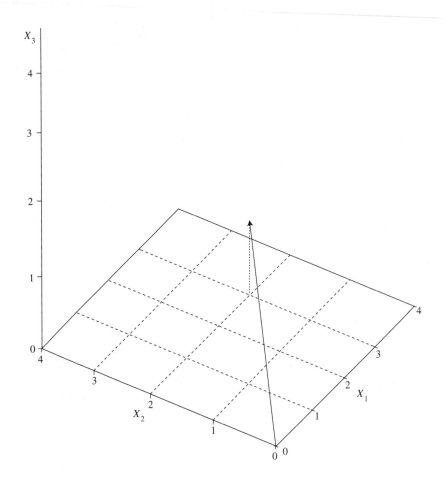

F I G. **4.7**
Vector in three di-
mensions

each other. The length of the football is in the direction of the eigenvector corresponding to the largest eigenvalue, and the variance in this direction is equal to λ_1. The axis between the two sides that are pressed together is in the direction of the eigenvector corresponding to the smallest eigenvalue, and the variance in this direction is equal to λ_3. The remaining axis is perpendicular to the preceding two axes and is in the direction of the eigenvector corresponding to λ_2; the variance in this direction is equal to λ_2.

It is interesting to consider the shape of the region in which trivariate normal data fall as the values of the eigenvalues of Σ change. For example, if λ_2 and λ_3 are equal, but smaller than λ_1, the shape of the region in which the data fall is similar to a traditional football shape. How long and how short the football becomes depends on relative sizes that the eigenvalues have with respect to each other. If λ_2 and λ_3 are both equal to zero, then the data fall onto a straight line because there is no variability in directions orthogonal to the line containing the data points.

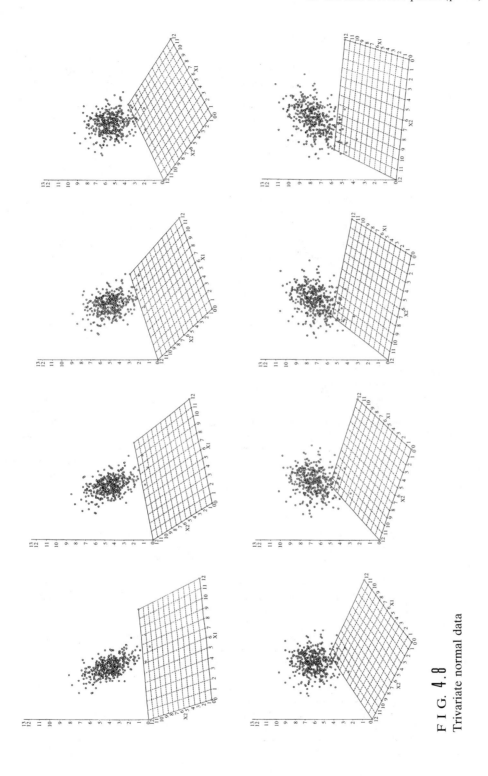

F I G. 4.8
Trivariate normal data

We can see that the shape of the ellipsoid changes as the relative values of these three eigenvalues change. Table 4.1 describes the geometric shape of ellipsoids for all of the various possibilities for relationships among λ_1, λ_2, and λ_3. To summarize, the values of the eigenvalues of a three-dimensional variance–covariance matrix give valuable information as to the general shape of the region in which three-dimensional multivariate normal data would tend to fall.

4.6
Geometric Descriptions ($p > 3$)

We cannot visualize geometric figures in more than three dimensions. Still, in cases for which $p > 3$, we can compute the eigenvalues of the variance–covariance matrix.

Suppose a researcher is measuring 15 variables. The data could be described in a 15-dimensional space. Suppose the researcher looked at the eigenvalues of the variance–covariance matrix and found that all but two were approximately equal to zero. This means that even though the researcher is

T A B L E   4.1

Relationships between the eigenvalues of Σ and the shape of a trivariate normal distribution

Eigenvalues	Geometric description
$\lambda_1 > \lambda_2 > \lambda_3 > 0$	A football with some of its air removed, the data are three dimensional.
$\lambda_1 > \lambda_2 > \lambda_3 = 0$	A football with all of its air removed, the data are two dimensional.
$\lambda_1 > \lambda_2 = \lambda_3 > 0$	A regular football, the data are three dimensional.
$\lambda_1 > \lambda_2 = \lambda_3 = 0$	The data falls on a straight line, the data are one dimensional.
$\lambda_1 = \lambda_2 = \lambda_3 > 0$	A basketball-shaped region, the data are three dimensional.
$\lambda_1 = \lambda_2 > \lambda_3 > 0$	A basketball with some of its air removed, the data are three dimensional.
$\lambda_1 = \lambda_2 > \lambda_3 = 0$	A basketball with all of its air removed, the data are two dimensional.
$\lambda_1 = \lambda_2 = \lambda_3 = 0$	The data all fall at a single point, the data are zero dimensional.

measuring 15 variables, a plot of the data in those 15 dimensions would not actually encompass more than two dimensions! Wouldn't it be great if we could go out into those two dimensions and have a look around? If we could look at a plot of the data in these two dimensions, we would be able to see whether there are outliers in the data set, whether there are clusters of points in the data, and whether the data tend to fall into an elliptically shaped region, that is, whether the data appear to be bivariate normal in these two dimensions.

While it is likely to be very difficult for a researcher to actually go out into the two dimensions in which the data fall, it is quite easy to bring those two dimensions back to the researcher so that he or she can examine plots of the data in these two dimensions.

The multivariate technique that allows us to determine the dimensionality of the space in which the data fall and examine the data in this reduced dimensional space is called *principal components analysis.* A formal development of principal components analysis is given in Chapter 5 and many of the examples throughout the rest of the book use principal components variables to visualize complex data sets in a reduced dimensional space.

Summary

A good deal of information about a population that is being sampled is contained in the eigenvalues and eigenvectors of a sample variance–covariance matrix. The eigenvalues generally describe the shape of the region in which the data fall; the eigenvectors show the directions of the major and minor axes that can be associated with the data points. If only a few of the eigenvalues are different than zero, we can look at higher dimensional data in the reduced dimensional space in which the data actually fall. The ability to do this often allows us to identify outliers in a data set and to examine distributional properties of the population being sampled.

Exercises

1 Verify that the two vectors,

$$\mathbf{a}_1 = \begin{bmatrix} 2/\sqrt{5} \\ 1/\sqrt{5} \end{bmatrix} \quad \text{and} \quad \mathbf{a}_2 = \begin{bmatrix} 1/\sqrt{5} \\ -2/\sqrt{5} \end{bmatrix}$$

are orthogonal to each other. Plot these two vectors on a two-dimensional plot and draw lines from the origin to the location of each of the vectors in your plot.

2 Consider the data matrix of Exercise 6 in Chapter 1. Complete each of the following parts. Your instructor will tell you whether you are to work these out by hand or use a program that allows for matrix manipulations such as SAS-IML, GAUSS, MATLAB, or S.

a Find the trace of both $\hat{\Sigma}$ and $\mathbf{R}$.

b Find the determinant of both $\hat{\Sigma}$ and $\mathbf{R}$.

c Find the eigenvalues and eigenvectors of both $\hat{\Sigma}$ and $\mathbf{R}$. Show that all pairs of eigenvectors are orthogonal to one another.

d Show that the sum of the eigenvalues of $\hat{\Sigma}$ is equal to the trace of $\hat{\Sigma}$. Also show that the product of the eigenvalues of $\hat{\Sigma}$ is equal to the determinant of $\hat{\Sigma}$.

e Compute $\sqrt{\hat{\boldsymbol{\mu}}'\hat{\boldsymbol{\mu}}}$, $\sqrt{\hat{\boldsymbol{\mu}}'\hat{\Sigma}^{-1}\hat{\boldsymbol{\mu}}}$, and $\sqrt{(\mathbf{x}_2 - \hat{\boldsymbol{\mu}})'\hat{\Sigma}^{-1}(\mathbf{x}_2 - \hat{\boldsymbol{\mu}})}$.

f Let

$$\mathbf{a} = \begin{bmatrix} 2 \\ -1 \\ 3 \end{bmatrix}$$

Compute $\mathbf{a}'\mathbf{x}_1$.

3 Make a photocopy of Figure 4.8. On this copy, sketch in the eigenvector (or axis) that shows the direction of where the most variability occurs in these data. Do this for each plot in Figure 4.8.

Principal Components Analysis 5

Data were collected by the U.S. Navy to attempt to estimate manpower needs for Bachelor Officers' Quarters. Myers (1990) uses these data to illustrate multiple regression techniques. These data are reproduced in Table 5.1.

Myers uses the data from Table 5.1 to build a multiple regression model that relates MMH to predictor variables ADO, MAC, WHR, CUA, WNGS, OBC, and RMS. The validity of the results of a multiple regression on these data depends on the validity of certain modeling assumptions. In particular, three important questions should be considered prior to attempting multiple regression analysis on these data:

1 Are there outliers in the data set? A single outlier may dramatically affect the results of a regression analysis.

2 Is there multicollinearity among the predictor variables? Note that multicollinearity exists if the dimension of the space in which the predictor variables lie is less than the number of predictor variables. The existence of multicollinearity greatly affects our interpretations of any fitted regression models.

3 Is the distribution of the predictor variables a multivariate normal distribution? If the distribution of the predictor variables is not multivariate normal, we may want to consider transformations on the predictor variables.

Principal component analysis can be used to help provide answers to these questions and to examine the data for other types of abnormalities that might be present.

5.1
Reasons for Using Principal Components Analysis

Principal components analysis (PCA) involves a mathematical procedure that transforms a set of correlated response variables into a smaller set of uncorrelated variables called *principal components*. By looking carefully at this new set of uncorrelated variables, answers to many important questions such as those given in the preceding section can be obtained. The answers to these questions will likely affect other analyses to be performed later.

There are many legitimate reasons for considering principal components analysis. Some are described in this section.

Data Screening

Principal components analysis is perhaps most useful for screening multivariate data. For almost all data analysis situations, PCA can be recommended as a first step. It can and should be performed on a set of data prior to performing any other kinds of multivariate analyses. Follow-up analyses on the principal

Data on bachelor officers' quarters

SITE	ADO	MAC	WHR	CUA	WNGS	OBC	RMS	MMH
1	2.00	4.00	4.00	1.26	1.00	6.00	6.00	180.23
2	3.00	1.58	40.00	1.25	1.00	5.00	5.00	182.61
3	16.60	23.78	40.00	1.00	1.00	13.00	13.00	164.38
4	7.00	2.37	168.00	1.00	1.00	7.00	8.00	284.55
5	5.30	1.67	42.50	7.79	3.00	25.00	25.00	199.92
6	16.50	8.25	168.00	1.12	2.00	19.00	19.00	267.38
7	25.89	3.00	40.00	.00	3.00	36.00	36.00	999.09
8	44.42	159.75	168.00	.60	18.00	48.00	48.00	1103.24
9	39.63	50.86	40.00	27.37	10.00	77.00	77.00	944.21
10	31.92	40.08	168.00	5.52	6.00	47.00	47.00	931.84
11	97.33	255.08	168.00	19.00	6.00	165.00	130.00	2268.06
12	56.63	373.42	168.00	6.03	4.00	36.00	37.00	1489.50
13	96.67	206.67	168.00	17.86	14.00	120.00	120.00	1891.70
14	54.58	207.08	168.00	7.77	6.00	66.00	66.00	1387.82
15	113.88	981.00	168.00	24.48	6.00	166.00	179.00	3559.92
16	149.58	233.83	168.00	31.07	14.00	185.00	202.00	3115.29
17	134.32	145.82	168.00	25.99	12.00	192.00	192.00	2227.76
18	188.74	937.00	168.00	45.44	26.00	237.00	237.00	4804.24
19	110.24	410.00	168.00	20.05	12.00	115.00	115.00	2628.32
20	96.83	677.33	168.00	20.31	10.00	302.00	210.00	1880.84
21	102.33	288.83	168.00	21.01	14.00	131.00	131.00	3036.63
22	274.92	695.25	168.00	46.63	58.00	363.00	363.00	5539.98
23	811.08	714.33	168.00	22.76	17.00	242.00	242.00	3534.49
24	384.50	1473.66	168.00	7.36	24.00	540.00	453.00	8266.77
25	95.00	368.00	168.00	30.26	9.00	292.00	196.00	1845.89

SITE, identifies different Bachelor Officers' Quarters; ADO, average daily occupancy; MAC, monthly average number of check-ins; WHR, number of hours per week service desk is in operation; CUA, total common use area in square feet; WNGS, number of building wings; OBC, operational berthing capacity; RMS, total number of rooms; MMH, monthly man-hours required.

components are useful for checking assumptions that a researcher might make about a set of multivariate data and for identifying and locating possible outliers in the data. If other abnormalities occur in a multivariate data set, PCA can help reveal them.

Clustering

Principal components analysis is also helpful whenever a researcher wants to group experimental units into subgroups of similar types. It can be used to help cluster experimental units into subgroups or for verifying the results of clustering programs. See Chapter 9 for examples of clustering.

Discriminant Analysis

Chatfield and Collins (1980) discuss a researcher who was attempting a discriminant analysis with a larger number of response variables. Unfortunately, the researcher had only a small sample of experimental units from each candidate population. Discriminant analysis programs require that an estimate of the variance–covariance matrix be inverted to develop a discrimination rule. If you have sampled fewer experimental units than there are response variables being measured, the estimated variance–covariance matrix cannot be inverted, and discriminant analysis programs will fail. A PCA revealed that a few principal components contained almost all of the information that was available in the original variables. Values for the principal components were obtained for each experimental unit, and these new variables were used as input variables to a discriminant analysis program. The estimated variance–covariance matrix of the new variables could be inverted, and the discriminant analysis program was able to produce a discrimination rule for classifying observations.

Regression

We have long known that multiple regression can be dangerous when predictor variables are highly correlated in some fashion. This has been referred to as *multicollinearity* among the predictor variables. Principal component analysis can help determine whether multicollinearity occurs among the predictor variables.

5.2
Objectives of Principal Components Analysis

Computer programs make it very easy to perform a PCA. This technique should be used primarily as an exploratory technique, and it should help researchers get a feel for a set of data. Sometimes PCA can help a researcher to better understand the correlation structure among the responses, and at times it may help the researcher generate hypotheses about the variables or the data.

The majority of the books on multivariate methods suggest that the primary objectives of a PCA are to (1) reduce the dimensionality of the data set and (2) identify new meaningful underlying variables.

In reality, objective 1 is not quite true. What we are really trying to do is discover the true dimensionality of the data. An important question is the following: "If the data are plotted in a p-dimensional space, will the data take up all p dimensions?" If not, then even though p variables are being measured, the actual dimensionality of the data is less than p. Principal components analysis can be used to assess the actual dimensionality of the data, and when the actual dimensionality is less than p, the original variables can be replaced by a smaller number of underlying variables without losing any information. This smaller number of variables can then be used in ensuing analyses.

With regard to objective 2, PCA will always identify new variables. However, we cannot guarantee that the new variables will be meaningful. Unfortunately, more often than not, they will not be meaningful. However, even though the new variables are not meaningful, the principal component variables are still useful. The new variables are useful for a variety of things including data screening, assumption checking, and cluster verifying.

There is a strong tendency among researchers to attempt to give meaning to newly created principal component variables. If interpretations are obvious, then go ahead and use them. Those few instances where the principal components can be interpreted can be considered a bonus because we usually do not expect to be able to interpret the principal component variables. It is really important to remember that a principal components analysis is very useful regardless of whether the principal components can be interpreted.

5.3
Principal Components Analysis
on the Variance–Covariance Matrix Σ

As noted earlier, one basic objective of a principal components analysis is to discover the true dimensionality of the space in which the data lie. In the process of doing this, new variables called *principal components* can be formed, in decreasing order of importance, so that

1 they are uncorrelated,

2 the first principal component accounts for as much of the variability in the data as possible, and

3 each succeeding component accounts for as much of the remaining variability as is possible.

Next a formal definition is given. In the following definitions, $\mathbf{x}$ represents a randomly selected observation from a population that has mean, $\boldsymbol{\mu}$, and variance–covariance matrix, Σ.

D E F I N I T I O N 5.1

The first principal component variable is defined by $y_1 = \mathbf{a}_1'(\mathbf{x} - \boldsymbol{\mu})$, where $\mathbf{a}_1$ is chosen so that the variance of $\mathbf{a}_1'(\mathbf{x} - \boldsymbol{\mu})$ is maximized over all vectors $\mathbf{a}_1$ satisfying $\mathbf{a}_1'\mathbf{a}_1 = 1$.

We can show that the maximum value of the variance of $\mathbf{a}_1'(\mathbf{x} - \boldsymbol{\mu})$ among all vectors $\mathbf{a}_1$ satisfying $\mathbf{a}_1'\mathbf{a}_1 = 1$ is equal to λ_1, the largest eigenvalue of Σ, and that this maximum occurs when $\mathbf{a}_1$ is an eigenvector of Σ corresponding to the eigenvalue λ_1 and satisfying $\mathbf{a}_1'\mathbf{a}_1 = 1$.

D E F I N I T I O N 5.2

The second principal component is defined by $y_2 = \mathbf{a}_2'(\mathbf{x} - \boldsymbol{\mu})$ where $\mathbf{a}_2$ is chosen so that the variance of $\mathbf{a}_2'(\mathbf{x} - \boldsymbol{\mu})$ is a maximum among all such linear combinations of $\mathbf{x}$ that are uncorrelated with the first principal component variable and have $\mathbf{a}_2'\mathbf{a}_2 = 1$.

We can show that the maximum value of the variance of $\mathbf{a}_2'(\mathbf{x} - \boldsymbol{\mu})$ among all linear combinations of $\mathbf{x}$ that have $\mathbf{a}_2'\mathbf{a}_2 = 1$ and that are uncorrelated with y_1 is equal to λ_2, the second largest eigenvalue of Σ, and that this maximum occurs when $\mathbf{a}_2$ is an eigenvector of Σ corresponding to the second largest eigenvalue, λ_2, and satisfying $\mathbf{a}_2'\mathbf{a}_2 = 1$.

In a similar manner, we can define additional principal components. The jth ($j = 3, 4, \ldots, p$) principal component is given by $y_j = \mathbf{a}_j'(\mathbf{x} - \boldsymbol{\mu})$ where $\mathbf{a}_j$ is chosen so that $\mathbf{a}_j'\mathbf{a}_j = 1$ and so that the variance of $\mathbf{a}_j'(\mathbf{x} - \boldsymbol{\mu})$ is a maximum among all such linear combinations of $\mathbf{x}$ that are uncorrelated with the first $j - 1$ principal components. The maximum value of the variance of $\mathbf{a}_j'(\mathbf{x} - \boldsymbol{\mu})$ among all linear combinations of $\mathbf{x}$ that are uncorrelated with y_1 through y_{j-1} and have $\mathbf{a}_j'\mathbf{a}_j = 1$ is equal to λ_j, and jth largest eigenvalue of Σ, and this maximum occurs when $\mathbf{a}_j$ is an eigenvector of Σ corresponding to the eigenvalue λ_j and satisfying $\mathbf{a}_j'\mathbf{a}_j = 1$.

Thus $\lambda_1 \geq \lambda_2 \geq \cdots \geq \lambda_p$ denote the ordered eigenvalues of Σ and $\mathbf{a}_1, \mathbf{a}_2, \ldots, \mathbf{a}_p$ denote corresponding normalized eigenvectors of Σ. We can show that two principal components are uncorrelated if and only if their defining eigenvectors are orthogonal to each other. This result allows us to place an orthogonal axis system within the principal component space in which the data fall.

The variance of the jth component, y_j, is λ_j, $j = 1, 2, \ldots, p$. Recall that $\text{tr}(\mathbf{\Sigma}) = \sigma_{11} + \sigma_{22} + \cdots + \sigma_{pp}$. Thus $\text{tr}(\mathbf{\Sigma})$, in some sense, measures the total variation in the original variables. Also recall that $\text{tr}(\mathbf{\Sigma}) = \lambda_1 + \lambda_2 + \cdots + \lambda_p$ and hence the total variation accounted for by all of the principal component variables is equal to the total amount of variation measured by the original variables. Because of this, to measure the importance of the jth principal component, it is customary to look at the ratio $\lambda_j/\text{tr}(\mathbf{\Sigma})$, for $j = 1, 2, \ldots, p$. This ratio measures the proportion of the total variability in the original variables that is accounted for by the jth principal component.

Principal Component Scores

To use the principal component variables in ensuing statistical analyses, it is necessary to compute principal component scores (values of the principal component variables) for each experimental unit in the data set. These scores provide the locations of the observations in a data set with respect to its principal component axes.

Let $\mathbf{x}_r$ be the vector of measured variables for the rth experimental unit. Then the value (score) of the jth principal component variable for the rth experimental unit is $y_{rj} = \mathbf{a}_j'(\mathbf{x}_r - \boldsymbol{\mu})$, for $j = 1, 2, \ldots, p$ and $r = 1, 2, \ldots, N$.

Component Loading Vectors

Note that the eigenvectors of the variance–covariance matrix that are being used to define the principal components are normalized to have length 1, that is, $\mathbf{a}_j'\mathbf{a}_j = 1$, for $j = 1, 2, \ldots, p$. This can sometimes be confusing when we are trying to interpret the principal components by examining the elements in the eigenvectors that define the principal components. A large element in one eigenvector may or may not be large in another eigenvector. That is, elements within an eigenvector are comparable to one another, but elements in different eigenvectors are not comparable. This is because the eigenvectors are normalized to have length 1, which requires that the sum of the squares of the elements in each vector must equal 1. Thus the more elements in a single vector that are actually different from 0, the smaller each element must be. For example, if eight elements in a vector were nonzero and if all were of similar magnitude, they would each have a value equal to $\pm 1/\sqrt{8} = \pm 0.3536$; but if only two elements in a vector were nonzero and were of the same magnitude, they would each have a value equal to $\pm 1/\sqrt{2} = \pm 0.7071$.

To make comparisons between eigenvectors, many researchers scale the eigenvectors by multiplying the elements in each vector by the square root of its corresponding eigenvalue. Let $\mathbf{c}_j = \lambda_j^{1/2}\mathbf{a}_j$, for $j = 1, 2, \ldots, p$. These new vectors are called *component loading vectors*. The elements in the vector $\mathbf{c}_j$ are called *component loadings* and are scaled so that they are generally larger than those of the less important components. The $\mathbf{c}_j$'s are still eigenvectors of $\boldsymbol{\Sigma}$, but they have lengths equal to $\sqrt{\lambda_j}$ rather than length 1. All of the elements in all of the $\mathbf{c}_j$'s are comparable to one another. The ith element in $\mathbf{c}_j$ gives the covariance between the ith original variable and the jth principal component.

5.4
Estimation of Principal Components

The preceding development assumed that both $\boldsymbol{\mu}$ and $\boldsymbol{\Sigma}$ are known. This is hardly ever the case so $\boldsymbol{\mu}$ and $\boldsymbol{\Sigma}$ will need to be estimated from sample data.

Suppose one has a random sample from a population with mean $\boldsymbol{\mu}$ and variance–covariance matrix $\boldsymbol{\Sigma}$. Recall that the estimators of $\boldsymbol{\mu}$ and $\boldsymbol{\Sigma}$ are $\hat{\boldsymbol{\mu}}$ and $\hat{\boldsymbol{\Sigma}}$, respectively, as indicated in Section 1.8.

The estimators of the λ_i's and the $\mathbf{a}_i$'s are always taken as the corresponding eigenvalues and eigenvectors of $\hat{\boldsymbol{\Sigma}}$. These are denoted by $\hat{\lambda}_i$ and $\hat{\mathbf{a}}_i$, respectively.

Estimation of Principal Component Scores

When working with sample data the principal component scores are estimated by

$$y_{rj} = \hat{\mathbf{a}}_j'(\mathbf{x}_r - \hat{\boldsymbol{\mu}}), \qquad \text{for } j = 1, 2, \ldots, p; \quad r = 1, 2, \ldots, N$$

5.5
Determining the Number of Principal Components

When performing a PCA, we need to determine the actual dimensionality of the space in which the data fall; that is, the number of principal components that have variances larger than zero. If several of the eigenvalues of $\hat{\boldsymbol{\Sigma}}$ are zero or sufficiently close to zero, then the actual dimensionality of the data is effectively that of the number of nonzero eigenvalues.

There are two methods for helping to choose the number of principal components to use when PCA is being applied to $\hat{\boldsymbol{\Sigma}}$. Both are based on the eigenvalues of $\hat{\boldsymbol{\Sigma}}$. Let d be the dimensionality of the space in which the data actually lie as you read about these two methods.

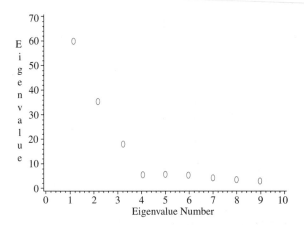

F I G. **5.1**

A SCREE plot

Method 1

Suppose we wish to account for $\gamma 100\%$ of the total variability in the original variables. One method of estimating d considers $V = (\lambda_1 + \lambda_2 + \cdots + \lambda_k)/\text{tr}(\mathbf{\Sigma})$ for successive values of $k = 1, 2, \ldots, p$. Then d is estimated by the smallest value for k at which V first exceeds γ. How much variability one desires to account for often depends on the type of population being sampled. For laboratory-type data, it may be quite easy to account for more than 95% of the total variability with only two or three principal components. On the other hand, for "people-type" data, five or six principal components may be required to account for more than 70 to 75% of the total variation. Unfortunately, the more principal components that are required, the less useful each one becomes.

Method 2

A second method of estimating d utilizes a SCREE plot of the eigenvalues. A SCREE plot is constructed by plotting the value of each eigenvalue against which one it is. That is, you plot the pairs $(1, \hat{\lambda}_1), (2, \hat{\lambda}_2), \ldots, (p, \hat{\lambda}_p)$. When the points on the graph tend to level off, these eigenvalues are usually close enough to zero that they can be ignored. At the very least, the smaller ones are probably measuring nothing but random noise, and you should not try to interpret noise. Thus, for this method, the dimensionality of the data space is assumed to be that corresponding to the smallest large eigenvalue. An example of a SCREE plot is shown in Figure 5.1. This SCREE plot would suggest that the actual dimensionality of the space in which the data lie is three and, hence, the appropriate number of principal components to use is also three.

In practice, researchers almost always consider simultaneously both of the methods presented. To illustrate the ideas discussed thus far, consider the following example.

E X A M P L E 5.1

Consider the applicant data introduced in Example 2.1 and consider performing a principal components analysis on these data. The applicant data were analyzed with the PRINCOMP procedure in SAS® using the following commands:

```
OPTIONS LINESIZE=66 PAGESIZE=60 NODATE;
TITLE 'EXAMPLE 5.1 - SAS PRINCOMP PROCEDURE';
DATA APPLICAN;
INPUT ID FL APP AA LA SC LC HON SMS EXP DRV AMB GSP
      POT KJ SUIT;
CARDS;
```

(The data in Table 2.1 were included here)

```
PROC PRINCOMP DATA=APPLICAN OUT=PCSCORES COVARIANCE;
 VAR FL--SUIT;
 TITLE2 'PC ANALYSIS IS ON RAW DATA AND THE SAMPLE
COVARIANCE    MATRIX';
 RUN;


PROC PRINT DATA=PCSCORES;
 VAR ID PRIN1-PRIN4;
 TITLE2 'VALUES OF THE FIRST 4 PRINCIPAL COMPONENT SCORES';
  RUN;
```

Some of the results from the preceding commands are shown in the partial computer printout of Figure 5.2. The first five pages of this output come from the PRINCOMP procedure, while page 6 comes from the PRINT procedure. On the upper half of page 1, the PRINCOMP procedure produces the means and standard deviations for each of the measured response variables. On the lower half of page 1 and continuing through page 3, the PRINCOMP procedure prints $\hat{\Sigma}$, the sample variance–covariance matrix. Note that the diagonal elements of this matrix, 7.1489 . . . , 3.8652 . . . , 10.8918 . . . , are the sample variances for the variables FL, APP, . . . , SUIT, respectively. The off-diagonal elements are the sample covariances between the pairs of variables that correspond to each row and column. For example, the estimated covariance between LA and SC is 2.0519.

In the middle of page 3, note the line that reads

```
Total Variance = 122.53501773
```

Principal Component Analysis

48 Observations
15 Variables
Simple Statistics

	FL	APP	AA	LA
Mean	6.000000000	7.083333333	7.083333333	6.145833333
Std	2.673749459	1.966023455	1.987549901	2.805690140

	SC	LC	HON	SMS
Mean	6.937500000	6.312500000	8.041666667	4.854166667
Std	2.418072470	3.170047654	2.534513536	3.439381336

	EXP	DRV	AMB	GSP
Mean	4.229166667	5.312500000	5.979166667	6.250000000
Std	3.308529167	2.947456531	2.935400827	3.035253854

	POT	KJ	SUIT
Mean	5.687500000	5.562500000	5.958333333
Std	3.183442871	2.657036023	3.300279378

Covariance Matrix = $\hat{\Sigma}$

	FL	APP	AA	LA
FL	7.14893617 $\hat{\gamma}_{11}$	1.25531915 $\hat{\gamma}_{12}$	0.23404255	2.29787234
APP	1.25531915 ·	3.86524823 $\hat{\gamma}_{22}$	0.48226950	2.09397163
AA	0.23404255 ·	0.48226950 ·	3.95035461	0.00886525
LA	2.29787234 ·	2.09397163 ·	0.00886525	7.87189716
SC	0.59574468	2.04787234 ·	0.00531915	2.05186170
LC	1.93617021	2.31382979	0.48404255	4.29388298
HON	-0.72340426	1.76241135	-0.15248227	4.58953901
SMS	2.48936170	3.31028369	0.37411348	3.48980496
EXP	4.85106383	0.91666667	1.74645390	1.30629433
DRV	2.72340426	1.97340426	0.54787234	3.25132979
AMB	2.23404255	3.17198582	0.25709220	2.85416667
GSP	2.74468085	3.02127660	1.19148936	4.28191489
POT	3.12765957	3.17553191	1.83510638	5.40824468
KJ	3.31914894	1.48404255	-1.70744681	5.10771277
SUIT	5.17021277	2.49290780	0.91843972	3.02748227

FIG. 5.2

Pages 1–4 and 6
of computer print-
out, Example 5.1

This is the trace of $\hat{\Sigma}$, which is the sum of the diagonal elements of the sample variance–covariance matrix and, in some sense, is a measure of the total variation in the response variables.

The lower part of page 3 contains four columns of numbers. The first

EXAMPLE 5.1 - SAS PRINCOMP PROCEDURE   2
PC ANALYSIS IS ON RAW DATA AND THE SAMPLE COVARIANCE MATRIX

Principal Component Analysis

Covariance Matrix

	SC	LC	HON	SMS
FL	0.59574468	1.93617021	-0.72340426	2.48936170
APP	2.04787234	2.31382979	1.76241135	3.31028369
AA	0.00531915	0.48404255	-0.15248227	0.37411348
LA	2.05186170	4.29388298	4.58953901	3.48980496
SC	5.84707447	6.19015957	2.51329787	6.65026596
LC	6.19015957	10.04920213	2.85904255	8.91888298
HON	2.51329787	2.85904255	6.42375887	2.09131206
SMS	6.65026596	8.91888298	2.09131206	11.82934397
EXP	0.12101064	1.54388298	-1.30762411	2.90647163
DRV	5.01994681	6.51728723	2.09308511	8.25930851
AMB	5.97739362	7.04920213	1.59663121	8.67774823
GSP	5.29255319	8.49468085	2.96808511	8.16489362
POT	5.17154255	7.84441489	3.35372340	8.25132979
KJ	3.09973404	4.43750000	3.01861702	5.14760638
SUIT	1.99734043	4.35372340	0.02304965	6.33421986

	EXP	DRV	AMB	GSP
FL	4.85106383	2.72340426	2.23404255	2.74468085
APP	0.91666667	1.97340426	3.17198582	3.02127660
AA	1.74645390	0.54787234	0.25709220	1.19148936
LA	1.30629433	3.25132979	2.85416667	4.28191489
SC	0.12101064	5.01994681	5.97739362	5.29255319
LC	1.54388298	6.51728723	7.04920213	8.49468085
HON	-1.30762411	2.09308511	1.59663121	2.96808511
SMS	2.90647163	8.25930851	8.67774823	8.16489362
EXP	10.94636525	3.28856383	1.89849291	3.00531915
DRV	3.28856383	8.68750000	6.75132979	6.38829787
AMB	1.89849291	6.75132979	8.61657801	6.98404255
GSP	3.00531915	6.38829787	6.98404255	9.21276596
POT	3.66888298	7.39760638	7.18484043	8.46276596
KJ	1.88962766	4.79920213	4.26728723	4.43085106
SUIT	7.56294326	6.05585106	4.21187943	5.28723404

F I G. 5.2

Pages 1–4 and 6
of computer print-
out, Example 5.1

column gives the eigenvalues of $\hat{\boldsymbol{\Sigma}}$, so we can see that $\hat{\lambda}_1 = 66.5364$, $\hat{\lambda}_2 = 18.1805$, $\hat{\lambda}_3 = 10.5910$, ... The third and fourth columns give the proportion of the total variation that is accounted for by successive principal components and the cumulative proportion of the total variation that is accounted for by that principal component and all of the previous ones. For this example, the first principal component accounts for 54.30% of the total variability, the second principal component accounts for 14.84% of the total variability, and the first two principal components together account for 69.14% of the total variability. The second column gives the difference between successive eigenvalues. This difference has limited use. A SCREE

Principal Component Analysis

Covariance Matrix

	POT	KJ	SUIT
FL	3.12765957	3.31914894	5.17021277
APP	3.17553191	1.48404255	2.49290780
AA	1.83510638	-1.70744681	0.91843972
LA	5.40824468	5.10771277	3.02748227
SC	5.17154255	3.09973404	1.99734043
LC	7.84441489	4.43750000	4.35372340
HON	3.35372340	3.01861702	0.02304965
SMS	8.25132979	5.14760638	6.33421986
EXP	3.66888298	1.88962766	7.56294326
DRV	7.39760638	4.79920213	6.05585106
AMB	7.18484043	4.26728723	4.21187943
GSP	8.46276596	4.43085106	5.28723404
POT	10.13430851	4.56250000	6.02925532
KJ	4.56250000	7.05984043	3.47074468
SUIT	6.02925532	3.47074468	10.89184397

Total Variance = 122.53501773 = $tr(\hat{\Sigma})$

Eigenvalues of the Covariance Matrix

	Eigenvalue		Difference	Proportion	Cumulative
PRIN1	66.5364	$\hat{\lambda}_1$	48.3559	0.542999	0.54300
PRIN2	18.1805	$\hat{\lambda}_2$	7.5895	0.148370	0.69137
PRIN3	10.5910	$\hat{\lambda}_3$	3.8230	0.086432	0.77780
PRIN4	6.7679	·	2.7823	0.055233	0.83303
PRIN5	3.9856	·	0.3580	0.032527	0.86556
PRIN6	3.6277	·	0.7119	0.029605	0.89517
PRIN7	2.9157		0.0802	0.023795	0.91896
PRIN8	2.8355		0.8804	0.023141	0.94210
PRIN9	1.9551		0.3416	0.015956	0.95806
PRIN10	1.6135		0.4770	0.013168	0.97123
PRIN11	1.1365		0.2636	0.009275	0.98050
PRIN12	0.8729		0.1662	0.007123	0.98762
PRIN13	0.7067		0.1981	0.005767	0.99339
PRIN14	0.5085		0.2072	0.004150	0.99754
PRIN15	0.3014		·	0.002459	1.00000

F I G. 5.2

Pages 1–4 and 6
of computer print-
out, Example 5.1

plot of these eigenvalues is shown in Figure 5.3. The SCREE plot shows a break between the third and fourth eigenvalues; and note that the first three principal components account for 77.8% of the total variability, so it would seem that the applicant data tend to fall within a 3-dimensional subspace of the 15-dimensional sample space.

The eigenvectors (normalized to have length 1) are shown on pages 4 and 5 of the computer output (however, only page 4 and the first 5 eigen-

EXAMPLE 5.1 - SAS PRINCOMP PROCEDURE   4
PC ANALYSIS IS ON RAW DATA AND THE SAMPLE COVARIANCE MATRIX

Principal Component Analysis

Eigenvectors

	PRIN1	PRIN2	PRIN3	PRIN4	PRIN5
FL	0.149129	0.371461	0.200481	-.277311	0.636939
APP	0.132250	-.029296	0.041918	0.134231	0.042210
AA	0.029611	0.101846	-.131030	0.603168	0.167474
LA	0.203126	-.093042	0.619733	0.126399	0.053473
SC	0.231436	-.235740	-.189273	-.072088	-.025117
LC	0.336870	-.195978	-.124714	0.052788	0.231817
HON	0.120238	-.300549	0.447178	0.255587	-.334369
SMS	0.379017	-.090010	-.281581	-.172303	-.177778
EXP	0.164016	0.636212	0.025043	0.166245	-.191487
DRV	0.316050	0.012486	-.113315	-.134844	-.338054
AMB	0.312106	-.122150	-.244517	-.147307	0.105416
GSP	0.338764	-.074347	-.050497	0.206271	0.258316
POT	0.357165	-.024920	0.041308	0.317232	0.108875
KJ	0.226076	-.044837	0.385206	-.459715	-.026846
SUIT	0.274483	0.470867	0.016815	-.015962	-.349972
	$\hat{\mathbf{a}}_1$	$\hat{\mathbf{a}}_2$	$\hat{\mathbf{a}}_3$	...	

F I G. 5.2
Pages 1–4 and 6
of computer print-
out, Example 5.1

vectors are shown in Figure 5.2). To illustrate, note that

$$\hat{\mathbf{a}}_1' = [0.149, 0.132, \ldots, 0.274] \quad \text{and} \quad \hat{\mathbf{a}}_2' = [0.371, -0.029, \ldots, 0.471]$$

The jth principal component score for the rth individual is computed by $y_{rj} = \hat{\mathbf{a}}_j'(\mathbf{x}_r - \hat{\boldsymbol{\mu}})$. To illustrate, the first individual in the applicant data set has FL=6, APP=7, AA=2, LA=5, SC=8, LC=7, HON=8, SMS=8, EXP=3, DRV=8, AMB=9, GSP=7, POT=5, KJ=7, and SUIT=10. So the first principal component score for this individual is

$$y_{11} = 0.149(6 - 6) + 0.132(7 - 7.083) + 0.030(2 - 7.083)$$
$$+ 0.203(5 - 6.146) + \cdots + 0.274(10 - 5.958) = 4.304$$

As a result of using the OUT=PCSCORES option, the PRINCOMP proce-dure automatically computes all 15 possible principal component scores for each individual in the data set and stores them in a second data set named PCSCORES. A printout of the first four principal component scores for each individual in the applicant data set is shown on page 6 of the computer out-put (Figure 5.2). Additional uses for these principal component scores are given later in this chapter in the continuation of Example 5.1.

In this example, there are no obvious interpretations for the first three principal components. Any such interpretation must come from an examina-tion of the eigenvectors on page 4 of the computer printout. The variables that tend to have strong relationships (i.e., have elements in the eigenvector that tend to be larger in absolute value than the others in the eigenvector)

EXAMPLE 5.1 - SAS PRINCOMP PROCEDURE   6
VALUES OF THE FIRST 4 PRINCIPAL COMPONENT SCORES

OBS	ID	y_1 PRIN1	y_2 PRIN2	y_3 PRIN3	y_4 PRIN4
1	1	4.3040	-0.3819	-1.7574	-5.61558
2	2	10.1416	0.4179	0.0874	-3.08784
3	3	6.5297	-0.1686	-0.3237	-4.52532
4	4	-1.3281	2.1759	1.0981	2.81460
5	5	1.4804	3.4916	3.2469	2.45018
6	6	2.3832	2.4176	1.0401	1.18088
7	7	9.5935	4.9223	0.3070	-0.09617
8	8	11.0723	4.4857	0.0833	0.09941
9	9	7.5936	4.6013	1.4609	-0.51860
10	10	8.2294	-1.2104	-8.8270	2.10878
11	11	4.3825	-0.2660	-11.4878	1.81192
12	12	5.9910	-3.0709	-6.3429	3.46936
13	13	-0.7646	1.4111	4.4261	1.70872
14	14	-1.9066	2.1314	3.7461	2.18465
15	15	-2.9391	2.0457	2.6388	4.05528
16	16	6.8485	2.7686	-1.0857	-0.41740
17	17	3.2896	2.1463	0.3424	-1.81086
18	18	-2.4699	-1.3436	-2.3240	-1.03799
19	19	-2.6360	-0.2087	-2.8799	-0.07141
20	20	4.9529	-2.8031	1.6593	0.92051
21	21	3.3098	-2.7063	0.6176	1.89694
22	22	10.4187	-1.8985	0.9701	-2.56450
23	23	10.6044	-2.7566	1.3076	-1.36303
24	24	9.8197	-2.3259	2.8235	-2.01274
25	25	-4.7128	-1.1959	2.6621	0.41022
26	26	-3.0377	0.7447	2.9647	0.63370
27	27	1.2640	-4.4574	0.7550	3.17409
28	28	-13.2762	-1.8724	-0.4129	-4.87070
29	29	-15.0505	-0.2647	-2.5238	-6.49352
30	30	-6.7359	-4.1335	2.5080	-2.76280
31	31	-4.0775	-3.4440	3.3333	-3.59605
32	32	-3.0224	-5.1850	1.4391	0.46134
33	33	-4.8156	-5.4946	1.5877	0.61968
34	34	-10.6381	-2.2224	0.9446	-1.04874
35	35	-12.6686	-0.3487	2.9770	-2.32254
36	36	-4.1900	-1.0116	2.2025	-1.52308
37	37	0.3559	-6.3630	-5.1620	-0.46163
38	38	1.5343	-6.5074	-3.7165	-1.06516
39	39	13.1802	3.8946	2.3557	0.11512
40	40	13.8044	3.6503	1.8667	-0.17950
41	41	-11.3814	13.5076	-3.0319	0.32818
42	42	-14.0307	14.6597	-3.5266	-0.99695
43	43	-7.8023	4.7752	1.2897	2.68144
44	44	5.5585	0.7543	-1.8583	-0.04394
45	45	-0.1003	-1.7368	3.1581	3.43819
46	46	2.8045	-3.2592	2.7652	4.21745
47	47	-15.6618	-4.2901	-1.0614	4.04622
48	48	-16.2003	-4.0747	-2.3428	3.65919

F I G. 5.2

Pages 1–4 and 6
of computer print-
out, Example 5.1

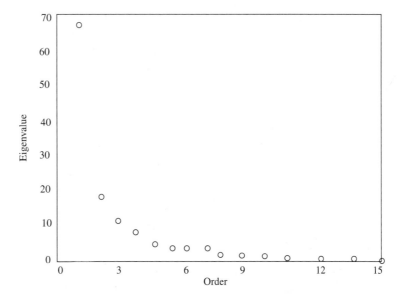

F I G. 5.3
SCREE plot of the eigenvalues from the applicant data

with the first principal component are LC, SMS, DRV, AMB, GSP, and POT. The variables that tend to have strong relationships with the second principal component are FL, EXP, and SUIT. The variables that tend to have strong relationships with the third principal component are LA, HON, and KJ. It is interesting to note that these three groups of variables are similar to the first three groups of variables obtained in Example 2.1. Recall that the relative sizes of the elements in one eigenvector cannot be compared to the sizes of elements in another eigenvector since each eigenvector is normalized to have length 1.

5.6 Caveats

Principal components analysis transforms a set of correlated variables into a new set of uncorrelated variables. If the original variables are already nearly uncorrelated, then nothing can be gained by carrying out a PCA. In this case, the actual dimensionality of the data is equal to the number of response variables measured and it is not possible to examine the data in a reduced dimensional space. When the measured variables have a multivariate normal distribution, a statistical test for testing the hypothesis that the original variables are uncorrelated is available. This test is described in Section 5.8.

Although PCA allows one to use a reduced number of variables in ensuing analyses, PCA cannot generally be used to eliminate variables. This is because all of the original variables are needed to score or evaluate the principal component variables for each of the individuals in a data set. Thus elimination of some of the original variables should not be a primary objective when using PCA.

Principal components analysis is appropriate only in those cases where all of the variables arise "on an equal footing." What does this mean? It means a couple of things: (1) The variables should all be measured in the same units, or at least in comparable units, and (2) the variables should have variances that are roughly similar in size.

To expand on point 1, if the response variables are not measured in the same units, then any change in the scale of measurement on one or more of the variables will have an effect on the principal components. Such a scaling could reverse the roles of important and unimportant variables. For example, suppose a variable that was originally measured in feet was changed to inches. This means that the variable's variance will be increased by a multiple of $12^2 = 144$. Since PCA is variance oriented, this variable would have a much greater influence on the principal components when measured in inches than it would when measured in feet.

To expand on point 2, the principal components are generally changed by a rescaling of the variables. Thus they are not a unique characteristic of the data. If one variable has a much larger variance than all the other variables, it will dominate the first principal component regardless of the covariance structure of the variables, and there is little point in performing a PCA in this case.

In those cases where the variables do not seem to be occurring on an equal footing, many researchers apply PCA to the correlation matrix of the responses rather than to the variance–covariance matrix. This is equivalent to applying PCA to Z scores or standardized data rather than applying it to raw data values. In this case, the principal components are defined by the eigenvalues and eigenvectors of $\mathbf{P}$, the correlation matrix, rather than by those of $\mathbf{\Sigma}$, the variance–covariance matrix. The eigenvalues and eigenvectors of $\mathbf{P}$ are different than those of $\mathbf{\Sigma}$, and no simple transformation exists for taking one set of values into the other set of values.

By default, most multivariate computing packages apply PCA to the correlation matrix rather than to the variance–covariance matrix, and many do not even give you an option for analyzing the variance–covariance matrix. This is unfortunate, because choosing to analyze $\mathbf{P}$ rather than $\mathbf{\Sigma}$ involves deciding that all of the response variables are equally important. Do not make this assumption arbitrarily.

Situations may exist where the variables are not in comparable units and where the researcher would prefer to not treat all variables as being equally important. Some statistical computing packages will allow the assignment of a weight to each variable. The data could be standardized and then variables that are thought to be more important than other variables could be given larger weights than those variables of lesser importance.

To illustrate the preceding comments about variables being equally important, consider the applicant data described in Example 2.1 and analyzed in Example 5.1. Recall that each applicant was evaluated on 15 different variables and that each variable takes on a value between 0 and 10. A primary goal of the analysis in Example 2.1 was to select six individuals to receive job offers. Consider two of the variables in the data set, namely, the variables ambition and appearance. It is likely that applicants vary more with respect to their ambition than they vary with respect to their appearance. That is, individuals may range from 0 to 10 in ambition, but might only range from 7 to 10 in appearance. In terms of selecting people to hire, ambition may be a good criterion for selection because people vary widely in ambition. At the same time, appearance may not be a good criterion to use because all candidates tend to look pretty good. Standardizing these two variables puts them into comparable units and, as a result, both variables vary equally. The result of this is that appearance will be just as important as ambition when selecting applicants. The equalization of these two variables is not likely to be desirable when choosing applicants to hire. Be sure to note that the applicant data were not standardized in Example 5.1. Standardization would not be desirable for these data because all variables are already on an equal footing with one another in that they were all evaluated on a 0 to 10 scale.

Every time a researcher considers applying PCA to a set of data, the researcher must make a conscientious decision about the units in which the variables will be analyzed. Researchers should not let a statistical computing package make this decision for them. If they do let the computing package make the decision, the data are likely to be standardized.

5.7
PCA on the Correlation Matrix **P**

As mentioned earlier, when variables do not occur on equal footings it is necessary to apply PCA methods to standardized data (Z scores). This is done by computing the eigenvalues and eigenvectors of the correlation matrix. The eigenvalues and eigenvectors of **P** are denoted by $\lambda_1^* \geq \lambda_2^* \geq \cdots \geq \lambda_p^*$ and $\mathbf{a}_1^*, \mathbf{a}_2^*, \ldots, \mathbf{a}_p^*$, respectively.

Convention   A notation that is used throughout the remainder of this book is to place asterisks (*) on quantities that come from Z scores or correlation matrices. Whenever asterisks do not appear, the reader can be assured that the functions being considered are from the raw data and/or variance–covariance matrices; and when asterisks do appear, the functions being considered come from Z scores and/or correlation matrices.

Principal Component Scores

When PCA is performed on a correlation matrix, principal component scores must be computed from the Z scores. In this case the jth principal component score for the rth experimental unit is defined by $y_{rj}^* = \mathbf{a}_j^{*'}\mathbf{z}_r$, for $j = 1, 2, \ldots, p$.

Component Correlation Vectors

When PCA has been performed on the correlation matrix, the elements of $\mathbf{c}_j^* = \lambda_j^{*1/2}\mathbf{a}_j^*$ give the correlations between the original variables and the jth principal component variable. These vectors are called *component correlation vectors*. Thus,

$$\mathrm{corr}(x_i, y_j^*) = c_{ij}^*$$

Note that c_{ij}^* is the ith element of the jth component loading vector. These correlations are called *component correlations*.

Sample Correlation Matrix

If we apply PCA to the sample correlation $\mathbf{R}$, the estimators of λ_j^* and $\mathbf{a}_j^*$ are denoted by $\hat{\lambda}_j^*$ and $\hat{\mathbf{a}}_j^*$, respectively, where the $\hat{\lambda}_j^*$'s and the $\hat{\mathbf{a}}_j^*$'s are the eigenvalues and eigenvectors of $\mathbf{R}$.

If the analysis is performed on $\mathbf{R}$, the principal component scores are computed from Z scores according to the formula

$$y_{rj}^* = \hat{\mathbf{a}}_j^{*'}\mathbf{z}_r, \qquad \text{for } j = 1, 2, \ldots, p; \quad r = 1, 2, \ldots, N$$

Determining the Number of Principal Components

Both methods described in Section 5.5 for determining the dimensionality of the space in which the standardized data actually lie can also be used when a PCA is being performed on a correlation matrix. When analyzing the correlation matrix, a third method can also be used.

The third method looks for eigenvalues that are greater than 1, and estimates the dimensionality of the sample space to be that of the number of eigenvalues that are greater than 1. The reason for comparing the eigenvalues to 1 is that when the analysis is being done on standardized data (i.e., the correlation matrix), the variance of each standardized variable is equal to 1. The belief is that if a principal component cannot account for more variation than a single variable can by itself, then it is probably not important; hence, components whose eigenvalues are less than 1 are often ignored. Comparing eigenvalues to 1 must never be considered when analyzing the raw data or, equivalently, the sample variance–covariance matrix.

When analyzing a correlation matrix, a good researcher usually looks at all three methods simultaneously. In all instances, the decision as to how many principal components to consider is a subjective one.

5.8
Testing for Independence of the Original Variables

If we believe the data come from a multivariate normal distribution, we can—and probably should—test whether the response variables are independent (i.e., uncorrelated) before performing a PCA. This can be done by testing $\mathbf{P} = \mathbf{I}$ or, equivalently, by testing that $\boldsymbol{\Sigma}$ is a diagonal matrix.

A likelihood ratio test statistic for testing $H_0: \mathbf{P} = \mathbf{I}$ is given by $V = |\mathbf{R}|$. For large values of N, we reject H_0 if

$$-a \log V > \chi^2_{\alpha, p(p-1)/2}$$

where $a = N - 1 - (2p + 5)/6$. If H_0 cannot be rejected, a PCA should not be performed.

Note The value of V can also be computed by using either of these formulas:

$$V = \hat{\lambda}_1^* \hat{\lambda}_2^* \ldots \hat{\lambda}_p^* \qquad \text{or} \qquad V = (\hat{\lambda}_1 \hat{\lambda}_2 \ldots \hat{\lambda}_p)/(\hat{\sigma}_{11} \hat{\sigma}_{22} \ldots \hat{\sigma}_{pp})$$

5.9
Structural Relationships

If some of the eigenvalues of $\boldsymbol{\Sigma}$ and/or $\mathbf{P}$ are zero, then some of the original variables are linearly dependent on others. Suppose we have determined that a PCA has d principal components. Then there are $p - d$ eigenvalues that are equal to zero. When there are $p - d$ eigenvalues equal to zero, their corresponding eigenvectors define $p - d$ linearly independent linear constraints on the response variables. The eigenvectors corresponding to the small eigenvalues define the *structural relationships* among the original variables. New variables created by these linear constraints are called *structural relationship variables*. Let $y_j = \mathbf{a}_j'(\mathbf{x} - \boldsymbol{\mu})$, where $j > d$. Then $\text{Var}(y_j) = \lambda_j \approx 0$. This implies that $\mathbf{a}_j'(\mathbf{x} - \boldsymbol{\mu}) \approx 0$ with probability 1. Thus the eigenvectors that correspond to eigenvalues near zero define the structural relationships among the original variables.

The structural relationships are never unique when $p - d > 1$, so you should never try to give physical interpretations to the structural relationships in cases where $p - d > 1$.

The structural relationship equations provide information about how the variables are related to one another, but they are not useful for multivariate data screening. They are most useful for examining multicollinearity relationships among a set of predictor variables in regression problems.

5.10
Statistical Computing Packages

Numerous statistical computing programs are available that allow us to perform PCAs on sets of multivariate data.

SAS[R] PRINCOMP Procedure

As already illustrated in Example 5.1, SAS has a procedure called PRINCOMP for performing a PCA. This procedure allows one to perform a PCA on raw data (i.e., computing eigenvalues and eigenvectors of the sample variance–covariance matrix) or on standardized data (i.e., computing eigenvalues and eigenvectors of the sample correlation matrix). This procedure also creates a new set of data that contains computed values of the principal component scores.

The basic commands to perform a PCA on raw data with the PRINCOMP procedure are

```
PROC PRINCOMP COVARIANCE OUT=PCSCORES;
 VAR X₁--Xₚ;
 TITLE 'PCA ON RAW DATA';
 RUN;
```

The basic commands to perform a PCA on standardized data (Z scores) with the PRINCOMP procedure are

```
PROC PRINCOMP OUT=PCSCORES;
 VAR X₁--Xₚ;
 TITLE 'PCA ON Z-SCORES';
 RUN;
```

The main difference in these two sets of commands is that the first set uses an option called the COVARIANCE option. The default in the PRINCOMP procedure is to analyze a correlation matrix. When one uses the COVARIANCE option, the sample variance–covariance matrix is used for the PCA. Both sets of commands will create a new data set called PCSCORES here. This new data set will contain all of the original data plus each of the principal component scores. The principal component scores will be labeled PRIN1, PRIN2, ... The first four principal component scores can be printed with these commands (assuming there is a variable in the original data set labeled ID that identifies each experimental unit):

```
PROC PRINT DATA=PCSCORES;
 VAR ID PRIN1-PRIN4;
 TITLE 'THE FIRST FOUR PRINCIPAL COMPONENT SCORES';
 RUN;
```

To create a scatter plot of the first two principal component scores, use these commands:

```
PROC PLOT;
 PLOT PRIN2*PRIN1='+';
 TITLE 'A SCATTER PLOT OF THE FIRST TWO PC SCORES';
 RUN;
```

The plot created by the preceding commands is useful for examining assumptions a researcher might make about the population under study and for checking the sample data for possible outliers or other abnormalities. If the population being sampled is multivariate normal, then each of the principal components will have a univariate normal distribution, and a pairwise scatter plot of the first two principal component scores should result in the data falling into an elliptically shaped region with the major and minor axes of the ellipse parallel to the PRIN1 and PRIN2 axes, respectively.

To create a three-dimensional plot of the first three principal component scores, use these commands:

```
PROC PLOT;
 PLOT PRIN2*PRIN1=PRIN3/CONTOUR=10;
 TITLE 'A CONTOUR PLOT OF THE FIRST THREE PC SCORES';
 RUN;
```

The plot created by these commands could be called a "poor man's" blob or bubble plot. Such plots were described in Chapter 3. The primary difference between the plot created by the preceding commands and a blob plot is that the symbol being plotted by the CONTOUR=10 option gets "darker" rather than "bigger" as the value of PRIN3 increases. This plot is also useful for examining assumptions about the population being studied and for checking the data set for possible abnormalities.

If the original data represent a random sample from a single multivariate normal population, then each set of principal component scores should be distributed normally. A test of univariate normality is available in the SAS UNIVARIATE procedure. To get normality tests for each of the first four principal components as well as stem-and-leaf plots, box plots, and normal probability plots, we can use the commands:

```
PROC UNIVARIATE DATA=PCSCORES PLOT NORMAL;
      VAR PRIN1-PRIN4;
      TITLE 'UNIVARIATE ANALYSES ON PC SCORES';
      RUN;
```

E X A M P L E 5.1 (continued)

Plots similar to those described here were obtained for the first three PC scores for the applicant data by using the following commands (these commands are also a part of the file labeled EX5_1.SAS on the enclosed disk):

```
PROC PLOT DATA=PCSCORES;
  PLOT PRIN1*PRIN2=PRIN3/CONTOUR=10 VAXIS = -20 TO 20 BY 5
    HAXIS = -15 TO 15 BY 5 VPOS=50 HPOS=72;
  TITLE2 'BLOB PLOT OF THE FIRST 3 PRINCIPAL COMPONENT
    SCORES';
  RUN;


PROC UNIVARIATE DATA=PCSCORES PLOT NORMAL;
  VAR PRIN1-PRIN3;
  TITLE2 'UNIVARIATE ANALYSES ON PC SCORES';
RUN;
```

Recall that there are no obvious interpretations for the first three principal components of these data. Nevertheless, the principal component scores can be examined to determine if there are abnormalities in the data. A blob-type plot of the first three principal component scores is given on page 7 of the computer printout shown in Figure 5.4. A three-dimensional plot is shown in Figure 5.5. There does not seem to be anything too unusual about these plots. The points seem to fall into an elliptically shaped region with axes parallel to the principal component axes. The two points in the lower right-hand corner of the plot on page 7 of the printout could be outliers. These two points correspond to individuals 41 (the $\bigcirc$ in the plot) and 42 (the $+$ in the plot). These are also the individuals shown at the far left of the plot in Figure 5.5.

Suppose we want to use the principal components to identify the best individuals to hire as was a primary objective of Example 2.1. An examination of the first three eigenvectors reveals that the best individuals to hire would be those who have large PRIN1, PRIN2, and PRIN3 scores. This is because the signs on the elements of each eigenvector that correspond to elements that are not close to zero are mostly positive. Therefore, individuals who have large values on the 15 original variables will also have large scores for the first three principal components and vice versa. From the plot on page 7 of the computer printout (Figure 5.4), the four best individuals are those in the upper right-hand corner. These are individuals 39 and 40 (the two *'s) and 7 and 8 (the two W's). Other individuals who are above average for all three of the first three principal components (i.e., those who have all of the first three principal components (i.e., those who have all of the first three principal component scores greater than zero) are individuals 2, 5, 6, 9, and 17. These individuals can be identified by looking at the printout of the PC scores on page 6 of Figure 5.2.

Pages 9 through 14 of the computer output of Figure 5.4 contain results produced by the UNIVARIATE procedure in SAS for each of the first three principal components. Note that the mean is zero for each of the principal components and that the variance is equal to the value of the corresponding eigenvalue. The line labeled W:Normal is a test for normality of the

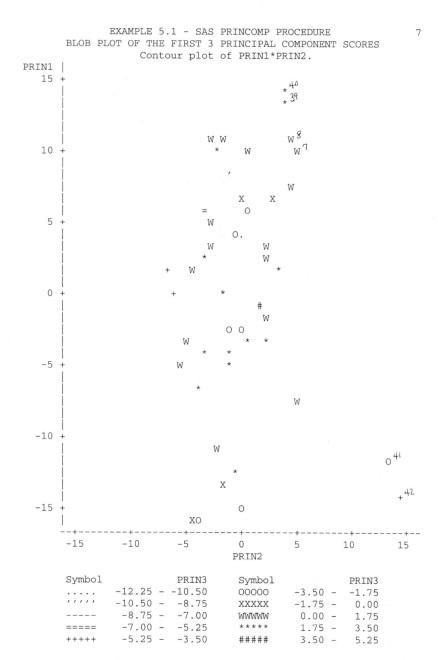

EXAMPLE 5.1 - SAS PRINCOMP PROCEDURE

BLOB PLOT OF THE FIRST 3 PRINCIPAL COMPONENT SCORES

Contour plot of PRIN1*PRIN2.

F I G. 5.4

Pages 7 and 9–14
of computer print-
out, Example 5.1

EXAMPLE 5.1 - SAS PRINCOMP PROCEDURE   9

UNIVARIATE ANALYSES ON PC SCORES

UNIVARIATE PROCEDURE

Variable=PRIN1 = y_1

Moments

N	48	Sum Wgts	48
Mean	0	Sum	0
Std Dev	8.156986	Variance	66.53642 $\hat{\lambda}_1$
Skewness	-0.35677	Kurtosis	-0.67107
USS	3127.212	CSS	3127.212
CV	.	Std Mean	1.17736
T:Mean=0	0	Prob>|T|	1.0000
Sgn Rank	19	Prob>|S|	0.8479
Num ^= 0	48		
W:Normal	0.951263	Prob<W	0.0743

Quantiles(Def=5)

100% Max	13.80439		99%	13.80439
75% Q3	6.260349		95%	11.07233
50% Med	0.80994		90%	10.41867
25% Q1	-4.45141		10%	-13.2762
0% Min	-16.2003		5%	-15.0505
			1%	-16.2003
Range	30.00472			
Q3-Q1	10.71176			
Mode	-16.2003			

Extremes

Lowest	Obs	Highest	Obs
-16.2003(	48)	10.41867(	22)
-15.6618(	47)	10.60439(	23)
-15.0505(	29)	11.07233(	8)
-14.0307(	42)	13.18018(	39)
-13.2762(	28)	13.80439(	40)

F I G. **5.4**

Pages 7 and 9–14
of computer print-
out, Example 5.1

principal component. The observed significance levels for the normality tests for each of the first three principal components are 0.0743 for the first, 0.0003 for the second, and 0.0001 for the third. One needs to interpret these significance levels very carefully for large data sets. For very large data sets, virtually any distribution can be rejected if you look only at statistical significance. In almost all cases, it is probably safer to look at the stem-and-leaf plots, the box plots, and the normal probability plots to determine whether departures from normality are likely to be of any practical concern.

The three plots for PRIN1 on page 10 do not reveal anything that would be of concern to an experienced data analyst. The plots for PRIN2

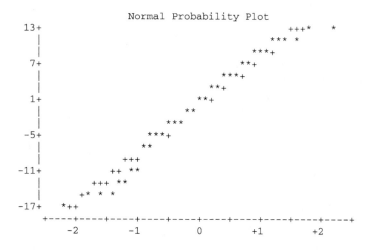

```
           EXAMPLE 5.1 - SAS PRINCOMP PROCEDURE              10
              UNIVARIATE ANALYSES ON PC SCORES

                     UNIVARIATE PROCEDURE

 Variable=PRIN1

        Stem Leaf                    #            Boxplot
          12 28                      2               |
          10 1461                    4               |
           8 268                     3               |
           6 0586                    4            +-----+
           4 3406                    4            |     |
           2 4833                    4            |     |
           0 4355                    4            *--+--*
          -0 9381                    4            |     |
          -2 00965                   5            |     |
          -4 8721                    4            +-----+
          -6 87                      2               |
          -8                                         |
         -10 46                      2               |
         -12 37                      2               |
         -14 710                     3               |
         -16 2                       1               |
             ----+----+----+----+
```

                          Normal Probability Plot
```
     13+                                          +++*     *
       |                                        *** *
       |                                      ***+
      7+                                     **+
       |                                  ***+
       |                                 **+
      1+                              **+
       |                            **
       |                          ***
     -5+                       ***+
       |                       **
       |                  +++
    -11+                ++ **
       |             +++ **
       |           +* * *
    -17+      *++
       +----+----+----+----+----+----+----+----+----+----+
          -2        -1         0        +1        +2
```

FIG. 5.4

Pages 7 and 9–14 of computer print-out, Example 5.1

on page 12 seem to identify a couple of possible outliers; these are the two large values that correspond to individuals 41 and 42. The plots for PRIN3 on page 14 reveals that individual 11 could possibly be an outlier and individuals 10 and 12 are slightly suspicious. Figures 5.6 through 5.8 show histograms with a normal curve overlay for each of the first three principal component variables.

To summarize, the primary outcome of an examination of the various plots of the principal component scores seems to reveal that individuals 41 and 42 are outliers in these data. It would be interesting to redo these analyses and

EXAMPLE 5.1 - SAS PRINCOMP PROCEDURE   11

UNIVARIATE ANALYSES ON PC SCORES

UNIVARIATE PROCEDURE

Variable=PRIN2

Moments

N	48	Sum Wgts	48
Mean	0	Sum	0
Std Dev	4.263864	Variance	18.18053 $\hat{\lambda}_2$
Skewness	1.402756	Kurtosis	3.319069
USS	854.4851	CSS	854.4851
CV	.	Std Mean	0.615436
T:Mean=0	0	Prob>|T|	1.0000
Sgn Rank	-65	Prob>|S|	0.5107
Num ^= 0	48		
W:Normal	0.897791	Prob<W	0.0003

Quantiles(Def=5)

100% Max	14.65972	99%	14.65972
75% Q3	2.161142	95%	4.922279
50% Med	-0.36531	90%	4.601323
25% Q1	-2.77987	10%	-4.4574
0% Min	-6.50738	5%	-5.49464
		1%	-6.50738
Range	21.1671		
Q3-Q1	4.941013		
Mode	-6.50738		

Extremes

Lowest	Obs	Highest	Obs
-6.50738(	38)	4.601323(	9)
-6.36295(	37)	4.775186(	43)
-5.49464(	33)	4.922279(	7)
-5.18499(	32)	13.50763(	41)
-4.45740(	27)	14.65970(	42)

those in Chapter 2 after removing these two individuals from the data set. You are asked to do this in one of the exercises at the end of this chapter.

Principal Components Analysis Using Factor Analysis Programs

Factor analysis (FA) is a multivariate procedure that is discussed in detail in the following chapter. Almost all computer programs that perform FA can be used to perform a principal components analysis on standardized data. As

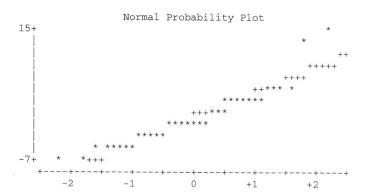

EXAMPLE 5.1 - SAS PRINCOMP PROCEDURE   12
          UNIVARIATE ANALYSES ON PC SCORES

```
                        UNIVARIATE PROCEDURE

Variable=PRIN2

              Stem Leaf                    #          Boxplot
                14 7                       1             0
                12 5                       1             0
                10
                 8
                 6
                 4 5689                    4             |
                 2 011248579              9          +-----+
                 0 4784                   4          |  +  |
                -0 9973220433322          13         *-----*
                -2 43188732               8          +-----+
                -4 525311                 6             |
                -6 54                     2             |
                   ----+----+----+----+
```

```
                        Normal Probability Plot
          15+                                              *
            |                                          *
            |                                               ++
            |                                        +++++
            |                                     ++++
            |                                 ++*** *
            |                              *******
            |                        +++***
            |                     *******
            |                *****
            |        *  *****
          -7+    *    *+++
            +----+----+----+----+----+----+----+----+----+----+
                -2        -1        0        +1        +2
```

F I G. 5.4

Pages 7 and 9–14
of computer print-
out, Example 5.1

a consequence, most statistical computing packages do not have a separate
routine for performing principal components analysis.

I find it unfortunate that many statistical computing packages require users
to perform PCA with FA programs. This is primarily because the outputs
from such programs are almost always labeled with FA notation rather than
PCA notation. This can be very confusing to inexperienced researchers. Another disadvantage of using FA programs to perform principal components
analysis is that many FA programs do not allow the use of the raw data values.
In other words, many programs standardize the data initially and then perform
principal components analysis on the standardized data.

Another consequence of using FA programs as a substitute for PCA pro-

EXAMPLE 5.1 - SAS PRINCOMP PROCEDURE   13

UNIVARIATE ANALYSES ON PC SCORES

UNIVARIATE PROCEDURE

Variable=PRIN3

Moments

N	48	Sum Wgts	48
Mean	0	Sum	0
Std Dev	3.254379	Variance	10.59099 $\hat{\lambda}_3$
Skewness	-1.5163	Kurtosis	2.804929
USS	497.7763	CSS	497.7763
CV	.	Std Mean	0.469729
T:Mean=0	0	Prob>|T|	1.0000
Sgn Rank	90	Prob>|S|	0.3614
Num ^= 0	48		
W:Normal	0.878361	Prob<W	0.0001

Quantiles(Def=5)

100% Max	4.426072		99%	4.426072
75% Q3	2.431861		95%	3.333284
50% Med	0.957372		90%	3.158135
25% Q1	-1.80781		10%	-3.71647
0% Min	-11.4878		5%	-6.3429
			1%	-11.4878
Range	15.91383			
Q3-Q1	4.239675			
Mode	-11.4878			

Extremes

Lowest	Obs	Highest	Obs
-11.4878(	11)	3.158135(	45)
-8.82701(	10)	3.246946(	5)
-6.3429(	12)	3.333284(	31)
-5.16203(	37)	3.746051(	14)
-3.71647(	38)	4.426072(	13)

FIG. 5.4

Pages 7 and 9–14 of computer print-out, Example 5.1

grams has to do with the output produced. For example, when analyzing a correlation matrix, most factor analysis programs produce estimates of the $\mathbf{c}_j^*$'s, the component correlation vectors (recall that these vectors are normalized to have lengths equal to the square roots of their corresponding eigenvalues), rather than estimates of the $\mathbf{a}_j^*$'s. (Recall that these eigenvectors are normalized to have length 1.) Actually, this may be an advantage, since the component correlation vectors give the correlations between the original variables and the newly defined principal component variables, and interpreting correlations is straightforward.

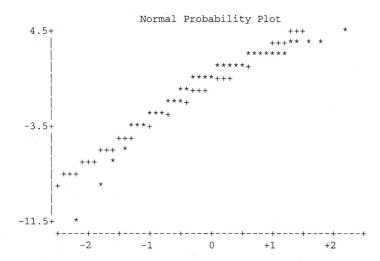

```
            EXAMPLE 5.1 - SAS PRINCOMP PROCEDURE              14
               UNIVARIATE ANALYSES ON PC SCORES

                     UNIVARIATE PROCEDURE

Variable=PRIN3

        Stem Leaf                          #        Boxplot
          4  4                             1           |
          3  002237                        6           |
          2  2456788                       7        +-----+
          1  0013345679                   10        *-----*
          0  1133689                       7        |  +  |
         -0  43                            2        |     |
         -1  9811                          4        +-----+
         -2  9533                          4           |
         -3  750                           3           |
         -4                                            |
         -5  2                             1           |
         -6  3                             1           |
         -7                                            |
         -8  8                             1           0
         -9                                            |
        -10                                            |
        -11  5                             1           0
             ----+----+----+----+
```

Normal Probability Plot

```
   4.5+                                          +++           *
      |                                       +++** * *
      |                                     *******
      |                               *****+
      |                           ****+++
      |                        **+++
      |                      ***+
      |                   ***+
  -3.5+                 ***+
      |               +++
      |            +++ *
      |         +++   *
      |      +++
      |    +
      |
 -11.5+    *
      +----+----+----+----+----+----+----+----+----+----+
          -2        -1         0        +1        +2
```

F I G. 5.4
Pages 7 and 9–14
of computer print-
out, Example 5.1

One big disadvantage of using FA programs to perform PCA is the way
in which FA programs compute principal component scores. FA programs
usually standardize the principal component scores so that they have standard
deviations equal to 1. This is extremely unfortunate, especially when one is
plotting the principal component scores to look for outliers in the data or
clusters of points in the data. For example, suppose that the first principal

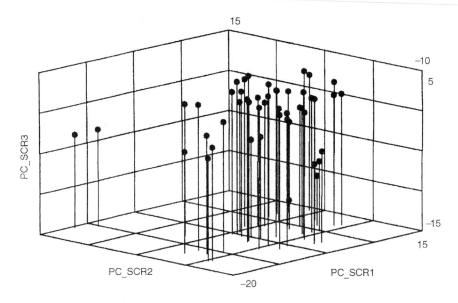

F I G. 5.5
Three-dimensional
plot of the first
three principal
component scores,
Example 5.1

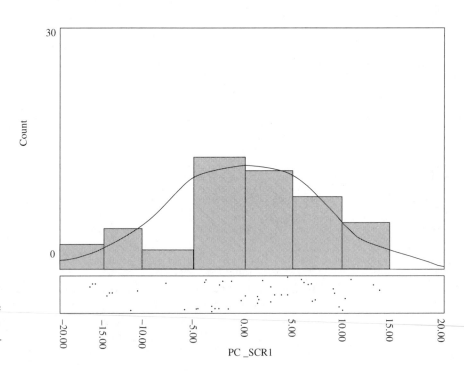

F I G. 5.6
Histogram for the
applicant data us-
ing the first princi-
pal component

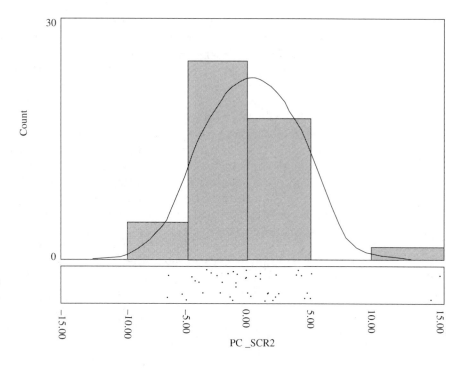

FIG. 5.7
Histogram for the applicant data using the second principal component

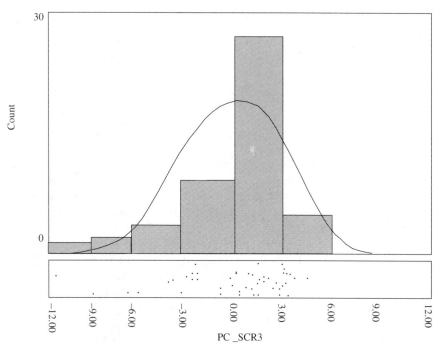

FIG. 5.8
Histogram for the applicant data using the third principal component

component accounts for 75% of the total variability and that the second principal component accounts for an additional 15% of the variability. In some sense, the first principal component is five times as important as the second principal component when looking for outliers and clusters of points in the data. However, if we plot standardized principal component scores, the two principal components will appear to be equally important, and this can be misleading to many unsuspecting researchers.

Obviously, principal component scores that have been standardized can be *un*standardized; you merely need to multiply each of the standardized PC scores by the square root of its corresponding eigenvalue. This should always be done prior to plotting principal component scores when you are looking for outliers and clusters of points in the principal component space. It is not quite so important if you are simply looking at the distributional properties of the principal component scores to see if they appear to be normally distributed because the standardized principal component scores will be normally distributed if and only if the unstandardized principal component scores are normally distributed.

PCA with SPSS's FACTOR Procedure

SPSS® contains a procedure called FACTOR that can be used to perform principal components analysis on standardized data (*Z* scores) by including EXTRACTION=PC and ROTATE=NOROTATE options. The FACTOR procedure in SPSS does not allow one to perform a PCA on the raw data values. SPSS's FACTOR procedure is illustrated in Example 5.2 using the data of Table 5.1.

E X A M P L E   5.2

The U.S. Navy collected data in an attempt to estimate manpower needs for Bachelor Officers' Quarters. Myers (1990) uses these data to illustrate multiple regression techniques and diagnostics. Here the data are used to look for possible problems in those variables Myers was using as predictors of monthly manpower needs. The data are shown in Table 5.1 near the beginning of the chapter. A description of each of the variables is also given in the table.

The variables in Table 5.1 are measurements that occur in many different kinds of measurement units. That is, the variables are not comparable to one another and thus do not occur on an equal footing, as is required for a PCA on raw data values. As a result, PCA should be performed on standardized data values. SPSS only performs PC analyses on standardized data. So SPSS can be used appropriately on these data. The data were analyzed with SPSS using the following commands:

```
SET WIDTH = 80
TITLE "EXAMPLE 5.2 - PCA ON U.S. NAVY BACHELOR OFFICERS'
QUARTERS"
DATA LIST LIST
 /SITE 1-2 ADO MAC WHR CUA WNGS OBC RMS MMH

VARIABLE LABELS ADO 'AVERAGE DAILY OCCUPANCY'/
                MAC 'AVERAGE NUMBER OF CHECK-INS PER MO.'/
                WHR 'WEEKLY HRS OF SERVICE DESK OPERATION'/
                CUA 'SQ FT OF COMMON USE AREA'/
                WNGS 'NUMBER OF BUILDING WINGS'/
                OBC 'OPERATIONAL BERTHING CAPACITY'/
                RMS 'NUMBER OF ROOMS'/
                MMH 'MONTHLY MAN-HOURS'

PRINT / SITE ADO MAC WHR CUA WNGS OBC RMS MMH

BEGIN DATA
```

(The data in Table 5.1 are placed here.)

```
END DATA

FACTOR  VARIABLES=ADO TO RMS
 /PLOT=EIGEN
 /CRITERIA=FACTORS(3)
 /EXTRACTION=PC
 /ROTATE=NOROTATE
 /PRINT=UNIVARIATE CORRELATION DET EXTRACTION FSCORE
 /SAVE REG (ALL PCSCR)

PLOT PLOT = PCSCR2 WITH PCSCR1

LIST VARIABLES=SITE PCSCR1 TO PCSCR3

COMPUTE PCSCRU1=PCSCR1*4.64302**.5
COMPUTE PCSCRU2=PCSCR2*.74021**.5
COMPUTE PCSCRU3=PCSCR3*.70630**.5

LIST VARIABLES = SITE PCSCRU1 TO PCSCRU3
```

The important and useful results from the preceding commands are shown on the partial computer printout pages shown in Figure 5.9. All of the pages of printout are available if the reader executes the SPSS file named EX5_2.SSX on the enclosed disk. Before describing the output, a few comments about the SPSS commands just given are in order.

1 The EXTRACTION=PC option is used to tell SPSS's FACTOR procedure that a principal components analysis is desired.

EXAMPLE 5.2 - PCA ON U.S. NAVY BACHELOR OFFICERS' QUARTERS   Page 3

- - - - - - - -   F A C T O R   A N A L Y S I S   - - - - - - - -

	MEAN	STD DEV	LABEL
ADO	118.35560	169.80118	AVERAGE DAILY OCCUPANCY
MAC	330.50560	382.80452	AVERAGE NUMBER OF CHECK-INS PER MO.
WHR	135.94000	58.62528	WEEKLY HRS OF SERVICE DESK OPERATION
CUA	15.71720	13.95082	SQ FT OF COMMON USE AREA
WNGS	11.12000	12.04270	NUMBER OF BUILDING WINGS
OBC	137.40000	134.15041	OPERATIONAL BERTHING CAPACITY
RMS	126.28000	116.49375	NUMBER OF ROOMS

NUMBER OF CASES =   25

CORRELATION MATRIX: = $\mathcal{R}$

	ADO	MAC	WHR	CUA	WNGS	OBC	RMS
ADO	1.00000						
MAC	.61918	1.00000					
WHR	.34730	.47139	1.00000				
CUA	.38744	.47319	.38829	1.00000			
WNGS	.48838	.55245	.38079	.68614	1.00000		
OBC	.62004	.84953	.47278	.59383	.67632	1.00000	
RMS	.67632	.86076	.49006	.66189	.75894	.97819	1.00000

DETERMINANT OF CORRELATION MATRIX =   .0004962 = $|\mathcal{R}|$ *Want this close to zero!*

EXTRACTION 1 FOR ANALYSIS 1, PRINCIPAL-COMPONENTS ANALYSIS (PC)

F I G. 5.9
Pages 3–5, 7, 10–12, 14, 17–19, 22, 24, and 26 of computer printout, Example 5.2

2 The next chapter demonstrates that a rotation of factor axes is often desirable when doing factor analysis. A default in the SPSS FACTOR procedure is to rotate the factor axes. Here the option ROTATE=NOROTATE is used to keep the procedure from rotating the PC axes.

3 The CRITERIA=FACTORS(3) option is used to tell the FACTOR procedure that three principal components are desired. This was done to override the SPSS's default option for choosing the number of principal components. SPSS's default option considers only those PCs that correspond to eigenvalues that are greater than 1.

The means and standard deviations of each of the measured variables are shown on the top of page 3 of the computer output. These are followed by the sample correlation matrix. Near the bottom of page 3, you can see that the determinant of the correlation matrix is equal to 0.0004962. Recall that you should only perform PCA in those cases where the determinant of the correlation matrix is close to zero because this indicates that linear dependencies exist among the response variables. Obviously, this correlation

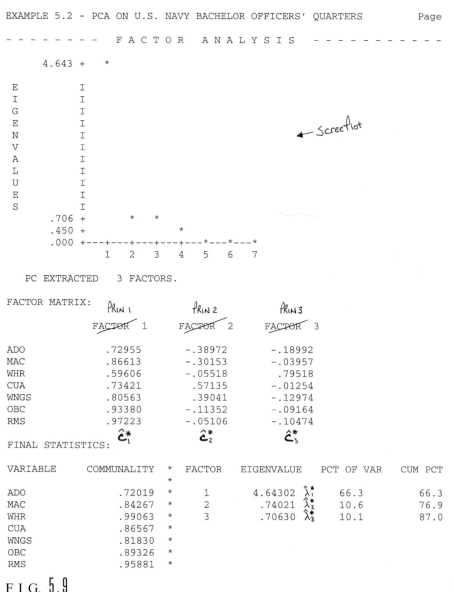

```
EXAMPLE 5.2 - PCA ON U.S. NAVY BACHELOR OFFICERS' QUARTERS            Page    4

- - - - - - - -   F A C T O R   A N A L Y S I S   - - - - - - - - - - -

        4.643 +    *

  E             I
  I             I
  G             I
  E             I
  N             I
  V             I
  A             I
  L             I
  U             I
  E             I
  S             I
         .706 +         *    *
         .450 +              *
         .000 +---+---+---+---+---*---*---*
                 1   2   3   4   5   6   7
```

← Screeplot

```
    PC EXTRACTED  3 FACTORS.

FACTOR MATRIX:     PRIN 1        PRIN 2         PRIN 3
                   FACTOR  1     FACTOR  2      FACTOR  3

  ADO              .72955        -.38972        -.18992
  MAC              .86613        -.30153        -.03957
  WHR              .59606        -.05518         .79518
  CUA              .73421         .57135        -.01254
  WNGS             .80563         .39041        -.12974
  OBC              .93380        -.11352        -.09164
  RMS              .97223        -.05106        -.10474
```

$\hat{c}_1^*$ $\hat{c}_2^*$ $\hat{c}_3^*$

```
FINAL STATISTICS:

  VARIABLE      COMMUNALITY  *  FACTOR   EIGENVALUE   PCT OF VAR   CUM PCT
                             *
  ADO              .72019    *    1       4.64302       66.3        66.3
  MAC              .84267    *    2        .74021       10.6        76.9
  WHR              .99063    *    3        .70630       10.1        87.0
  CUA              .86567    *
  WNGS             .81830    *
  OBC              .89326    *
  RMS              .95881    *
```

Eigenvalue column labels: $\hat{\lambda}_1^*$, $\hat{\lambda}_2^*$, $\hat{\lambda}_3^*$

F I G. 5.9

Pages 3–5, 7, 10–12, 14, 17–19, 22, 24, and 26 of computer printout, Example 5.2

matrix has a determinant that is close to zero. See the next paragraph for a statistical measure of closeness to zero.

In Section 5.8 a procedure was given for testing the hypothesis that the population correlation matrix is equal to the identity matrix, that is, that all variables are uncorrelated, when the data are multivariate normal. The

EXAMPLE 5.2 - PCA ON U.S. NAVY BACHELOR OFFICERS' QUARTERS   Page   5

- - - - - - - - - -   F A C T O R   A N A L Y S I S   - - - - - - - - - - -

FACTOR SCORE COEFFICIENT MATRIX:

	FACTOR 1	FACTOR 2	FACTOR 3
ADO	.15713	-.52650	-.26889
MAC	.18654	-.40736	-.05602
WHR	.12838	-.07455	1.12583
CUA	.15813	.77188	-.01776
WNGS	.17351	.52744	-.18369
OBC	.20112	-.15336	-.12975
RMS	.20940	-.06898	-.14829

$$\frac{1}{\hat{\lambda}_1^*}\hat{a}_1^* \qquad \frac{1}{\hat{\lambda}_2^*}\hat{a}_2^* \qquad \frac{1}{\hat{\lambda}_3^*}\hat{a}_3^*$$

EXAMPLE 5.2 - PCA ON U.S. NAVY BACHELOR OFFICERS' QUARTERS   Page   7

	y_1^*	y_2^*	y_3^*
SITE	PCSCR1	PCSCR2	PCSCR3
1	-1.27858	-.14568	-1.84874
2	-1.20341	-.19080	-1.15638
3	-1.15647	-.28431	-1.19877
4	-.91348	-.38470	1.28983
5	-1.02689	.21353	-1.19566
6	-.84829	-.39021	1.23291
7	-1.06469	-.29865	-1.29120
8	-.42838	-.01634	.85817
9	-.48240	1.35759	-1.55343
10	-.61870	-.10185	1.07449
11	.02550	.02833	.70250
12	-.49089	-.57407	1.03981
13	.01820	.42656	.64630
14	-.42820	-.25830	.96874
15	.54625	-.52244	.49974
16	.47497	.84162	.37447
17	.32409	.61185	.45445
18	1.33053	1.21239	-.08663
19	.10937	.21036	.63398
20	.65247	-.37578	.34447
21	.13544	.47677	.59667
22	2.18240	2.45113	-.95962
23	1.42764	-2.13805	-.88461
24	2.15526	-2.63448	-.96421
25	.55826	.48551	.42271

Standardized Principal Component Scores ←

NUMBER OF CASES READ =   25   NUMBER OF CASES LISTED =   25

F I G. 5.9

Pages 3–5, 7, 10–12, 14, 17–19, 22, 24, and 26 of computer printout, Example 5.2

```
* * * * * * * * * * * * * * * *  P L O T  * * * * * * * * * * * * * * * *

EXAMPLE 5.2 - PCA ON U.S. NAVY BACHELOR OFFICERS' QUARTERS          Page   10

                        PLOT OF PCSCR2 WITH PCSCR1
        ++----+----+----+----+----+----+----+----+----+----+----+----+----++
  R    3+                                                                  +
  E     I                                                                  I
  G     I                                                     1 22         I
  R     I                                                                  I
        I                                                                  I
  F    2+                                                                  +
  A     I                                                                  I
  C     I                                                                  I
  T     I                      1 9                                         I
  O     I                                          1 18                     I
  R    1+                                                                  +
        I                            1                                     I
  S     I                            1                                     I
  C     I                         11    1                                  I
  O     I                  1       1                                       I
  R    0+                      1   1                                       +
  E     I             121    1 1                                           I
        I                11               1                                I
        I                   1           1                                  I
        I                                                                  I
  2   -1+                                                                  +
        I                                                                  I
  F     I                                                                  I
  O     I                                                                  I
  R   -2+                                                                  +
  A     I                                1 23                              I
  N     I                                                                  I
  A     I                                         1 24                     I
  L     I                                                                  I
  Y   -3+                                                                  +
  S     I                                                                  I
  I     I                                                                  I
  S     I                                                                  I
      -4+                                                                  +
        I                                                                  I
        I                                                                  I
  1     I                                                                  I
        I                                                                  I
      -5+                                                                  +
        ++----+----+----+----+----+----+----+----+----+----+----+----+----++
        -3        -2        -1         0         1         2         3

                    REGR FACTOR SCORE    1 FOR ANALYSIS    1
```

This plot does not accurately reflect distances between points.

    25 cases plotted.

F I G. 5.9

Pages 3–5, 7, 10–12, 14, 17–19, 22, 24, and 26 of computer printout, Example 5.2

EXAMPLE 5.2 - PCA ON U.S. NAVY BACHELOR OFFICERS' QUARTERS   Page 11

PLOT OF PCSCR3 WITH PCSCR1

```
         ++----+----+----+----+----+----+----+----+----+----+----+----+----++
   R   1.5+                                                                  +
   E     I                                                                   I
   G     I                    1                                              I
   R     I                    1                                              I
         I                     1                                             I
   F    1+                      11                                           +
   A     I                      1                                            I
   C     I                                                                   I
   T     I                          1                                        I
   O     I                          12                                       I
   R    .5+                          1 1        '                            +
         I                            11                                     I
   S     I                             1                                     I
   C     I                                                                   I
   O     I                                                                   I
   R    0+                                      1                            +
   E     I                                                                   I
         I                                                                   I
         I                                                                   I
         I                                                                   I
   3   -.5+                                                                  +
         I                                                                   I
   F     I                                                                   I
   O     I                                                                   I
   R     I                                     1                             I
       -1+                                          2                        +
   A     I                                                                   I
   N     I              2 1                                                  I
   A     I              1                                                    I
   L     I                                                                   I
   Y  -1.5+                                                                  +
   S     I                  1                                                I
   I     I                                                                   I
   S     I          1                                                        I
         I                                                                   I
       -2+                                                                   +
         I                                                                   I
         I                                                                   I
   1     I                                                                   I
         I                                                                   I
     -2.5+                                                                   +
         ++--+----+----+----+----+----+----+----+----+----+----+----+----++
          -3        -2        -1         0         1         2         3
```

Plot may be misleading.

REGR FACTOR SCORE   1 FOR ANALYSIS   1

25 cases plotted.

F I G. 5.9
Pages 3–5, 7, 10–12, 14, 17–19, 22, 24, and 26 of computer printout, Example 5.2

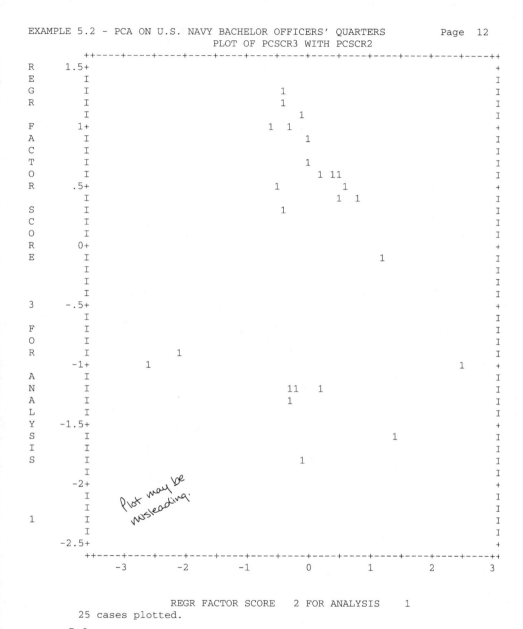

EXAMPLE 5.2 - PCA ON U.S. NAVY BACHELOR OFFICERS' QUARTERS   Page 12
                    PLOT OF PCSCR3 WITH PCSCR2

Plot may be misleading.

REGR FACTOR SCORE   2 FOR ANALYSIS   1
25 cases plotted.

F I G. 5.9

Pages 3–5, 7, 10–12, 14, 17–19, 22, 24, and 26 of computer printout, Example 5.2

EXAMPLE 5.2 - PCA ON U.S. NAVY BACHELOR OFFICERS' QUARTERS   Page   14

SITE   PCSCRU1   PCSCRU2   PCSCRU3

SITE	PCSCRU1	PCSCRU2	PCSCRU3
1	-2.76	-.13	-1.55
2	-2.59	-.16	-.97
3	-2.49	-.24	-1.01
4	-1.97	-.33	1.08
5	-2.21	.18	-1.00
6	-1.83	-.34	1.04
7	-2.29	-.26	-1.09
8	-.92	-.01	.72
9	-1.04	1.17	-1.31
10	-1.33	-.09	.90
11	.05	.02	.59
12	-1.06	-.49	.87
13	.04	.37	.54
14	-.92	-.22	.81
15	1.18	-.45	.42
16	1.02	.72	.31
17	.70	.53	.38
18	2.87	1.04	-.07
19	.24	.18	.53
20	1.41	-.32	.29
21	.29	.41	.50
22	4.70	2.11	-.81
23	3.08	-1.84	-.74
24	4.64	-2.27	-.81
25	1.20	.42	.36

← Unstandardized Principal Component Scores

NUMBER OF CASES READ =   25   NUMBER OF CASES LISTED =   25

F I G. **5.9**

Pages 3–5, 7, 10–12, 14, 17–19, 22, 24, and 26 of computer printout, Example 5.2

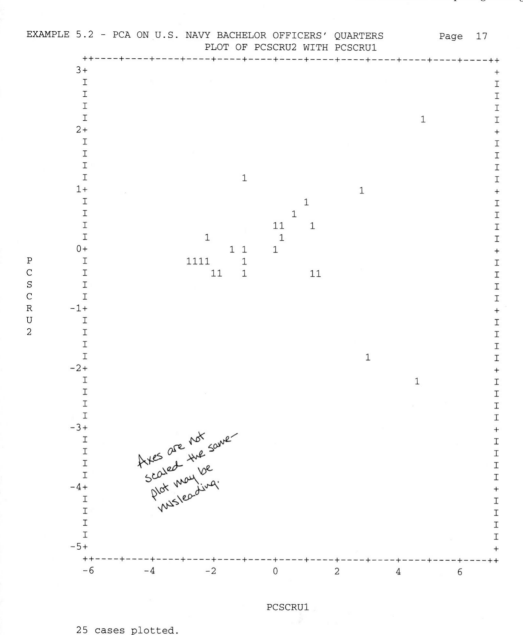

EXAMPLE 5.2 - PCA ON U.S. NAVY BACHELOR OFFICERS' QUARTERS   Page   17
                  PLOT OF PCSCRU2 WITH PCSCRU1

PCSCRU1

25 cases plotted.

F I G. 5.9

Pages 3–5, 7, 10–12, 14, 17–19, 22, 24, and 26 of computer printout, Example 5.2

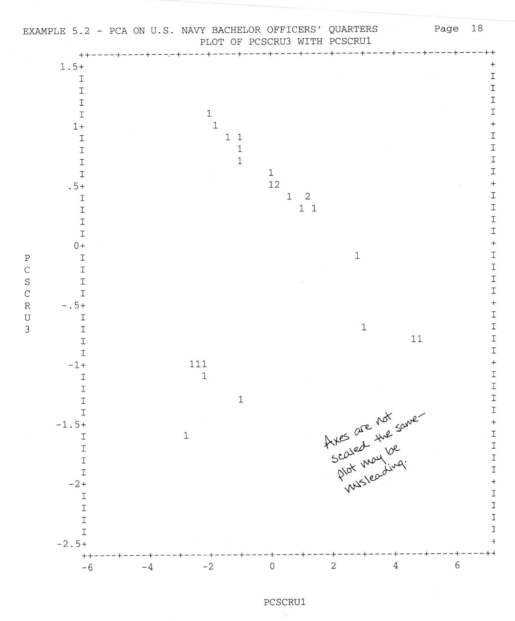

EXAMPLE 5.2 - PCA ON U.S. NAVY BACHELOR OFFICERS' QUARTERS   Page   18
                      PLOT OF PCSCRU3 WITH PCSCRU1

PCSCRU1

25 cases plotted.

FIG. 5.9

Pages 3–5, 7, 10–12, 14, 17–19, 22, 24, and 26 of computer printout, Example 5.2

EXAMPLE 5.2 - PCA ON U.S. NAVY BACHELOR OFFICERS' QUARTERS   Page 19

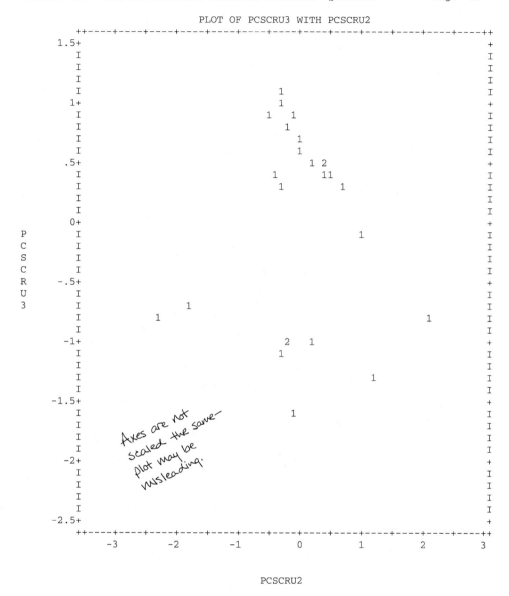

25 cases plotted.

FIG. 5.9
Pages 3–5, 7, 10–12, 14, 17–19, 22, 24, and 26 of computer printout, Example 5.2

```
EX 5.2 - PCA ON U.S. NAVY BACHELOR OFFICERS' QUARTERS          Page   22

PCSCRU1
     COUNT   MIDPOINT    ONE SYMBOL EQUALS APPROXIMATELY   .10 OCCURRENCES

        0     -3.0           .
        3     -2.6     *******:*******************
        2     -2.2     **********:*********
        2     -1.8     ************:*******
        1     -1.4     **********       .
        4     -1.0     ****************:**********************
        0      -.6                    .
        0      -.2                    .
        4       .2     ****************:********************
        1       .6     **********       .
        2      1.0     ****************:***
        2      1.4     ***************:*****
        0      1.8                .
        0      2.2              .
        0      2.6             .
        2      3.0     ******:*************
        0      3.4       .
        0      3.8      .
        0      4.2      .
        2      4.6     *:*****************
        0      5.0      .
              I....+....I....+....I....+....I....+....I....+....I
              0         1         2         3         4         5

                         HISTOGRAM FREQUENCY
```

MEAN	.000	STD ERR	.431	MEDIAN	.039
MODE	-2.755	STD DEV	2.155	VARIANCE	4.643
KURTOSIS	-.071	S E KURT	.902	SKEWNESS	.759
S E SKEW	.464	RANGE	7.458	SUM	.000
MAXIMUM	4.703			MINIMUM	-2.755

$\hat{\lambda}_1^*$

F I G. 5.9

Pages 3–5, 7, 10–12, 14, 17–19, 22, 24, and 26 of computer printout, Example 5.2

EX 5.2 - PCA ON U.S. NAVY BACHELOR OFFICERS' QUARTERS   Page 24

PCSCRU2

COUNT	MIDPOINT	ONE SYMBOL EQUALS APPROXIMATELY .20 OCCURRENCES
0	-2.60	
1	-2.35	*****
0	-2.10	.
1	-1.85	:****
0	-1.60	.
0	-1.35	.
0	-1.10	.
0	-.85	.
1	-.60	***** .
6	-.35	************:***************
6	-.10	**************:**************
2	.15	********** .
3	.40	************:**
2	.65	**********.
0	.90	.
2	1.15	*****:****
0	1.40	.
0	1.65	.
0	1.90	.
1	2.15	:****
0	2.40	

```
                    I....+....I....+....I....+....I....+....I....+....I
                    0         2         4         6         8        10
                              HISTOGRAM FREQUENCY
```

MEAN	.000	STD ERR	.172	MEDIAN	-.088 $\hat{\lambda}_2^*$
MODE	-2.267	STD DEV	.860	VARIANCE	.740
KURTOSIS	2.568	S E KURT	.902	SKEWNESS	-.384
S E SKEW	.464	RANGE	4.375	MINIMUM	-2.267
MAXIMUM	2.109	SUM	.000		

F I G. 5.9

Pages 3–5, 7, 10–12, 14, 17–19, 22, 24, and 26 of computer printout, Example 5.2

EX 5.2 - PCA ON U.S. NAVY BACHELOR OFFICERS' QUARTERS   Page 26

PCSCRU3

```
   COUNT   MIDPOINT     ONE SYMBOL EQUALS APPROXIMATELY     .10 OCCURRENCES
      0     -1.75    .
      1     -1.60    **:*******
      0     -1.45       .
      1     -1.30    ****:*****
      1     -1.15    ******:***
      3     -1.00    *******:********************
      2      -.85    **********:*********
      1      -.70    **********    .
      0      -.55              .
      0      -.40              .
      0      -.25              .
      1      -.10    **********    .
      0       .05              .
      0       .20              .
      5       .35    ***************:*********************************
      3       .50    **************:****************
      2       .65    ************:*******
      2       .80    **********:*********
      1       .95    ********:*
      2      1.10    ******:**************
      0      1.25       .
              I....+....I....+....I....+....I....+....I....+....I
              0         1         2         3         4         5
                        HISTOGRAM FREQUENCY
```

MEAN	.000	STD ERR	.168	MEDIAN	.355 $\hat{\lambda}_3^*$
MODE	-1.554	STD DEV	.840	VARIANCE	.706
KURTOSIS	-1.360	S E KURT	.902	SKEWNESS	-.465
S E SKEW	.464	RANGE	2.638	MINIMUM	-1.554
MAXIMUM	1.084	SUM	.000		

F I G. 5.9

Pages 3–5, 7, 10–12, 14, 17–19, 22, 24, and 26 of computer printout, Example 5.2

value of the test statistic for these data is

$$-a \cdot \log(V) = -(N - 1 - (2p + 5)/6) \log(|\mathbf{R}|)$$
$$= -(25 - 1 - (2 \cdot 7 + 5)/6) \log(0.0004962)$$
$$= -20.83333 \cdot (-7.6085)$$
$$= 158.5$$

and the critical point of the chi-square distribution with $p(p - 1)/2 = 21$ degrees of freedom for $\alpha = 0.0001$ is 53.96. Clearly, the hypothesis will be rejected at the 0.0001 significance level because $158.5 > 53.96$.

From the lower right-hand corner of page 4 (Figure 5.9), you can see that the three largest eigenvalues of the sample correlation matrix are $\hat{\lambda}_1^* = 4.64302$, $\hat{\lambda}_2^* = 0.74021$, $\hat{\lambda}_3^* = 0.70630$. The first principal component accounts for 66.3% of the total variation, while second and third each account for approximately 10% of the total variation. The first three together account for 87.0% of the total variation. A SCREE plot of the eigenvalues is shown at the top of page 4 of the computer printout. The SCREE plot shows that the value of the fourth eigenvalue is equal to 0.450. The remaining three eigenvalues are not actually equal to zero as suggested by the SCREE plot, but they are extremely close to zero. Only one of the three eigenvalues is greater than 1. You could certainly make a valid argument for using only the first principal component based on both the SCREE plot and from the fact that only the first eigenvalue is greater than 1 (remember the correlation matrix is being analyzed). Still, the first principal component only accounts for 66.3% of the total variation. Since the second and third eigenvalues are somewhat close to 1, I have chosen to use each of the first three principal components.

Try to visualize what the standardized data must look like in the three-dimensional principal component subspace of the seven-dimensional sample space. (You might find it useful to review Sections 4.4 through 4.6.) For this example and if the data were multivariate normal, the standardized data should appear to fall into a geometric region that could be described by a somewhat elongated football whose length is about 2.5 times longer than its diameter. Since the two smaller eigenvalues are nearly similar in size, the data would likely be symmetrical about the axis corresponding to the largest eigenvalue.

The vectors shown in the middle of page 4 under the label FACTOR MA-TRIX are the first three component correlation vectors. Thus, the first column of numbers is $\hat{\mathbf{c}}_1^*$, and the second and third columns are $\hat{\mathbf{c}}_2^*$ and $\hat{\mathbf{c}}_3^*$, respectively. Recall that $\hat{\mathbf{c}}_i^* = \sqrt{\hat{\lambda}_i^*} \hat{\mathbf{a}}_i^*$ for each i. These vectors contain the correlations between the original variables and the principal component variables. For example, the correlation between variable ADO and y_1, the first principal component, is 0.73, and the correlation between ADO and y_2 is -0.39.

Each of the original variables is highly correlated with the first principal component, so it would seem that the first principal component can be interpreted as some sort of an aggregate or composite measure of size. You should not try to interpret the second and third principal components for this example since the near equality of their corresponding eigenvalues indicates that the sample region is nearly symmetrical in directions orthogonal to the major (first) axis. As a result, an infinite number of possibilities exists for placing second and third axes in the principal component space, and you should not be overly enthusiastic about the two directions picked by the PCA procedure.

The column of numbers in the lower left-hand corner of page 4 is labeled COMMUNALITY. *Communality* is a factor analysis term. In PCA we can interpret it in the same way we would interpret R^2 in the multiple regression. Recall that in regression terminology, R^2 measures the proportion of variability in a dependent variable that is accounted for by the predictor variables in a multiple regression model.

Suppose we were to build a multiple regression model with ADO as the variable to be predicted and with y_1, y_2, and y_3, the first three principal components, as predictors. Then the value of R^2 for this regression model will be equal to 0.720. If we were to build a model to predict MAC from y_1, y_2, and y_3, the value of R^2 for this model is equal to 0.843. Similarly, R^2 values for predicting the variables WHR, CUA, WNGS, OBC, and RMS from the first three principal components are 0.991, 0.866, 0.818, 0.893, and 0.959, respectively. These R^2 values indicate how well we would be able to use the first three principal component variables to predict each of the original variables. Generally, we hope that each of these R^2 values will be somewhat close to 1 as this indicates that little information is lost.

The vectors shown on page 5 of the Figure 5.9 computer printout are the ones used by SPSS to compute principal component scores. For example, the first principal component score for a particular SITE would be computed by

$$y = 0.157 \cdot Z_{ADO} + 0.187 \cdot Z_{MAC} + 0.128 \cdot Z_{WHR} + 0.158 \cdot Z_{CUA} + 0.173 \cdot Z_{WNGS}$$
$$+ 0.201 \cdot Z_{OBC} + 0.209 \cdot Z_{RMS}$$

where Z_{xxx} represents the Z score for the variable named xxx at the chosen SITE. As noted earlier, these vectors produce principal component scores that are standardized. Generally, standardized PC scores are not desirable. In SPSS unstandardized PC scores can be obtained from the standardized ones by using a COMPUTE command.

Page 7 of the computer output shows the first three standardized principal component scores, and pages 10–12 show pairwise plots of the three (standardized) principal component scores. You must be extremely careful when using these plots to identify outliers because the distances between sites in standardized units are not realistic. You can safely use the plots to

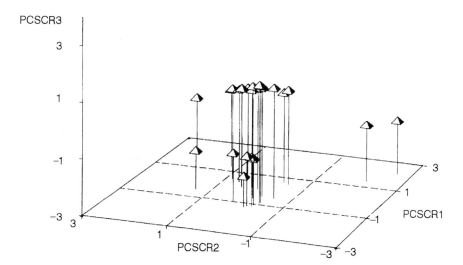

FIG. **5.10**
Three-dimensional
plot of the first
three standard-
ized principal
components

assess multivariate normality of the principal component variables and, consequently, to assess the multivariate normality of the original data. If the
original data were multivariate normal, pairwise plots of the standardized
principal component scores should fall into circular-shaped regions. Obviously, these data do not appear to be multivariate normal.

A listing of the unstandardized PC scores is shown on page 14 of the
printout (Figure 5.9). This latter set of PC scores was obtained by the
COMPUTE commands in the SPSS program used. Pairwise plots of these unstandardized PC scores are shown on pages 17–19. Generally, a plot of the
unstandardized PC scores can be used to reveal outliers in a data set.
These plots must be interpreted carefully, however, since the scales of measurement used on each of the principal component axes are not the same.

Figure 5.10 shows a three-dimensional plot of the first three standardized principal component scores, and Figure 5.11 shows a three-dimensional
plot of the unstandardized principal component scores. Note that the measurement units on each of the axes of the plot in Figure 5.11 are the same.
This allows us to assess accurately the relative locations of the individual
SITEs with respect to one another. Do you get a different perception from
the plots in Figures 5.10 and 5.11? Which gives the most accurate represen-
tation for interpretation purposes? Your answer should be Figure 5.11.

Pages 22, 24, and 26 of the computer printout of Figure 5.9 show histograms that were created by SPSS using the following commands:

```
FREQUENCIES VARIABLES=PCSCRU1 PCSCRU2 PCSCRU3
     /HISTOGRAM = NORMAL /STATISTICS = ALL
```

A normal curve is superimposed on each plot. From these plots, it also
seems clear that there are some abnormalities in these data.

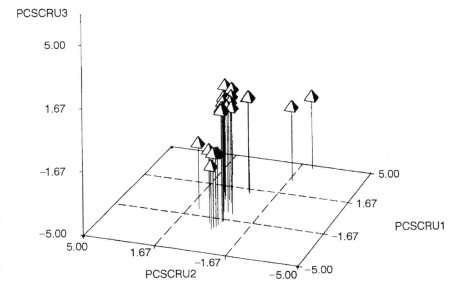

F I G. 5.11
Three-dimensional
plot of the first
three unstan-
dardized principal
components

Summary

To summarize the results of the PCA on the U.S. Navy data, it seems clear that the first principal component can be interpreted as a composite measurement of size. The histogram on page 22 of the computer printout shown in Figure 5.9 reveals that four sites are much larger than the remaining sites. These four sites are likely to be very influential on any regression models that might be used to model these data. Since some of the eigenvalues of the correlation matrix are very close to zero, there is a great deal of multicollinearity among these variables. This will have a significant impact on the interpretability of any fitted regression model. See Myers' (1990) analyses of these data. An examination of all of the plots seems to indicate that these variables do not have a multivariate normal distribution. Transformations of some of the variables may allow new variables to be created that could be more normally distributed.

Exercises

1   Each of the following questions refers to Example 5.1 and its SAS analyses in Figures 5.2 and 5.4.

   a   What are estimates of the mean, variance, and standard deviation of HON?

b What is the sample covariance between KJ and LA?

c Recall that the determinant of a symmetric matrix is equal to the product of its eigenvalues. What is the determinant of $\hat{\Sigma}$?

d Use the test described in Section 5.8 to test the hypothesis that the variables in the applicant data set are independent.

e Which of the original response variables has the largest mean? Which has the smallest variance? Which pair of variables has the greatest covariance between them?

f How many principal components would be required to explain at least 90% of the total variation in the original variables?

g What is the value of $\hat{\lambda}_4$? What proportion of the total variability does this eigenvalue account for?

h What is the value of $\hat{\mathbf{a}}_3$?

i Compute $\hat{\mathbf{c}}_1$, the component loading vector corresponding to the first principal component.

j Compute $\hat{\mathbf{c}}_2$ and $\hat{\mathbf{c}}_3$. Consider the first three component loading vectors. For each of the original variables, identify the principal component that each associates with most strongly.

k What is the value of y_{43}, the third principal component score for the fourth individual?

l Describe how y_{43} in part k can be computed by hand.

m List the five individuals who have the largest values for their first principal component score. List the five who have the smallest values for their first principal component score. (*Hint:* Use results from the UNIVARIATE procedure on pages 9–14 of Figure 5.4.)

n What is the variance of the first principal component? Get this answer from page 9 of the computer printout of Figure 5.4. How does this compare to $\hat{\lambda}_1$?

o Suppose we were going to hire individuals from the applicant data set. Which individuals are below average with respect to all three of the first three principal components?

2 Remove applicants 41 and 42 from the applicant data set and redo the principal component analysis. Also perform follow-up analyses similar to those done in Example 5.1. Write a short report that discusses what you have learned from this reanalysis. Also compare the results of this analysis with those of the first analysis.

3 The following questions refer to Example 5.2 and its SPSS analyses in Figure 5.9.

a What are the mean and standard deviations for the variables CUA, MAC, and WHR?

b What is the correlation between OBC and CUA? Construct a 95% confidence interval for the true value of the population correlation coefficient between these two variables.

c Which two variables are most highly correlated with one another?

d Use the test described in Section 5.8 to test the hypothesis that the variables in this data set are independent.

e Make a rough sketch of what an ellipsoid of concentration must look like in the space defined by the first three principal components.

f What proportion of the variance of MAC is accounted for by the first three principal components?

g What is the value of $\hat{c}_2^*$, the component correlation vector for the second principal component?

h Compute $\hat{a}_1^*$, the eigenvector corresponding to the first principal component of the correlation matrix.

i Consider the first three component loading vectors. For each of the original variables, identify the principal component that each associates with most strongly.

j What is the correlation between WHR and the third principal component?

k Page 7 of the computer printout of Figure 5.9 gives a listing of the standardized principal component scores for each of the sites. Compute the values of the first three unstandardized principal component scores for sites 8 and 16. How do these compare with the values printed on page 14 of the computer printout?

4 In the U.S. Navy data set (see Table 5.1), create new variables by taking the logs of each of the original variables. Perform a PCA on these seven new variables. Do these new variables appear to have a multivariate normal distribution? Explain your answer. Do multicollinearities exist among the new variables? Explain your answer.

5 Perform a PCA on the data you provided for this class in Exercise 1 of Chapter 1. Are there any abnormalities in your data? If so, what are they? Do your important principal components appear to be distributed normally? Are there any outliers in your data? Explain your answers and write a short report summarizing what you have learned from these analyses.

6 Consider the data from the Big 8 football conference described in Exercise 7 in Chapter 1.

a Should these data be standardized prior to performing a PCA? Explain your answer.

b Perform a PCA on these data without using the variables WINS or GAMES.

c How many principal components are required to describe the space in which the data actually fall? Explain your answer.

d Create a scatter plot of the first two (unstandardized) principal component scores and discuss any abnormalities that might be in this data set.

7 Consider the world temperature data described in Exercise 8 of Chapter 2.

   a Should these data be standardized prior to performing a PCA? Explain your answer.

   b Perform a PCA on these data using all three of the temperature variables.

   c How many principal components are required to describe the space in which the data actually fall? Explain your answer.

   d Perform some univariate analyses on the first principal component variable. Does this variable appear to be normally distributed? Why or why not?

   e Do there appear to be any abnormalities in this data set? Explain your answer.

8 Consider the responses to the first 15 questions on the survey given to 975 Pizzazz employees described in Appendix B. Note that these 15 questions are related to organizational commitment.

   a Is it necessary to standardize these data prior to performing a PCA? Explain your answer.

   b Perform a PCA on this subset of Pizzazz data.

   c How many principal components are required to describe adequately the space in which these data actually fall? Explain your answer.

   d Perform some univariate analyses on the first three principal component variables. Do these variables appear to be normally distributed? Why or why not?

   e Create a three-dimensional blob plot using the first three principal component variables.

   f Do there appear to be any abnormalities related to organizational commitment in this data set? Explain your answer.

9 Repeat Exercise 8 using the responses to questions 34–53. These 20 questions are related to job security.

10 Repeat Exercise 8 using the responses to questions 61–80. These 20 questions are related to job satisfaction.

11 Consider the survey of 304 married adults described in Appendix C and their responses to the 60-item Family Assessment Device.

   a Explain why it is not necessary or desirable to standardize these responses prior to performing a principal component analysis on these data.

    b Perform a principal components analysis on these 60 items.

    c How many principal components are required to describe adequately the space in which these data actually fall? Explain your answer.

    d Perform some univariate analyses on the first three principal component variables. Do these principal component variables appear to be normally distributed? Why or why not?

    e Create a three-dimensional blob plot of these data using the first three principal component variables.

    f Do there appear to be any abnormalities with respect to these 60 items in this data set? Explain your answer.

12 Repeat Exercise 11 using the responses to the Family Adaptability Cohesion Evaluation Scales.

13 Repeat Exercise 11 using the responses to the Kansas Marital Satisfaction Scale and the Kansas Family Life Satisfaction Scale simultaneously.

Factor Analysis

Factor analysis (FA) is another variable-directed technique. Principal components analysis (PCA) produces an orthogonal transformation of the variables and does not depend on an underlying model. Factor analysis, however, does depend on a reasonable statistical model. Factor analysis is more concerned with explaining the covariance and/or correlation structure among the measured variables, whereas PCA is more concerned with explaining the variability in the variables.

One basic objective of factor analysis is to determine whether the p response variables exhibit patterns of relationship with each other such that the variables can be partitioned into, say, m subsets, each consisting of a group of variables tending to be more highly related to others within the subset than to those in other subsets.

6.1 Objectives of Factor Analysis

Another basic purpose of a factor analysis is to derive, create, or develop a new set of uncorrelated variables, called *underlying factors* or *underlying characteristics,* with the hope that these new variables will give a better understanding of the data being analyzed. These new variables can then be used in future analyses of the data.

Whenever a large number of variables is measured on each experimental unit, the variables are often related to each other in many different ways. The factor analysis model assumes there is a smaller set of uncorrelated variables that, in some sense, drives or controls the values of the variables that are actually being measured.

In order for these new variables to give a better understanding of the data, we need to give reasonable interpretations to the new variables being created. Whether reasonable interpretations exist is almost always determined, subjectively, by the researcher rather than, objectively, by statistical methods.

To use these new variables, we need to evaluate or score them for each experimental unit in the data set. Factor analysis may allow a researcher to eliminate some of the original variables, but this often depends on how we decide to score the new variables.

To summarize, the goals of factor analysis include the following:

1 To determine whether a smaller set of uncorrelated variables exists that will explain the relationships that exist between the original variables,

2 to determine the number of underlying variables,

3 to interpret these new variables,

4 to evaluate individuals or experimental units in the data set on these new variables, and

5 to use these new variables in other statistical analyses of the data.

In addition to the examples found in this chapter, FA and regression are used in Chapter 11 as an alternative to canonical correlation analysis.

6.2
Caveats

Factor analysis creates a new set of uncorrelated variables from a set of correlated variables. Hence, if the original variables are already uncorrelated, then there is little reason to consider carrying out a FA. Factor analysis attempts to explain the correlations between the original variables and, when the original variables are already uncorrelated, there is nothing to explain.

Many researchers and statisticians believe that FA is not a valid and useful statistical technique. In fact, Hills (1977) stated that "FA is not worth the time necessary to understand it and carry it out." He went on to say that FA is "an elaborate way of doing something which can only ever be crude, namely picking out clusters of inter-related variables, and then finding some sort of average of the variables in a cluster in spite of the fact that the variables may be measured on different scales." I do not agree with this assessment, but researchers who use FA should be prepared for criticism and they must be knowledgeable enough about FA methods to be able to defend their decision to use it. This book attempts to present all of the important issues, and each reader will be required to make his or her own conclusions about the acceptability and validity of FA methods.

Most of the criticism that FA receives is due to the nonuniqueness of FA solutions and the subjectivity involved with numerous aspects of FA. Subjective decisions are made when a researcher determines the number of underlying factors, determines how they are created, determines how they should be interpreted, and determines how individuals in the sample should be evaluated on these new variables. Because of these numerous instances in which subjectivity can play a role, critics of FA suspect that a researcher may be able to show anything that he or she wishes to show. Practitioners of FA methods, on the other hand, believe that things are not nearly as bad or as shaky as they might at first appear to be.

6.3
Some History of Factor Analysis

Spearman (1904) published a paper that is thought to be where FA originated. Spearman was working with examination scores of schoolchildren, and noticed certain systematic effects in the matrix of correlations between the scores the children made on different tests.

Table 6.1 shows the correlations among test scores at which Spearman was looking. Spearman noted that when considering the upper triangular portion of this correlation matrix, the correlations decreased in size in each row, and decreased in size more or less to the same extent. He pointed out that such an effect could be accounted for if the score on each of the variables was composed of two parts. He proposed that, for a randomly selected child, each of his or her measured variables (in a standardized form) could be modeled by the following equations:

$$\text{Classics} = \lambda_1 f + \eta_1$$
$$\text{French} = \lambda_2 f + \eta_2$$
$$\text{Eng} = \lambda_3 f + \eta_3$$
$$\text{Math} = \lambda_4 f + \eta_4$$
$$\text{Discr} = \lambda_5 f + \eta_5$$
$$\text{Music} = \lambda_6 f + \eta_6$$

where f is a random component that is common to all measured variables, η_j is a random component that is specific to the jth measured variable, and with the η_j's independent of each other and independent of f.

Spearman believed that a child's performance on any test was the sum of a general factor, f, which could be interpreted as general intelligence, and an ability, η_j, specific to the particular test being measured.

This should seem reasonable. Most readers will have known classmates who were generally thought of as being A students, others who were thought to be B students, etc. This general classification might be identified with general intelligence. Still each student may have extra-special talents in some specific areas and deficiencies in other specific areas, so that general intelligence does not explain all that goes into a student's score on a particular test.

T A B L E 6.1

Correlations among test scores of schoolchildren

	Classics	French	Eng	Math	Discr	Music
Classics	—	.83	.78	.70	.66	.63
French	.83	—	.67	.67	.65	.57
Eng	.78	.67	—	.64	.54	.51
Math	.70	.67	.64	—	.45	.51
Discr	.66	.65	.54	.45	—	.40
Music	.63	.57	.51	.51	.40	—

6.4
The Factor Analysis Model

Suppose one observes a p-variate response vector $\mathbf{x}$ from a population that has mean $\boldsymbol{\mu}$ and variance–covariance matrix $\boldsymbol{\Sigma}$. The general FA model assumes there are m underlying factors (certainly we want $m < p$) denoted by $f_1, f_2, \ldots, f_m$ such that

$$x_j = \mu_j + \lambda_{j1} f_1 + \lambda_{j2} f_2 + \cdots + \lambda_{jm} f_m + \eta_j \qquad \text{for } j = 1, 2, \ldots, p$$

Assumptions

In the preceding model, we assume that

1 the f_k's are independently and identically distributed with mean 0 and variance 1 for $k = 1, 2, \ldots, m$;

2 the η_j's are independently distributed with mean 0 and variance ψ_j for $j = 1, 2, \ldots, p$; and

3 f_k and η_j have independent distributions for all combinations of k and j, $k = 1, 2, \ldots, m$ and $j = 1, 2, \ldots, p$.

The assumptions that the f's and η's have zero means can be made without any loss of generality. This is because if the means are not zero, their contribution could be described by the μ_j's, and the f's and η's could be replaced by a new set of f's and η's that would have zero means. Also, the assumption that the f's have variance 1 can be made without any loss of generality. This is because if the variances were not all equal to 1, the λ's could be changed to create a new set of f's that would have variance 1. Thus, the only real restriction in the assumptions for the FA model is that all f's and η's be independent.

Also, without any loss of generality, it is almost always assumed that $\mu_j = 0$ and that the $\text{var}(x_j) = 1$ for every j. This can always be the case if one simply standardizes the measured variables before beginning a factor analysis. This is the default in almost all statistical computing packages. However, factor analysis concepts are introduced for nonstandardized x's simply to remind readers that it is not actually necessary to standardize the data even though the data are usually standardized by most researchers.

Hence, the FA model becomes

$$x_j = \lambda_{j1} f_1 + \lambda_{j2} f_2 + \cdots + \lambda_{jm} f_m + \eta_j, \qquad \text{for } j = 1, 2, \ldots, p$$

where the x's have been centered about their means.

Matrix Form of the Factor Analysis Model

In matrix form, the model is

$$\mathbf{x} = \mathbf{\Lambda f} + \boldsymbol{\eta}$$

where **x** has been centered, and

$$\mathbf{x} = [x_1, x_2, \ldots, x_p]'$$
$$\mathbf{f} = [f_1, f_2, \ldots, f_m]'$$
$$\boldsymbol{\eta} = [\eta_1, \eta_2, \ldots, \eta_p]'$$
$$\mathbf{\Lambda} = \begin{bmatrix} \lambda_{11} & \lambda_{12} & \cdots & \lambda_{1m} \\ \lambda_{21} & \lambda_{22} & \cdots & \lambda_{2m} \\ \vdots & \vdots & \cdots & \vdots \\ \lambda_{p1} & \lambda_{p2} & \cdots & \lambda_{pm} \end{bmatrix}$$

In matrix form the factor analysis model assumptions become

1 $\mathbf{f} \sim (\mathbf{0}, \mathbf{I})$,
2 $\boldsymbol{\eta} \sim (\mathbf{0}, \mathbf{\Psi})$, where $\mathbf{\Psi} = \mathrm{diag}(\psi_1, \ldots, \psi_p)$, and
3 $\mathbf{f}$ and $\boldsymbol{\eta}$ are independent.

Definitions of Factor Analysis Terminology

The new variables $f_1, f_2, \ldots, f_m$ are called *common factors*, and $\eta_1, \eta_2, \ldots, \eta_p$ are called *specific factors*. The quantity η_j describes the residual variation specific to the *j*th response variable, and ψ_j is called the specific variance of the *j*th response variable.

The multipliers, the λ_{jk}'s, are called *factor loadings*. Each λ_{jk} measures the contribution of the *k*th common factor to the *j*th response variable. It is said that λ_{jk} is the loading of the *j*th response variable on the *k*th factor.

6.5
Factor Analysis Equations

Next note that

$$\mathbf{x} = \mathbf{\Lambda f} + \boldsymbol{\eta} \qquad \text{implies that}$$
$$\mathbf{\Sigma} = \mathrm{Cov}(\mathbf{x})$$
$$= \mathrm{Cov}(\mathbf{\Lambda f} + \boldsymbol{\eta})$$
$$= \mathbf{\Lambda} \cdot \mathrm{Cov}(\mathbf{f}) \cdot \mathbf{\Lambda}' + \mathbf{\Psi}$$
$$= \mathbf{\Lambda I \Lambda}' + \mathbf{\Psi}$$
$$= \mathbf{\Lambda \Lambda}' + \mathbf{\Psi}$$

Thus, to determine if $\mathbf{f}$, $\boldsymbol{\Lambda}$, and $\boldsymbol{\eta}$ exist such that $\mathbf{x} = \boldsymbol{\Lambda}\mathbf{f} + \boldsymbol{\eta}$, we instead try to find $\boldsymbol{\Lambda}$ and $\boldsymbol{\Psi}$ so that

$$\boldsymbol{\Sigma} = \boldsymbol{\Lambda}\boldsymbol{\Lambda}' + \boldsymbol{\Psi} \tag{6.1}$$

The relationships described in Eq. (6.1) are called the *factor analysis equations*.

Remarks

1 If $\boldsymbol{\Lambda}$ and $\boldsymbol{\Psi}$ exist so that $\boldsymbol{\Sigma} = \boldsymbol{\Lambda}\boldsymbol{\Lambda}' + \boldsymbol{\Psi}$, then the common factors explain the covariances among the response variables exactly. This follows because $\boldsymbol{\Psi}$ is a diagonal matrix.

2 The variance of x_j can be partitioned as $\sigma_{jj} = \sum_{k=1}^{m} \lambda_{jk}^2 + \psi_j$, and the proportion of the variance of x_j that is explained by the common factors, $(\sum_{k=1}^{m} \lambda_{jk}^2)/\sigma_{jj}$, is called the *communality* of the jth response variable.

3 The covariance between x_j and $x_{j'}$ is

$$\text{cov}(x_j, x_{j'}) = \sum_{k=1}^{m} \lambda_{jk}\lambda_{j'k}$$

4 The covariance between x_j and f_k is λ_{jk}, the loading of the jth response variable on the kth factor.

The preceding description of the main ideas of factor analysis assumes that we are factor analyzing the variance–covariance matrix $\boldsymbol{\Sigma}$. Actually, factor analysis procedures are almost always applied to Z scores and to the correlation matrix $\mathbf{P}$.

If the factor analysis has been applied to the correlation matrix $\mathbf{P}$, then $\boldsymbol{\Lambda}$ is the matrix of correlations between the z_j's and the f_k's. That is, $\text{corr}(z_j, f_k) = \lambda_{jk}$. Also in this case, $\sum_{k=1}^{m} \lambda_{jk}^2 + \psi_j = 1$, and the communality of the jth (Z score) variable is simply $\sum_{k=1}^{m} \lambda_{jk}^2$. In the remainder of this chapter, we assume that FA methods are being applied to $\mathbf{P}$, the correlation matrix.

Nonuniqueness of the Factors

If $m > 1$, the factor loading matrix is not unique. That is, if $\boldsymbol{\Lambda}$ and $\boldsymbol{\Psi}$ exist so that

$$\mathbf{P} = \boldsymbol{\Lambda}\boldsymbol{\Lambda}' + \boldsymbol{\Psi}, \quad \text{then}$$
$$= \boldsymbol{\Lambda}\mathbf{T}\mathbf{T}'\boldsymbol{\Lambda}' + \boldsymbol{\Psi}$$

for every orthogonal matrix $\mathbf{T}$. (Recall that $\mathbf{T}$ is an orthogonal matrix if $\mathbf{T}\mathbf{T}' = \mathbf{I}$.) It then follows that

$$\mathbf{P} = (\boldsymbol{\Lambda}\mathbf{T})(\boldsymbol{\Lambda}\mathbf{T})' + \boldsymbol{\Psi}$$

Thus, if Λ is a loading matrix, then $\Lambda \mathbf{T}$ is also a loading matrix for every orthogonal matrix $\mathbf{T}$.

This result can also be illustrated with the FA model. If

$$\mathbf{z} = \Lambda \mathbf{f} + \boldsymbol{\eta}, \qquad \text{then}$$
$$\mathbf{z} = (\Lambda \mathbf{T})(\mathbf{T}'\mathbf{f}) + \boldsymbol{\eta}$$
$$= \Lambda^* \mathbf{f}^* + \boldsymbol{\eta} \text{ (say)}$$

Thus, if Λ is a loading matrix, then $\Lambda^* = \Lambda \mathbf{T}$ is also a loading matrix for any orthogonal matrix $\mathbf{T}$.

Hence, the factor loading matrix Λ is not unique. This is a serious dilemma for many statisticians and researchers. Others see this as a great opportunity. Those that see it as an opportunity feel that if we are not able to interpret the initial factor solution, it is probably because we have the wrong solution, and another solution, obtainable by multiplying by an orthogonal matrix (called a *rotation*), may exist that can be easily interpreted.

Since the factor loading matrix is not unique, the custom has developed of rotating (transforming by multiplying by an orthogonal matrix) the factors obtained in order to select factors in such a way that they make "sense." Critics of FA then ask "Who determines when a solution makes sense?" The fear is that the solution makes sense when the results agree with many of the biases and prejudices held by the researcher who is analyzing the data. Thus, it should be apparent that transforming factor solutions can seem attractive to some researchers and potentially dangerous to others.

6.6
Solving the Factor Analysis Equations

In this section solutions to the factor analysis equations are considered. We assume here that the correlation matrix $\mathbf{P}$ is being analyzed. The techniques described here can also be applied to a variance–covariance matrix.

To determine whether a set of m underlying factors exist, one determines if Λ and $\boldsymbol{\Psi}$ could possibly exist such that

$$\mathbf{P} = \Lambda \Lambda' + \boldsymbol{\Psi}$$

The number of unknown quantities in Λ and $\boldsymbol{\Psi}$ is $pm + p = p(m + 1)$. The number of known quantities in $\mathbf{P}$ is $p(p + 1)/2$ (since $\mathbf{P}$ is symmetric). Thus, the FA equations give rise to $p(p + 1)/2$ equations in $p(m + 1)$ unknowns that must be solved. For example, if $p = 20$ and $m = 5$, then there are 210 equations in 120 unknowns that must be solved.

If $p(m + 1) > p(p + 1)/2$ or, equivalently, if $m > (p - 1)/2$, then there are more unknowns than there are equations, and no unique solution exists.

However, even when $m \leq (p - 1)/2$, no unique solution exists for $m \geq 2$, since any solution can be rotated to an infinite number of other solutions simply by multiplying the initial solution by any of an infinite number of orthogonal matrices.

One question that would seem to be important is this: "Do solutions exist that are not rotations of each other?" To answer this question, suppose $p = 2$ and $m = 1$. In this case the FA equations give rise to three equations in four unknowns. An infinite number of solutions are possible and since there are only two possible orthogonal 1×1 matrices, only pairs of these solutions can be rotations of each other.

As a second example, suppose $p = 3$ and $m = 2$. In this case, the FA equations give rise to six equations in nine unknowns. Again it is possible to show that many solutions exist that are not rotations of each other.

Let's look at an actual solution to a factor analysis problem. Suppose $p = 3$ and $m = 1$. The FA equations give rise to six equations in six unknowns. In this case, the actual equations are

$$1 = \lambda_{11}^2 + \psi_1, \qquad \rho_{12} = \lambda_{11}\lambda_{21}$$
$$1 = \lambda_{21}^2 + \psi_2, \qquad \rho_{13} = \lambda_{11}\lambda_{31}$$
$$1 = \lambda_{31}^2 + \psi_3, \qquad \rho_{23} = \lambda_{21}\lambda_{31}$$

where ρ_{12}, ρ_{13}, and ρ_{23} are the pairwise correlations between the three original variables. These equations imply that

$$\lambda_{21} = \rho_{12}/\lambda_{11} \qquad \text{and} \qquad \lambda_{31} = \rho_{13}/\lambda_{11}$$

which imply that $\rho_{23} = \rho_{12}\rho_{13}/\lambda_{11}^2$ and, hence, that $\lambda_{11}^2 = \rho_{12}\rho_{13}/\rho_{23}$. Without any loss of generality, we can take λ_{11} to be the positive square root of $\rho_{12}\rho_{13}/\rho_{23}$. Then $\lambda_{21} = \rho_{21}/\lambda_{11}$ and $\lambda_{31} = \rho_{13}/\lambda_{11}$, and $\psi_1 = 1 - \lambda_{11}^2$, $\psi_2 = 1 - \lambda_{21}^2$, and $\psi_3 = 1 - \lambda_{31}^2$.

If

$$\mathbf{P} = \begin{bmatrix} 1 & 0.3 & -0.3 \\ 0.3 & 1 & 0.3 \\ -0.3 & 0.3 & 1 \end{bmatrix}$$

then $\lambda_1^2 = -0.3$, which is clearly impossible, and a realistic solution to the factor analysis equations does not exist when $m = 1$.

Suppose

$$\mathbf{P} = \begin{bmatrix} 1 & 0.84 & 0.60 \\ 0.84 & 1 & 0.35 \\ 0.60 & 0.35 & 1 \end{bmatrix}$$

In this case one obtains $\lambda_{11} = 1.2$, $\lambda_{21} = 0.7$, $\lambda_{31} = 0.5$, $\psi_1 = -0.44$, $\psi_2 = 0.51$, and $\psi_3 = 0.75$. Here a real solution exists, but the solution still does not make sense since ψ_1 is a specific variance that is negative, and it is not possible to have a negative variance. Also, $\lambda_{11} = 1.2$, which supposedly measures the correlation between the first variable and the underlying factor. Correlations cannot be greater than 1.

In each of the preceding two cases, we would need to increase m in order to get a solution that would be mathematically possible. For the latter example, we might argue that an increase in the number of underlying factors is necessary because all three variables are not highly correlated with one another. There is a relatively high correlation between variables 1 and 2, but the correlations between these and variable 3 are much lower. Thus, while variables 1 and 2 seem to have something in common, variable 3 has little in common with variables 1 and 2.

The preceding two examples, where there are six equations in six unknowns, are given, partially, to illustrate how difficult it might be to solve the FA equations. Think about how you might solve these equations when there are 210 equations in 120 unknowns. Fortunately, computer algorithms are available that provide approximate solutions to the FA equations for these more complicated cases.

The preceding sections assumed, perhaps without the reader realizing it, that true values of Σ and/or P were known. In practice, these will need to be estimated by the sample variance–covariance matrix $\hat{\Sigma}$ and the sample correlation matrix R, respectively. Then FA methods are applied to one of these two matrices.

6.7
Choosing the Appropriate Number of Factors

Before we actually begin to solve the FA equations, we should try to estimate or guess at the value of m. That is, guess how many underlying characteristics or factors there really are that are driving the values of the variables actually being measured.

One fairly good method that can be used to make an initial guess as to the number of underlying factors is to begin with a PCA and determine how many principal components would be required by using the methods recommended in Chapter 5 for PCA. Then use this as the number of factors that are required. See Section 5.7 for a review of methods for choosing the number of principal components.

An initial guess as to the number of factors will not always agree with a final determination. Still, we have to start somewhere, and this may be as good a place to start as any. To expand on this, see the next section.

Subjective Criteria

Several subjective criteria can help you choose a final number of underlying factors, and the process of selecting a final number of factors may be as much of an art as it is a science. Here are some things that should be considered:

1   Do not include *trivial* factors. Trivial factors are defined as factors that have one and only one of the original variables loading on the factor. Variables that load on only one factor are generally uncorrelated with all the other variables in the data set, and such variables, by themselves, are underlying characteristics. In these instances, it is probably best to remove such variables and restart the FA procedure. This does not mean that these variables are not important or not useful, it just means that these variables are measuring characteristics of the population that are independent of the characteristics being measured by the other variables. It does not make sense to create factors for such variables when the variables themselves can be used

2   Some researchers believe that the η_j's (or equivalently, the ψ_j's) in the FA model should all be close to zero. However, there is nothing in the FA model or its assumptions that requires this. Trying to get the ψ_j's close to zero will require larger numbers of underlying factors (many of which are likely to be trivial factors) and tends to take FA toward PCA where one is trying to explain the variability of the variables rather than their covariances and/or correlations. Note that these remarks could be made in terms of the communalities as well as the specific variances since the communalities approach one if and only if the specific variances approach zero. Thus, you do not need to expect all communalities to approach one.

3   Many computing programs will produce matrices of differences between the observed correlations between variables and those that are reproduced by the FA solution. If these differences are small, you might be able to reduce the number of factors; on the other hand, if some differences are large (many greater than 0.25 or a few greater than 0.40, perhaps), then the number of factors might need to be increased.

4   Some programs produce partial correlations among the variables after adjusting for the common factors. If any of the partial correlations is large, then you should consider increasing the number of factors. If all partial correlations are small, then you might try to reduce the number of factors.

Objective Criteria

There are also objective methods for choosing the number of underlying factors.

One of the methods proposed for conducting a factor analysis is based on a maximum likelihood approach, which is based on the data having a multivariate normal distribution. Although a discussion of maximum likelihood methods is beyond the scope of this book, one nice advantage of this approach is that it allows you to compute a likelihood ratio test (LRT) statistic that can be used to test the adequacy of the chosen number of factors. This is a good method to consider when you believe that the data come from a multivariate normal distribution.

To use this method, a researcher will usually want to try several possibilities for m, the number of underlying factors. If we test for the adequacy of m and reject, then the number of factors should be increased, but if we do not reject, then we could consider reducing the number of factors. For example, if six factors are adequate, it might also be the case that four or five would be adequate. But if six factors are not adequate, then the number of factors would need to be increased.

Many computer programs print a statistic called Akaike's information criterion (AIC) (Akaike, 1973; Akaike, 1974). Supposedly, the number of factors that provides the minimum value for AIC is considered to be the best choice. Use of this criterion requires the data analyst to try several different choices for m in order to see which choice produces the minimum value for AIC.

Other programs print a statistic called Schwarz's Bayesian criterion (SBC) (Schwarz, 1978). As with the AIC, the choice for m that makes SBC the smallest is considered best.

The test procedure based on maximum likelihood methods and Akaike's information criterion tends to produce factors that are statistically significant but inconsequential for practical purposes. That is, these procedures may tend to produce trivial factors. Trivial factors are less likely to be produced by Schwarz's Bayesian criterion. If trivial factors are produced, the corresponding variables should be removed and the process of solving the FA equations restarted as recommended earlier.

6.8
Computer Solutions of the Factor Analysis Equations

In the early days of factor analysis, a number of interactive methods were used to solve the FA equations. Most of these methods required the user to make some subjective judgments, such as guessing the communalities for each of the response variables. Because of this, many different researchers could guess different communality values when analyzing a data set and as a result could possibly find entirely different factors. This is still true today.

Many different methods have been proposed for solving the FA equations, and many of the books on factor analysis describe different computer algorithms that could be used. Among the methods that have been proposed are

1 principal factoring with or without iteration,

2 Rao's canonical factoring,

3 alpha factoring,

4 image factoring,

5 maximum likelihood,

6 unweighted Least-Squares factor analysis, and

7 Harris factoring.

Which of these methods is best is not known. Method 1 may be the most popular of these. Two advantages that the maximum likelihood method holds over the others are (1) it is invariant to the units in which the variables are measured and (2) there are objective ways (LRT, AIC, and SBC) to help a researcher estimate an appropriate number of underlying factors. If a researcher feels that the data being analyzed are multivariate normal, then the maximum likelihood procedure should be given strong consideration.

FA solution methods 1, 3, 5, and 6 require prior estimates of the communalities for each response variable. How should these prior estimates be obtained? When a user has no idea about what to choose for the communalities, the communality estimate for a variable should be taken as the squared multiple correlation of that variable with all remaining variables. This is the default in many computing packages. Recall that the squared multiple correlation between one variable and a set of variables is given by the proportion of the variability in the first variable that is accounted for by the other variables. This is usually called R^2 in multiple regression problems.

Another possibility, not often used, is to take the prior communality estimate for each variable to be its maximum correlation with any of the remaining variables.

As previously stated, several acceptable methods are available for solving the FA equations. The easiest one of these to explain is the principal factor method. For the interested reader, this solution method is described next. The uninterested reader may skip this section.

Principal Factor Method on **R**

1 This method initially requires suitable estimates of the communalities or, equivalently, estimates of the specific variances, $\psi_1, \psi_2, \ldots, \psi_p$.

2 Next, since $\mathbf{R} = \mathbf{\Lambda}\mathbf{\Lambda}' + \mathbf{\Psi}$ must be satisfied, we must have $\mathbf{\Lambda}\mathbf{\Lambda}' = \mathbf{R} - \mathbf{\Psi}$.

3 To obtain a unique solution for $\mathbf{\Lambda}$, we can force $\mathbf{\Lambda}'\mathbf{\Lambda}$ to be a diagonal matrix. Let $\mathbf{\Lambda}'\mathbf{\Lambda} = \mathbf{D}$.

4 Then $\mathbf{\Lambda}\mathbf{\Lambda}' = \mathbf{R} - \mathbf{\Psi}$ implies $\mathbf{\Lambda}\mathbf{\Lambda}'\mathbf{\Lambda} = (\mathbf{R} - \mathbf{\Psi})\mathbf{\Lambda}$, which implies $\mathbf{\Lambda}\mathbf{D} = (\mathbf{R} - \mathbf{\Psi})\mathbf{\Lambda}$, which, by looking at the columns of both sides of this matrix equation, implies that

$$[d_1\boldsymbol{\lambda}_1, \ldots, d_m\boldsymbol{\lambda}_m] = [(\mathbf{R} - \mathbf{\Psi})\boldsymbol{\lambda}_1, \ldots, (\mathbf{R} - \mathbf{\Psi})\boldsymbol{\lambda}_m]$$

That is, $(\mathbf{R} - \mathbf{\Psi})\boldsymbol{\lambda}_k = d_k \boldsymbol{\lambda}_k$ where $\boldsymbol{\lambda}_k$ is the kth column of $\mathbf{\Lambda}$ for $k = 1, 2, \ldots, m$.

5   The only way the equations in item 4 can be true is if the diagonal elements of $\mathbf{D}$ are eigenvalues of $\mathbf{R} - \mathbf{\Psi}$ and if the columns of $\mathbf{\Lambda}$ are their corresponding eigenvectors.

6   Any subset of m eigenvalues and eigenvectors would solve the FA equations, but the vectors corresponding to the m largest eigenvalues are chosen since the elements of $\mathbf{D}$ are communalities and can be expressed as

$$d_k = \sum_{j=1}^{p} \lambda_{jk}^2, \qquad \text{for } k = 1, 2, \ldots, m$$

By choosing vectors corresponding to the m largest eigenvalues, we can maximize the communalities, which, in turn, tends to maximize the λ_{jk}'s (the factor loadings). Hence, these vectors should correspond to the most important factors.

Remarks   If we have chosen the communalities wisely, there will be m non-negative eigenvalues in $\mathbf{R} - \mathbf{\Psi}$. If we have chosen them badly, a nonsensical solution will result, and we must reduce m or reestimate the communalities and start over.

If the communalities are all taken to be equal to 1, then $\mathbf{\Psi} = \mathbf{0}$, and the principal factor method reduces to a PCA on the correlation matrix. Obviously, this is why most statistical computing packages allow users to perform a PCA with the use of their FA program.

Principal Factor Method with Iteration

Another method of solving the FA equations is called the *principal factor method with iteration*. This procedure begins in the same way as the principal factor method. However, after an initial solution is found, the communalities corresponding to this solution are determined. Then these communalities are used as an initial guess, and the procedure starts all over again. This cycle is repeated until either all estimates converge or a nonsensical result is obtained. Experience has shown that these possibilities appear to be equally likely when working with real data sets. Note also that each iteration of this process actually produces a solution to the FA equations.

E X A M P L E   **6**.**1**

The data in Table 6.2 is taken from Gunst and Mason (1980) and consist of anthropometric and physical fitness measurements that were taken on 50 white male applicants to the police department of a major metropolitan city. The variables include reaction time in seconds to a visual stimulus

T A B L E 6.2
Police Department applicant data

ID	REACT	HEIGHT	WEIGHT	SHLDR	PELVIC	CHEST	THIGH	PULSE	DIAST	CHNUP	BREATH	RECVR	SPEED	ENDUR	FAT
1	0.310	179.6	74.20	41.7	27.3	82.4	19.0	64	64	2	158	108	5.5	4.0	11.91
2	0.345	175.6	62.04	37.5	29.1	84.1	5.5	88	78	20	166	108	5.5	4.0	3.13
3	0.293	166.2	72.96	39.4	26.8	88.1	22.0	100	88	7	167	116	5.5	4.0	16.89
4	0.254	173.8	85.92	41.2	27.6	97.6	19.5	64	62	4	220	120	5.5	4.0	19.59
5	0.384	184.8	65.88	39.8	26.1	88.2	14.5	80	68	9	210	120	5.5	5.0	7.74
6	0.406	189.1	102.26	43.3	30.1	101.2	22.0	60	68	4	188	91	6.0	4.0	30.42
7	0.344	191.5	84.04	42.8	28.4	91.0	18.0	64	48	1	272	110	6.0	3.0	13.70
8	0.321	180.2	68.34	41.6	27.3	90.4	5.5	74	64	14	193	117	5.5	4.0	3.04
9	0.425	183.8	95.14	42.3	30.1	100.2	13.5	80	78	4	199	105	5.5	4.0	20.26
10	0.385	163.1	54.28	37.2	24.2	80.5	7.0	84	78	13	157	113	6.0	4.0	3.04
11	0.317	169.6	75.92	39.4	27.2	92.0	16.5	65	78	6	180	110	5.0	5.0	12.83
12	0.353	171.6	71.70	39.1	27.0	86.2	25.5	68	72	0	193	105	5.5	4.0	15.95
13	0.413	180.0	80.68	40.8	28.3	87.4	17.5	73	88	4	218	109	5.0	4.2	11.86
14	0.392	174.6	70.40	39.8	25.9	83.9	16.5	104	78	6	190	129	5.0	4.0	9.93
15	0.312	181.8	91.40	40.6	29.5	95.1	32.0	92	88	1	206	139	5.0	3.5	32.63
16	0.342	167.4	65.74	39.7	26.4	86.0	13.0	80	86	6	181	120	5.5	4.0	6.64
17	0.293	173.0	79.28	41.2	26.9	96.1	11.5	72	68	6	184	111	5.5	3.9	11.57
18	0.317	179.8	92.06	40.0	29.8	100.9	15.0	60	78	0	205	92	5.0	4.0	24.21
19	0.333	176.8	87.96	41.2	28.4	100.8	20.5	76	90	1	228	147	4.0	3.5	22.39
20	0.317	179.3	77.66	41.4	31.6	90.1	9.5	58	86	15	198	98	5.5	4.1	6.29
21	0.427	193.5	98.44	41.6	29.2	95.7	21.0	54	74	0	254	110	5.5	3.8	23.63
22	0.266	178.8	65.42	39.3	27.1	83.0	16.5	88	72	7	206	121	5.5	4.0	10.53
23	0.311	179.6	97.04	43.8	30.1	100.8	22.0	100	74	3	194	124	5.0	4.0	20.62
24	0.284	172.6	81.72	40.9	27.3	91.5	22.0	74	76	4	201	113	5.5	5.1	18.39
25	0.259	171.5	69.60	40.4	27.8	87.7	15.5	70	72	10	175	110	5.5	3.0	11.14

ID	REACT	HEIGHT	WEIGHT	SHLDR	PELVIC	CHEST	THIGH	PULSE	DIAST	CHNUP	BREATH	RECVR	SPEED	ENDUR	FAT
26	0.317	168.9	63.66	39.8	26.7	83.9	6.0	68	70	7	179	119	5.5	5.0	5.16
27	0.263	183.1	87.24	43.2	28.3	95.7	11.0	88	74	7	245	115	5.5	4.0	9.60
28	0.336	163.6	64.86	37.5	26.6	84.0	15.5	64	64	6	146	115	5.0	4.4	11.93
29	0.267	184.3	84.68	40.3	29.0	93.2	8.5	64	76	2	213	109	5.5	5.0	8.55
30	0.271	181.0	73.78	42.8	29.7	90.3	8.5	56	88	11	181	109	6.0	5.0	4.94
31	0.264	180.2	75.84	41.4	28.7	88.1	13.5	76	76	9	192	144	5.5	3.6	10.62
32	0.357	184.1	70.48	42.0	28.9	81.3	14.0	84	72	5	231	123	5.5	4.5	8.46
33	0.259	178.9	86.90	42.5	28.7	95.0	16.0	54	68	12	186	118	6.0	4.0	13.47
34	0.221	170.0	76.68	39.7	27.7	93.6	15.0	50	72	4	178	108	5.5	4.5	12.81
35	0.333	180.6	77.32	42.1	27.3	89.5	16.0	88	72	11	200	119	5.5	4.6	13.34
36	0.359	179.0	79.90	40.8	28.2	90.3	26.5	80	80	3	201	124	5.5	3.7	24.57
37	0.314	186.6	100.36	42.5	31.5	100.3	27.0	62	76	2	208	120	5.5	4.1	28.35
38	0.295	181.4	91.66	41.9	28.9	96.6	25.5	68	78	2	211	125	6.0	3.0	26.12
39	0.296	176.5	79.00	40.7	29.1	86.5	20.5	60	66	5	210	117	5.5	4.2	15.21
40	0.308	174.0	69.10	40.9	27.0	88.1	18.0	92	74	5	161	140	5.0	5.5	12.51
41	0.327	178.2	87.78	42.9	27.2	100.3	16.5	72	72	4	189	115	5.5	3.5	20.50
42	0.303	177.1	70.18	39.4	27.6	85.5	16.0	72	74	14	201	110	6.0	4.8	10.67
43	0.297	180.0	67.66	40.9	28.7	86.1	15.0	76	76	5	177	110	5.5	4.5	10.76
44	0.244	176.8	86.12	41.3	28.2	92.7	12.5	76	68	7	181	110	5.5	4.0	14.55
45	0.282	176.3	65.00	39.0	26.0	83.3	7.0	88	72	12	167	127	5.5	5.0	5.27
46	0.285	192.4	99.14	43.7	28.7	96.1	20.5	64	68	4	174	105	6.0	4.0	17.94
47	0.299	175.2	75.70	39.4	27.3	90.8	19.0	56	76	7	174	111	5.5	4.5	12.64
48	0.280	175.9	78.62	43.4	29.3	90.7	18.0	64	72	7	170	117	5.5	3.7	10.81
49	0.268	174.6	64.88	42.3	29.2	82.6	3.5	72	80	11	199	113	6.0	4.5	2.01
50	0.362	179.0	71.00	41.2	27.3	85.6	16.0	68	90	5	150	108	5.5	5.0	10.00

(REACT), the applicant's height in centimeters (HEIGHT), the applicant's weight in kilograms (WEIGHT), the applicant's shoulder width in centimeters (SHLDR), the applicant's pelvic width in centimeters (PELVIC), the applicant's minimum chest circumference in centimeters (CHEST), the applicant's thigh skinfold thickness in millimeters (THIGH), the applicant's resting pulse rate (PULSE), the applicant's diastolic blood pressure (DIAST), the number of chin-ups the applicant was able to complete (CHNUP), the applicant's maximum breathing capacity in liters (BREATH), the applicant's pulse rate after 5 minutes of recovery from treadmill running (RECVR), the applicant's maximum treadmill speed (SPEED), the applicant's treadmill endurance time in minutes (ENDUR), and the applicant's total body fat measurement (FAT).

An analysis of the data in Table 6.2, using the principal factor method with iteration, was initially attempted with SPSS's FACTOR procedure by using its EXTRACTION=PAF option. The commands actually used follow:

```
FACTOR    VARIABLES=REACT,HEIGHT,WEIGHT,SHLDR,PELVIC,CHEST,
             THIGH,PULSE,DIAST,CHNUP,BREATH,RECVR,SPEED,ENDUR,FAT
          /WIDTH=78

          /EXTRACTION=PAF

          /ROTATION=VARIMAX

          /PRINT=UNIVARIATE INITIAL CORRELATION DET EXTRACTION

             REPR ROTATION FSCORE

          /FORMAT SORT
```

Some of the important results from the preceding commands are shown on pages 4–8 of the computer output shown in Figure 6.1. Interested readers can see all of the output by executing the file named EX6_1.SSX on the enclosed disk.

The top of page 4 shows the means and standard deviations for each of the 15 response variables. Beginning in the middle of page 4 and continuing through the first half of page 5 is a listing of the lower half of the sample correlation matrix. Following the correlation matrix is its determinant, which is equal to 0.0000173.

Recall that if the determinant of the correlation matrix is close to 1, then we should not attempt a factor analysis because this would be an indication that the variables are all uncorrelated with one another and, as a result, there would be no underlying common factors. For this example, the determinant of the correlation matrix is very close to zero, which is an indication that there are linear dependencies among the response variables and that there are likely to be underlying common factors.

If we believe the data are normal, we can also test the hypothesis that

EX 6.1 - FITNESS AND SIZE MEASUREMENTS ON POLICE APPLICANTS   Page   4
- - - - - - - -   F A C T O R   A N A L Y S I S   - - - - - - - - -
ANALYSIS NUMBER 1 LISTWISE DELETION OF CASES WITH MISSING VALUES

```
                MEAN       STD DEV    LABEL
REACT          .31620       .04821
HEIGHT      177.90600      6.70313
WEIGHT       78.35240     11.46837
SHLDR        40.95200      1.58607
PELVIC       28.10600      1.43718
CHEST        90.62000      5.97092
THIGH        16.13000      6.10186
PULSE        73.08000     12.91075
DIAST        74.60000      8.11398
CHNUP         6.28000      4.38010
BREATH      193.34000     25.45473
RECVR       115.54000     11.17981
SPEED         5.48000       .36365
ENDUR         4.17400       .56417
FAT          13.78240      7.34763
```

NUMBER OF CASES =   50

CORRELATION MATRIX: R

```
            REACT      HEIGHT     WEIGHT     SHLDR      PELVIC     CHEST      THIGH
REACT      1.00000
HEIGHT      .22235    1.00000
WEIGHT      .05623     .63534    1.00000
SHLDR     -.09378     .65429     .66562    1.00000
PELVIC    -.05589     .58589     .64702     .58244    1.00000
CHEST     -.03182     .42592     .88869     .55449     .52205    1.00000
THIGH      .13240     .22319     .55422     .20457     .20749     .39780    1.00000
PULSE      .16311    -.18151    -.26384    -.16824    -.32295    -.24628    -.00622
DIAST      .14732    -.18655    -.05170    -.14488     .14774    -.00842     .04868
CHNUP     -.15846    -.27601    -.57578    -.27387    -.15816    -.45359    -.66953
BREATH     .15953     .58783     .45020     .36765     .35357     .34730     .20652
RECVR     -.12957    -.12126    -.12188    -.02394    -.20686    -.08323     .20313
SPEED     -.14934     .21563    -.05264     .20175     .04124    -.16241    -.20804
ENDUR     -.05255    -.18916    -.36798    -.24478    -.22935    -.32754    -.33573
FAT        .16481     .36511     .80951     .33063     .41322     .72462     .84421
```

F I G. **6.1**
Pages 4–8 of computer printout, Example 6.1

the correlation matrix is equal to an identity matrix by using the procedure discussed in Section 5.8.

The principal axis factoring (PAF) method needs initial estimates of the communalities of each response variable. The default in SPSS is to select the initial communality estimates to be equal to the squared multiple correlation that each variable has with all of the other response variables. This was the recommendation made earlier in this chapter.

The initial communality estimates (shown under the column labeled COMMUNALITY at the bottom of page 5 of Figure 6.1), the initial eigenvalues of the matrix $\mathbf{R} - \mathbf{\Psi}$ (shown under the column labeled EIGEN-

EX 6.1 - FITNESS AND SIZE MEASUREMENTS ON POLICE APPLICANTS   Page   5

- - - - - - - - - - -   F A C T O R   A N A L Y S I S   - - - - - - - - - -

	PULSE	DIAST	CHNUP	BREATH	RECVR	SPEED	ENDUR
PULSE	1.00000						
DIAST	.23409	1.00000					
CHNUP	.15478	.05375	1.00000				
BREATH	-.06436	-.16483	-.35762	1.00000			
RECVR	.50389	.15384	-.06025	.09128	1.00000		
SPEED	-.29958	-.32092	.32390	-.03122	-.43652	1.00000	
ENDUR	.00730	.13366	.21278	-.31457	-.07280	-.02248	1.00000
FAT	-.09483	.04465	-.69117	.29870	.06652	-.20551	-.40546

	FAT
FAT	1.00000

DETERMINANT OF CORRELATION MATRIX =   .0000173 |R|

EXTRACTION 1 FOR ANALYSIS 1, PRINCIPAL AXIS FACTORING (PAF)

INITIAL STATISTICS:

VARIABLE	COMMUNALITY	*	FACTOR	EIGENVALUE	PCT OF VAR	CUM PCT
REACT	.43327	*	1	5.21853	34.8	34.8
HEIGHT	.74643	*	2	2.40680	16.0	50.8
WEIGHT	.94074	*	3	1.31268	8.8	59.6
SHLDR	.73989	*	4	1.23108	8.2	67.8
PELVIC	.72887	*	5	1.20385	8.0	75.8
CHEST	.89251	*	6	.84790	5.7	81.5
THIGH	.84259	*	7	.70475	4.7	86.2
PULSE	.47264	*	8	.57841	3.9	90.0
DIAST	.35876	*	9	.39355	2.6	92.7
CHNUP	.71587	*	10	.36819	2.5	95.1
BREATH	.50650	*	11	.32659	2.2	97.3
RECVR	.58341	*	12	.18687	1.2	98.5
SPEED	.62236	*	13	.13881	.9	99.5
ENDUR	.28693	*	14	.04388	.3	99.7
FAT	.93751	*	15	.03813	.3	100.0

   PAF ATTEMPTED TO EXTRACT   5 FACTORS.
IN ITERATION   21 COMMUNALITY OF A VARIABLE EXCEEDED 1.0.

Ignore these columns for f.a.

F I G. **6**.1

Pages 4–8 of computer printout, Example 6.1

VALUE), the proportion of the total variation for which each accounts (under the column labeled PCT OF VAR), and the cumulative proportion of the total variation (under the column labeled CUM PCT) are not very interesting when performing an FA. These were interesting in PCA, but are not really interesting or useful when doing a FA. Recall that in PCA, you are trying to account for as much of the variation in the response variables

EX 6.1 - FITNESS AND SIZE MEASUREMENTS ON POLICE APPLICANTS   Page 6

- - - - - - - - - - F A C T O R   A N A L Y S I S - - - - - - - - - -

*** EXTRACTION TERMINATED ***

EXTRACTION 2 FOR ANALYSIS 1, MAXIMUM LIKELIHOOD (ML)   *May be ignored*

INITIAL STATISTICS:

VARIABLE	COMMUNALITY	*	FACTOR	EIGENVALUE	PCT OF VAR	CUM PCT
REACT	.43327	*	1	5.21853	34.8	34.8
HEIGHT	.74643	*	2	2.40680	16.0	50.8
WEIGHT	.94074	*	3	1.31268	8.8	59.6
SHLDR	.73989	*	4	1.23108	8.2	67.8
PELVIC	.72887	*	5	1.20385	8.0	75.8
CHEST	.89251	*	6	.84790	5.7	81.5
THIGH	.84259	*	7	.70475	4.7	86.2
PULSE	.47264	*	8	.57841	3.9	90.0
DIAST	.35876	*	9	.39355	2.6	92.7
CHNUP	.71587	*	10	.36819	2.5	95.1
BREATH	.50650	*	11	.32659	2.2	97.3
RECVR	.58341	*	12	.18687	1.2	98.5
SPEED	.62236	*	13	.13881	.9	99.5
ENDUR	.28693	*	14	.04388	.3	99.7
FAT	.93751	*	15	.03813	.3	100.0

   ML EXTRACTED 5 FACTORS.   9 ITERATIONS REQUIRED.

CHI-SQUARE STATISTIC: 53.7867, D.F.: 40, SIGNIFICANCE: 0.0713

Tests H_0: 5 factors are adequate. vs. H_1: more than 5 factors are needed.

FACTOR MATRIX: $\hat{\Lambda}$ *the initial loading matrix*

	FACTOR 1	FACTOR 2	FACTOR 3	FACTOR 4	FACTOR 5
RECVR	.99949	.00240	.00450	-.00012	-.00011
PULSE	.50504	-.19335	-.08477	.04875	.27563
SPEED	-.43717	-.13061	.16898	.29861	-.38960
WEIGHT	-.12509	.95862	.18869	-.04543	-.01043
FAT	.06572	.91331	-.31177	-.04854	.06576
CHEST	-.08644	.85351	.24073	-.34980	.05074
THIGH	.20414	.72625	-.58269	.16388	-.09597
CHNUP	-.06010	-.66109	.29232	-.06014	-.07071
SHLDR	-.02761	.59165	.49093	.18250	-.24903

FIG. 6.1
Pages 4–8 of computer printout, Example 6.1

EX 6.1 - FITNESS AND SIZE MEASUREMENTS ON POLICE APPLICANTS Page 7

- - - - - - - - - - - F A C T O R   A N A L Y S I S - - - - - - - - - - -

	FACTOR 1	FACTOR 2	FACTOR 3	FACTOR 4	FACTOR 5
HEIGHT	-.12472	.58612	.45384	.56874	.05023
PELVIC	-.20978	.58043	.33669	.10627	-.09787
BREATH	.08895	.44315	.29541	.32277	.13001
ENDUR	-.07202	-.40673	.04292	-.01658	.03689
REACT	-.12896	.09789	-.16910	.33628	.67879
DIAST	.15454	-.02047	-.12127	-.16471	.25161

FINAL STATISTICS:

VARIABLE	COMMUNALITY	*	FACTOR	EIGENVALUE	PCT OF VAR	CUM PCT
		*				
REACT	.62865	*	1	1.63181	10.9	10.9
HEIGHT	.89105	*	2	4.90279	32.7	43.6
WEIGHT	.97237	*	3	1.34425	9.0	52.5
SHLDR	.68715	*	4	.86154	5.7	58.3
PELVIC	.51515	*	5	.86543	5.8	64.0
CHEST	.91884	*				
THIGH	.94470	*				
PULSE	.37798	*				
DIAST	.12945	*				
CHNUP	.53473	*				
BREATH	.41265	*				
RECVR	.99900	*				
SPEED	.47769	*				
ENDUR	.17409	*				
FAT	.94234	*				

Ignore these.

final communalities

REPRODUCED CORRELATION MATRIX:

	REACT	HEIGHT	WEIGHT	SHLDR	PELVIC
REACT	.62865*	.00028	.00053	.03542	-.05213
HEIGHT	.22206	.89105*	-.00140	-.01001	.01120
WEIGHT	.05570	.63674	.97237*	.00805	.00464
SHLDR	-.12921	.66431	.65756	.68715*	.02417
PELVIC	-.00376	.57469	.64238	.55827	.51515*
CHEST	-.02920	.42389	.88979	.54907	.55245
THIGH	.13327	.22414	.55426	.19180	.20933
PULSE	.13377	-.17321	-.26961	-.22970	-.26851
DIAST	.11398	-.16735	-.05698	-.16863	-.12726
CHNUP	-.17462	-.28507	-.56759	-.23934	-.27216
BREATH	.17875	.57282	.45341	.43129	.35960

F I G. 6.1

Pages 4–8 of computer printout, Example 6.1

EX 6.1 - FITNESS AND SIZE MEASUREMENTS ON POLICE APPLICANTS Page 8

- - - - - - - - - - F A C T O R   A N A L Y S I S - - - - - - - - - - -

	REACT	HEIGHT	WEIGHT	SHLDR	PELVIC
RECVR	-.12953	-.12128	-.12187	-.02396	-.20676
SPEED	-.14903	.20492	-.04814	.16927	.14265
ENDUR	-.01832	-.21751	-.37242	-.22980	-.21189
FAT	.16197	.36131	.80998	.36026	.39976

	CHEST	THIGH	PULSE	DIAST	CHNUP
REACT	-.00261	-.00087	.02934	.03334	.01616
HEIGHT	.00203	-.00095	-.00830	-.01920	.00907
WEIGHT	-.00110	-.00004	.00577	.00528	-.00820
SHLDR	.00542	.01278	.06146	.02375	-.03453
PELVIC	-.03040	-.00184	-.05444	.27501	.11401
CHEST	.91884*	-.00195	-.01412	-.01878	.01764
THIGH	.39975	.94470*	.00017	.01248	-.00374
PULSE	-.23215	-.00639	.37798*	.08048	.10451
DIAST	.01036	.03620	.15361	.12945*	.09284
CHNUP	-.47123	-.66579	.05027	-.03909	.53473*
BREATH	.33535	.20828	-.01423	-.05160	-.24057
RECVR	-.08323	.20314	.50390	.15386	-.06033
SPEED	-.15724	-.19624	-.30269	-.23259	.17161
ENDUR	-.32292	-.34135	.04799	.00401	.28415
FAT	.71910	.84411	-.10121	.05381	-.70060

	BREATH	RECVR	SPEED	ENDUR	FAT
REACT	-.01922	-.00005	-.00031	-.03423	.00284
HEIGHT	.01502	.00002	.01071	.02835	.00380
WEIGHT	-.00321	-.00001	-.00450	.00444	-.00048
SHLDR	-.06364	.00002	.03249	-.01498	-.02963
PELVIC	-.00603	-.00010	-.10142	-.01746	.01345
CHEST	.01195	.00000	-.00517	-.00462	.00551
THIGH	-.00176	-.00002	-.01180	.00563	.00010
PULSE	-.05013	-.00001	.00311	-.04069	.00638
DIAST	-.11323	-.00002	-.08833	.12965	-.00916
CHNUP	-.11706	.00008	.15229	-.07137	.00943
BREATH	.41265*	.00003	-.03010	-.14005	-.01266
RECVR	.09125	.99900*	-.00002	-.00003	.00004
SPEED	-.00112	-.43649	.47769*	-.09502	.03531
ENDUR	-.17453	-.07277	.07254	.17409*	-.01911
FAT	.31137	.06648	-.24082	-.38635	.94234*

THE LOWER LEFT TRIANGLE CONTAINS THE REPRODUCED CORRELATION MATRIX; THE
DIAGONAL, COMMUNALITIES; AND THE UPPER RIGHT TRIANGLE, RESIDUALS BETWEEN
THE OBSERVED CORRELATIONS AND THE REPRODUCED CORRELATIONS.

THERE ARE 19 (18.0%) RESIDUALS (ABOVE DIAGONAL) THAT ARE > 0.05

F I G. **6.1**

Pages 4–8 of computer printout, Example 6.1

as possible, while in FA, you are trying to account for as much of the correlation between the response variables as possible.

In spite of the remark in the previous paragraph, the default in the SPSS PAF extraction method is to choose the number of factors to be equal to the number of eigenvalues of $\mathbf{R} - \mathbf{\Psi}$ that are greater than 1. Note that this is not one of the methods recommended in this book! The book does suggest looking initially at the eigenvalues of $\mathbf{R}$, but not $\mathbf{R} - \mathbf{\Psi}$. Still this may be all right as an initial guess, but every data analyst must look at this number suspiciously and should be willing to consider other possibilities for the number of underlying factors.

There is a note on the bottom of page 5 of the computer printout (see Figure 6.1) indicating that the PAF method attempted to extract five (the number of eigenvalues that are greater than 1) factors. However, in the iteration process, the PAF procedure ran into a snag. On the 21st iteration (see the very bottom of page 5), the communality of one of the response variables exceeded 1. This is ridiculous, since this would mean that more than 100% of the variability in that variable could be accounted for by the common factors, which is an impossibility. Equivalently, this would imply that the specific variance of one of the variables is negative. Because this solution would not make any sense, the FACTOR procedure quit. There is a message indicating this near the top of page 6. We could see the solution that was available at the end of iteration number 20 by inserting a

```
/CRITERIA=ITERATE(20)
```

option before the

```
/EXTRACTION=PAF
```

option in the SPSS command stream.

Next the maximum likelihood method was attempted by adding the following SPSS commands to the previous command stream:

```
/EXTRACTION=ML
/CRITERIA=ITERATE(20) RCONVERGE(.01)
/ROTATION=VARIMAX
/PRINT=UNIVARIATE INITIAL CORRELATION DET EXTRACTION
      REPR ROTATION FSCORE
/FORMAT SORT
```

The results from these commands begin on page 6 of the computer printout shown in Figure 6.1. The maximum likelihood procedure also requires prior estimates of communalities. These were once again chosen to be equal to the squared multiple correlation of each variable with all other variables. SPSS once again selected five factors for the same reason as before. The maximum likelihood procedure required nine iterations to converge as stated near the middle of page 6.

Following this message is a CHI-SQUARE STATISTIC, which is equal to 53.7867 and is based on 40 degrees of freedom. This statistic is the test of the hypothesis that the selected number of factors (five in this example) is adequate. The significance level of this test statistic is 0.0713, and we would not reject the adequacy of the selected number of factors at the 5% significance level since 0.0713 > 0.05. This indicates that at the 5% significance level it is not necessary to consider more than five factors; however, remember that fewer than five factors may also be adequate.

The initial loading matrix is shown at the bottom of page 6 and continues onto page 7 (see Figure 6.1). This matrix contains the correlations between the original variables and the factors. We could try to interpret the factors by examining the correlations shown here, but this is usually a wasted effort. We can generally interpret the factors more easily after performing a rotation of the factor loading matrix. Hence, interpretations are not discussed for this example until after performing a rotation. Rotations are considered in the next section.

The final communalities (the proportions of variability in each of the variables that are accounted for by the underlying factors) are shown near the middle of page 7 under the heading FINAL STATISTICS and the column labeled COMMUNALITY. Note that the communalities for the variables DIAST and ENDUR are quite low, 0.12945 and 0.17407, respectively. This is an indication that these two variables are likely to be uncorrelated with the other variables in the data set and with each other. It is likely that the only way to increase the communalities for these two variables would be to increase the number of factors selected. This is neither desirable nor necessary. This is not to say that these two variables are not important; only that the characteristics of police department applicants being measured by these two variables are independent of characteristics being measured by the other variables.

The eigenvalues printed to the right of the final communality estimates are not useful and can be ignored.

At the bottom of page 7 and continuing onto page 8, SPSS prints the REPRODUCED CORRELATION MATRIX. This matrix contains three different pieces of information. The elements below the diagonal show the correlations between the variables that would be reproduced by this five-factor solution. Mathematically, these are the elements below the diagonal of $\hat{\Lambda}\hat{\Lambda}'$. Note that these elements do not change as a result of performing an orthogonal rotation of the loading matrix. The elements on the diagonal (identified by the asterisks) are the communalities of each of the variables. These are the diagonal elements of $\hat{\Lambda}\hat{\Lambda}'$. The most useful parts of this matrix are the elements above the diagonal. These elements are the differences between the observed sample correlations and those reproduced by the five-factor solution.

If the number of factors selected is adequate, then we would expect the numbers above the diagonal to be close to zero. SPSS counts the number of differences that are greater than 0.05 and provides a message at the bottom of page 8 that states "There are 19 (18.0%) residuals (above diagonal) that are >0.05." An examination of the residuals reveals none that are greater than 0.40 and only one that is greater than 0.25. Thus there seems to be no reason, based on these differences, to increase the number of factors for this analysis. However, remember that the number of factors could still be decreased.

6.9
Rotating Factors

As stated before, when a set of factors is derived, they are not always easy to interpret. If you want to rotate the factors in hopes of finding a set that is easy to interpret, you should at least proceed with some kind of plan or objective in mind. Do not try to interpret underlying factors until you have performed a factor rotation.

Most rotation procedures try to make as many factor loadings as possible near zero and to maximize as many of the others as possible. In addition, since the factors are independent, it would be nice, but not critical, if response variables were not loaded heavily on more than one factor.

T A B L E 6.3

Factor loadings of six variables on two factors

Response variable	Initial solution		Rotated (57°)		Rotated (7°)	
x_1	.90	−.04	.52	.73	.90	.07
x_2	.88	−.15	.61	.66	.89	−.04
x_3	.93	−.10	.59	.73	.94	.01
x_4	.61	.40	.00	.73	.56	.47
x_5	.55	.35	.01	.66	.50	.41
X_6	.75	.41	.06	.85	.69	.50
	f_1	f_2	f_1'	f_2'	f_1''	f_2''

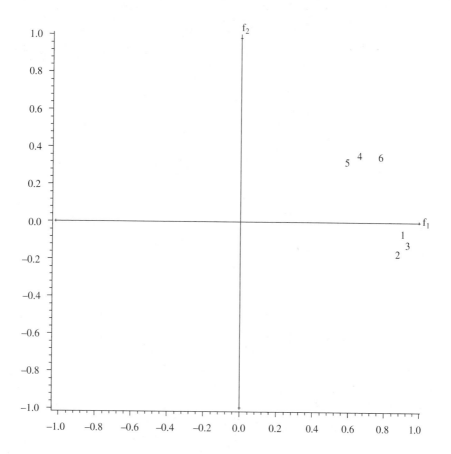

F I G. **6.2**

Initial factor loadings plotted on factor axes

Examples (*m* = 2)

Suppose *m* = 2 and *p* = 6. Consider the initial factor loadings shown in Table 6.3. When *m* = 2 the correlations that each variable has with each factor can be plotted on a two-dimensional axis system (Figure 6.2). This plot is somewhat different than most of the plots you might consider since variables are located in the plot according to the correlations that each variable has with respect to the two factor axes.

When there are two underlying factors, multiplying the original factor loading matrix by an orthogonal matrix is geometrically equivalent to rotating the axes in Figure 6.2 about the origin. As the axes are rotated to new factor axes, the correlations that the variables have with each of the rotated factors will change. To see what these correlations are, you must examine the coordinates of the variables with respect to the new factor axes.

If you rotate the axes 57 degrees in a clockwise direction, the correlations that variables x_4, x_5, and x_6 have with factor 1 will be, in some sense, minimized,

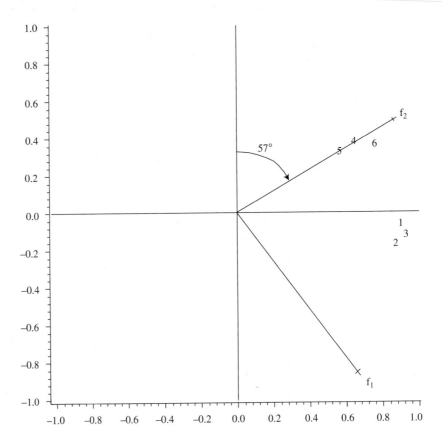

F I G. **6.3**

Factor axes rotated 57 degrees

whereas the correlations that these variables have with the second factor will be, in some sense, maximized. The actual correlations that the variables have with these rotated factor axes are given by the second two columns in Table 6.3 with such a rotation illustrated in Figure 6.3.

If you rotate the original axes 7 degrees in a clockwise direction, the correlations that variables x_1, x_2, and x_3 have with factor 2 will be, in some sense, minimized, whereas the correlations that these variables have with the first factor will be, in some sense, maximized. The actual correlations that the variables would have with these rotated factor axes are given by the last two columns in Table 6.3 with this rotation illustrated in Figure 6.4.

Rotation Methods

Many different orthogonal rotation algorithms have been developed. Orthogonal rotation procedures keep the factors uncorrelated whenever you start with a set of uncorrelated factors. Some orthogonal rotation methods that have

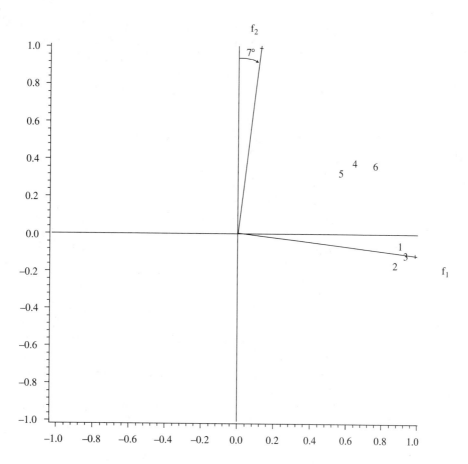

F I G. 6.4

Factor axes rotated 7 degrees

been proposed are Quartimax, Varimax, Transvarimax, Equamax, Ratiomax, and Parsimax. The most popular of these is the Varimax rotation procedure, which is described in the next subsection.

The Varimax Rotation Method

Suppose $\mathbf{B} = \Lambda\mathbf{T}$ where $\mathbf{T}$ is an orthogonal matrix. Kaiser (1958) proposed the sum of the variances of the squared loadings within each column of the factor matrix as a measure of simple structure. His (raw) Varimax criteria is to maximize

$$V^* = \sum_{q=1}^{m} \left(\frac{\left[\sum_{j=1}^{p} b_{jq}^4 - \left(\sum_{j=1}^{p} b_{jq}^2 \right)^2 \Big/ p \right]}{p} \right)$$

Note that the quantity within the larger parentheses in this expression is the variance of the squared loadings within the qth column of B. Since the squared loadings are all between 0 and 1, trying to maximize the variance of the squared loadings within a column is somewhat equivalent to trying to spread out the squared loadings within a column, that is, forcing as many of the loadings as possible toward 0 and forcing the others toward 1. Kaiser then sums the variances of the squared loadings within a column across the columns. The orthogonal matrix **T** that produces a maximum to this sum of column variances results in Kaiser's (raw) Varimax rotation of the factor loading matrix Λ.

The preceding criterion gives equal weight to response variables having both large and very small communalities; because of this, Kaiser suggested that it would be better to divide the factor loadings for each variable by the variable's own communality, and then maximize the sum of the variances of the squared ratios within a column as before. Thus, Kaiser would actually maximize

$$V = \frac{1}{p^2} \sum_{q=1}^{m} \left[p \sum_{j=1}^{p} \frac{b_{jq}^4}{h_j^4} - \left(\sum_{j=1}^{p} \frac{b_{jq}^2}{h_j^2} \right)^2 \right]$$

where h_j^2 is the communality of the jth response variable, $j = 1, 2, \ldots, p$. The orthogonal matrix **T** that maximizes the preceding sum produces the Varimax rotation of the factor loading matrix. This adjustment gives more weight to the variables having the larger communalities and less weight to the variables that have small communalities (i.e., less weight to those variables that have less in common with the other variables).

Note

$$h_j^2 = \lambda_{j1}^2 + \lambda_{j2}^2 + \cdots + \lambda_{jm}^2 = b_{j1}^2 + b_{j2}^2 + \cdots + b_{jm}^2$$

That is, the communalities are not changed because of the rotation and, thus, they remain constant. Every orthogonal rotation has this property. That is, orthogonal rotations of factor loading matrices do not affect the communalities of the response variables. Consequently, orthogonal rotations do not affect the specific variances of the variables.

6.10
Oblique Rotation Methods

Consider once again the plot shown in Figure 6.2. Obviously, it is not possible to rotate the axes so that one axis goes through each cluster of variables while keeping the axes orthogonal (perpendicular) to one another. In such cases, many researchers have suggested using oblique rotations. For the previous example, an oblique rotation would occur if you rotated the first axis 7 degrees

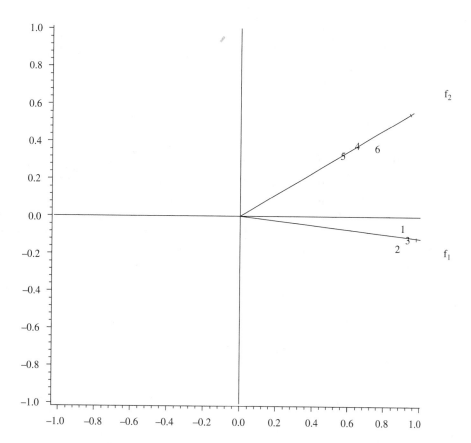

in a clockwise direction and the second factor axis 57 degrees in a clockwise direction. The results of such a rotation are illustrated in Figure 6.5.

In practice, such rotations can be achieved by multiplying Λ by a matrix $\mathbf{Q}$ where $\mathbf{Q}$ is not an orthogonal matrix. Oblique rotations do not produce new factors that remain uncorrelated, which is a contradiction of the initial FA assumptions. From a philosophical point of view, these kinds of rotations do not make much sense to this author.

In the initial development of a FA model, we assume that there exists an uncorrelated set of underlying factors that drive or control the variables being measured. Allowing oblique rotations seems to say that we did not really believe the assumptions of the initial model. What would happen if we eliminated the assumptions of uncorrelated factors from the factor analysis assumptions initially? Although we do not know for sure what would happen, it seems unlikely that it would result in the same thing that happens when we allow oblique rotations at the very end of the analysis.

This book does not recommend oblique rotation methods. I believe that the independence requirement imposed on the factors actually makes them easier to interpret and that obliquely rotated factors are much harder to interpret. Nevertheless, several oblique rotation methods are available in computing packages. Some that exist are Quartimin, Covarimin, Oblimin, Direct Oblimin, Orthoblique, Second Generation Little Jiffy, Oblimax, Promax, and Maxplane.

E X A M P L E 6 . 1 (continued)

The results of a Varimax rotation of the factors in Example 6.1 can be found on page 9 of the SPSS computer output shown in Figure 6.6, under the label ROTATED FACTOR MATRIX. While it is not generally useful to examine the orthogonal matrix that produced the Varimax rotation, it is shown at the bottom of page 9 and continues onto the top of page 10 of Figure 6.6.

Variables that are correlated (absolute correlations that are greater than 0.40) with the five rotated factor axes and their correlations are as follows:

Factor 1: THIGH (0.96), FAT (0.90), CHNUP (−0.69), WEIGHT (0.61), and CHEST (0.49)

Factor 2: HEIGHT (0.89), SHLDR (0.75), WEIGHT (0.62), BREATH (0.60), PELVIC (0.59), and CHEST (0.46)

Factor 3: RECVR (0.95), PULSE (0.57), and SPEED (−0.53)

Factor 4: WEIGHT (0.42) and CHEST (0.66)

Factor 5: REACT (0.78)

The next step in factor analysis is to attempt to give physical interpretations to the rotated factors. Interpretation of the rotated factors requires researchers to possess knowledge, experience, discretion, and wisdom, while remaining objective, and while suppressing, as much as possible, their own biases and prejudices. Researchers must carefully consider the population being sampled when making interpretations and should always keep in mind that the underlying factors are measuring unique and independent characteristics of the population that was sampled.

An examination of the correlations that the original variables have with rotated factors 1, 2, and 4 reveals that each of these three factors has something to do with body size. But how do they differ? Keeping in mind that the three measures of body size are independent (and, hence, must be independent in our interpretations), it seems that the first is a measure of body mass or lack of it. This factor is likely a measure of the obesity of police department applicants. The primary variables used in this interpretation are THIGH, FAT, and CHNUP. Obviously THIGH and FAT should be highly correlated with each other, and it is reasonable that the ability to do chin-ups is lessened in people who tend to be obese.

EX 6.1 - FITNESS AND SIZE MEASUREMENTS ON POLICE APPLICANTS Page 9

- - - - - - - - - - F A C T O R A N A L Y S I S - - - - - - - - - - -

VARIMAX ROTATION 1 FOR EXTRACTION 2 IN ANALYSIS 1 - KAISER NORMALIZATION.

 VARIMAX CONVERGED IN   5 ITERATIONS.

ROTATED FACTOR MATRIX: $\hat{\Lambda}^* = B$

Trivial factor

	FACTOR 1	FACTOR 2	FACTOR 3	FACTOR 4	FACTOR 5
THIGH	.95650 $\hat{\lambda}_{11}$	.05936	.10374	-.11693	.04300
FAT	.89518 $\hat{\lambda}_{12}$	.24680	.01455	.27120	.07951
CHNUP	-.69005	-.17518	-.02580	-.10835	-.12433
ENDUR	-.35390	-.19812	-.02695	-.08598	.03843
HEIGHT	.17586	.88717	-.16275	-.10250	.18990
SHLDR	.19251	.74751	-.14639	.09309	-.24744
WEIGHT	.61466	.61825	-.19005	.41767	-.04198
BREATH	.16627	.59744	.08179	.00840	.14596
PELVIC	.23807	.58660	-.27274	.18855	-.06658
RECVR	.10226	.05829	.95125	-.12144	-.25596
PULSE	-.07857	-.11544	.57435	-.08284	.14746
SPEED	-.19194	.16477	-.53039	-.33280	-.14708
DIAST	.03783	-.16505	.22761	.17002	.14163
CHEST	.48837	.46242	-.11724	.66136	-.12393
REACT	.09227	.07459	.05873	-.00754	.78171

factor loadings are the correlations between the original variables and the factors

FACTOR TRANSFORMATION MATRIX: $= T$

	FACTOR 1	FACTOR 2	FACTOR 3	FACTOR 4	FACTOR 5
FACTOR 1	.10319	.05389	.95192	-.12361	-.25500
FACTOR 2	.77493	.53356	-.07354	.33062	-.00844
FACTOR 3	-.61037	.71085	.00598	.27941	-.20986
FACTOR 4	.03120	.45505	-.03250	-.80511	.37775
FACTOR 5	-.12374	-.00517	.29556	.38619	.86497

T may be ignored.

F I G. 6.6
Pages 9 and 10 of computer printout, Example 6.1

EX 6.1 - FITNESS AND SIZE MEASUREMENTS ON POLICE APPLICANTS Page 10

- - - - - - - - - - - F A C T O R   A N A L Y S I S - - - - - - - - - - -

FACTOR SCORE COEFFICIENT MATRIX:

	FACTOR 1	FACTOR 2	FACTOR 3	FACTOR 4	FACTOR 5
REACT	-.07040	.04313	.23057	.17696	.75694
HEIGHT	-.13228	.66439	.02623	-.52156	.44212
WEIGHT	.06569	.49486	-.07602	.46274	-.35922
SHLDR	-.00606	.14488	-.10742	-.17232	-.28896
PELVIC	-.00723	.06339	-.02861	-.04385	-.07340
CHEST	-.08156	-.09737	.09427	.80393	-.08720
THIGH	.73563	-.23701	-.25523	-.86060	-.30247
PULSE	-.02052	-.00451	.05789	.05979	.17469
DIAST	-.01046	-.02288	.03854	.07157	.09943
CHNUP	-.03656	.01370	-.01734	-.00122	-.07555
BREATH	-.02422	.07441	.02520	-.02353	.11083
RECVR	-.07378	.30414	.93658	-.01940	-.37332
SPEED	.02574	.06125	-.09999	-.19820	-.25057
ENDUR	-.01017	-.00247	.00632	.00916	.01487
FAT	.35182	-.24167	.13475	.26517	.46131

   5 BARTLETT FACTOR SCORES WILL BE SAVED WITH ROOTNAME: BARTFSC

FACTOR SCORE COEFFICIENT MATRIX:

	FACTOR 1	FACTOR 2	FACTOR 3	FACTOR 4	FACTOR 5
REACT	-.04249	.03597	.15938	.10151	.53752
HEIGHT	-.11597	.59017	.01220	-.44864	.34528
WEIGHT	.07784	.48376	-.06222	.43741	-.29284
SHLDR	-.01481	.13172	-.07565	-.12231	-.20094
PELVIC	-.00865	.05807	-.02050	-.03024	-.05141
CHEST	-.05987	-.06104	.06628	.68159	-.11593
THIGH	.66875	-.23286	-.18406	-.69801	-.14179
PULSE	-.01367	-.00454	.04051	.03877	.12277
DIAST	-.00600	-.02004	.02702	.05317	.06774

F I G. 6.6
Pages 9 and 10 of computer printout, Example 6.1

However, the second factor seems to be a measure of frame size or skeletal structure. The primary variables used in this interpretation are HEIGHT, SHLDR, and PELVIC.

Although both of the first two factors are measures of body size, they are getting at different measures of body size. It is reasonable that the two are independent measures, because there are tall people who are obese and tall people who are not obese, as well as short people who are obese and short people who are not obese.

The fourth factor also seems to measure body size. But how does it differ from the first two? This is perhaps the hardest factor to interpret. The variable most highly correlated with the factor is CHEST, and WEIGHT is also correlated with this factor somewhat. The easy way out would be to call this factor the chest size factor, but this does not seem to be a characteristic that is clearly independent of obesity and skeletal structure. One interpretation that has been put forth for this factor is that it may be a measure of upper body strength and may be related to whether an individual applicant lifts weights or not. One can argue that this interpretation is independent of the other two measures of body size.

Note that both WEIGHT and CHEST are correlated with all three of these factors. This is not unreasonable; we would expect these two variables to be correlated with obesity, skeletal structure, and upper body strength even though obesity, skeletal structure, and upper body strength are not themselves correlated.

Next consider the third factor. The variables correlated with this factor are RECVR, PULSE, and SPEED. This factor might be interpreted as a measure of aerobic fitness or cardiovascular fitness. The primary variables used in this interpretation are RECVR and PULSE. Is it reasonable to conclude that this factor is independent of the other three? We could argue that it is, since this factor may be related to whether an individual exercises regularly or not. If obese people exercise, they are likely to be cardiovascularly fit, and if nonobese people do not exercise, they are not likely to be cardiovascularly fit. In addition, cardiovascular fitness may have little to do with upper body strength or skeletal structure.

The fifth and last factor has only one variable that is highly correlated with it, namely, REACT. This factor is called a trivial factor. It could be interpreted as measuring an individual's reaction time. As indicated earlier, we generally want to reduce the number of common factors and not include trivial factors in the final analysis. This is done later in this chapter for this example.

Two of the variables do not correlate with any of the five factors. These are the variables ENDUR and DIAST. Recall that their communalities with the underlying factors were low. These two variables are measuring characteristics that are uncorrelated with those characteristics measured by the five common factors. Does this seem reasonable? While it seems plausible that DIAST is not correlated with obesity, skeletal structure, cardiovascular fitness, upper body strength, and reaction time, it seems a bit unusual that ENDUR is not correlated with cardiovascular fitness.

The reason for this could be that cardiovascular fitness is likely determined by whether an individual exercises or not, and there are various kinds of exercise that can improve cardiovascular fitness such as swimming, tennis, handball, etc. ENDUR measures the length of time an individual is willing to run on a treadmill. An individual can be very fit cardiovascularly, but if he or she did not get fit through jogging or running, then he or she may not be able to run for long periods of time on a treadmill. ENDUR might also be related to how well an individual reacts to pain. People who do not run regularly often quit running because their muscles get sore and tighten up even though they may not be gasping for air.

To summarize, the variables in this data set seem to be determined by seven underlying characteristics. These underlying characteristics are obesity (factor 1), skeletal structure (factor 2), cardiovascular fitness (factor 3), upper body strength (factor 4), reaction time (factor 5), endurance, and diastolic blood pressure.

A client once performed a factor analysis on data collected from recently divorced individuals. Upon examining the rotated factor matrix, the researcher wanted to label one of the factors as a measure of "connectedness" and another one as a measure of "disconnectedness." Such labeling does not make sense because these two variables would not be independent. Each factor has both high and low values. The "connectedness" factor would have values that are very high (indicating an individual who is very connected) and values that are very low (indicating an individual who is very disconnected).

One final note: Factor analysis does not distinguish variables as to their relative importance with respect to one another. It simply identifies underlying variables that are measuring independent and/or uncorrelated characteristics of the population being sampled.

6.11
Factor Scores

Factor analysis is frequently used to reduce a large number of responses to a smaller set of uncorrelated variables. If this new set of variables is going to be used in subsequent statistical analyses, it is necessary to assign a score or value for each of the new variables for each experimental unit in the data set.

Evaluation of factor scores is not simple because the model for each individual is $\mathbf{x} = \Lambda\mathbf{f} + \boldsymbol{\eta}$ with $\boldsymbol{\eta}$ unknown and Λ estimated. Thus, for a given observation vector $\mathbf{x}$, it is not actually possible to determine $\mathbf{f}$ explicitly, although things should be improved if we consider only nontrivial factors. Two formal methods have been proposed for estimating $\mathbf{f}$ for a given individual. One of these is known as *Bartlett's method* or the *weighted least-squares method*, and the other is known as *Thompson's method* or the *regression method*.

There are also two ad hoc methods that can be recommended, and these ad hoc methods are preferred to the formal methods by many researchers.

Bartlett's Method or the Weighted Least-Squares Method

After solving the FA model, we have $\mathbf{z} = \Lambda\mathbf{f} + \boldsymbol{\eta}$ where $\boldsymbol{\eta} \sim (\mathbf{0}, \boldsymbol{\Psi})$. Bartlett suggested that the next step be to find the $\mathbf{f}$ that minimizes

$$(\mathbf{z}_r - \hat{\Lambda}\mathbf{f})\hat{\boldsymbol{\Psi}}^{-1}(\mathbf{z}_r - \hat{\Lambda}\mathbf{f})$$

where $\mathbf{z}_r$ is the standardized data vector for the rth individual. For a given $\mathbf{z}_r$, the preceding expression is minimized when

$$\mathbf{f}_r = (\hat{\Lambda}'\hat{\boldsymbol{\Psi}}^{-1}\hat{\Lambda})^{-1}\hat{\Lambda}\hat{\boldsymbol{\Psi}}^{-1}\mathbf{z}_r$$

Then $\mathbf{f}_r$ is taken as the vector of estimated factor scores for the rth individual, $r = 1, 2, \ldots, N$.

Thompson's Method or the Regression Method

Thompson noted that for normally distributed data the joint distribution of $\mathbf{x}$ (standardized) and $\mathbf{f}$ is

$$\begin{bmatrix} \mathbf{z} \\ \mathbf{f} \end{bmatrix} \sim N\left(\begin{bmatrix} \mathbf{0} \\ \mathbf{0} \end{bmatrix}, \begin{bmatrix} \mathbf{P} & \Lambda \\ \Lambda' & \mathbf{I} \end{bmatrix}\right)$$

This implies that the conditional expectation of $\mathbf{f}$ given that $\mathbf{z} = \mathbf{z}^*$ is

$$E[\mathbf{f}|\mathbf{z} = \mathbf{z}^*] = \Lambda'\mathbf{P}^{-1}\mathbf{z}^*$$

Therefore, Thompson's method estimates the vector of factor scores for the rth individual as $\mathbf{f}_r = \hat{\Lambda}'\mathbf{R}^{-1}\mathbf{z}_r$.

Note Some software packages may use

$$\mathbf{f}_r = \hat{\Lambda}'(\hat{\Lambda}\hat{\Lambda}' + \hat{\boldsymbol{\Psi}})^{-1}\mathbf{z}_r$$

instead of the formula just given. In this formula $\mathbf{R}$ in the previous formula is replaced by its factor analysis reproduction $\hat{\Lambda}\hat{\Lambda}' + \hat{\boldsymbol{\Psi}}$.

Ad Hoc Methods

Two other methods are used for scoring rotated factors. Even though both of these are ad hoc procedures, they hold some advantages over the formal procedures described. One advantage they hold over the formal procedures suggested by Bartlett and Thompson is that it is easy to explain what each means and it is very clear how each can be interpreted. These methods assume that only orthogonal rotations have been used.

When scoring a selected factor, all you really need is some variable that is (1) highly correlated with the selected factor and (2) uncorrelated with all the other factors. Any procedure that produces factor scores having these two properties should be acceptable.

One way to score factors that would generally have the two properties described in the previous paragraph would be to take an (adjusted) average of all of the variables that have high correlations with the factor. The word *adjusted* is used here to mean that variables that have high positive correlations with the factor would be added, whereas those that have high negative correlations would be subtracted. For example, if z_1, z_2, and z_3 had correlations of 0.96, 0.81, and -0.93, respectively, with the first factor and if all other z_i's had low correlations with the first factor, then scores for the first factor would be computed by the formula $f_1 = (z_1 + z_2 - z_3)/3$.

A second ad hoc way to score factors is to take the variable that has the highest correlation with the factor and use its value for the score of the factor. For the example in the previous paragraph, the score for f_1 would be given by z_1 only.

E X A M P L E   6 . 1 (continued)

In the SPSS analysis of Example 6.1, factor scores were computed and saved for subsequent analyses that one might want to do by including the following options in the SPSS command stream. The first computes factor scores by Bartlett's method and the second computes factor scores by the regression method.

```
/SAVE BART (ALL BARTFSC)

/SAVE REG (ALL REGFSC)
```

The resulting factor scores as well as the original data were printed with the aid of the following commands:

```
LIST VARIABLES = ID REACT TO THIGH FAT

LIST VARIABLES = ID PULSE TO ENDUR

LIST VARIABLES = ID BARTFSC1 TO BARTFSC5

LIST VARIABLES = ID REGFSC1 TO REGFSC5
```

The results of these commands can be found on pages 16–22 of the complete computer printout available by executing the file EX6_1.SSX on the enclosed disk; of these results, only pages 16, 18, 20, and 22 are shown in Figure 6.7, along with pages 10–12, which provide different results as discussed next.

EX 6.1 - FITNESS AND SIZE MEASUREMENTS ON POLICE APPLICANTS Page 10

- - - - - - - - - - - F A C T O R   A N A L Y S I S - - - - - - - - - - -

         FACTOR 5

FACTOR 1	-.25500
FACTOR 2	-.00844
FACTOR 3	-.20986
FACTOR 4	.37775
FACTOR 5	.86497

FACTOR SCORE COEFFICIENT MATRIX: For Bartlett's Method

	FACTOR 1	FACTOR 2	FACTOR 3	FACTOR 4	FACTOR 5
REACT	-.07040	.04313	.23057	.17696	.75694
HEIGHT	-.13228	.66439	.02623	-.52156	.44212
WEIGHT	.06569	.49486	-.07602	.46274	-.35922
SHLDR	-.00606	.14488	-.10742	-.17232	-.28896
PELVIC	-.00723	.06339	-.02861	-.04385	-.07340
CHEST	-.08156	-.09737	.09427	.80393	-.08720
THIGH	.73563	-.23701	-.25523	-.86060	-.30247
PULSE	-.02052	-.00451	.05789	.05979	.17469
DIAST	-.01046	-.02288	.03854	.07157	.09943
CHNUP	-.03656	.01370	-.01734	-.00122	-.07555
BREATH	-.02422	.07441	.02520	-.02353	.11083
RECVR	-.07378	.30414	.93658	-.01940	-.37332
SPEED	.02574	.06125	-.09999	-.19820	-.25057
ENDUR	-.01017	-.00247	.00632	.00916	.01487
FAT	.35182	-.24167	.13475	.26517	.46131

  5 BARTLETT FACTOR SCORES WILL BE SAVED WITH ROOTNAME: BARTFSC

FACTOR SCORE COEFFICIENT MATRIX: For the Regression Method

	FACTOR 1	FACTOR 2	FACTOR 3	FACTOR 4	FACTOR 5
REACT	-.04249	.03597	.15938	.10151	.53752
HEIGHT	-.11597	.59017	.01220	-.44864	.34528
WEIGHT	.07784	.48376	-.06222	.43741	-.29284
SHLDR	-.01481	.13172	-.07565	-.12231	-.20094
PELVIC	-.00865	.05807	-.02050	-.03024	-.05141
CHEST	-.05987	-.06104	.06628	.68159	-.11593
THIGH	.66875	-.23286	-.18406	-.69801	-.14179
PULSE	-.01367	-.00454	.04051	.03877	.12277
DIAST	-.00600	-.02004	.02702	.05317	.06774

F I G. 6.7

Pages 10–12, 16, 18, 20, and 22 of computer printout, Example 6.1

EX 6.1 - FITNESS AND SIZE MEASUREMENTS ON POLICE APPLICANTS   Page   11

- - - - - - - - - - -F A C T O R   A N A L Y S I S- - - - - - - - - - -

	FACTOR 1	FACTOR 2	FACTOR 3	FACTOR 4	FACTOR 5
CHNUP	-.03673	.01281	-.01151	.00319	-.05615
BREATH	-.01948	.06645	.01693	-.02484	.08080
RECVR	-.06131	.28812	.93957	-.02299	-.35485
SPEED	.01409	.05316	-.07020	-.14902	-.16960
ENDUR	-.00918	-.00236	.00454	.00642	.00997
FAT	.35146	-.21267	.08958	.19693	.33177

COVARIANCE MATRIX FOR ESTIMATED REGRESSION FACTOR SCORES:

	FACTOR 1	FACTOR 2	FACTOR 3	FACTOR 4	FACTOR 5
FACTOR 1	.96089				
FACTOR 2	.02261	.93064			
FACTOR 3	.01221	.00279	.97228		
FACTOR 4	.02670	.03735	-.03820	.85373	
FACTOR 5	.02198	-.01299	-.07570	-.05348	.74847

  5 REGRESSION FACTOR SCORES WILL BE SAVED WITH ROOTNAME: REGFSC

OBLIMIN ROTATION 2 FOR EXTRACTION 2 IN ANALYSIS 1 - KAISER NORMALIZATION.

   OBLIMIN CONVERGED IN   9 ITERATIONS.

PATTERN MATRIX:

	FACTOR 1	FACTOR 2	FACTOR 3	FACTOR 4	FACTOR 5
RECVR	1.02323	.09099	.11785	-.02102	-.26129
PULSE	.53904	-.06616	-.04571	-.05612	.14865
THIGH	.08483	1.08397	-.18243	.17340	.01963
FAT	-.08523	.82601	.10492	-.20804	.02065
CHNUP	.03728	-.67728	-.05044	.06357	-.09143
ENDUR	-.02648	-.31062	-.14768	.04641	.05589
HEIGHT	-.00851	.02955	.90999	.23491	.23999
SHLDR	.01041	.01228	.76599	.04861	-.23479
BREATH	.15230	.03061	.63962	.04016	.16525

F I G. 6.7
Pages 10–12, 16, 18, 20, and 22 of computer printout, Example 6.1

EX 6.1 - FITNESS AND SIZE MEASUREMENTS ON POLICE APPLICANTS   Page   12

- - - - - - - - - - F A C T O R   A N A L Y S I S - - - - - - - - -

	FACTOR 1	FACTOR 2	FACTOR 3	FACTOR 4	FACTOR 5
PELVIC	-.21093	.06909	.58863	-.05890	-.07032
WEIGHT	-.21776	.38464	.58330	-.29087	-.08935
CHEST	-.22444	.18445	.49273	-.58283	-.20561
SPEED	-.34890	-.08739	.09572	.47557	-.08392
DIAST	.11046	-.00122	-.12499	-.25361	.10986
REACT	-.08881	.07006	.10775	-.05855	.79185

STRUCTURE MATRIX: Correlations between variables and obliquely rotated factors

	FACTOR 1	FACTOR 2	FACTOR 3	FACTOR 4	FACTOR 5
RECVR	.94707	.13704	-.05710	-.19984	-.07649
PULSE	.58826	-.07292	-.21837	-.12819	.26159
FAT	-.08353	.94611	.49004	-.49094	.00848
THIGH	.09130	.94232	.26119	-.21737	.04355
CHNUP	.03007	-.72215	-.34627	.30515	-.08357
ENDUR	.01606	-.39098	-.28459	.16101	.06389
HEIGHT	-.21788	.34329	.88959	.17584	.11757
WEIGHT	-.33305	.74394	.82319	-.40537	-.17483
SHLDR	-.22586	.32863	.79369	.03386	-.32308
PELVIC	-.35679	.34886	.67921	-.06489	-.17286
CHEST	-.29152	.60729	.67032	-.61367	-.25990
BREATH	.02422	.29323	.59643	-.02972	.11914
SPEED	-.46239	-.20907	.13406	.56488	-.19804
DIAST	.20177	.03241	-.15554	-.27475	.16455
REACT	.04590	.13892	.07206	-.13330	.76710

FACTOR CORRELATION MATRIX: Correlations between factors

		FACTOR 1	FACTOR 2	FACTOR 3	FACTOR 4	FACTOR 5
FACTOR	1	1.00000				
FACTOR	2	-.01236	1.00000			
FACTOR	3	-.23930	.43560	1.00000		
FACTOR	4	-.15888	-.35255	-.03483	1.00000	
FACTOR	5	.19204	.00022	-.11304	-.07629	1.00000

FIG. 6.7

Pages 10–12, 16, 18, 20, and 22 of computer printout, Example 6.1

EX 6.1 - FITNESS AND SIZE MEASUREMENTS ON POLICE APPLICANTS   Page   16

ID	REACT	HEIGHT	WEIGHT	SHLDR	PELVIC	CHEST	THIGH	FAT
1	.310	179.6	74.20	41.7	27.3	82.4	19.0	11.91
2	.345	175.6	62.04	37.5	29.1	84.1	5.5	3.13
3	.293	166.2	72.96	39.4	26.8	88.1	22.0	16.89
4	.254	173.8	85.92	41.2	27.6	97.6	19.5	19.59
5	.384	184.8	65.88	39.8	26.1	88.2	14.5	7.74
6	.406	189.1	102.26	43.3	30.1	101.2	22.0	30.42
7	.344	191.5	84.04	42.8	28.4	91.0	18.0	13.70
8	.321	180.2	68.34	41.6	27.3	90.4	5.5	3.04
9	.425	183.8	95.14	42.3	30.1	100.2	13.5	20.26
10	.385	163.1	54.28	37.2	24.2	80.5	7.0	3.04
11	.317	169.6	75.92	39.4	27.2	92.0	16.5	12.83
12	.353	171.6	71.70	39.1	27.0	86.2	25.5	15.95
13	.413	180.0	80.68	40.8	28.3	87.4	17.5	11.86
14	.392	174.6	70.40	39.8	25.9	83.9	16.5	9.93
15	.312	181.8	91.40	40.6	29.5	95.1	32.0	32.63
16	.342	167.4	65.74	39.7	26.4	86.0	13.0	6.64
17	.293	173.0	79.28	41.2	26.9	96.1	11.5	11.57
18	.317	179.8	92.06	40.0	29.8	100.9	15.0	24.21
19	.333	176.8	87.96	41.2	28.4	100.8	20.5	22.39
20	.317	179.3	77.66	41.4	31.6	90.1	9.5	6.29
21	.427	193.5	98.44	41.6	29.2	95.7	21.0	23.63
22	.266	178.8	65.42	39.3	27.1	83.0	16.5	10.53
23	.311	179.6	97.04	43.8	30.1	100.8	22.0	20.62
24	.284	172.6	81.72	40.9	27.3	91.5	22.0	18.39
25	.259	171.5	69.60	40.4	27.8	87.7	15.5	11.14
26	.317	168.9	63.66	39.8	26.7	83.9	6.0	5.16
27	.263	183.1	87.24	43.2	28.3	95.7	11.0	9.60
28	.336	163.6	64.86	37.5	26.6	84.0	15.5	11.93
29	.267	184.3	84.68	40.3	29.0	93.2	8.5	8.55
30	.271	181.0	73.78	42.8	29.7	90.3	8.5	4.94
31	.264	180.2	75.84	41.4	28.7	88.1	13.5	10.62
32	.357	184.1	70.48	42.0	28.9	81.3	14.0	8.46
33	.259	178.9	86.90	42.5	28.7	95.0	16.0	13.47
34	.221	170.0	76.68	39.7	27.7	93.6	15.0	12.81
35	.333	180.6	77.32	42.1	27.3	89.5	16.0	13.34
36	.359	179.0	79.90	40.8	28.2	90.3	26.5	24.57
37	.314	186.6	100.36	42.5	31.5	100.3	27.0	28.35
38	.295	181.4	91.66	41.9	28.9	96.6	25.5	26.12
39	.296	176.5	79.00	40.7	29.1	86.5	20.5	15.21
40	.308	174.0	69.10	40.9	27.0	88.1	18.0	12.51
41	.327	178.2	87.78	42.9	27.2	100.3	16.5	20.50
42	.303	177.1	70.18	39.4	27.6	85.5	16.0	10.67
43	.297	180.0	67.66	40.9	28.7	86.1	15.0	10.76
44	.244	176.8	86.12	41.3	28.2	92.7	12.5	14.55
45	.282	176.3	65.00	39.0	26.0	83.3	7.0	5.27
46	.285	192.4	99.14	43.7	28.7	96.1	20.5	17.94
47	.299	175.2	75.70	39.4	27.3	90.8	19.0	12.64
48	.280	175.9	78.62	43.4	29.3	90.7	18.0	10.81
49	.268	174.6	64.88	42.3	29.2	82.6	3.5	2.01
50	.362	179.0	71.00	41.2	27.3	85.6	16.0	10.00

F I G. 6.7

Pages 10–12, 16, 18, 20, and 22 of computer printout, Example 6.1

EX 6.1 - FITNESS AND SIZE MEASUREMENTS ON POLICE APPLICANTS Page 18

ID	PULSE	DIAST	CHNUP	BREATH	RECVR	SPEED	ENDUR
1	64	64	2	158	108	5.5	4.0
2	88	78	20	166	108	5.5	4.0
3	100	88	7	167	116	5.5	4.0
4	64	62	4	220	120	5.5	4.0
5	80	68	9	210	120	5.5	5.0
6	60	68	4	188	91	6.0	4.0
7	64	48	1	272	110	6.0	3.0
8	74	64	14	193	117	5.5	4.0
9	80	78	4	199	105	5.5	4.0
10	84	78	13	157	113	6.0	4.0
11	65	78	6	180	110	5.0	5.0
12	68	72	0	193	105	5.5	4.0
13	73	88	4	218	109	5.0	4.2
14	104	78	6	190	129	5.0	4.0
15	92	88	1	206	139	5.0	3.5
16	80	86	6	181	120	5.5	4.0
17	72	68	6	184	111	5.5	3.9
18	60	78	0	205	92	5.0	4.0
19	76	90	1	228	147	4.0	3.5
20	58	86	15	198	98	5.5	4.1
21	54	74	0	254	110	5.5	3.8
22	88	72	7	206	121	5.5	4.0
23	100	74	3	194	124	5.0	4.0
24	74	76	4	201	113	5.5	5.1
25	70	72	10	175	110	5.5	3.0
26	68	70	7	179	119	5.5	5.0
27	88	74	7	245	115	5.5	4.0
28	64	64	6	146	115	5.0	4.4
29	64	76	2	213	109	5.5	5.0
30	56	88	11	181	109	6.0	5.0
31	76	76	9	192	144	5.5	3.6
32	84	72	5	231	123	5.5	4.5
33	54	68	12	186	118	6.0	4.0
34	50	72	4	178	108	5.5	4.5
35	88	72	11	200	119	5.5	4.6
36	80	80	3	201	124	5.5	3.7
37	62	76	2	208	120	5.5	4.1
38	68	78	2	211	125	6.0	3.0
39	60	66	5	210	117	5.5	4.2
40	92	74	5	161	140	5.0	5.5
41	72	72	4	189	115	5.5	3.5
42	72	74	14	201	110	6.0	4.8
43	76	76	5	177	110	5.5	4.5
44	76	68	7	181	110	5.5	4.0
45	88	72	12	167	127	5.5	5.0
46	64	68	4	174	105	6.0	4.0
47	56	76	7	174	111	5.5	4.5
48	64	72	7	170	117	5.5	3.7
49	72	80	11	199	113	6.0	4.5
50	68	90	5	150	108	5.5	5.0

F I G. 6.7

Pages 10–12, 16, 18, 20, and 22 of computer printout, Example 6.1

EX 6.1 - FITNESS AND SIZE MEASUREMENTS ON POLICE APPLICANTS   Page   20

ID	BARTFSC1	BARTFSC2	BARTFSC3	BARTFSC4	BARTFSC5
1	.47342	-.18450	-1.06249	-2.06021	-.18981
2	-1.84661	-.56138	-.04028	.32850	1.46376
3	1.09792	-2.00749	-.03171	.01399	-.30942
4	.80526	-.33651	.00639	.86046	-1.69722
5	-.82872	.50806	.85380	-.96602	1.85633
6	1.38100	.99732	-1.94436	.94747	1.68352
7	.04395	1.95695	-.74483	-1.79225	.48028
8	-1.95794	.68966	.35235	.36393	-.00012
9	-.25959	1.06468	-.14749	2.36735	2.15357
10	-1.37553	-2.16609	.43243	.32794	1.30376
11	.15064	-1.39660	-.25560	1.13619	.18823
12	1.47731	-1.75442	-.98014	-1.28598	.84007
13	-.06548	.22009	-.00016	-.03024	2.32072
14	-.29121	-.32858	1.74221	-.27720	1.70852
15	2.56092	.20154	1.92902	-.34560	.20234
16	-.58194	-1.26628	.65188	.27559	.21371
17	-.54008	-.46241	-.32271	1.59289	-.69061
18	.43997	-.41010	-1.49618	2.64025	1.49157
19	.53623	.38661	3.30188	2.42429	.42647
20	-1.15233	.35702	-1.47344	.34825	.10749
21	.69732	2.04339	-.02812	.09724	2.53582
22	-.08398	-.29871	.29984	-1.74990	-.07094
23	.85439	.83284	.63614	1.38616	-.72633
24	1.10151	-.90241	-.50516	-.06364	-.72716
25	.05465	-1.14447	-.80714	-.43661	-1.09948
26	-1.45311	-.83753	.56343	.43713	-.08812
27	-.94157	1.45779	-.17137	.76870	-.86626
28	.16365	-2.37357	.23223	.36758	.40800
29	-1.16642	1.13419	-.50369	.82265	-.18362
30	-1.32792	.75340	-.90414	-.33608	-1.27283
31	-.62614	1.16640	2.12709	-.66040	-1.70816
32	-.71207	1.16575	.73117	-1.95982	1.01936
33	.03527	.70467	-.45979	.03620	-2.40435
34	.18773	-1.28569	-1.06125	.76467	-1.88839
35	-.20068	.47665	.34725	-.40948	.30033
36	1.65224	-.31856	.74131	-.92807	.82737
37	1.82136	1.19568	.02625	.15670	-.53117
38	1.68745	.53643	.33697	-.33863	-1.09116
39	.72639	-.14168	-.29160	-1.22558	-.75142
40	.03363	-.36164	2.14204	-.35581	-.48754
41	.30894	.16994	.06620	1.66254	-.19250
42	-.10508	-.43478	-.66015	-1.19794	-.08857
43	-.24464	-.23805	-.55790	-1.19024	.27042
44	-.20864	.11592	-.73624	.87594	-1.26108
45	-1.51723	-.02862	1.17265	-.21654	-.15439
46	.68603	1.94677	-1.55835	-.79835	-1.01111
47	.42538	-.82604	-.56637	-.23497	-.42228
48	.19494	.07014	-.47130	-.69336	-1.84024
49	-1.93132	.26135	-.42583	-.60409	-1.12023
50	-.17924	-.34714	-.48726	-.84180	1.07288

← Factor scores computed by Bartlett's Method

F I G. 6.7

Pages 10–12, 16, 18, 20, and 22 of computer printout, Example 6.1

EX 6.1 - FITNESS AND SIZE MEASUREMENTS ON POLICE APPLICANTS   Page   22

ID	REGFSC1	REGFSC2	REGFSC3	REGFSC4	REGFSC5
1	.37857	-.23844	-.93471	-1.70238	.06135
2	-1.74662	-.57106	-.18663	.13343	1.04776
3	1.00276	-1.83898	-.00014	-.01595	-.17973
4	.75189	-.24077	.11072	.83406	-1.29475
5	-.75937	.39628	.71781	-.95977	1.35162
6	1.38811	.96747	-2.03446	.86725	1.37399
7	.04009	1.74697	-.68609	-1.45308	.48726
8	-1.85174	.61213	.30671	.27072	-.09822
9	-.11660	1.04500	-.39707	1.94437	1.47691
10	-1.32802	-2.05044	.28639	.07610	.92344
11	.14452	-1.25706	-.30822	.92156	.12092
12	1.35202	-1.66101	-.95430	-1.13145	.82700
13	-.00774	.17207	-.17487	-.14345	1.73432
14	-.23582	-.34006	1.57070	-.41462	1.15957
15	2.48408	.23532	1.90526	-.30365	.07758
16	-.56780	-1.18229	.59647	.13612	.09952
17	-.50600	-.37500	-.33021	1.37747	-.58353
18	.49852	-.29665	-1.66425	2.22788	1.10346
19	.63841	.46613	3.09309	1.94952	-.05365
20	-1.10551	.31370	-1.46711	.33041	.14340
21	.77425	1.88806	-.20881	.04340	1.88371
22	-.13207	-.34348	.36189	-1.51500	.01982
23	.86861	.85737	.63330	1.25187	-.65796
24	1.01416	-.80926	-.42275	-.00044	-.46667
25	-.01905	-1.06813	-.68738	-.32440	-.72241
26	-1.39860	-.79326	.51771	.28631	-.15305
27	-.87239	1.37487	-.13783	.73844	-.71614
28	.12520	-2.19617	.17625	.19885	.30256
29	-1.08336	1.06085	-.51833	.74259	-.18367
30	-1.30693	.67258	-.78400	-.19164	-.90523
31	-.60460	1.07466	2.21735	-.52989	-1.43286
32	-.67886	.98440	.70316	-1.73109	.78163
33	-.00768	.68789	-.26401	.20431	-1.77509
34	.11726	-1.14215	-.91938	.75136	-1.35313
35	-.18215	.42083	.32941	-.36647	.20979
36	1.58287	-.30244	.71286	-.83266	.65324
37	1.76998	1.16676	.08533	.25430	-.38341
38	1.60466	.53985	.44527	-.17852	-.79397
39	.64197	-.15225	-.17135	-.98089	-.45698
40	.03007	-.33678	2.13257	-.37213	-.50260
41	.34167	.22991	.01968	1.44172	-.23342
42	-.15280	-.45243	-.59188	-1.01181	.05108
43	-.27310	-.27659	-.52109	-1.02472	.30600
44	-.21113	.15021	-.65257	.84201	-.94135
45	-1.45340	-.06376	1.14151	-.26299	-.22574
46	.64065	1.80624	-1.39431	-.47695	-.60632
47	.36759	-.76400	-.50684	-.17587	-.24055
48	.12418	.06638	-.28986	-.46770	-1.30123
49	-1.89583	.19035	-.32899	-.48136	-.81977
50	-.18492	-.37385	-.52598	-.77519	.88549

← Factor scores computed by the Regression Method

F I G. 6.7

Pages 10–12, 16, 18, 20, and 22 of computer printout, Example 6.1

The way in which factor scores are computed by Bartlett's method is shown by the FACTOR SCORE COEFFICIENT MATRIX printed in the middle of page 10 of Figure 6.7. To illustrate, the first factor score (obesity) for an individual would be computed by the following formula:

$$\text{Obesity} = -0.070 \cdot Z_{\text{REACT}} - 0.132 \cdot Z_{\text{HEIGHT}} + 0.066 \cdot Z_{\text{WEIGHT}}$$
$$+ \cdots + 0.352 \cdot Z_{\text{FAT}}$$

where $Z_{\text{REACT}}, Z_{\text{HEIGHT}}, Z_{\text{WEIGHT}}, \ldots, Z_{\text{FAT}}$ are the Z scores for the individual being evaluated. The values of the factor scores computed by Bartlett's method are shown on page 20 of Figure 6.7.

The way in which factor scores are computed by the regression method is shown by the FACTOR SCORE COEFFICIENT MATRIX printed at the bottom of page 10 and at the top of page 11. The values of the factor scores computed by the regression method are shown on page 22 of Figure 6.7.

In the middle of page 11, you will find the sample variance–covariance matrix for the factor scores computed by the regression method. Note that while the covariances and, hence, correlations between the factor scores are close to zero, they are not quite equal to zero.

For illustration purposes only, an oblique rotation was performed by the OBLIMIN method using the option:

/ROTATION=OBLIMIN

Although such a rotation cannot be recommended, the results are shown in Figure 6.7 beginning at the bottom of page 11 and continuing through page 12. The matrix labeled the PATTERN MATRIX can be ignored. It is the matrix obtained after performing the oblique rotation and cannot be interpreted. The matrix labeled the STRUCTURE MATRIX is much more interesting and useful. This matrix contains the correlations between the original variables and the new obliquely rotated factors. This matrix would be used by researchers attempting to make physical interpretations of the obliquely rotated factors. We make no attempt to interpret these factors here. After performing an oblique rotation the factors are no longer uncorrelated. The correlations between the obliquely rotated factors are shown at the bottom of page 12.

E X A M P L E 6.2

The data in Table 6.2 were also analyzed using SAS's FACTOR procedure. The steps taken in this example illustrate a recommended approach to performing a factor analysis from start to finish. This example also illustrates the type of output available from the SAS FACTOR procedure. The interested reader can reproduce the complete computer output shown for this example by executing the file named EX6_2.SAS on the enclosed computer disk.

A first step that is recommended when performing a factor analysis is to perform a PCA on the data to obtain an initial guess at an appropriate number of factors using methods described in Chapter 5 for determining the number of required principal components. PCA will usually produce an upper bound on the number of factors required. That is, the number of factors required will usually be less than the number of principal components required if you are willing to eliminate trivial factors from consideration as recommended.

While SAS's `PRINCOMP` procedure could be used to perform the PCA, SAS's `FACTOR` procedure is used instead. The `FACTOR` procedure is used because it can produce a SCREE plot of the eigenvalues of the correlation matrix and this is often helpful in choosing an appropriate number of principal components. The commands (assuming the data is in a SAS data file named `POLICE`) used in this first step were

```
OPTIONS LINESIZE=75 PAGESIZE=66 NODATE;
   TITLE1 'EX6.2 - ANTHROPOMETRIC AND PHYSICAL FITNESS
     MEASUREMENTS';
   TITLE2 'OF POLICE DEPARTMENT APPLICANTS';

PROC FACTOR DATA=POLICE METHOD=PRINCIPAL SCREE;
   VAR REACT -- FAT;
    RUN;
```

The results from these commands can be found on pages 1–3 of the computer printout, of which only pages 1 and 2 are shown in Figure 6.8. The eigenvalues of the correlation matrix are shown on page 1 and a SCREE plot of the eigenvalues is shown on page 2. There are five eigenvalues greater than 1 and these five account for 75.8% of the total variability. The SCREE plot does not level off quite as fast as we would like, but there is a gap between the fifth and sixth eigenvalues. Thus, for an initial guess, we could try taking the number of factors equal to five.

Next a FA is attempted using the maximum likelihood method for solving the factor equations by using the following commands:

```
PROC FACTOR DATA=POLICE METHOD=ML NFACTOR=5 ROTATE=VARIMAX
      S C EV RES REORDER SCORE OUT=SCORES;
   VAR REACT -- FAT;
   RUN;
```

The results of these commands are shown on pages 4–6 of the computer output of Figure 6.8. Page 4 contains the means and standard deviations of each of the variables. The correlation matrix is printed at the bottom of page 4 and continues through page 5.

The maximum likelihood method in SAS `FACTOR` also requires prior communality estimates, and the default is to select communality estimates equal to the squared multiple correlation of each variable with all re-

```
EX6.2 - ANTHROPOMETRIC AND PHYSICAL FITNESS MEASUREMENTS OF        1
                    POLICE DEPARTMENT APPLICANTS

Initial Factor Method: Principal Components

                 Prior Communality Estimates: ONE

      Eigenvalues of the Correlation Matrix:   Total = 15  Average = 1

                        1         2         3         4         5
      Eigenvalue   5.218525  2.406798  1.312678  1.231075  1.203846
      Difference   2.811727  1.094120  0.081603  0.027230  0.355941
      Proportion     0.3479    0.1605    0.0875    0.0821    0.0803
      Cumulative     0.3479    0.5084    0.5959    0.6779    0.7582

                        6         7         8         9        10
      Eigenvalue   0.847904  0.704748  0.578409  0.393547  0.368192
      Difference   0.143156  0.126339  0.184862  0.025355  0.041606
      Proportion     0.0565    0.0470    0.0386    0.0262    0.0245
      Cumulative     0.8147    0.8617    0.9003    0.9265    0.9510

                       11        12        13        14        15
      Eigenvalue   0.326585  0.186870  0.138806  0.043881  0.038135
      Difference   0.139715  0.048064  0.094925  0.005746
      Proportion     0.0218    0.0125    0.0093    0.0029    0.0025
      Cumulative     0.9728    0.9853    0.9945    0.9975    1.0000
         5 factors will be retained by the MINEIGEN criterion.
```

F I G. **6.8**

Pages 1, 2, 4–6, and 9–17 of computer printout, Example 6.2

maining variables. This is called the *SMC option*. The values of the prior communality estimates are shown at the top of page 6. The number of factors was taken to be five because the NFACTOR=5 option was used in the SAS command stream.

The maximum likelihood method requires an iterative process to solve the maximum likelihood equations. Near the bottom of page 6 (see Figure 6.8), the output indicates the new set of communalities obtained at the end of each iteration. For example, 0.27784, 0.90750, ..., 0.95775 are the communalities of REACT, HEIGHT, ..., FAT after the first iteration. The column labeled Change is used to decide when to stop the iteration process. This column gives the largest difference between the communalities in the previous step and the communalities in this step. For example, 0.32160 = 0.90501 − 0.58341 is the difference between the communality of RECVR at the conclusion of the first iteration and its initial communality estimate. This difference is the largest of all the possible differences between pairs of communality estimates at the end of the first iteration and their initial estimates. Whenever this maximum change gets to be smaller than 0.001, SAS stops the iteration process and claims that the procedure has converged.

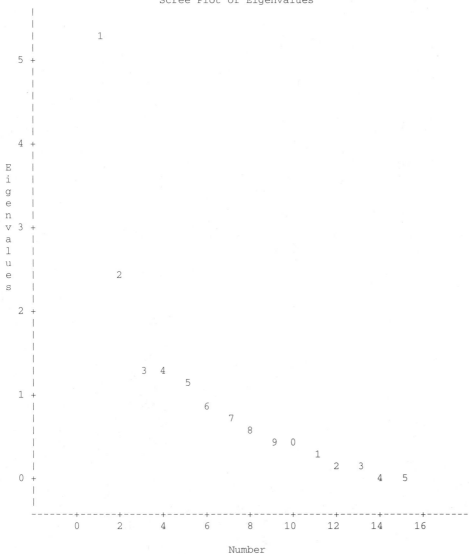

```
   EX6.2 - ANTHROPOMETRIC AND PHYSICAL FITNESS MEASUREMENTS OF        2
                    POLICE DEPARTMENT APPLICANTS

Initial Factor Method: Principal Components

                        Scree Plot of Eigenvalues
     |
     |
     |          1
     |
   5 +
     |
     |
     |
     |
     |
     |
   4 +
 E   |
 i   |
 g   |
 e   |
 n   |
 v 3 +
 a   |
 l   |
 u   |
 e   |          2
 s   |
     |
   2 +
     |
     |
     |
     |
     |          3   4
     |                 5
   1 +
     |
     |            6
     |              7
     |                8
     |                  9  0
     |                        1
     |                          2   3
   0 +                              4    5
     |
     |
     -------+------+------+------+------+------+------+------+------+-------
            0      2      4      6      8     10     12     14     16

                                  Number
```

F I G. **6.8**
Pages 1, 2, 4–6, and 9–17 of computer printout, Example 6.2

EX6.2 - ANTHROPOMETRIC AND PHYSICAL FITNESS MEASUREMENTS OF   4
POLICE DEPARTMENT APPLICANTS

Means and Standard Deviations from 50 observations

	REACT	HEIGHT	WEIGHT	SHLDR	PELVIC
Mean	0.3162	177.906	78.3524	40.952	28.106
Std Dev	0.04821466	6.7031339	11.4683699	1.58606688	1.43717552

	CHEST	THIGH	PULSE	DIAST	CHNUP
Mean	90.62	16.13	73.08	74.6	6.28
Std Dev	5.9709159	6.10186489	12.9107454	8.1139839	4.38010344

	BREATH	RECVR	SPEED	ENDUR	FAT
Mean	193.34	115.54	5.48	4.174	13.7824
Std Dev	25.4547297	11.1798105	0.36365492	0.56417178	7.34763118

Correlations

R matrix

	REACT	HEIGHT	WEIGHT	SHLDR	PELVIC
REACT	1.00000	0.22235	0.05623	-0.09378	-0.05589
HEIGHT	0.22235	1.00000	0.63534	0.65429	0.58589
WEIGHT	0.05623	0.63534	1.00000	0.66562	0.64702
SHLDR	-0.09378	0.65429	0.66562	1.00000	0.58244
PELVIC	-0.05589	0.58589	0.64702	0.58244	1.00000
CHEST	-0.03182	0.42592	0.88869	0.55449	0.52205
THIGH	0.13240	0.22319	0.55422	0.20457	0.20749
PULSE	0.16311	-0.18151	-0.26384	-0.16824	-0.32295
DIAST	0.14732	-0.18655	-0.05170	-0.14488	0.14774
CHNUP	-0.15846	-0.27601	-0.57578	-0.27387	-0.15816
BREATH	0.15953	0.58783	0.45020	0.36765	0.35357
RECVR	-0.12957	-0.12126	-0.12188	-0.02394	-0.20686
SPEED	-0.14934	0.21563	-0.05264	0.20175	0.04124
ENDUR	-0.05255	-0.18916	-0.36798	-0.24478	-0.22935
FAT	0.16481	0.36511	0.80951	0.33063	0.41322

	CHEST	THIGH	PULSE	DIAST	CHNUP
REACT	-0.03182	0.13240	0.16311	0.14732	-0.15846
HEIGHT	0.42592	0.22319	-0.18151	-0.18655	-0.27601
WEIGHT	0.88869	0.55422	-0.26384	-0.05170	-0.57578
SHLDR	0.55449	0.20457	-0.16824	-0.14488	-0.27387
PELVIC	0.52205	0.20749	-0.32295	0.14774	-0.15816
CHEST	1.00000	0.39780	-0.24628	-0.00842	-0.45359
THIGH	0.39780	1.00000	-0.00622	0.04868	-0.66953
PULSE	-0.24628	-0.00622	1.00000	0.23409	0.15478
DIAST	-0.00842	0.04868	0.23409	1.00000	0.05375
CHNUP	-0.45359	-0.66953	0.15478	0.05375	1.00000
BREATH	0.34730	0.20652	-0.06436	-0.16483	-0.35762
RECVR	-0.08323	0.20313	0.50389	0.15384	-0.06025
SPEED	-0.16241	-0.20804	-0.29958	-0.32092	0.32390
ENDUR	-0.32754	-0.33573	0.00730	0.13366	0.21278
FAT	0.72462	0.84421	-0.09483	0.04465	-0.69117

F I G. 6.8

Pages 1, 2, 4–6, and 9–17 of computer printout, Example 6.2

EX6.2 - ANTHROPOMETRIC AND PHYSICAL FITNESS MEASUREMENTS OF   5
POLICE DEPARTMENT APPLICANTS

Correlations

	BREATH	RECVR	SPEED	ENDUR	FAT
REACT	0.15953	-0.12957	-0.14934	-0.05255	0.16481
HEIGHT	0.58783	-0.12126	0.21563	-0.18916	0.36511
WEIGHT	0.45020	-0.12188	-0.05264	-0.36798	0.80951
SHLDR	0.36765	-0.02394	0.20175	-0.24478	0.33063
PELVIC	0.35357	-0.20686	0.04124	-0.22935	0.41322
CHEST	0.34730	-0.08323	-0.16241	-0.32754	0.72462
THIGH	0.20652	0.20313	-0.20804	-0.33573	0.84421
PULSE	-0.06436	0.50389	-0.29958	0.00730	-0.09483
DIAST	-0.16483	0.15384	-0.32092	0.13366	0.04465
CHNUP	-0.35762	-0.06025	0.32390	0.21278	-0.69117
BREATH	1.00000	0.09128	-0.03122	-0.31457	0.29870
RECVR	0.09128	1.00000	-0.43652	-0.07280	0.06652
SPEED	-0.03122	-0.43652	1.00000	-0.02248	-0.20551
ENDUR	-0.31457	-0.07280	-0.02248	1.00000	-0.40546
FAT	0.29870	0.06652	-0.20551	-0.40546	1.00000

F I G. 6.8

Pages 1, 2, 4–6, and 9–17 of computer printout, Example 6.2

EX6.2 - ANTHROPOMETRIC AND PHYSICAL FITNESS MEASUREMENTS OF   6
POLICE DEPARTMENT APPLICANTS

Initial Factor Method: Maximum Likelihood

Prior Communality Estimates: SMC

REACT	HEIGHT	WEIGHT	SHLDR	PELVIC
0.433273	0.746427	0.940737	0.739887	0.728866

CHEST	THIGH	PULSE	DIAST	CHNUP
0.892508	0.842592	0.472639	0.358759	0.715872

BREATH	RECVR	SPEED	ENDUR	FAT
0.506498	0.583405	0.622358	0.286925	0.937507

5 factors will be retained by the NFACTOR criterion.

Iter	Criterion	Ridge	Change	Communalities
1	1.44621	0.000	0.32160	0.27784 0.90750 0.97656 0.67692 0.49074
				0.90980 0.88450 0.34842 0.29435 0.53165
				0.37336 0.90501 0.58936 0.19221 0.95775
2	1.36345	0.000	0.21891	0.49675 0.89863 0.97288 0.70362 0.50535
				0.91077 0.91795 0.34980 0.15698 0.53569
				0.41517 1.02633 0.55784 0.19001 0.94478

ERROR: Communality greater than 1.0. *This try stopped!*

F I G. 6.8

Pages 1, 2, 4–6, and 9–17 of computer printout, Example 6.2

```
         EX6.2 - ANTHROPOMETRIC AND PHYSICAL FITNESS MEASUREMENTS OF          9

Iter Criterion    Ridge    Change    Communalities

  1    1.44621    0.000    0.32160    0.27784 0.90750 0.97656 0.67692 0.49074
                                      0.90980 0.88450 0.34842 0.29435 0.53165
                                      0.37336 0.90501 0.58936 0.19221 0.95775

  2    1.37279    0.000    0.17140    0.44924 0.90056 0.97368 0.69783 0.50217
                                      0.91056 0.91069 0.34950 0.18679 0.53482
                                      0.40610 1.00000 0.56468 0.19049 0.94759

  3    1.35475    0.000    0.10111    0.55035 0.90113 0.97335 0.70023 0.50795
                                      0.91400 0.93346 0.36290 0.14267 0.53867
                                      0.41749 1.00000 0.51945 0.18087 0.94224

  4    1.35106    0.000    0.03815    0.58850 0.89787 0.97319 0.69362 0.51216
                                      0.91573 0.94370 0.37064 0.13328 0.53673
                                      0.41591 1.00000 0.49421 0.17575 0.94147

  5    1.35039    0.000    0.02037    0.60887 0.89477 0.97281 0.69046 0.51399
                                      0.91723 0.94544 0.37426 0.13018 0.53581
                                      0.41459 1.00000 0.48406 0.17443 0.94167

  6    1.35026    0.000    0.00998    0.61886 0.89296 0.97259 0.68871 0.51469
                                      0.91805 0.94531 0.37607 0.12947 0.53527
                                      0.41366 1.00000 0.48035 0.17415 0.94198

  7    1.35023    0.000    0.00496    0.62382 0.89201 0.97247 0.68790 0.51495
                                      0.91847 0.94503 0.37694 0.12937 0.53500
                                      0.41315 1.00000 0.47882 0.17410 0.94216

  8    1.35022    0.000    0.00240    0.62622 0.89154 0.97241 0.68750 0.51506
                                      0.91866 0.94486 0.37735 0.12939 0.53486
                                      0.41288 1.00000 0.47814 0.17409 0.94226

  9    1.35022    0.000    0.00116    0.62738 0.89131 0.97239 0.68731 0.51511
                                      0.91876 0.94476 0.37755 0.12940 0.53479
                                      0.41275 1.00000 0.47783 0.17409 0.94230
```

F I G. 6.8
Pages 1, 2, 4–6, and 9–17 of computer printout, Example 6.2

EX6.2 - ANTHROPOMETRIC AND PHYSICAL FITNESS MEASUREMENTS OF   10
                POLICE DEPARTMENT APPLICANTS

Initial Factor Method: Maximum Likelihood

Final communality estimates ←

10   1.35022   0.000   0.00056   0.62794 0.89120 0.97237 0.68722 0.51513
                                    0.91881 0.94471 0.37765 0.12941 0.53476
                                    0.41269 1.00000 0.47768 0.17409 0.94233

                    Convergence criterion satisfied.

        Significance tests based on 50 observations:

            Test of H0: No common factors.
                vs HA: At least one common factor.

                                                ⟸ $H_0 : \rho = I$ *vs.*
                                                   $H_1 : \rho \neq I$

            Chi-square = 473.196   df = 105   Prob>chi**2 = 0.0000

            Test of H0: 5 Factors are sufficient.
                vs HA: More factors are needed.

            Chi-square = 53.784   df = 40   Prob>chi**2 = 0.0714

        Akaike's Information Criterion = 227.51111709
        Schwarz's Bayesian Criterion = 190.23647876

F I G. 6.8
Pages 1, 2, 4–6, and 9–17 of computer printout, Example 6.2

EX6.2 - ANTHROPOMETRIC AND PHYSICAL FITNESS MEASUREMENTS OF   11
                POLICE DEPARTMENT APPLICANTS

                    Factor Pattern = $\hat{\Lambda}$

	FACTOR1	FACTOR2	FACTOR3	FACTOR4	FACTOR5
RECVR	1.00000	-0.00000	0.00000	0.00000	-0.00000
PULSE	0.50389	-0.19460	-0.08711	0.04878	0.27561
SPEED	-0.43652	-0.12948	0.17123	0.29858	-0.38970
WEIGHT	-0.12188	0.95896	0.18902	-0.04564	-0.01046
FAT	0.06652	0.91307	-0.31229	-0.04848	0.06581
CHEST	-0.08323	0.85373	0.24075	-0.34999	0.05062
THIGH	0.20313	0.72563	-0.58376	0.16414	-0.09586
SHLDR	-0.02394	0.59185	0.49095	0.18223	-0.24918
HEIGHT	-0.12126	0.58658	0.45458	0.56854	0.05041
PELVIC	-0.20686	0.58103	0.33755	0.10606	-0.09779
BREATH	0.09128	0.44303	0.29501	0.32258	0.13023
ENDUR	-0.07280	-0.40654	0.04336	-0.01653	0.03696
CHNUP	-0.06025	-0.66088	0.29276	-0.06020	-0.07086
REACT	-0.12957	0.09818	-0.16829	0.33604	0.67845
DIAST	0.15384	-0.02090	-0.12208	-0.16473	0.25156

F I G. 6.8
Pages 1, 2, 4–6, and 9–17 of computer printout, Example 6.2

Initial Factor Method: Maximum Likelihood

Residual Correlations With Uniqueness on the Diagonal

	REACT	HEIGHT	WEIGHT	SHLDR	PELVIC
REACT	0.37203	0.00029	0.00054	0.03545	-0.05223
HEIGHT	0.00029	0.10880	-0.00140	-0.00999	0.01118
WEIGHT	0.00054	-0.00140	0.02763	0.00805	0.00464
SHLDR	0.03545	-0.00999	0.00805	0.31281	0.02419
PELVIC	-0.05223	0.01118	0.00464	0.02419	0.48486
CHEST	-0.00264	0.00204	-0.00110	0.00541	-0.03040
THIGH	-0.00089	-0.00095	-0.00004	0.01277	-0.00183
PULSE	0.02947	-0.00829	0.00576	0.06156	-0.05447
DIAST	0.03344	-0.01918	0.00528	0.02381	0.27499
CHNUP	0.01619	0.00907	-0.00819	-0.03458	0.11401
BREATH	-0.01925	0.01496	-0.00320	-0.06354	-0.00602
RECVR	0.00000	0.00000	0.00000	0.00000	0.00000
SPEED	-0.00031	0.01071	-0.00450	0.03235	-0.10141
ENDUR	-0.03429	0.02831	0.00444	-0.01497	-0.01747
FAT	0.00286	0.00379	-0.00047	-0.02963	0.01345

	CHEST	THIGH	PULSE	DIAST	CHNUP
REACT	-0.00264	-0.00089	0.02947	0.03344	0.01619
HEIGHT	0.00204	-0.00095	-0.00829	-0.01918	0.00907
WEIGHT	-0.00110	-0.00004	0.00576	0.00528	-0.00819
SHLDR	0.00541	0.01277	0.06156	0.02381	-0.03458
PELVIC	-0.03040	-0.00183	-0.05447	0.27499	0.11401
CHEST	0.08119	-0.00195	-0.01411	-0.01878	0.01765
THIGH	-0.00195	0.05529	0.00019	0.01248	-0.00374
PULSE	-0.01411	0.00019	0.62230	0.08057	0.10450
DIAST	-0.01878	0.01248	0.08057	0.87058	0.09285
CHNUP	0.01765	-0.00374	0.10450	0.09285	0.46525
BREATH	0.01195	-0.00175	-0.05007	-0.11323	-0.11705
RECVR	0.00000	0.00000	0.00000	0.00000	0.00000
SPEED	-0.00520	-0.01182	0.00294	-0.08835	0.15226
ENDUR	-0.00461	0.00563	-0.04074	0.12963	-0.07136
FAT	0.00552	0.00010	0.00635	-0.00917	0.00944

	BREATH	RECVR	SPEED	ENDUR	FAT
REACT	-0.01925	0.00000	-0.00031	-0.03429	0.00286
HEIGHT	0.01496	0.00000	0.01071	0.02831	0.00379
WEIGHT	-0.00320	0.00000	-0.00450	0.00444	-0.00047
SHLDR	-0.06354	0.00000	0.03235	-0.01497	-0.02963
PELVIC	-0.00602	0.00000	-0.10141	-0.01747	0.01345
CHEST	0.01195	0.00000	-0.00520	-0.00461	0.00552
THIGH	-0.00175	0.00000	-0.01182	0.00563	0.00010
PULSE	-0.05007	0.00000	0.00294	-0.04074	0.00635
DIAST	-0.11323	0.00000	-0.08835	0.12963	-0.00917
CHNUP	-0.11705	0.00000	0.15226	-0.07136	0.00944
BREATH	0.58735	0.00000	-0.03009	-0.14009	-0.01269
RECVR	0.00000	0.00000	0.00000	0.00000	0.00000
SPEED	-0.03009	0.00000	0.52235	-0.09498	0.03534
ENDUR	-0.14009	0.00000	-0.09498	0.82591	-0.01911
FAT	-0.01269	0.00000	0.03534	-0.01911	0.05767

F I G. 6.8

Pages 1, 2, 4–6, and 9–17 of computer printout, Example 6.2

```
EX6.2 - ANTHROPOMETRIC AND PHYSICAL FITNESS MEASUREMENTS OF    13
              POLICE DEPARTMENT APPLICANTS

Initial Factor Method: Maximum Likelihood

Root Mean Square Off-diagonal Residuals: Over-all = 0.05055719

          REACT      HEIGHT     WEIGHT     SHLDR      PELVIC
       0.023597   0.011664   0.004377   0.031550   0.087331

          CHEST      THIGH      PULSE      DIAST      CHNUP
       0.012114   0.006072   0.045944   0.096671   0.073938

         BREATH      RECVR      SPEED      ENDUR       FAT
       0.062417   0.000000   0.061993   0.062762   0.014883


         Partial Correlations Controlling Factors

              REACT      HEIGHT     WEIGHT     SHLDR      PELVIC

REACT       1.00000    0.00144    0.00528    0.10391   -0.12297
HEIGHT      0.00144    1.00000   -0.02546   -0.05416    0.04867
WEIGHT      0.00528   -0.02546    1.00000    0.08656    0.04005
SHLDR       0.10391   -0.05416    0.08656    1.00000    0.06211
PELVIC     -0.12297    0.04867    0.04005    0.06211    1.00000
CHEST      -0.01520    0.02168   -0.02323    0.03395   -0.15323
THIGH      -0.00618   -0.01220   -0.00111    0.09710   -0.01119
PULSE       0.06124   -0.03188    0.04391    0.13952   -0.09915
DIAST       0.05876   -0.06232    0.03404    0.04563    0.42325
CHNUP       0.03891    0.04029   -0.07228   -0.09065    0.24004
BREATH     -0.04117    0.05918   -0.02511   -0.14823   -0.01128
RECVR       0.00000    0.00000    0.00000    0.00000    0.00000
SPEED      -0.00069    0.04491   -0.03743    0.08003   -0.20150
ENDUR      -0.06186    0.09443    0.02938   -0.02945   -0.02760
FAT         0.01955    0.04785   -0.01188   -0.22059    0.08042

              CHEST      THIGH      PULSE      DIAST      CHNUP

REACT      -0.01520   -0.00618    0.06124    0.05876    0.03891
HEIGHT      0.02168   -0.01220   -0.03188   -0.06232    0.04029
WEIGHT     -0.02323   -0.00111    0.04391    0.03404   -0.07228
SHLDR       0.03395    0.09710    0.13952    0.04563   -0.09065
PELVIC     -0.15323   -0.01119   -0.09915    0.42325    0.24004
CHEST       1.00000   -0.02913   -0.06279   -0.07063    0.09081
THIGH      -0.02913    1.00000    0.00102    0.05689   -0.02332
PULSE      -0.06279    0.00102    1.00000    0.10946    0.19420
DIAST      -0.07063    0.05689    0.10946    1.00000    0.14590
CHNUP       0.09081   -0.02332    0.19420    0.14590    1.00000
BREATH      0.05471   -0.00968   -0.08281   -0.15835   -0.22391
RECVR       0.00000    0.00000    0.00000    0.00000    0.00000
SPEED      -0.02523   -0.06955    0.00516   -0.13102    0.30886
ENDUR      -0.01781    0.02634   -0.05683    0.15288   -0.11511
FAT         0.08064    0.00180    0.03354   -0.04093    0.05764
```

F I G. 6.8
Pages 1, 2, 4–6, and 9–17 of computer printout, Example 6.2

```
    EX6.2 - ANTHROPOMETRIC AND PHYSICAL FITNESS MEASUREMENTS OF    14
                    POLICE DEPARTMENT APPLICANTS

Initial Factor Method: Maximum Likelihood

          Partial Correlations Controlling Factors

            BREATH     RECVR      SPEED     ENDUR       FAT
   REACT   -0.04117   0.00000  -0.00069  -0.06186   0.01955
   HEIGHT   0.05918   0.00000   0.04491   0.09443   0.04785
   WEIGHT  -0.02511   0.00000  -0.03743   0.02938  -0.01188
   SHLDR   -0.14823   0.00000   0.08003  -0.02945  -0.22059
   PELVIC  -0.01128   0.00000  -0.20150  -0.02760   0.08042
   CHEST    0.05471   0.00000  -0.02523  -0.01781   0.08064
   THIGH   -0.00968   0.00000  -0.06955   0.02634   0.00180
   PULSE   -0.08281   0.00000   0.00516  -0.05683   0.03354
   DIAST   -0.15835   0.00000  -0.13102   0.15288  -0.04093
   CHNUP   -0.22391   0.00000   0.30886  -0.11511   0.05764
   BREATH   1.00000   0.00000  -0.05433  -0.20114  -0.06895
   RECVR    0.00000   0.00000   0.00000   0.00000   0.00000
   SPEED   -0.05433   0.00000   1.00000  -0.14461   0.20363
   ENDUR   -0.20114   0.00000  -0.14461   1.00000  -0.08757
   FAT     -0.06895   0.00000   0.20363  -0.08757   1.00000


Root Mean Square Off-diagonal Partials: Over-all = 0.09757115

          REACT     HEIGHT     WEIGHT      SHLDR     PELVIC
       0.054011   0.046151   0.039335   0.101184   0.156092

          CHEST      THIGH      PULSE      DIAST      CHNUP
       0.062306   0.037796   0.084883   0.146018   0.147381

         BREATH      RECVR      SPEED      ENDUR        FAT
       0.107320   0.000000   0.129232   0.094542   0.094286
```

F I G. **6.8**

Pages 1, 2, 4–6, and 9–17 of computer printout, Example 6.2

EX6.2 - ANTHROPOMETRIC AND PHYSICAL FITNESS MEASUREMENTS OF 15
POLICE DEPARTMENT APPLICANTS

Rotation Method: Varimax

Orthogonal Transformation Matrix

	1	2	3	4	5
1	0.10224	0.05828	0.95184	-0.12100	-0.25596
2	0.77459	0.53376	-0.07600	0.33057	-0.00794
3	-0.61096	0.71079	0.00151	0.27926	-0.20862
4	0.03136	0.45438	-0.03195	-0.80543	0.37791
5	-0.12370	-0.00495	0.29530	0.38650	0.86492

Rotated Factor Pattern $= \hat{\Lambda}^*$

	FACTOR1	FACTOR2	FACTOR3	FACTOR4	FACTOR5
THIGH	0.95650	0.05928	0.10377	-0.11698	0.04315
FAT	0.89519	0.24691	0.01443	0.27105	0.07947
ENDUR	-0.35393	-0.19811	-0.02688	-0.08587	0.03854
CHNUP	-0.69006	-0.17518	-0.02568	-0.10832	-0.12444
HEIGHT	0.17583	0.88722	-0.16259	-0.10291	0.19001
SHLDR	0.19258	0.74751	-0.14643	0.09256	-0.24765
WEIGHT	0.61471	0.61842	-0.19024	0.41725	-0.04215
BREATH	0.16627	0.59741	0.08181	0.00831	0.14612
PELVIC	0.23810	0.58668	-0.27281	0.18814	-0.06658
RECVR	0.10224	0.05828	0.95184	-0.12100	-0.25596
PULSE	-0.07855	-0.11562	0.57411	-0.08238	0.14756
DIAST	0.03785	-0.16505	0.22738	0.17029	0.14159
SPEED	-0.19198	0.16476	-0.53001	-0.33327	-0.14719
CHEST	0.48846	0.46268	-0.11761	0.66098	-0.12418
REACT	0.09224	0.07457	0.05856	-0.00729	0.78129

Loadings are the correlations between the original variables and the factors.

F I G. 6.8
Pages 1, 2, 4–6, and 9–17 of computer printout, Example 6.2

```
     EX6.2 - ANTHROPOMETRIC AND PHYSICAL FITNESS MEASUREMENTS OF     16
                   POLICE DEPARTMENT APPLICANTS


     Scoring Coefficients Estimated by Regression

Rotation Method: Varimax

            Standardized Scoring Coefficients

          FACTOR1    FACTOR2    FACTOR3    FACTOR4    FACTOR5

  THIGH    0.66883   -0.23335   -0.18435   -0.69832   -0.14139
  FAT      0.35126   -0.21238    0.08967    0.19726    0.33217
  ENDUR   -0.00919   -0.00233    0.00463    0.00643    0.00998
  CHNUP   -0.03671    0.01280   -0.01157    0.00312   -0.05626
  HEIGHT  -0.11613    0.59026    0.01241   -0.44885    0.34632
  SHLDR   -0.01471    0.13143   -0.07615   -0.12254   -0.20120
  WEIGHT   0.07794    0.48389   -0.06145    0.43747   -0.29403
  BREATH  -0.01946    0.06629    0.01668   -0.02473    0.08105
  PELVIC  -0.00865    0.05805   -0.02032   -0.03026   -0.05148
  RECVR   -0.06136    0.28841    0.94104   -0.02268   -0.35554
  PULSE   -0.01362   -0.00475    0.03979    0.03887    0.12306
  DIAST   -0.00599   -0.02005    0.02688    0.05324    0.06784
  SPEED    0.01407    0.05328   -0.06948   -0.14921   -0.17009
  CHEST   -0.05974   -0.06072    0.06552    0.68122   -0.11606
  REACT   -0.04247    0.03604    0.15956    0.10172    0.53650
```

F I G. 6.8
Pages 1, 2, 4–6, and 9–17 of computer printout, Example 6.2

EX6.2 - ANTHROPOMETRIC AND PHYSICAL FITNESS MEASUREMENTS OF   17
POLICE DEPARTMENT APPLICANTS

OBS	ID	FACTOR1	FACTOR2	FACTOR3	FACTOR4	FACTOR5
1	1	0.37841	-0.23915	-0.93428	-1.70276	0.06223
2	2	-1.74676	-0.57083	-0.18711	0.13405	1.04804
3	3	1.00293	-1.83968	-0.00199	-0.01531	-0.17913
4	4	0.75203	-0.24044	0.11085	0.83370	-1.29483
5	5	-0.75960	0.39627	0.71814	-0.95946	1.35235
6	6	1.38810	0.96808	-2.03538	0.86575	1.37237
7	7	0.03996	1.74639	-0.68543	-1.45403	0.48799
8	8	-1.85167	0.61238	0.30644	0.27043	-0.09872
9	9	-0.11640	1.04569	-0.39839	1.94420	1.47436
10	10	-1.32810	-2.05020	0.28707	0.07742	0.92126
11	11	0.14466	-1.25693	-0.30950	0.92194	0.12085
12	12	1.35197	-1.66185	-0.95516	-1.13121	0.82777
13	13	-0.00767	0.17145	-0.17612	-0.14270	1.73395
14	14	-0.23579	-0.34069	1.57025	-0.41297	1.15908
15	15	2.48388	0.23530	1.90613	-0.30268	0.07892
16	16	-0.56766	-1.18249	0.59635	0.13702	0.09833
17	17	-0.50574	-0.37459	-0.33147	1.37708	-0.58436
18	18	0.49853	-0.29591	-1.66731	2.22761	1.10477
19	19	0.63859	0.46676	3.09274	1.95118	-0.05392
20	20	-1.10537	0.31355	-1.46832	0.32966	0.14324
21	21	0.77401	1.88854	-0.20754	0.04333	1.88293
22	22	-0.13236	-0.34413	0.36198	-1.51449	0.02274
23	23	0.86912	0.85703	0.63113	1.25156	-0.65876
24	24	1.01426	-0.80962	-0.42349	-0.00049	-0.46617
25	25	-0.01900	-1.06855	-0.68841	-0.32469	-0.72140
26	26	-1.39869	-0.79289	0.51897	0.28721	-0.15422
27	27	-0.87206	1.37448	-0.13980	0.73783	-0.71538
28	28	0.12508	-2.19592	0.17687	0.20008	0.30187
29	29	-1.08345	1.06150	-0.51793	0.74224	-0.18257
30	30	-1.30690	0.67280	-0.78338	-0.19281	-0.90544
31	31	-0.60476	1.07513	2.22088	-0.52939	-1.43333
32	32	-0.67906	0.98352	0.70415	-1.73041	0.78224
33	33	-0.00756	0.68845	-0.26210	0.20304	-1.77661
34	34	0.11730	-1.14166	-0.91948	0.75079	-1.35235
35	35	-0.18210	0.42045	0.32880	-0.36647	0.20982
36	36	1.58269	-0.30270	0.71348	-0.83213	0.65347
37	37	1.76995	1.16723	0.08658	0.25355	-0.38371
38	38	1.60458	0.54020	0.44715	-0.17900	-0.79449
39	39	0.64185	-0.15261	-0.16990	-0.98084	-0.45671
40	40	0.03004	-0.33679	2.13364	-0.37111	-0.50260
41	41	0.34190	0.23044	0.01850	1.44117	-0.23472
42	42	-0.15298	-0.45263	-0.59118	-1.01200	0.05180
43	43	-0.27328	-0.27703	-0.52177	-1.02490	0.30795
44	44	-0.21103	0.15043	-0.65320	0.84156	-0.94096
45	45	-1.45365	-0.06332	1.14308	-0.26212	-0.22517
46	46	0.64067	1.80638	-1.39376	-0.47895	-0.60612
47	47	0.36753	-0.76375	-0.50635	-0.17619	-0.24022
48	48	0.12443	0.06591	-0.28986	-0.46857	-1.30193
49	49	-1.89584	0.19005	-0.32839	-0.48148	-0.82001
50	50	-0.18501	-0.37407	-0.52621	-0.77524	0.88549

Factor scores computed by the Regression Method

F I G. 6.8
Pages 1, 2, 4–6, and 9–17 of computer printout, Example 6.2

In this analysis, the SAS FACTOR procedure hit a snag in the second iteration step. In this step, the communality of one of the variables (RECVR) was greater than 1. This makes no sense, so the procedure stopped.

The SAS FACTOR procedure offers an option called the *HEYWOOD option* that can be used in this instance. Because communalities greater than 1 make no sense, this option replaces such communality estimates by 1 and the process continues to iterate. This option was selected and the analysis process restarted by using the following commands:

```
PROC FACTOR DATA=POLICE METHOD=ML NFACTOR=5 ROTATE=VARIMAX
     S C EV RES REORDER SCORE OUT=SCORES HEYWOOD;
VAR REACT -- FAT;
RUN;
```

The new and important results from these commands are shown on pages 7–16 of the computer output shown in Figure 6.8. The means and standard deviations of the variables and the correlations between variable pairs are once again printed on pages 7 and 8; these pages are not reproduced in Figure 6.8. The top portion of page 9 is identical to that on page 6 so it is not reproduced here either.

Page 9 shows information concerning the iteration process. Note that in the second iteration step, the communality for RECVR that was greater than 1 in the previous step has now been replaced by 1, and the iteration process continues. The information on the iteration process continues on the top of page 10. In iteration step 10, the maximum change in a communality was less than 0.001, so the convergence criterion is satisfied and the process stops.

In the middle of page 10, note two statistical tests that are of interest. These tests assume the data are multivariate normal. The first is a test of

$$H_0: \text{There are no common factors}$$

versus

$$H_a: \text{There is at least one common factor}$$

This is equivalent to testing the hypothesis that $\mathbf{P} = \mathbf{I}$ versus $\mathbf{P} \neq \mathbf{I}$; that is, it tests the hypothesis that all of the original variables are independent. This test was discussed in Section 5.8. If this null hypothesis cannot be rejected, a FA should not be performed. The p value is very small in this case, the hypothesis of independence among the variables would be rejected, and a factor analysis seems appropriate.

The second test in the middle of page 10 tests

$$H_0: \text{Five factors are sufficient}$$

versus the alternative

$$H_a: \text{More factors are needed}$$

This test is not significant at the 0.05 level since 0.0714 > 0.05. So it would seem that five factors are adequate. Remember that a fewer number of factors than five might also be adequate.

The SAS FACTOR procedure also prints Akaike's information criterion (AIC = 227.51) and Schwarz's Bayesian criterion (SBC = 190.24). Recall that the number of factors that produces minimum values for one or both of these criteria is often considered a best choice for the number of factors. To use these criteria, one would need to rerun the FACTOR procedure selecting different numbers of factors each time to see what number of factors produces the minimum values for these two criteria.

The matrix labeled Factor Pattern on page 11 of Figure 6.8 is the estimate of Λ, the initial (unrotated) loading matrix. Page 12 shows the differences between the true sample correlations and those that are reproduced by this five-factor solution. The numbers on the diagonal are the communalities of the variables subtracted from 1. This matrix is simply $\mathbf{R} - \hat{\Lambda}\hat{\Lambda}'$.

Again we would hope that these differences would be small when we have an appropriate number of factors. Readers might recall that SPSS counted the number of residuals that were greater than 0.05. SAS computes the root mean squares of the off-diagonal residuals. SAS does this for all off-diagonal residuals and for the residuals occurring for each variable. The results are printed on the top of page 13.

The root mean square of the off-diagonal residuals is computed by taking the mean of the squares of the off-diagonal residuals, and then taking the square root of this result. In some sense, it measures an average sized absolute residual. We again would hope that the root mean squares of the off-diagonal residuals would be small. In our example the largest of these is 0.096671 for DIAST, and there seems to be no evidence that the number of factors should be increased.

On the bottom of page 13 and continuing on to page 14 (see Figure 6.8), SAS computes the partial correlations between the variables adjusted for the newly created underlying factors. To illustrate partial correlation, suppose you were to regress each of the original variables on the underlying factors and then compute the residuals from these regressions. The correlation between the two sets of residuals for a pair of variables is the partial correlation between the variables after adjusting for the underlying factors.

We would expect there to be little correlation left between the variables when we have an appropriate number of factors. And if the number of chosen factors is adequate, we would hope that the partial correlations would be small. SAS also computes the root mean squares of the off-diagonal partials. These are shown at the bottom of page 14.

Near the middle of page 15, you can see the Varimax-rotated factor matrix. This is the same as the one given by SPSS and can be interpreted the same way as before. On the top of page 15 is the orthogonal matrix that is used to obtain the rotated factor matrix.

On page 16, the way in which factor scores would be computed by the regression method is shown by the Standardized Scoring Coefficients. The values of the factor scores computed by the regression method were printed by the commands:

```
PROC PRINT DATA=SCORES;
 VAR ID FACTOR1-FACTOR5;
 RUN;
```

and the results are shown on page 17.

E X A M P L E 6.3

This example continues to analyze the police department applicant data. This example strives to "fine-tune" the factor analysis process. The reader might recall that among the five factors in the previous analyses one was a trivial factor with only REACT loading on that factor. There were also two variables, DIAST and ENDUR, that did not load on any of the factors. As a result, it was suggested that there were basically seven underlying characteristics: obesity, skeletal structure, cardiovascular fitness, upper body strength, reaction time, endurance, and diastolic blood pressure. To substantiate these conjectures, another pair of analyses is performed. The first factor analyzes the data after removing the variables REACT, ENDUR, and DIAST. We would expect to see four factors: one that could be identified as a measure of obesity, one that is a measure of skeletal structure, one that is a measure of cardiovascular fitness, and one that is a measure of upper body strength.

Next each of these factors is scored and added to the data set. Then another factor analysis is performed on these four new variables along with the three variables that were held out. We expect to find no common factors in this analysis.

The commands required to perform these two analyses are

```
TITLE1 'EX6.3 - FURTHER ANALYSES ON POLICE DEPARTMENT
    APPLICANTS';

PROC FACTOR METHOD=ML NFACT=4 ROTATE=VARIMAX RES REORDER
    DATA=POLICE SCORE OUT=SCORES2 HEYWOOD;
 VAR HEIGHT--PULSE CHNUP BREATH RECVR SPEED FAT;
 RUN;
DATA SCORES2; SET SCORES2; DROP FACTOR1-FACTOR4;
 SKEL=FACTOR1;   OBESITY=FACTOR2;
 CARDVASC=FACTOR3; UBSTRGTH=FACTOR4;
```

```
        EX6.3 - FURTHER ANALYSES ON POLICE DEPARTMENT APPLICANTS              2

Initial Factor Method: Maximum Likelihood

    Significance tests based on 50 observations:

      Test of H0: No common factors.
           vs HA: At least one common factor.

      Chi-square = 429.357   df = 66   Prob>chi**2 = 0.0000

      Test of H0: 4 Factors are sufficient.
           vs HA: More factors are needed.

      Chi-square = 39.931   df = 24   Prob>chi**2 = 0.0218

     Akaike's Information Criterion = 156.10954809
     Schwarz's Bayesian Criterion = 129.67939519
     Tucker and Lewis's Reliability Coefficient = 0.8794297195
```

F I G. 6.9

Pages 2, 4–6, 8, 10, and 13 of computer printout, Example 6.3

```
        PROC FACTOR METHOD=ML S C EV RES REORDER DATA=SCORES2
            HEYWOOD;
         VAR REACT DIAST ENDUR SKEL OBESITY CARDVASC UBSTRGTH;

        PROC PRINT DATA=SCORES2;
          VAR ID REACT RECVR DIAST OBESITY SKEL UBSTRGTH CARDVASC;
          RUN;
```

Some of the basic results from the preceding commands can be found on the computer printout pages shown in Figure 6.9. The interested reader can get all of the output by executing the file named EX6_3.SAS on the enclosed computer disk. The test of "H_0: 4 Factors are sufficient" versus "H_a: More factors are needed" is rejected at the 0.0218 significance level as shown from the excerpt of page of the computer printout.

This rejection seems a little strange, but remember that the maximum likelihood method has a tendency of adding trivial factors, so you should hold off on making any conclusions until you look at other parts of the output. In particular, an examination of the residual correlations on page 4 and the root mean squares of the off-diagonal residuals would seem to indicate that there is no need for additional factors. The partial correlations among the variables after adjusting for the common factors on page 5 also indicate that there is no need to add additional factors.

The rotated factor matrix is shown on page 6 of the Figure 6.9 computer printout. In this case FACTOR1 seems to be a measure of skeletal structure, FACTOR2 seems to be a measure of obesity, FACTOR3 seems to be a measure of cardiovascular fitness, and FACTOR4 seems to be a measure of upper body strength.

EX6.3 - FURTHER ANALYSES ON POLICE DEPARTMENT APPLICANTS   4

Initial Factor Method: Maximum Likelihood

Residual Correlations With Uniqueness on the Diagonal

	HEIGHT	WEIGHT	SHLDR	PELVIC	CHEST	THIGH
HEIGHT	0.24786	-0.00320	-0.02102	-0.00452	0.00006	-0.01531
WEIGHT	-0.00320	0.03681	0.00691	0.00699	-0.00036	-0.00168
SHLDR	-0.02102	0.00691	0.31938	0.01698	0.00286	0.04065
PELVIC	-0.00452	0.00699	0.01698	0.47353	-0.00614	-0.01842
CHEST	0.00006	-0.00036	0.00286	-0.00614	0.02089	-0.00031
THIGH	-0.01531	-0.00168	0.04065	-0.01842	-0.00031	0.12922
PULSE	0.02960	0.00286	0.01056	-0.07287	-0.00435	-0.03038
CHNUP	0.00998	-0.01105	-0.02807	0.12925	0.00329	-0.01758
BREATH	0.10160	-0.00892	-0.09989	-0.02674	0.00476	-0.01421
RECVR	-0.00026	-0.00005	0.00065	0.00016	-0.00001	0.00008
SPEED	0.01197	-0.00959	0.10270	-0.09164	-0.00006	0.01075
FAT	0.00637	0.00000	-0.01710	0.01207	0.00051	0.00074

	PULSE	CHNUP	BREATH	RECVR	SPEED	FAT
HEIGHT	0.02960	0.00998	0.10160	-0.00026	0.01197	0.00637
WEIGHT	0.00286	-0.01105	-0.00892	-0.00005	-0.00959	0.00000
SHLDR	0.01056	-0.02807	-0.09989	0.00065	0.10270	-0.01710
PELVIC	-0.07287	0.12925	-0.02674	0.00016	-0.09164	0.01207
CHEST	-0.00435	0.00329	0.00476	-0.00001	-0.00006	0.00051
THIGH	-0.03038	-0.01758	-0.01421	0.00008	0.01075	0.00074
PULSE	0.69453	0.09078	0.00091	-0.00048	-0.08924	0.01477
CHNUP	0.09078	0.47423	-0.12649	0.00112	0.18881	0.01026
BREATH	0.00091	-0.12649	0.64945	-0.00011	-0.06564	-0.00369
RECVR	-0.00048	0.00112	-0.00011	0.00918	-0.00003	0.00008
SPEED	-0.08924	0.18881	-0.06564	-0.00003	0.65777	0.01427
FAT	0.01477	0.01026	-0.00369	0.00008	0.01427	0.03553

Root Mean Square Off-diagonal Residuals: Over-all = 0.04642011

HEIGHT	WEIGHT	SHLDR	PELVIC	CHEST	THIGH
0.033290	0.006117	0.046862	0.053944	0.002996	0.018535

PULSE	CHNUP	BREATH	RECVR	SPEED	FAT
0.046388	0.084225	0.061529	0.000431	0.078291	0.009631

F I G. 6.9
Pages 2, 4–6, 8, 10, and 13 of computer printout, Example 6.3

The results of the second analysis start on page 8. Look carefully at the correlation matrix printed in the middle of page 8 and note that the correlations among the seven variables in this analysis are very low. Also note that on page 10, the test of "H_0: No common factors" versus "H_a: At least one common factor" was not significant with Prob>chi**2 = 0.5558.

A portion of the printout of the final data set that was created is shown on page 13 of Figure 6.9. This data set is now ready for any additional analyses a researcher might wish to perform.

EX6.3 - FURTHER ANALYSES ON POLICE DEPARTMENT APPLICANTS   5

Initial Factor Method: Maximum Likelihood

Partial Correlations Controlling Factors

	HEIGHT	WEIGHT	SHLDR	PELVIC	CHEST	THIGH
HEIGHT	1.00000	-0.03353	-0.07470	-0.01320	0.00085	-0.08558
WEIGHT	-0.03353	1.00000	0.06374	0.05298	-0.01295	-0.02436
SHLDR	-0.07470	0.06374	1.00000	0.04367	0.03507	0.20008
PELVIC	-0.01320	0.05298	0.04367	1.00000	-0.06170	-0.07447
CHEST	0.00085	-0.01295	0.03507	-0.06170	1.00000	-0.00593
THIGH	-0.08558	-0.02436	0.20008	-0.07447	-0.00593	1.00000
PULSE	0.07135	0.01787	0.02242	-0.12706	-0.03608	-0.10141
CHNUP	0.02912	-0.08362	-0.07212	0.27275	0.03306	-0.07101
BREATH	0.25324	-0.05766	-0.21934	-0.04821	0.04087	-0.04906
RECVR	-0.00551	-0.00271	0.01206	0.00236	-0.00082	0.00225
SPEED	0.02965	-0.06165	0.22407	-0.16421	-0.00050	0.03687
FAT	0.06790	0.00003	-0.16055	0.09305	0.01887	0.01089

	PULSE	CHNUP	BREATH	RECVR	SPEED	FAT
HEIGHT	0.07135	0.02912	0.25324	-0.00551	0.02965	0.06790
WEIGHT	0.01787	-0.08362	-0.05766	-0.00271	-0.06165	0.00003
SHLDR	0.02242	-0.07212	-0.21934	0.01206	0.22407	-0.16055
PELVIC	-0.12706	0.27275	-0.04821	0.00236	-0.16421	0.09305
CHEST	-0.03608	0.03306	0.04087	-0.00082	-0.00050	0.01887
THIGH	-0.10141	-0.07101	-0.04906	0.00225	0.03687	0.01089
PULSE	1.00000	0.15817	0.00136	-0.00603	-0.13203	0.09405
CHNUP	0.15817	1.00000	-0.22792	0.01704	0.33806	0.07904
BREATH	0.00136	-0.22792	1.00000	-0.00140	-0.10043	-0.02427
RECVR	-0.00603	0.01704	-0.00140	1.00000	-0.00040	0.00439
SPEED	-0.13203	0.33806	-0.10043	-0.00040	1.00000	0.09335
FAT	0.09405	0.07904	-0.02427	0.00439	0.09335	1.00000

Root Mean Square Off-diagonal Partials: Over-all = 0.10224824

HEIGHT	WEIGHT	SHLDR	PELVIC	CHEST	THIGH
0.090349	0.045924	0.128895	0.113910	0.029639	0.081322

PULSE	CHNUP	BREATH	RECVR	SPEED	FAT
0.087893	0.162749	0.129542	0.007022	0.145767	0.076238

F I G. **6.9**

Pages 2, 4–6, 8, 10, and 13 of computer printout, Example 6.3

Rotation Method: Varimax

Orthogonal Transformation Matrix

	1	2	3	4
1	-0.10975	-0.09925	0.96609	-0.21161
2	0.49056	0.62315	0.24217	0.55892
3	0.32895	-0.77181	0.07609	0.53881
4	0.79944	-0.07843	-0.04729	-0.59373

Rotated Factor Pattern

	FACTOR1	FACTOR2	FACTOR3	FACTOR4
HEIGHT	0.82924	0.17660	-0.17918	-0.03476
SHLDR	0.79155	0.07968	-0.06739	0.20777
WEIGHT	0.63613	0.54414	-0.17665	0.48086
PELVIC	0.62348	0.19947	-0.24966	0.18875
BREATH	0.56677	0.15536	0.05304	0.04867
THIGH	0.12468	0.91342	0.14417	0.01043
FAT	0.26128	0.88225	0.01004	0.34314
CHNUP	-0.20828	-0.68161	-0.01382	-0.13266
RECVR	0.05475	0.05860	0.99178	-0.02776
PULSE	-0.14207	-0.03377	0.51449	-0.13946
SPEED	0.17819	-0.19027	-0.44637	-0.27392
CHEST	~ 0.44146	0.38373	-0.10882	0.79066

Variance explained by each factor

	FACTOR1	FACTOR2	FACTOR3	FACTOR4
Weighted	28.912644	45.134335	109.72126	40.008617
Unweighted	2.805832	2.662875	1.613478	1.169442

Final Communality Estimates and Variable Weights
Total Communality: Weighted = 223.77686   Unweighted = 8.251627

	HEIGHT	WEIGHT	SHLDR	PELVIC	CHEST	THIGH
Communality	0.752141	0.963189	0.680617	0.526472	0.979114	0.870784
Weight	4.034966	27.165540	3.130381	2.111795	47.878202	7.738762

	PULSE	CHNUP	BREATH	RECVR	SPEED	FAT
Communality	0.305473	0.525766	0.350547	0.990823	0.342229	0.964472
Weight	1.439838	2.108859	1.539983	108.96350	1.520287	28.144745

F I G. 6.9
Pages 2, 4–6, 8, 10, and 13 of computer printout, Example 6.3

EX6.3 - FURTHER ANALYSES ON POLICE DEPARTMENT APPLICANTS   8

Means and Standard Deviations from 50 observations

	REACT	DIAST	ENDUR	SKEL
Mean	0.3162	74.6	4.174	0
Std Dev	0.04821466	8.1139839	0.56417178	0.94355339

	OBESITY	CARDVASC	UBSTRGTH
Mean	0	0	0
Std Dev	0.9746088	0.99514238	0.95907906

Correlations

	REACT	DIAST	ENDUR	SKEL
REACT	1.00000	0.14732	-0.05255	-0.00146
DIAST	0.14732	1.00000	0.13366	-0.14714
ENDUR	-0.05255	0.13366	1.00000	-0.22327
SKEL	-0.00146	-0.14714	-0.22327	1.00000
OBESITY	0.22532	0.05772	-0.33103	0.02784
CARDVASC	-0.14428	0.16188	-0.04637	0.00389
UBSTRGTH	-0.16509	0.06912	-0.15049	0.06296

	OBESITY	CARDVASC	UBSTRGTH
REACT	0.22532	-0.14428	-0.16509
DIAST	0.05772	0.16188	0.06912
ENDUR	-0.33103	-0.04637	-0.15049
SKEL	0.02784	0.00389	0.06296
OBESITY	1.00000	0.00324	0.02183
CARDVASC	0.00324	1.00000	-0.00738
UBSTRGTH	0.02183	-0.00738	1.00000

F I G. 6.9

Pages 2, 4–6, 8, 10, and 13 of computer printout, Example 6.3

EX6.3 - FURTHER ANALYSES ON POLICE DEPARTMENT APPLICANTS   10

Initial Factor Method: Maximum Likelihood

Significance tests based on 50 observations:

Test of H0: No common factors.
    vs HA: At least one common factor.

Chi-square = 19.458   df = 21   Prob>chi**2 = 0.5558

Test of H0: 2 Factors are sufficient.
    vs HA: More factors are needed.

Chi-square = 3.726   df = 8   Prob>chi**2 = 0.8810

Akaike's Information Criterion = 44.186266728
Schwarz's Bayesian Criterion = 41.213363418
Tucker and Lewis's Reliability Coefficient = 0

F I G. 6.9

Pages 2, 4–6, 8, 10, and 13 of computer printout, Example 6.3

OBS	ID	REACT	RECVR	DIAST	OBESITY	SKEL	UBSTRGTH	CARDVASC
1	1	0.310	108	64	0.54473	0.01385	-1.97446	-0.76015
2	2	0.345	108	78	-1.40703	-0.92442	-0.25918	-0.54702
3	3	0.293	116	88	0.97752	-1.54439	-0.12289	0.07893
4	4	0.254	120	62	0.44216	-0.18150	1.34749	0.41604
5	5	0.384	120	68	-0.74946	-0.19949	-0.18133	0.45896
6	6	0.406	91	68	1.74664	0.78434	0.65819	-2.32292
7	7	0.344	110	48	-0.01471	1.59506	-0.95210	-0.60530
8	8	0.321	117	64	-1.99044	0.22029	0.70287	0.25711
9	9	0.425	105	78	0.09497	0.82196	1.39776	-0.95246
10	10	0.385	113	78	-0.95581	-2.22608	-0.43187	-0.06402
11	11	0.317	110	78	0.01700	-1.18478	0.85784	-0.40187
12	12	0.353	105	72	1.31408	-1.43023	-0.95134	-0.95535
13	13	0.413	109	88	0.14559	0.34231	-0.91900	-0.62622
14	14	0.392	129	78	-0.04497	-0.34734	-0.94706	1.21302
15	15	0.312	139	88	2.63848	0.25430	-0.16553	1.92544
16	16	0.342	120	86	-0.65204	-1.04910	-0.03053	0.49581
17	17	0.293	111	68	-0.75550	-0.48141	1.65219	-0.28718
18	18	0.317	92	78	0.88570	-0.58358	1.76013	-2.06750
19	19	0.333	147	90	0.41567	0.26766	2.12106	2.83723
20	20	0.317	98	86	-1.21702	0.49223	0.01186	-1.52533
21	21	0.427	110	74	1.12544	1.66310	-0.40411	-0.67180
22	22	0.266	121	72	0.07540	-0.46576	-1.33502	0.48053
23	23	0.311	124	74	0.37697	1.11171	1.43171	0.72637
24	24	0.284	113	76	0.90720	-0.52005	0.02256	-0.24570
25	25	0.259	110	72	-0.11070	-0.94765	-0.12171	-0.43590

F I G. 6.9

Pages 2, 4–6, 8, 10, and 13 of computer printout, Example 6.3

Summary

Factor analysis is a useful technique for determining a set of underlying new variables, called *factors,* that appear to be driving the measurements that are being made with the original variables. Factor analysis has received criticism from those who do not like the nonuniqueness of the FA solution to the FA equations and to the nonuniqueness of the rotation scheme that is being used. While the solutions are not unique in the mathematical sense, they do seem to be unique in a practical sense. To illustrate, the police department applicant data have been factor analyzed in a large number of different ways by using different schemes for solving the factor analysis equations and different schemes for rotating the initial solution to the factor analysis equations. Although no solutions are mathematically identical, they all seem to produce factors that can be identified as measures of obesity, skeletal structure, cardiovascular fitness, and upper body strength. Furthermore, reaction time, diastolic blood pressure, and endurance are always uncorrelated with these four underlying factors.

Exercises

1 Consider the correlations among children's test scores shown in Table 6.1. An approximate solution to the factor analysis equations gives

$$\hat{\Lambda} = \begin{bmatrix} 0.95 \\ 0.89 \\ 0.83 \\ 0.77 \\ 0.71 \\ 0.65 \end{bmatrix}$$

(Note that this approximate solution was obtained by following Spearman's beliefs. The elements in this factor loading matrix are equally spaced between 0.95 and 0.65.)

a For the loading matrix given, what is the matrix of reproduced correlations?

b What is the matrix of residual differences between the original correlation matrix and those reproduced by this one-factor solution?

c Determine the communalities of each of the original variables.

d What are the specific variances for each of the original variables?

e Use a statistical computing package on the sample correlation matrix to find a one-factor solution to the factor analysis equations using the principal factor method. Use prior communality estimates equal to the squared multiple correlation of each variable with all of the remaining variables. How do the factor loadings in your solution compare to the ones given above?

2 Consider Example 6.1 where the determinant of the correlation matrix was equal to 0.0000173. Use the procedure described in Section 5.8 to test H_0: $\mathbf{P} = \mathbf{I}$ versus H_1: $\mathbf{P} \neq \mathbf{I}$.

3 Consider the applicant data, with individuals 41 and 42 removed, originally discussed in Example 2.1. Using the principal axis factoring method with a Varimax rotation and concepts that are described in Chapter 6, determine whether there is a set of underlying characteristics that summarizes the information in the original variables. Write a short report indicating the processes you went through to accomplish the following:

a Determine the number of underlying characteristics.

b Identify and/or interpret the underlying characteristics.

c Score each of the applicants in the data set with respect to each of the underlying characteristics identified in part b.

4   Repeat Exercise 3 using the maximum likelihood method for solving the factor analysis equations. How do these results compare to those obtained in Exercise 3?

5   Repeat Exercise 3 using a third method of your choice to solve the factor analysis equations and a different method of rotating factors. Discuss and compare these results to those found in Exercises 2 through 4. Be prepared to discuss your results with those found by your classmates.

6   Repeat Exercises 2 through 5 with another data set of your choice—perhaps the one you provided in Exercise 1 of Chapter 1. Write a short report indicating what you have learned about the data you are analyzing.

7   Consider the portion of the Pizzazz survey data described in Appendix B that comes from employees who have no supervisory responsibilities. That is, use only the data coming from employees whose positions are either kitchen crew, delivery driver, or waiter/waitress. Use the principal axis factoring method with a Varimax rotation and concepts described in this chapter to determine whether there is a set of underlying characteristics that summarizes the information in the organizational commitment variables (questions 1–15). Write a short report indicating the processes you went through to accomplish the following:

    a   Determine the number of underlying characteristics.

    b   Identify and/or interpret the underlying characteristics.

8   Repeat Exercise 7 for those employees with supervisory responsibilities.

9   Repeat Exercises 7 and 8 for the job security variables (questions 34–53).

10   Repeat Exercises 7 and 8 for the job satisfaction variables (questions 61–80).

11   Create factor scores for the underlying factors obtained in Exercise 7. Then use multiple regression to study the effects of these newly created underlying variables on each of the three job involvement variables (questions 16–18). Write a short report indicating what you have learned from this regression modeling.

12   Repeat Exercise 11 for the employees with supervisory responsibilities and the underlying factors found in Exercise 8.

13   Consider the survey data from 304 married adults described in Appendix C. Use only the data from the females and their responses to the 60-item Family Assessment Device questionnaire.

    a   Choose one of the factor analysis methods and one of the orthogonal rotation procedures described in this chapter to determine if there is a set of underlying characteristics that explains the responses of these adults to the 60-item Family Assessment Device. Write a short report describing what you have done and why.

      b Repeat part a using a different factor analysis method and a different orthogonal rotation procedure.

      c Repeat part a using a third factor analysis method and a third orthogonal rotation procedure.

      d Write a short report that contrasts and compares the results from the three different analyses performed in parts a–c.

14 Repeat Exercise 13 for the males in the survey data described in Appendix C.

15 Repeat Exercise 13 for females' responses to the 30-item Family Adaptability Cohesion Evaluation Scale questionnaire of Appendix C.

16 Repeat Exercise 15 for the males in the survey data described in Appendix C.

Discriminant Analysis 7

Suppose an anesthesiologist needs to determine whether an anesthetic is safe for a person who is having a heart operation. The anesthesiologist may know certain things about the patient such as age, gender, race, blood pressure, weight, etc. Based on these kinds of criteria, the anesthesiologist would like to know the following: (1) Can this knowledge be used to construct a rule that will classify new patients as to whether they are going to be safe or unsafe recipients of the anesthetic? (2) What is the rule and can the rule be used to classify new patients? (3) What are the chances of making mistakes when using the rule?

Note that, in the preceding example, mistakes occur whenever a new patient is classified into the wrong population. Thus, an error occurs when a patient who is really safe for the anesthetic is predicted to be unsafe or when a patient who is unsafe is predicted to be safe. Note also that these two kinds of errors are probably not equally serious. It may not be too serious to classify a patient as being unsafe for the anesthetic when the patient is really safe, provided that there are alternative anesthetics; but it would be an extremely serious mistake to classify a patient as safe for an anesthetic when, in fact, the anesthetic is really unsafe for that patient. *Discriminant analysis* is a multivariate technique that can be used to build rules that can classify patients into the appropriate population.

Discriminant analysis is similar to regression analysis except that the dependent variable is categorical rather than continuous. In regression analysis, we want to be able to predict the value of a variable of interest based on a set of predictor variables. In discriminant analysis, we want to be able to predict class membership of an individual observation based on a set of predictor variables.

Discriminant analysis is sometimes known as *classification analysis*. Suppose we have several populations from which observations may come. Suppose, also, we have a new observation that is known to come from one of these populations, but it is not known from which population. The basic objective of discriminant analysis is to produce a rule or a classification scheme that will enable a researcher to predict the population from which an observation is most likely to have come.

7.1
Discrimination for Two Multivariate Normal Populations

Suppose there are two multivariate normal populations, say, Π_1 that is $N_p(\boldsymbol{\mu}_1, \boldsymbol{\Sigma}_1)$ and Π_2 that is $N_p(\boldsymbol{\mu}_2, \boldsymbol{\Sigma}_2)$. Suppose a new observation vector $\mathbf{x}$ is known to come from either Π_1 or Π_2. A rule is needed that can be used to predict from which of the two populations $\mathbf{x}$ is most likely to have come. Four different ways of looking at this problem are considered. For many cases, these four ways of developing a discrimination rule are equivalent.

A Likelihood Rule

For mathematical statisticians, a reasonable rule might be:

Choose Π_1 if $L(\mathbf{x}; \boldsymbol{\mu}_1, \boldsymbol{\Sigma}_1) > L(\mathbf{x}; \boldsymbol{\mu}_2, \boldsymbol{\Sigma}_2)$, and choose Π_2 otherwise, where $L(\mathbf{x}; \boldsymbol{\mu}_i, \boldsymbol{\Sigma}_i)$ is the likelihood function for the ith population evaluated at $\mathbf{x}$, $i = 1, 2$.

Note that the likelihood function for $\mathbf{x}$ is simply the multivariate normal probability density function, given in Eq. (1.1), evaluated at the observation vector $\mathbf{x}$.

The Linear Discriminant Function Rule

When two multivariate normal populations have equal variance–covariance matrices (i.e., when $\boldsymbol{\Sigma}_1 = \boldsymbol{\Sigma}_2$), the likelihood rule simplifies to:

Choose Π_1 if $\mathbf{b}'\mathbf{x} - k > 0$ and choose Π_2 otherwise, where $\mathbf{b} = \boldsymbol{\Sigma}^{-1}(\boldsymbol{\mu}_1 - \boldsymbol{\mu}_2)$ and $k = (\frac{1}{2})(\boldsymbol{\mu}_1 - \boldsymbol{\mu}_2)'\boldsymbol{\Sigma}^{-1}(\boldsymbol{\mu}_1 + \boldsymbol{\mu}_2)$.

The function $\mathbf{b}'\mathbf{x}$ is called the *linear discriminant function* of $\mathbf{x}$. It is the single linear function of the elements in $\mathbf{x}$ that summarizes all of the information in $\mathbf{x}$ that is available for effective discrimination between two multivariate normal populations that have equal variance–covariance matrices.

A Mahalanobis Distance Rule

When two populations have equal variance–covariance matrices, the likelihood rule is also equivalent to:

Choose Π_1 when $d_1 < d_2$ where $d_i = (\mathbf{x} - \boldsymbol{\mu}_i)'\boldsymbol{\Sigma}^{-1}(\mathbf{x} - \boldsymbol{\mu}_i)$ for $i = 1, 2$.

The quantity d_i is, in some sense, a measure of how far $\mathbf{x}$ is from $\boldsymbol{\mu}_i$, and d_i is called the *Mahalanobis squared distance* between $\mathbf{x}$ and $\boldsymbol{\mu}_i$, for $i = 1, 2$. This distance measure takes the variances and covariances of the measured variables into account. The Mahalanobis squared distance rule classifies an observation into the population to whose mean it is "closest."

A Posterior Probability Rule

When the variance–covariance matrices are equal, the quantity $P(\Pi_i \mid \mathbf{x})$ defined by

$$P(\Pi_i \mid \mathbf{x}) = \exp[(-\tfrac{1}{2})d_i]/\{\exp[(-\tfrac{1}{2})d_1] + \exp[(-\tfrac{1}{2})d_2]\}$$

is called the *posterior probability* of population Π_i given $\mathbf{x}$, for $i = 1, 2$.

The posterior probability is not actually a true probability because no random event is under consideration. The observation either belongs to one population or the other. The uncertainty comes with a researcher's ability to choose the correct population. The major benefit of the posterior probability is that it gives an indication of how confident one might feel that he or she is making a correct decision when **x** is being assigned to one of the two populations. For example, if the posterior probability for Π_1 is 0.53 and that for Π_2 is 0.47, then we might not feel too confident that a correct decision is actually being made when **x** is assigned to the first population. However, if the posterior probability for Π_1 is 0.96 and that for Π_2 is 0.04, then we would feel very confident that a correct decision is being made when **x** is classified into the first population.

As suggested by the previous paragraph, a discriminant rule based on posterior probabilities is:

Choose Π_1 if $P(\Pi_1|\mathbf{x}) > P(\Pi_2|\mathbf{x})$ and choose Π_2 otherwise.

Remark When the variance–covariance matrices for both populations are equal, all four discriminant rules just described are equivalent. That is, all four will assign new observations into exactly the same groups.

Sample Discriminant Rules

The preceding descriptions of the four equivalent discriminant rules assume knowledge of the true values of $\boldsymbol{\mu}_1$, $\boldsymbol{\mu}_2$, $\boldsymbol{\Sigma}_1$, and $\boldsymbol{\Sigma}_2$. In practice, this will never be the case; instead, we will need to produce discriminant rules based on sample estimates of $\boldsymbol{\mu}_1$, $\boldsymbol{\mu}_2$, $\boldsymbol{\Sigma}_1$, and $\boldsymbol{\Sigma}_2$.

When we have random samples from each of the two populations of interest, unbiased estimates of $\boldsymbol{\mu}_1$, $\boldsymbol{\mu}_2$, $\boldsymbol{\Sigma}_1$, and $\boldsymbol{\Sigma}_2$ are given by $\hat{\boldsymbol{\mu}}_1$, $\hat{\boldsymbol{\mu}}_2$, $\hat{\boldsymbol{\Sigma}}_1$, and $\hat{\boldsymbol{\Sigma}}_2$, respectively. If we believe that the two variance–covariance matrices are equal (a test for equality of variance–covariance matrices is given in Section 10.1), then a pooled estimate of $\boldsymbol{\Sigma}$, the common variance–covariance matrix, is given by

$$\hat{\boldsymbol{\Sigma}} = \frac{(N_1 - 1)\hat{\boldsymbol{\Sigma}}_1 + (N_2 - 1)\hat{\boldsymbol{\Sigma}}_2}{N_1 + N_2 - 2}$$

where N_1 and N_2 are the sizes of the random samples taken from Π_1 and Π_2, respectively.

Discriminant rules based on samples from each population can then be formed exactly like those based on population values simply by substituting sample estimates for the parameters in the discriminant rules described earlier.

Estimating Probabilities of Misclassification

When performing a discriminant analysis, we need to be able to determine or estimate the probabilities of correct classifications of new observations. Obviously, a rule that classifies correctly 95% of the time would be preferred to a rule that only classifies correctly 75% of the time.

Unfortunately, estimating the probabilities of correct classification is not easy. Three basic methods can be used as explained in the following subsections.

Resubstitution Estimates

One simple method called the resubstitution method is to apply a discriminant rule to the data used to develop the rule and observe how often the rule correctly classifies these observations. The biggest drawback to this method is that it overestimates the probabilities of correct classification, although for extremely large sample sizes the bias is not too bad. Obviously, a rule is likely to do better on the data used to build the rule than it might do on any other data set. Nevertheless, almost all computer programs that perform discriminant analyses will produce a "classification summary matrix" that shows where observations in the sample data would be classified by the discriminant rule developed. If the discriminant rule does not work well for the data used to construct the rule, then we should expect the rule to be even worse for new data sets.

Estimates from Holdout Data

A second method for estimating the probabilities of correct classification, called the *holdout method,* uses a *holdout data set,* also sometimes called a *test data set.* A holdout data set is one for which we know where the observations should be classified, but the holdout data are not used to develop the discriminant rule. Then a discriminant rule developed from other data (often called *calibration data*) can be applied to the holdout data set to see how well observations in the holdout data set are classified according to the rule. This method has been shown to produce unbiased estimates of the probabilities of correct classification.

The biggest drawback to using a holdout data set is that you may not actually be getting the "best" possible discrimination rule because all the data are not being used to obtain the rule. For example, suppose you have 250 observations from each of two populations, for a total of 500 observations. You might form a discriminant rule based on 150 of the observations from each of the two samples, and then test the rule on the remaining 200 observations (100 from each population). But do you get the best discriminant rule when you use only 300 of the observations rather than all 500 observations? Probably not. Obviously, this method presents some problems for researchers who are faced with limited amounts of data, but it can be recommended to researchers who have large amounts of data.

Cross-Validation Estimates

A third method, which is usually preferred to the first two, is known as *cross-validation*. This method was initially proposed by Lachenbruch (1968). It can be described in the following manner: Remove the first observation vector from the data set, form a discriminant rule based on all of the remaining data, use this rule to classify the first observation, and note whether the observation is correctly classified or not. Next, replace the first observation and remove the second observation from the data set, form a discriminant rule based on all of the remaining data, use this rule to classify the second observation, and note whether the observation is correctly classified or not. Continue this process through the entire data set, removing one observation at a time, and noting whether that particular observation would be correctly classified by a rule formed from all of the remaining data. Finally, create a summary matrix for these cross-validated estimates. These estimates have been shown to be nearly unbiased estimates of the true probabilities of correct and incorrect classifications. This method is also referred to as "jackknifing" by some authors.

To illustrate the basic ideas of a discriminant analysis, consider the following example.

E X A M P L E 7.1

Dr. Michael Finnegan, professor of anthropology at Kansas State University, was interested in determining whether wild turkeys could be distinguished from domestic turkeys by considering measurements of certain bones. A man had been arrested for stealing many of his neighbor's turkeys. The man claimed that the turkey meat in his freezer actually came from wild turkeys. Bone measurements could be made on turkeys confiscated from the suspect, and the state needed to prove (in a court trial) that the turkeys in the man's freezer were domestic turkeys and not wild turkeys as the man had claimed.

Initially, bone measurements were made on 158 male and female turkeys, some of which were known to be domestic turkeys, and some of which were known to be wild turkeys. The data are in the file labeled EX7_1.SAS on the enclosed disk. In addition to variables that identify each bird according to its sex and type (wild or domestic), there are variables that correspond to measurements made on bones. All bone measurements are in millimeters. The bone measurement variables are length of humerus (HUM), length of radius (RAD), length of ulna (ULN), length of femur (FEM), length of tibiotarsus (TIB), length of tibiotarsus measured to the nutrient foramen (TIN), length of carpometacarpus (CAR), length of digit III to phalanx I (D3P), sternal length (STL), sternal breadth (STB), coracoid length (COR), pelvic length (PEL), tibiotarsus maximum diameter at the nutrient foramen (MAX), tibiotarsus minimum diameter at the nutrient foramen (MIN), and scapula length (SCA).

Unfortunately, not all bone measurements were available on all turkeys, so much of the data are missing. A portion of the data is shown on page 1 of the computer printout of Figure 7.1.

The analysis discussed here considers only male turkeys and uses only the bone measurements that were available on a majority of the birds. This portion of the data was analyzed with the following SAS commands:

```
TITLE 'Ex. 7.1 - Bone Measurements on Turkeys';

DATA; SET TURKEY;  IF SEX='MALE';
 KEEP ID TYPE HUM RAD ULN FEMUR TIN CAR D3P COR SCA;

PROC PRINT;
 RUN;

PROC DISCRIM LIST CROSSLIST;
 CLASSES TYPE;
 VAR HUM--SCA;
 RUN;
```

A portion of the results from these commands is shown on the pages of computer printout shown in Figure 7.1. All of the output can be viewed by executing the file called EX7_1.SAS on the enclosed disk.

There were 82 male turkeys, but complete data on the variables being used in this analysis were available for only 33 turkeys. Nineteen of these were domestic turkeys and 14 were wild turkeys. A portion of the data being analyzed is shown on page 1 of the computer printout.

On page 5 we see that the Mahalanobis squared distance between the two sample means is 13.18039. Section 10.3 gives a method for comparing the means of two populations that is based on the Mahalanobis squared distance between two sample means. If we use this test to compare wild turkeys to domestic turkeys, we find that there is a significant difference in the means of the two types of turkeys ($p < 0.0001$).

The linear discriminant function, $\mathbf{b}'\mathbf{x}$, can be obtained by subtracting the two vectors shown on page 6 (see Figure 7.1) from each other. These vectors actually define what are called *classification functions*. Classification

Ex. 7.1 - Bone Measurements on Turkeys   1

OBS	ID	HUM	RAD	ULN	FEMUR	TIN	CAR	D3P	COR	SCA	TYPE
1	K766	.	.	.	.	.	.	.	.	.	WILD
2	N399	153	138	153	139	162	810	307	.	.	WILD
3	NEX1	.	.	.	.	.	.	.	.	.	WILD
4	NEX2	.	.	.	.	.	.	.	.	.	WILD
5	NEX3	.	.	.	.	.	.	.	.	.	WILD
6	NEB4	.	.	.	.	.	.	.	.	.	WILD
7	NEB5	.	.	.	.	.	.	.	.	.	WILD
8	NEB6	.	.	.	.	.	.	.	.	.	WILD
9	NEB7	.	.	.	.	.	.	.	.	.	WILD
10	NE16	.	.	.	.	.	.	.	.	.	WILD
11	B396	.	.	.	140	.	.	.	107	.	WILD
12	B457	.	.	.	143	151	.	.	108	132	WILD
13	B710	153	140	147	142	151	817	305	102	128	WILD
14	B790	156	137	151	146	155	814	305	111	137	WILD
15	B791	.	132	148	138	145	775	.	106	128	WILD
16	B795	151	134	151	144	.	789	292	116	126	WILD
17	B819	158	135	151	146	152	790	289	111	125	WILD
18	B081	.	135	149	.	149	789	.	111	123	WILD
19	B085	148	129	146	139	147	767	287	106	123	WILD
20	B089	157	140	154	140	159	818	301	116	136	WILD
21	B090	153	138	153	141	151	822	312	115	133	WILD
22	B091	156	138	156	145	150	835	310	118	133	WILD
23	B093	151	133	148	139	152	793	290	105	.	WILD
24	B097	153	135	150	144	158	772	276	102	123	WILD
25	B099	152	140	151	144	158	792	303	111	122	WILD
26	B101	.	.	.	.	151	814	304	.	.	WILD
27	B102	153	.	.	145	156	.	.	115	136	WILD
28	B103	149	133	149	139	151	.	.	107	132	WILD
29	B105	.	.	.	145	160	.	.	116	134	WILD
30	B106	147	130	144	136	145	765	289	108	131	WILD
31	B111	154	138	155	142	153	827	315	111	128	WILD
32	B114	154	138	154	138	155	802	287	111	132	WILD
33	B116	161	131	150	140	151	816	301	112	134	WILD
34	B117	.	142	159	146	156	824	298	113	136	WILD
35	L641	150	.	.	133	141	.	.	111	122	WILD
36	L690	140	126	137	123	141	770	290	95	118	WILD
37	L783	154	.	.	144	157	.	.	113	135	WILD
38	L902	.	126	145	.	.	790	300	.	.	WILD
39	L905	169	.	.	154	163	.	.	122	143	WILD
40	L637	153	137	.	140	.	820	300	104	130	DOMESTIC
41	L674	142	131	140	128	131	800	250	95	122	DOMESTIC
42	L678	148	130	145	133	141	800	290	98	124	DOMESTIC
43	L679	148	131	144	133	135	820	300	103	123	DOMESTIC
44	L682	145	131	140	129	136	790	300	100	123	DOMESTIC
45	L684	153	135	146	138	146	840	310	107	125	DOMESTIC

F I G. 7.1
Pages 1, 5–7, 9, 10, and 12 of computer printout, Example 7.1

Ex. 7.1 - Bone Measurements on Turkeys   5

DISCRIMINANT ANALYSIS

Pairwise Generalized Squared Distances Between Groups

$$D^2(i|j) = (\bar{X}_i - \bar{X}_j)' \, COV^{-1} \, (\bar{X}_i - \bar{X}_j)$$

Generalized Squared Distance to TYPE

From
TYPE   DOMESTIC   WILD

DOMESTIC   0   13.18039
WILD   13.18039   0

Ex. 7.1 - Bone Measurements on Turkeys   6

DISCRIMINANT ANALYSIS   LINEAR DISCRIMINANT FUNCTION

$$Constant = -.5 \, \bar{X}_j' \, COV^{-1} \, \bar{X}_j \qquad Coefficient \; Vector = COV^{-1} \, \bar{X}_j$$

TYPE

	DOMESTIC	WILD
CONSTANT	-1071	-1086
HUM	2.45681	2.35200
RAD	0.61660	0.53661
ULN	5.23163	4.61215
FEMUR	2.47758	2.27022
TIN	-0.21695	0.69567
CAR	0.94512	0.91261
D3P	0.80872	0.70556
COR	-7.74922	-6.92758
SCA	3.42495	3.41294

Classification
functions

F I G. 7.1
Pages 1, 5–7, 9,
10, and 12 of com-
puter printout,
Example 7.1

Ex. 7.1 - Bone Measurements on Turkeys   7

DISCRIMINANT ANALYSIS

Classification Results for Calibration Data: WORK.TURKEY2

Resubstitution Results using Linear Discriminant Function

Generalized Squared Distance Function:

$$D^2_j(X) = (X-\bar{X}_j)' \, COV^{-1} \, (X-\bar{X}_j)$$

Posterior Probability of Membership in each TYPE:

$$Pr(j|X) = \exp(-.5 \, D^2_j(X)) \, / \, SUM_k \, \exp(-.5 \, D^2_k(X))$$

| | | | Posterior Probability of Membership in TYPE: | |
Obs	From TYPE	Classified into TYPE	DOMESTIC	WILD
13	WILD	DOMESTIC *	0.5662	0.4338
14	WILD	WILD	0.0006	0.9994
17	WILD	WILD	0.0008	0.9992
19	WILD	WILD	0.0038	0.9962
20	WILD	WILD	0.0000	1.0000
21	WILD	WILD	0.0022	0.9978
22	WILD	WILD	0.0115	0.9885
24	WILD	WILD	0.0002	0.9998
25	WILD	WILD	0.0000	1.0000
30	WILD	WILD	0.0009	0.9991
31	WILD	WILD	0.0632	0.9368
32	WILD	WILD	0.0001	0.9999
33	WILD	WILD	0.0012	0.9988
36	WILD	WILD	0.3388	0.6612
41	DOMESTIC	DOMESTIC	0.9999	0.0001
42	DOMESTIC	DOMESTIC	0.9978	0.0022
43	DOMESTIC	DOMESTIC	0.9998	0.0002
44	DOMESTIC	DOMESTIC	0.9961	0.0039
45	DOMESTIC	DOMESTIC	0.5250	0.4750
46	DOMESTIC	DOMESTIC	0.9992	0.0008
48	DOMESTIC	DOMESTIC	0.9996	0.0004
49	DOMESTIC	DOMESTIC	0.9997	0.0003
50	DOMESTIC	DOMESTIC	0.9998	0.0002
51	DOMESTIC	DOMESTIC	0.9932	0.0068
52	DOMESTIC	WILD *	0.1934	0.8066
53	DOMESTIC	DOMESTIC	1.0000	0.0000

F I G. 7.1
Pages 1, 5–7, 9, 10, and 12 of computer printout, Example 7.1

DISCRIMINANT ANALYSIS

Classification Summary for Calibration Data: WORK.TURKEY2

Resubstitution Summary using Linear Discriminant Function

Generalized Squared Distance Function:

$$D_j^2(X) = (X - \bar{X}_j)' \, COV^{-1} \, (X - \bar{X}_j)$$

Posterior Probability of Membership in each TYPE:

$$Pr(j|X) = \exp(-.5 \, D_j^2(X)) \, / \, \text{SUM}_k \exp(-.5 \, D_k^2(X))$$

Number of Observations and Percent Classified into TYPE:

From TYPE	DOMESTIC	WILD	Total
DOMESTIC	18	1	19
	94.74	5.26	100.00
WILD	1	13	14
	7.14	92.86	100.00
Total	19	14	33
Percent	57.58	42.42	100.00
Priors	0.5000	0.5000	

Error Count Estimates for TYPE:

	DOMESTIC	WILD	Total
Rate	0.0526	0.0714	0.0620
Priors	0.5000	0.5000	

F I G. 7.1
Pages 1, 5–7, 9, 10, and 12 of computer printout, Example 7.1

```
Ex. 7.1 - Bone Measurements on Turkeys                10
                    DISCRIMINANT ANALYSIS

     Classification Results for Calibration Data: WORK.TURKEY2

     Cross-validation Results using Linear Discriminant Function

        Generalized Squared Distance Function:

        2        _              -1    _
       D (X) = (X-X    )'  COV      (X-X    )
        j          (X)j       (X)       (X)j

        Posterior Probability of Membership in each TYPE:

                           2                    2
       Pr(j|X) = exp(-.5 D (X)) / SUM exp(-.5 D (X))
                         j         k           k

                              Posterior Probability of Membership in TYPE:
           Obs      From          Classified
                    TYPE          into TYPE      DOMESTIC          WILD

           13       WILD          DOMESTIC *     0.9999          0.0001
           14       WILD          WILD           0.0019          0.9981
           17       WILD          WILD           0.0024          0.9976
           19       WILD          WILD           0.0104          0.9896
           20       WILD          WILD           0.0000          1.0000
           21       WILD          WILD           0.0047          0.9953
           22       WILD          WILD           0.0996          0.9004
           24       WILD          WILD           0.0002          0.9998
           25       WILD          WILD           0.0000          1.0000
           30       WILD          WILD           0.0029          0.9971
           31       WILD          WILD           0.4310          0.5690
```

F I G. 7.1
Pages 1, 5–7, 9,
10, and 12 of com-
puter printout,
Example 7.1

functions are described in detail in Section 7.3. For this example, the linear discriminant rule is:

Choose the domestic classification if $\mathbf{b}'\mathbf{x} > k$ where

$$\mathbf{b} = \begin{bmatrix} 2.4568 \\ 0.6166 \\ 5.2316 \\ 2.4776 \\ -0.2170 \\ 0.9451 \\ 0.8087 \\ -7.7492 \\ 3.4250 \end{bmatrix} - \begin{bmatrix} 2.3520 \\ 0.5366 \\ 4.6122 \\ 2.2702 \\ 0.6957 \\ 0.9126 \\ 0.7056 \\ -6.9276 \\ 3.4129 \end{bmatrix} = \begin{bmatrix} 0.1048 \\ 0.0800 \\ 0.6195 \\ 0.2074 \\ -0.9126 \\ 0.0325 \\ 0.1032 \\ -0.8216 \\ 0.0120 \end{bmatrix} \quad \text{and}$$

$$k = -1086 - (-1071) = -15$$

and choose the wild classification otherwise.

DISCRIMINANT ANALYSIS

Classification Summary for Calibration Data: WORK.TURKEY2

Cross-validation Summary using Linear Discriminant Function

Generalized Squared Distance Function:

$$D_j^2(X) = (X-\bar{X}_{(X)j})' COV_{(X)}^{-1} (X-\bar{X}_{(X)j})$$

Posterior Probability of Membership in each TYPE:

$$Pr(j|X) = exp(-.5\ D_j^2(X))\ /\ \underset{k}{SUM}\ exp(-.5\ D_k^2(X))$$

Number of Observations and Percent Classified into TYPE:

From TYPE	DOMESTIC	WILD	Total
DOMESTIC	17	2	19
	89.47	10.53	100.00
WILD	2	12	14
	14.29	85.71	100.00
Total	19	14	33
Percent	57.58	42.42	100.00
Priors	0.5000	0.5000	

Error Count Estimates for TYPE:

	DOMESTIC	WILD	Total
Rate	0.1053	0.1429	0.1241
Priors	0.5000	0.5000	

F I G. 7.1

Pages 1, 5–7, 9, 10, and 12 of computer printout, Example 7.1

Pages 7 and 8 (only a portion of page 7 is shown in Figure 7.1; both pages can be seen by executing the file name EX7_1.SAS on the enclosed disk) show a listing of where each of the turkeys in the sample data would be classified if the sample discriminant rule were applied to the sample data. Those assignments that result in misclassifications are flagged with an asterisk (*). The last two columns on page 7 show the posterior probabilities for each of the two possible classifications. For example, for Obs=13, the posterior probability for DOMESTIC is 0.5662 and that for WILD is 0.4338. Because 0.5662 > 0.4338, observation 13 gets classified into the

DOMESTIC group; this is a misclassification because observation 13 is known to be from a wild turkey, so the observation is flagged with an asterisk.

Page 9 of the printout shown in Figure 7.1 summarizes the classifications that result when the sample discriminant rule is applied to the data used to form the rule. Note that 18 of the 19 (94.7%) DOMESTIC turkeys are classified correctly by the rule, and that 13 of the 14 (92.9%) WILD turkeys are classified correctly by the rule. Remember that these ratios are likely to be overestimates of the true probabilities of correct classification by the discriminant rule since they are obtained by using the rule on the data used to build the rule.

Pages 10 and 11 (only a portion of page 10 is shown) are similar to pages 7 and 8 except that these classifications are obtained through the use of the cross-validation method. Again misclassifications are flagged with an asterisk. On these two pages, when a turkey is being classified, it is removed from the data before building a rule for that turkey. Page 12 summarizes the classifications made using the cross-validation method. From page 12 we can see that 89.47% (17 of 19) of the DOMESTIC turkeys are classified correctly and that 85.71% (12 of 14) of the WILD turkeys are classified correctly.

When the discriminant rule was applied to the turkeys in the accused's freezer, all were classified as being domestic turkeys. Upon seeing the results of this statistical analysis the man accused of stealing his neighbor's turkeys agreed to pay his neighbor for the turkeys (even though the man never admitted that he was guilty of stealing the turkeys). As a result of his willingness to pay his neighbor for the turkeys, the neighbor dropped the charges against him. It is now believed that the state fish and game department is after the man for possessing more wild turkeys than the law allows him to possess.

7.2
Cost Functions and Prior Probabilities (Two Populations)

The discriminant rules given in Section 7.1 do not take into account the relative risks of making errors of misclassification. When there are only two competing populations, these rules have the property that the probability of misclassifying an observation into the first population when it really comes from the second is equal to the probability of misclassifying an observation into the second population when it really comes from the first.

However, a patient who is expecting to receive an anesthetic during surgery may not be too enthused about having these two probabilities of misclassification equal. A patient would not want to be given the anesthetic if there is any real chance that the anesthetic would be unsafe. If the anesthesiologist is going to err, the patient would prefer that she or he err on the side of

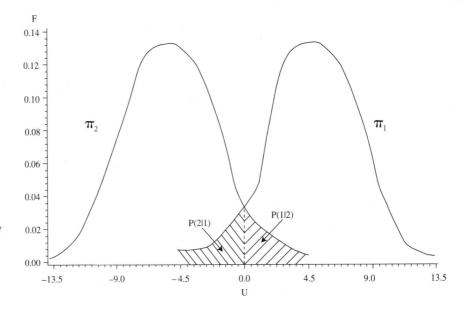

FIG. 7.2

Distribution of U for two populations, equal costs of misclassification

denying the anesthetic whenever there is a small chance that the anesthetic could be dangerous.

To see how the probabilities of misclassification can be changed and to see the effects of such changes, let

$$U = (\boldsymbol{\mu}_1 - \boldsymbol{\mu}_2)'\boldsymbol{\Sigma}^{-1}\mathbf{x} - \tfrac{1}{2}(\boldsymbol{\mu}_1 - \boldsymbol{\mu}_2)'\boldsymbol{\Sigma}^{-1}(\boldsymbol{\mu}_1 + \boldsymbol{\mu}_2)$$

Note that $U = \mathbf{b}'\mathbf{x} - k$ where $\mathbf{b}$ and k were defined earlier. We can show that if $\mathbf{x}$ comes from Π_1, then U will be distributed $N(\tfrac{1}{2}\delta, \delta)$, and if $\mathbf{x}$ comes from Π_2, then U is distributed $N(-\tfrac{1}{2}\delta, \delta)$ where

$$\delta = (\boldsymbol{\mu}_1 - \boldsymbol{\mu}_2)'\boldsymbol{\Sigma}^{-1}(\boldsymbol{\mu}_1 - \boldsymbol{\mu}_2)$$

Note that δ measures the Mahalanobis squared distance between the two population means.

The four discriminant rules described in Section 7.1 are also equivalent to this rule:

Choose Π_1 if $U > 0$, and choose Π_2 otherwise.

Figure 7.2 shows the distribution of U under the two possibilities from which $\mathbf{x}$ might come when $\delta = 9$. $P(2|1)$ is the probability of misclassifying an observation into population 2 when it comes from population 1 and $P(1|2)$ is the probability of misclassifying an observation into population 1 when it comes from population 2.

We can reduce the probability of making an error in one direction (how-

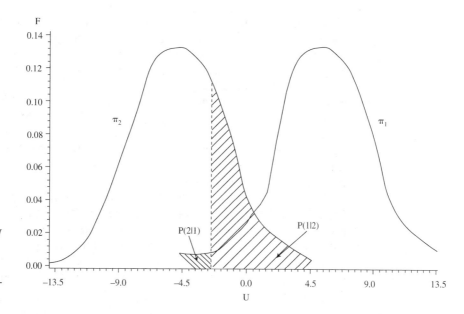

F I G. 7.3
Distribution of U
for two popula-
tions, unequal
costs of misclassi-
fication

ever, it causes an increase in the probability of making an error in the other
direction) simply by taking a discriminant rule of the form:

> Choose Π_1 if $U > u$, and choose Π_2 otherwise, where u is some nonzero con-
> stant.

If we want the probability of misclassifying an observation into the second
population when it comes from the first to be at most α, then we would choose
$u = \frac{1}{2}\delta - z_\alpha\sqrt{\delta}$ where z_α is the upper $\alpha \cdot 100\%$ critical point of the standard
normal probability distribution.

For example, suppose that $\delta = 9$, and that we want the probability of
misclassifying an observation into Π_2 when it comes from Π_1 to be at most
0.01. Then we would take $u = 4.5 - (2.326)(3) = -2.478$. This value of u
makes the probability of misclassification in the other direction equal to 0.2503.
This situation is illustrated in Figure 7.3. When $\delta = 9$ and $u = 0$, the probability
of a misclassification into either population when it comes from the other
is 0.0668.

7.3
A General Discriminant Rule (Two Populations)

In this section, a very general procedure for developing discriminant rules is
given. These general rules allow researchers to take into account the fact that
errors in one direction may be much more serious than errors in the other

direction by assigning relative costs to these two kinds of errors. These general rules also allow researchers to use prior information about the relative frequency with which the two groups generally occur whenever that relative frequency is known or can be estimated. For example, if a credit card company knew that 75% of credit card applicants are good credit risks without knowing anything else about the applicant, then that company might want to take this information into account when developing a discriminant rule.

The rules given in this section require that the probability density functions be known for each of the groups or, at least, that the densities can be estimated. The rules do not require the groups to have probability distributions that belong to the same general class.

Suppose Π_1 is distributed according to the probability density function $f_1(\mathbf{x}; \boldsymbol{\theta}_1)$, which depends on some parameters $\boldsymbol{\theta}_1$, and suppose Π_2 is distributed according to the probability density function $f_2(\mathbf{x}; \boldsymbol{\theta}_2)$, which depends on some other parameters $\boldsymbol{\theta}_2$. A general discriminant rule must divide the p-dimensional sample space (the space into which data vectors fall) into two parts, R_1 and R_2, so that when $\mathbf{x}$ falls in R_1, Π_1 is chosen, and when $\mathbf{x}$ falls in R_2, π_2 is chosen.

A Cost Function

Let $C(i|j)$ represent the cost of misclassifying an observation from Π_j into Π_i. Without any loss of generality, there is no reward for classifying an observation correctly, only a penalty if an observation is classified incorrectly. Also let $P(i|j)$ represent the probability of misclassifying an observation from Π_j into Π_i.

Prior Probabilities

In some cases, a researcher may have prior knowledge as to how likely it is that a randomly selected observation would come from each of the two groups. For example, the anesthesiologist may know that, in the absence of any other information about the patients, 80% of them are safe for the anesthetic. A good discriminant rule should be able to take this information into account.

Let p_i (called the *prior probability* for group i) represent the probability that a randomly selected observation comes from Π_i, for $i = 1, 2$.

Average Cost of Misclassification

We can show that the average cost of the misclassification of a randomly selected observation is

$$p_1 \cdot C(2|1) \cdot P(2|1) + p_2 \cdot C(1|2) \cdot P(1|2)$$

A Bayes Rule

Given prior probabilities p_1 and p_2 for each population, a rule that minimizes the average cost of misclassification of a randomly selected observation is called a Bayes rule with respect to prior probabilities p_1 and p_2.

We can show that the Bayes rule is:

Choose Π_1 if $p_2 \cdot f_2(\mathbf{x};\, \boldsymbol{\theta}_2) \cdot C(1|2) < p_1 \cdot f_1(\mathbf{x};\, \boldsymbol{\theta}_1) \cdot C(2|1)$, and choose Π_2 otherwise.

Note   If $p_1 = p_2$ and $C(1|2) = C(2|1)$, then the Bayes rule simplifies to the likelihood rule.

Note   If Π_1 and Π_2 are multivariate normal populations with equal variance–covariance matrices, the Bayes rule says to:

Choose Π_1 if $d_1^* < d_2^*$ where

$$d_i^* = \tfrac{1}{2}(\mathbf{x} - \boldsymbol{\mu}_i)'\boldsymbol{\Sigma}^{-1}(\mathbf{x} - \boldsymbol{\mu}_i) - \log[p_i \cdot C(j|i)], \qquad \text{for } i \neq j = 1, 2 \qquad (7.1)$$

Classification Functions

When the costs of classification errors in the two directions are equal and when the variance–covariance matrices in both populations are equal, we can compute functions, called *classification functions,* for each group. The classification function for the ith group is defined by

$$c_i = \boldsymbol{\mu}_i'\boldsymbol{\Sigma}^{-1}\mathbf{x} - \tfrac{1}{2}\boldsymbol{\mu}_i'\boldsymbol{\Sigma}^{-1}\boldsymbol{\mu}_i + \ln p_i, \qquad \text{for } i = 1, 2$$

We can show that

$$d_1^* < d_2^*$$

if and only if $c_1 > c_2$. Thus, we could compute the value for each group's classification function for an observed data vector, and assign the data vector to the population that produces the largest value for the classification function.

Unequal Covariance Matrices

Suppose $\boldsymbol{\Sigma}_1 \neq \boldsymbol{\Sigma}_2$. In this case the Bayes rule becomes:

Choose Π_1 if $d_1^{**} < d_2^{**}$ where

$$d_i^{**} = \tfrac{1}{2}(\mathbf{x} - \boldsymbol{\mu}_i)'\boldsymbol{\Sigma}_i^{-1}(\mathbf{x} - \boldsymbol{\mu}_i) + \tfrac{1}{2}\log(|\boldsymbol{\Sigma}_i|) - \log[p_i \cdot C(j|i)]$$

For equal costs of misclassification in each direction, the Bayes rule simplifies to:

Choose Π_1 if $d_1^{***} < d_2^{***}$ where

$$d_i^{***} = \tfrac{1}{2}(\mathbf{x} - \boldsymbol{\mu}_i)'\boldsymbol{\Sigma}_i^{-1}(\mathbf{x} - \boldsymbol{\mu}_i) + \tfrac{1}{2}\log(|\boldsymbol{\Sigma}_i|) - \log(p_i)$$

Some authors refer to the two rules based on d_i^{**} and d_i^{***} as *quadratic discriminant rules* since they involve quadratic functions of the elements in $\mathbf{x}$. Note that d_i^{**} and d_i^{***} are not true measures of distance. In fact, quite often d_i^{**} and d_i^{***} are actually negative numbers. This is because $|\mathbf{\Sigma}_i|$ can be less than 1, and in these cases, the $\log(|\mathbf{\Sigma}_i|)$ will be negative.

As described in previous sections of this chapter, if you do not know $\boldsymbol{\mu}_1$, $\boldsymbol{\mu}_2$, $\mathbf{\Sigma}_1$, and $\mathbf{\Sigma}_2$, then they are replaced by their respective estimates before using the preceding formulas.

Comment   Section 10.1 introduces a statistical test that can be used to test $H_0: \mathbf{\Sigma}_1 = \mathbf{\Sigma}_2$ versus $H_a: \mathbf{\Sigma}_1 \neq \mathbf{\Sigma}_2$, and several statistical computing packages contain an option to test for the equality of the population variance–covariance matrices. However, many statisticians have observed that the hypothesis of equal variance–covariance matrices is almost always rejected in practice when large sample sizes are used. When we have extremely large sample sizes, almost all tested hypotheses will be rejected. So a more useful question to consider when the hypothesis of equal variance–covariance matrices is rejected is "Will the inequality of the variance–covariance matrix have any practical implications?" This question is not directly answerable. To help determine whether the inequality of the variance–covariance matrices has any practical impact, we can analyze a set of data in two ways—one by pooling the variance–covariance matrices and one by not pooling—and then compare the classification summary matrices for the two options. We would then choose the option that tends to classify observations correctly more often.

Tricking Computing Packages

Many statistical computing packages allow users to choose different prior probabilities for each group, but none seems to allow users to take different cost functions into account. However, those packages that allow users to use prior probabilities can be tricked into using cost functions when there are only two competing populations. Let

$$p_1^* = \frac{p_1 C(2|1)}{p_1 C(2|1) + p_2 C(1|2)} \quad \text{and} \quad p_2^* = \frac{p_2 C(1|2)}{p_1 C(2|1) + p_2 C(1|2)}$$

Note that p_1^* and p_2^* are both nonnegative and they add to one, thus they possess the same properties that probabilities possess. Also note that $d_1^{**} < d_2^{**}$, if and only if $d_1^* < d_2^*$, where

$$d_i^* = \tfrac{1}{2}(\mathbf{x} - \boldsymbol{\mu}_i)\,'\mathbf{\Sigma}^{-1}\,(\mathbf{x} - \boldsymbol{\mu}_i) - \ln(p_i^*)$$

Thus, to trick those computing packages that allow use of prior probabilities into using unequal cost functions, we simply tell the computing package that the prior probabilities are p_1^* and p_2^*.

7.4
Discriminant Rules (More Than Two Populations)

The previous sections discussed general discriminant rules for cases when there are just two populations under consideration. Often, a researcher will have several populations into which observations must be classified. For example, a bank issuing credit cards might wish to classify applicants into several different risk categories. Individuals might be offered differing credit limits according to perceived risks based on information available about the applicant. As a result, some individuals may not be offered credit cards at all, while others may be offered a card but are subjected to a $1000 credit limit. Others may be offered a $3000 credit limit, others a $10,000 credit limit, etc. Thus, there are several different populations into which applicants could be classified.

Except for being able to obtain a single linear discriminant rule, all of the proposals for discriminating described in Section 7.1 can be extended to cases where there are more than two populations. In cases where there are more than two populations in which to classify observations, one can still

1 compute the Mahalanobis squared distance between an observation and each of the population means, and then classify the observation into the population to whose mean it is closest;

2 compute the posterior probability of an observation for each of the competing populations, and classify the observation into the population that gives the largest posterior probability; or

3 compute the value of each population's classification function, and classify an observation into the population that gives the largest value for the classification function.

Each of these alternatives is equivalent. It is simply a matter of how the user of discriminant analysis prefers to view a solution to the discrimination problem.

The only concept for discriminating between two populations that does not generalize to discriminating between more than two populations is that of being able to take the relative costs of misclassification errors into account. The case of more than two populations is considered in the next example.

E X A M P L E *7.2*

A researcher wanted to determine if it might be possible to develop a classification rule that could be used to discriminate among different groups of wheat kernels. She had four groups of wheat kernels. Groups 1 and 2 were grown in one location and groups 3 and 4 were grown in another location. In addition, groups 1 and 3 were of a variety called *Arkan* while groups 2 and 4 were of a variety called *Arthur*.

To understand why this is an important problem, you can read the following two paragraphs. For those who do not care, skip these paragraphs.

Arthur is a soft wheat variety and Arkan is a hard wheat variety. Arkan was developed by crossing a hard wheat variety called Kansas with a soft wheat variety called Arthur. The resultant variety (Arkan) looks like a soft wheat variety but has baking characteristics similar to those possessed by hard wheat varieties. In particular, Arkan, like other hard wheat varieties, has a higher level of protein than soft wheat varieties. Because of their higher protein content, hard wheat varieties are used for most bread products, while soft wheat varieties are generally used for cookie and cake products. Hard wheat varieties generally have a higher market value than the soft wheat varieties.

There is some concern about whether an unscrupulous person might try mixing Arthur with Arkan, and then try to sell the combination of the two varieties as being all Arkan, the high-protein variety. If successful, this person would then receive a hard wheat price for the Arthur that is being bought at soft wheat prices. While this is perhaps an easy way to make a tidy profit, it's hardly ethical. Unfortunately, such unscrupulous people exist. In fact, a few years ago, it was discovered that a grain elevator operator had mixed 300 tons of dirt into a shipment of wheat that was going to be sent overseas. It seems that the wheat was cleaner than the law requires, so this operator added dirt to the wheat to get the shipment up to the level of uncleanliness allowed by law. Thus he was able to buy dirt at dirt prices and sell it at wheat prices, making a tidy, but not ethical, profit. At the time this was discovered, adding dirt was not illegal, so no real penalty could be applied. Now it is illegal to make wheat dirtier by adding foreign material to it.

Although it is possible to identify these two varieties accurately by grinding wheat kernels and performing chemical analyses on the flour obtained, this researcher was looking for easier ways to identify varieties. This researcher wanted to use physical measurements of wheat kernels to try to discriminate among groups of kernels. Thus, she took measurements of the area, perimeter, length, and breadth of each kernel.

Each wheat kernel has a crease in it, and she took measurements on the kernels with the crease to the right and then again with the crease down. A portion of the data to be considered in this analysis is shown on page 1 of the computer printout of Figure 7.4. The variables measured when the crease was down are the square root of the area of the kernel (DOWN_A), the perimeter of the kernel (DOWN_P), the length of the kernel (DOWN_P) and the breadth of the kernel (DOWN_B). The variables RIGHT_A, RIGHT_P, RIGHT_L, and RIGHT_B are defined similarly, except that the kernel was measured with the crease to the right. All measurements were taken electronically by an "image analyzer." Groups 1 and 2 contain 36 observations each, and groups 3 and 4 contain 50 observations each.

Ex. 7.2 - SIZE MEASUREMENTS ON WHEAT KERNELS   1-4

O B S	L O C	V A R I E T Y	G R P	D O W N — A	D O W N — P	D O W N — L	D O W N — B	R I G H T — A	R I G H T — P	R I G H T — L	R I G H T — B
1	MAS0	ARKAN	1	54.4518	219	89	43	56.6039	226	89	47
2	MAS0	ARKAN	1	55.1453	221	91	46	56.2583	224	91	46
3	MAS0	ARKAN	1	53.9166	223	90	44	55.0908	223	91	44
4	MAS0	ARKAN	1	52.2303	212	87	41	53.5444	215	88	44
5	MAS0	ARKAN	1	51.5558	207	78	42	52.9811	211	81	44
6	MAS0	ARKAN	1	50.4282	203	82	41	51.2152	207	82	42
7	MAS0	ARKAN	1	50.5074	208	84	40	52.7636	211	84	42
8	MAS0	ARKAN	1	52.7636	218	91	41	52.3450	216	90	41
9	MAS0	ARKAN	1	53.7308	216	81	44	56.8155	221	86	50
37	MAS0	ARTHUR	2	50.4975	205	85	41	50.8724	215	86	42
38	MAS0	ARTHUR	2	52.4118	212	89	42	54.0185	217	91	44
39	MAS0	ARTHUR	2	52.0384	210	85	42	51.8266	209	85	42
40	MAS0	ARTHUR	2	56.3205	222	94	46	54.9181	223	92	45
41	MAS0	ARTHUR	2	56.2406	224	90	46	55.2178	223	91	46
42	MAS0	ARTHUR	2	56.4535	229	93	46	52.7447	221	99	39
43	MAS0	ARTHUR	2	52.7162	215	87	42	51.4782	213	87	40
44	MAS0	ARTHUR	2	54.5711	219	89	45	54.0925	215	88	44
45	MAS0	ARTHUR	2	56.0803	225	92	47	55.5608	229	92	45
73	VLAD12	ARKAN	3	52.4690	217	92	40	54.7175	221	93	44
74	VLAD12	ARKAN	3	56.7803	234	89	45	56.7891	230	95	46
75	VLAD12	ARKAN	3	58.4722	234	96	47	58.0086	235	95	46
76	VLAD12	ARKAN	3	54.8635	224	92	47	56.6569	224	92	47
77	VLAD12	ARKAN	3	54.7631	228	96	42	56.4092	232	96	46
78	VLAD12	ARKAN	3	54.7814	223	95	42	55.0727	228	95	44
79	VLAD12	ARKAN	3	55.0636	228	95	43	56.2406	232	95	46
80	VLAD12	ARKAN	3	56.5066	225	93	46	55.4166	228	92	44
81	VLAD12	ARKAN	3	51.0000	213	94	39	52.2972	216	89	41
123	VLAD12	ARTHUR	4	53.9907	220	100	43	50.1996	218	93	37
124	VLAD12	ARTHUR	4	56.6480	220	88	48	53.4509	213	87	44
125	VLAD12	ARTHUR	4	53.1695	213	88	43	51.3128	212	89	41
126	VLAD12	ARTHUR	4	56.3738	228	95	45	55.0999	225	92	47
127	VLAD12	ARTHUR	4	57.1752	229	96	46	54.2218	223	92	45
128	VLAD12	ARTHUR	4	56.3383	220	89	48	52.8772	217	89	44
129	VLAD12	ARTHUR	4	54.4794	216	88	45	52.6308	212	88	44
130	VLAD12	ARTHUR	4	56.2139	223	92	45	54.3047	223	91	42
131	VLAD12	ARTHUR	4	56.1694	223	88	47	56.8243	221	89	51

F I G. 7.4
Pages 1–9, 11, and 20 of computer printout, Example 7.2

Ex. 7.2 - SIZE MEASUREMENTS ON WHEAT KERNELS   5

DISCRIMINANT ANALYSIS

172 Observations   171 DF Total
  8 Variables   168 DF Within Classes
  4 Classes   3 DF Between Classes

Class Level Information

F I G. 7.4

Pages 1–9, 11,
and 20 of com-
puter printout,
Example 7.2

GRP	Frequency	Weight	Proportion	Prior Probability
1	36	36.0000	0.209302	0.209302
2	36	36.0000	0.209302	0.209302
3	50	50.0000	0.290698	0.290698
4	50	50.0000	0.290698	0.290698

Ex. 7.2 - SIZE MEASUREMENTS ON WHEAT KERNELS   6

DISCRIMINANT ANALYSIS   WITHIN COVARIANCE MATRIX INFORMATION

F I G. 7.4

Pages 1–9, 11,
and 20 of com-
puter printout,
Example 7.2

GRP	Covariance Matrix Rank	Natural Log of Determinant of the Covariance Matrix
1	8	12.18400
2	8	17.94579
3	8	11.59084
4	8	10.67200
Pooled	8	15.74996

Basic Discrimination

The wheat kernel data were initially analyzed with the following SAS commands:

```
OPTIONS PAGESIZE=54 LINESIZE=75 NODATE;
TITLE 'Ex. 7.2- SIZE MEASUREMENTS ON WHEAT KERNELS';

PROC PRINT;
 VAR LOC VARIETY GRP DOWN_A--DOWN_B RIGHT_A--RIGHT_B;

PROC DISCRIM DATA=FINAL POOL=TEST CROSSVALIDATE CROSSLIST;
 CLASSES GRP;
 VAR DOWN_A--DOWN_B RIGHT_A--RIGHT_B;
 PRIORS PROP;   RUN;
```

```
              Ex. 7.2 - SIZE MEASUREMENTS ON WHEAT KERNELS                    7

DISCRIMINANT ANALYSIS      TEST OF HOMOGENEITY OF WITHIN COVARIANCE MATRICES

      Notation: K     = Number of Groups

                P     = Number of Variables

                N     = Total Number of Observations - Number of Groups

                N(i) = Number of Observations in the i'th Group - 1


                      __                              N(i)/2
                      ||  |Within SS Matrix(i)|
             V     = ----------------------------------
                                                N/2
                      |Pooled SS Matrix|


                    _                     _   2
                    |       1        1    | 2P + 3P - 1
           RHO = 1.0 - |  SUM -----  -  ---  | -------------
                    |_       N(i)       N  _| 6(P+1)(K-1)


           DF    = .5(K-1)P(P+1)


                               |  _                      _ |
                               |       PN/2               |
                               |    N            V        |
Under null hypothesis:  -2 RHO ln | ------------------- |
                               |    __          PN(i)/2   |
                               |_   || N(i)              _|

is distributed approximately as chi-square(DF)


Test Chi-Square Value =   457.642902
with      108 DF       Prob > Chi-Sq = 0.0001
```

Tests Ho: $\Sigma_1 = \Sigma_2 = \Sigma_3 = \Sigma_4$

```
Since the chi-square value is significant at the  0.1000 level,
the within covariance matrices will be used in the discriminant function.

Reference: Morrison, D.F. (1976)    Multivariate Statistical Methods p252.
```

F I G. 7.4
Pages 1–9, 11, and 20 of computer printout, Example 7.2

Before looking at the output created by the preceding commands, we should discuss some of the options used in this command stream. The POOL=TEST option produces a statistical test of the hypothesis that the variance–covariance matrices of the four groups are equal. If the hypothesis is accepted, the DISCRIM procedure pools the four sample variance–covariance matrices into a single one, and then assigns an observation into

```
       Ex. 7.2 - SIZE MEASUREMENTS ON WHEAT KERNELS            8

                    DISCRIMINANT ANALYSIS

           Pairwise Generalized Squared Distances Between Groups

       2           _   _      -1   _   _
      D (i|j) = (X - X )' COV   (X - X ) + ln |COV | - 2 ln PRIOR
                  i   j     j    i   j          j               j

              Generalized Squared Distance to GRP
```

F I G. 7.4

Pages 1–9, 11, and 20 of computer printout, Example 7.2

From GRP	1	2	3	4
1	15.31195	26.04731	16.71814	23.50097
2	27.98156	21.07374	26.73456	30.59429
3	18.10906	27.76505	14.06178	20.64051
4	26.57096	22.20930	27.04606	13.14294

the population for which

$$d_i^* = \tfrac{1}{2}(\mathbf{x} - \hat{\boldsymbol{\mu}}_i)'\hat{\boldsymbol{\Sigma}}^{-1}(\mathbf{x} - \hat{\boldsymbol{\mu}}_i) - \log(p_i), \qquad \text{for } i = 1, 2, 3, 4$$

is minimized. If the hypothesis of equal variance–covariance matrices is rejected, then the DISCRIM procedure assigns an observation into the population for which

$$d_i^{**} = \tfrac{1}{2}(\mathbf{x} - \hat{\boldsymbol{\mu}}_i)'\hat{\boldsymbol{\Sigma}}_i^{-1}(\mathbf{x} - \hat{\boldsymbol{\mu}}_i) + \tfrac{1}{2}\log(|\hat{\boldsymbol{\Sigma}}_i|) - \log(p_i), \qquad \text{for } i = 1, 2, 3, 4$$

is minimized.

Because both alternatives produce classification summary matrices, some researchers prefer to run the DISCRIM procedure twice—once by pooling variance–covariance matrices and once by not pooling—and then compare the resulting classification summary matrices. This can be done by running the DISCRIM procedure first with a POOL=YES option, and then by running it again with a POOL=NO option. Running the DISCRIM procedures both ways is probably a good idea, especially when sample sizes are large (50 or more observations in each group). The reason for this is that in large sample size cases, the test of equal variance–covariance matrices is so powerful that it almost always rejects equality of the matrices. However, the differences between the covariance matrices may not be large enough to be of any real practical importance. The only way you can tell whether it is better to pool the covariance matrices or not is to try discriminating both ways. The advantages of not pooling is that simpler discriminant rules are possible.

The CROSSVALIDATE option produces estimates of the probabilities of misclassification according to Lachenbruch's cross-validation procedure, and the CROSSLIST option provides a listing of how each of the data vectors in the calibration samples would be classified by the cross-validation method. Those that are misclassified are flagged with an asterisk (*).

```
      Ex. 7.2 - SIZE MEASUREMENTS ON WHEAT KERNELS                    9

                    DISCRIMINANT ANALYSIS

        Classification Summary for Calibration Data: WORK.FINAL

     Resubstitution Summary using Quadratic Discriminant Function

       Generalized Squared Distance Function:
```

$$D_j^2(X) = (X-\bar{X}_j)'\ COV_j^{-1}\ (X-\bar{X}_j) + \ln\ |COV_j| - 2 \ln PRIOR_j$$

```
       Posterior Probability of Membership in each GRP:
```

$$Pr(j|X) = \exp(-.5\ D_j^2(X))\ /\ SUM_k\ \exp(-.5\ D_k^2(X))$$

Number of Observations and Percent Classified into GRP:

From GRP	1	2	3	4	Total
1	24 66.67	1 2.78	9 25.00	2 5.56	36 100.00
2	2 5.56	8 22.22	2 5.56	24 66.67	36 100.00
3	5 10.00	1 2.00	41 82.00	3 6.00	50 100.00
4	0 0.00	2 4.00	1 2.00	47 94.00	50 100.00
Total Percent	31 18.02	12 6.98	53 30.81	76 44.19	172 100.00
Priors	0.2093	0.2093	0.2907	0.2907	

FIG. 7.4

Pages 1–9, 11, and 20 of computer printout, Example 7.2

The PRIORS PROP option gives prior probabilities (for each group) equal to the ratio of the observed number of observations in each group's sample to the total number of observations in all of the samples. This option generally is used when we expect the relative frequencies of the observations in each group to be proportional to what it actually will be in the population to which the discriminant rule will be applied. This is not the case for this example, and we cannot make a very good argument for using such prior probabilities here. However, we cannot make a very good argument for excluding prior probability information either.

Ex. 7.2 - SIZE MEASUREMENTS ON WHEAT KERNELS   11

Discriminant Analysis   Classification Results for Calibration Data: WORK.FINAL

Cross-validation Results using Quadratic Discriminant Function

Generalized Squared Distance Function:

$$D^2_j(X) = (X-\bar{X}_{(X)j})' COV^{-1}_{(X)j} (X-\bar{X}_{(X)j}) + \ln |COV_{(X)j}| - 2 \ln PRIOR_j$$

Posterior Probability of Membership in each GRP:

$$Pr(j|X) = \exp(-.5 D^2_j(X)) / SUM_k \exp(-.5 D^2_k(X))$$

			Posterior Probability of Membership in GRP:			
Obs	From GRP	Classified into GRP	1	2	3	4
1	1	1	0.7120	0.0041	0.2688	0.0151
2	1	3 *	0.1705	0.0076	0.7630	0.0589
3	1	3 *	0.2002	0.0013	0.7978	0.0007
4	1	1	0.4697	0.0025	0.3834	0.1445
37	2	4 *	0.0269	0.0002	0.0001	0.9728
38	2	4 *	0.2810	0.0004	0.3177	0.4008
39	2	1 *	0.5081	0.0164	0.1334	0.3420
40	2	4 *	0.0007	0.0119	0.0027	0.9847
73	3	3	0.2464	0.0001	0.7130	0.0405
74	3	3	0.0070	0.1223	0.8627	0.0080
75	3	3	0.0010	0.0010	0.9095	0.0886
76	3	3	0.3682	0.0010	0.6308	0.0000
123	4	2 *	0.0000	0.9965	0.0001	0.0034
124	4	4	0.0000	0.0014	0.0000	0.9986
125	4	4	0.0033	0.0229	0.0009	0.9729
126	4	4	0.0000	0.0406	0.0000	0.9594

F I G. 7.4

Pages 1–9, 11, and 20 of computer printout, Example 7.2

The primary reason the PRIORS PROP option was included here is that the only data available for testing the discriminant rule are the data that are being used to form the rule, and these data sets do not have equal numbers in each of the four groups. As a consequence, to be fair to the rule, we should weigh classifications in favor of those groups that had more observations. This is accomplished by including the PRIORS PROP option. I

```
                Ex. 7.2 - SIZE MEASUREMENTS ON WHEAT KERNELS                    20

  Discriminant Analysis    Classification Summary for Calibration Data: WORK.FINAL

       Cross-validation Summary using Quadratic Discriminant Function

    Generalized Squared Distance Function:
```

$$
D^2_j(X) = (X - \bar{X}_{(X)j})' \, COV^{-1}_{(X)j} \, (X - \bar{X}_{(X)j}) + \ln |COV_{(X)j}| - 2 \ln PRIOR_j
$$

```
    Posterior Probability of Membership in each GRP:
```

$$
Pr(j|X) = \exp(-.5 \, D^2_j(X)) \, / \, SUM_k \, \exp(-.5 \, D^2_k(X))
$$

```
      Number of Observations and Percent Classified into GRP:
```

From GRP	1	2	3	4	Total
1	18	4	12	2	36
	50.00	11.11	33.33	5.56	100.00
2	2	7	2	25	36
	5.56	19.44	5.56	69.44	100.00
3	8	2	35	5	50
	16.00	4.00	70.00	10.00	100.00
4	0	4	3	43	50
	0.00	8.00	6.00	86.00	100.00
Total	28	17	52	75	172
Percent	16.28	9.88	30.23	43.60	100.00
Priors	0.2093	0.2093	0.2907	0.2907	

FIG. 7.4
Pages 1–9, 11, and 20 of computer printout, Example 7.2

must emphasize, however, that this is a weak argument for this particular example.

Whenever possible, prior probabilities should be taken to be equal to whatever they are in the population to which the discrimination rule is actually going to be applied. If these probabilities are not known, then you should make the prior probabilities equal. In SAS's DISCRIM procedure this can be done by including a PRIORS EQUAL option or by leaving the PRIORS option out altogether.

Some of the results from the preceding SAS commands are printed on pages 1–21 of the computer output pages. A few of these pages are shown in Figure 7.4; all of the output can be seen by executing the file named `EX7_2.SAS` on the enclosed floppy disk.

Pages 1–4 contain a printout of the data being analyzed. Only nine lines of data from each of the four groups are shown. Page 5 shows some summary information about each of the groups and is helpful for determining if the information agrees with the user's preconceptions about the data being analyzed.

Page 6 shows some of the information that is used to develop the test for equal variance–covariance matrices; it can safely be ignored by most data analysts. In the middle of page 7 we find a statistical test of the hypothesis that the population variance–covariance matrices are equal. The results are summarized in the lines:

```
Test Chi-Square Value =   457.642902
with     108 DF     Prob > Chi-Sq = 0.0001
```

In this case, the hypothesis is rejected at the 0.0001 level. As a result, the `DISCRIM` procedure uses the discriminant rule that is weighted by unequal covariance matrices and by unequal prior probabilities.

While page 8 can safely be ignored, you might note that it supposedly contains distances between groups. When you have equal variance–covariance matrices and equal prior probabilities, this is true; the distances printed are the Mahalanobis squared distances between the estimates of the group means. When the variance–covariance matrices are unequal and/or the prior probabilities are unequal, the generalized distance between groups 1 and 2 is weighted differently than the distance between groups 2 and 1. As a result the printed matrix is not symmetric. In those cases where the determinant of the individual sample variance–covariance matrices is less than 1, the printed generalized distances can even be negative. This is because the logarithms of numbers less than 1 are negative. For the data being analyzed here, note that the generalized distance from group 1 to group 2 is not equal to the generalized distance from group 2 to group 1.

Page 9 of the printout shown in Figure 7.4 summarizes how the kernels in the calibration data would be classified by the resubstitution method. You can see that 66.67% of the observations in group 1, 82% of those in group 3, and 94% of those in group 4 are classified correctly by the discriminant rule, while only 22.22% of those in group 2 are classified correctly. Remember that these are likely to be overestimates of the true probabilities of success, since they are obtained by applying the rule to data that were used to form the rule.

Recalling that groups 1 and 3 were of the same variety and that groups 2 and 4 were of the same variety, we can see that most of the errors have to do with being able to classify locations correctly rather than classifying varieties correctly. If we consider the estimated probabilities of correct variety classification, then group 1 is classified into the correct variety 92.67%

(*Note:* 92.67 = 67.67 + 25) of the time, group 2 is classified correctly 88.89% of the time, group 3 is classified correctly 92% of the time, and group 4 is classified correctly 98% of the time.

Pages 11–19 (only page 11 is shown in Figure 7.4; the first four observations in each group are shown; the remaining pages can be seen by executing the file called EX7_2.SAS on the enclosed disk) provide a listing of the cross-validations for each kernel. These pages show how each kernel would be classified by a rule obtained by using all of the data except that for the kernel being classified. These pages also show the posterior probabilities of the four groups for each kernel.

The classification results are summarized on page 20. These are likely to be nearly unbiased estimates of the true probabilities of successful classification and, as a consequence, are better estimates than those obtained by the resubstitution method. From this page, we can see that group 1 is classified into the correct variety 50% + 33.33% = 83.33% of the time, group 2 is classified into the correct variety 19.44% + 69.44% = 88.89% of the time, group 3 is classified into the correct variety 16% + 70% = 86% of the time, and group 4 is classified into the correct variety 8% + 86% = 94% of the time.

7.5
Variable Selection Procedures

When several variables are being considered for discrimination purposes, you might ask questions such as (1) are all the variables really necessary for effective discrimination and (2) which variables are the best discriminators?

Variable selection procedures have been proposed that can provide some guidance to researchers wishing to select a subset of the measured variables to use for discrimination purposes. Most existing variable selection procedures are somewhat similar to variable selection procedures used for multiple regression problems. The methods discussed in this book are (1) a forward selection procedure, (2) a backward elimination procedure, and (3) a stepwise procedure that is a combination of 1 and 2.

Forward Selection Procedure

The forward selection procedure begins by choosing the variable that is expected to be the best discriminator among all of the available variables. This is usually accomplished by testing a hypothesis of equal group means for each prospective discriminator. The variable that produces the largest F value in a one-way analysis of variance (ANOVA) is the first variable selected—provided that its F value is statistically significant at a level specified by the researcher. If no variable produces a significant *F* value, then we can conclude that the prospective variables will not discriminate between the groups.

If one of the variables is selected, then choosing a second variable is considered. This is usually accomplished by testing a hypothesis of equal group means for each remaining variable after adjusting for the variable already selected. The variable selected next is the one that produces the largest *F* value in a one-way analysis of covariance (ANCOVA) using the first variable selected as a covariate—provided that the second variable's *F* value is statistically significant at the chosen level. If no second variable is statistically significant, the procedure stops and the first variable is the only one selected.

If a second variable is selected, then choosing a third variable is considered, etc. During each step, an ANCOVA that includes all previously selected variables as covariates is used to select the next variable. The forward procedure stops when none of the remaining variables is statistically significant.

Backward Elimination Procedure

The backward elimination procedure begins by including all prospective variables as discriminators, after which it removes the variable that seems to be the least useful for discriminating. The variable selected for removal is identified by performing an ANCOVA for each variable using all other variables as covariates. If all variables are statistically significant at a chosen significance level, the procedure stops; but if some variables are not statistically significant then the procedure removes the variable that is least significant.

If a variable is removed, then the backward elimination procedure looks for a second variable to remove. The variable to be removed next is identified by performing an ANCOVA on each remaining variable using all other remaining variables as covariates. If all remaining variables are statistically significant, the procedure stops. If some remaining variables are not statistically significant then the procedure removes the variable that is least significant.

If a second variable is removed, then the procedure looks for a third variable, etc. Within each step, an ANCOVA, using all remaining variables as covariates, is used to select the next variable for removal. The backward elimination procedure stops when all remaining variables are statistically significant.

Stepwise Selection Procedure

The stepwise procedure uses combinations of the forward selection procedure and the backward elimination procedure. This procedure selects variables for inclusion within each step in exactly the same way that the forward selection procedure selects variables. Where the stepwise procedure differs from the forward selection procedure is that at each step, before choosing a new variable to include, it checks to see if all of the variables previously selected remain significant. In some instances, a variable may seem very useful early in the selection process, but after a few additional variables are included, one selected earlier may no longer be useful. This procedure would remove such a variable, while the forward selection procedure never removes any variables. This procedure stops when no other variables meet the criteria for entry or when the variable to be included next is one that was just removed.

Recommendations

Many statistical computing packages allow the use of variable selection procedures. I believe that the best selection procedure is the backward elimination procedure provided that the number of prospective variables is at most 15. When the number of variables exceeds 15, then the stepwise procedure is recommended. The forward selection procedure is not recommended since it may produce sets of discriminating variables for which not every variable is significant.

When using variable selection procedures, a researcher can usually choose the significance levels (or F values) at which variables enter into the discriminating set of variables or are removed from this set.

When using the backward elimination procedure, I recommend that a significance level for a variable to remain in the model be set at $\alpha = 0.01$. When using the stepwise selection procedure, I recommend that the significance level for entry of a variable into the discriminating set be fixed at an α somewhere between 0.25 and 0.50, and the significance level for a variable remaining in the discriminating set be fixed at $\alpha = 0.15$.

When using statistical computing packages, researchers should try to see the results of a large number of possible subsets of discriminating variables. The recommendations given earlier will force the computer to provide information about a large number of possibilities, and a researcher can subjectively choose from among these possibilities. There is no law, written or unwritten, that says a researcher must use the set of variables occurring on the last page of the computer printout.

Caveats

While selection procedures select variables, they usually do not evaluate how well the selected variables actually discriminate. To see how well they actually discriminate, you will often have to run a discriminant analysis program using the selected variables as discriminators. Contrary to what might be expected, a subset of well-chosen variables will often do a better job of discriminating between groups than you can do by using all possible variables.

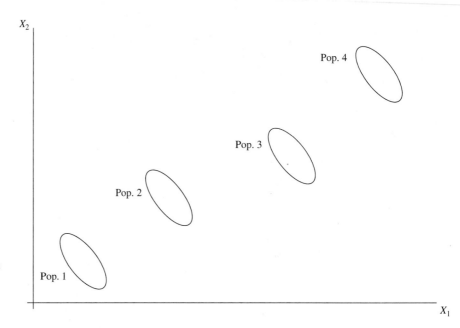

F I G. 7.5
Case where both x_1 and x_2 are statistically significant; the variable x_1 will discriminate among the populations

It is important to note that the selection methods are based on tests about the means of the groups on the candidate variables, and populations can have different means, but still overlap. Whether variables discriminate well depends not on whether the means are different, but on whether the populations overlap. Also just because populations have differing means on more than one variable, it does not necessarily follow that all variables will be required in order to discriminate adequately.

Figures 7.5 and 7.6 show the locations of the ellipsoids of concentration for distinct bivariate normal populations. Figure 7.5 illustrates a situation where both x_1 and x_2 are likely to be highly significant; however, only x_1 is actually needed. To see this, sketch some vertical lines midway between the different ellipsoids. Clearly, if we knew the value of x_1 for a new observation vector, we would know where to assign the new observation. Figure 7.6 illustrates a situation where x_1 is likely to be the most significant variable. However, x_2 will be a better discriminator. The variable x_1 will discriminate population 4 from the others, but it will not discriminate between the other three populations. On the other hand, x_2 should effectively discriminate all four populations from one another.

Because of the preceding remarks, all statistically significant variables selected by a variable selection procedure may not actually be required, the variables selected may not actually be the best discriminators, and the variables selected may not actually be good discriminators. Thus the results of any variable selection procedure must be taken with a grain of salt.

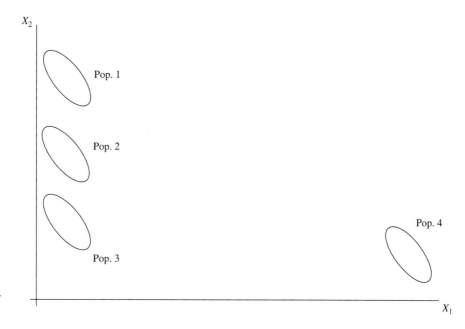

F I G. 7.6
Case where x_2 is a better discriminator than x_1

On the positive side, the selection procedures do provide some guidance, in the absence of any other alternatives. For example, if you have 20 candidate variables, then there are $2^{20} - 1 = 1,048,575$ possible subsets of discriminating variables. This is surely too many possibilities to consider. The variable selection procedures give, at least, a reduced number of possible subsets to consider and try.

E X A M P L E 7.2 (continued)

SAS has a procedure called STEPDISC that can be used to help select subsets of variables that might have chances at being good discriminators. This procedure was applied to the data in Example 7.2 by using the following commands:

```
PROC STEPDISC DATA=FINAL METHOD=STEPWISE SLE=.40 SLS=.15;
  CLASSES GRP;
  VAR DOWN_A--DOWN_B RIGHT_A--RIGHT_B;
  RUN;
```

The METHOD=STEPWISE option tells the procedure to use the stepwise method for variable selection. Another option that can be used here is the METHOD=BACKWARD for the backward elimination procedure. The forward selection procedure can also be used, but it is not recommended in this book. The SLE=.40 option tells the STEPDISC procedure to enter variables into the discriminating set only if they are significant at the 0.40 level. SLE stands for "significance level for entry." The SLS=.15 option tells the procedure to remove variables from the set if they are not significant at the 0.15 level. SLS stands for "significance level for staying."

The results of the individual steps in the stepwise procedure can be found on pages 22–28 of the computer output. A portion of these pages is shown in Figure 7.7. All pages of output are available by executing the file called EX7_2.SAS on the enclosed disk. The main results of the steps shown on pages 22–28 are summarized on page 29.

The *F* statistics shown on the top of page 23 are the *F*'s that would be obtained by performing a one-way analysis of variance on each of the candidate variables: DOWN_A, DOWN_P, ..., RIGHT_B. The variable producing the largest *F* value is selected first, provided that it is significant at the 0.40 level. Thus, the first variable selected is RIGHT_A, which has F=22.183 and an observed significance level that is less than 0.0001.

At the beginning of step 2 on the bottom of page 23, the procedure checks to see if RIGHT_A can be removed from the set of selected variables, and since RIGHT_A is significant at the 0.15 level, it is not removed.

On the top of page 24, the procedure looks for the next variable to be selected. The *F* statistics shown are the *F*'s that would be obtained by performing a one-way analysis of covariance on each of the remaining candidate variables using RIGHT_A as a covariate. The variable producing the largest *F* value (F=43.344) is DOWN_A. This variable is also significant at the 0.40 level, so it is appended to the set of discriminating variables.

At the beginning of step 3 on the bottom of page 24, the procedure checks to see if either of the selected variables can be eliminated. When checking DOWN_A, the procedure uses RIGHT_A as a covariate, and when checking RIGHT_A, the procedure uses DOWN_A as a covariate. Since the resulting *F* statistics (F=43.344, F=51.614, respectively) are both significant at the 0.15 level, neither variable is removed from the discriminating set.

In the middle of page 24, we see some multivariate test statistics for the two selected variables. These are multivariate analysis variance (MANOVA) test statistics for testing the null hypothesis that the four populations of wheat kernels have equal means on the variables DOWN_A and RIGHT_A simultaneously versus the alternative hypothesis that the means of the four populations differ with respect to at least one of the two variables. MANOVA is discussed in detail in Chapter 11.

Just below the multivariate test statistics, we find the line

```
Average Squared Canonical Correlation = 0.23170874
```

This is computed by dividing Pillai's trace (a multivariate analysis of variance test statistic that is discussed in Chapter 11) by one less than the number of populations under study. [*Note:* ASCC = 0.2317 = 0.6951/(4 − 1).] This statistic is useful when comparing different subsets of discriminating

Ex. 7.2 - SIZE MEASUREMENTS ON WHEAT KERNELS   23

STEPWISE DISCRIMINANT ANALYSIS

Stepwise Selection: Step 1

Statistics for Entry, DF = 3, 168   *one-way ANOVA F's*

Variable	R**2	F	Prob > F	Tolerance
DOWN_A	0.2239	16.156	0.0001	1.0000
DOWN_P	0.2103	14.915	0.0001	1.0000
DOWN_L	0.1388	9.022	0.0001	1.0000
DOWN_B	0.0848	5.188	0.0019	1.0000
RIGHT_A	0.2837	22.183	0.0001	1.0000
RIGHT_P	0.2633	20.017	0.0001	1.0000
RIGHT_L	0.2776	21.524	0.0001	1.0000
RIGHT_B	0.0454	2.665	0.0496	1.0000

Variable RIGHT_A will be entered

The following variable(s) have been entered:
RIGHT_A

Multivariate Statistics

Wilks' Lambda = 0.71627052   F(3, 168) =   22.183   Prob > F = 0.0001
Pillai's Trace =   0.283729   F(3, 168) =   22.183   Prob > F = 0.0001

Average Squared Canonical Correlation = 0.09457649

Stepwise Selection: Step 2

Statistics for Removal, DF = 3, 168

Variable	R**2	F	Prob > F
RIGHT_A	0.2837	22.183	0.0001

No variables can be removed

FIG. 7.7

Pages 23, 24, and 27–29 of computer printout, Example 7.2

variables. Values of ASCC that are close to 1 for a subset of variables indicate that the selected variables should do a good job of discriminating among all of the populations under consideration.

The stepwise process continues through the next several pages of computer output (see Figure 7.7). On page 27, we see that the variables DOWN_A, DOWN_P, RIGHT_A, RIGHT_L, and RIGHT_B have all been selected, and on the bottom of page 27, we can see that none of these variables can be removed.

Ex. 7.2 - SIZE MEASUREMENTS ON WHEAT KERNELS   24

STEPWISE DISCRIMINANT ANALYSIS

Stepwise Selection: Step 2

Statistics for Entry, DF = 3, 167

One-way ANOCOVAR F's

Variable	Partial R**2	F	Prob > F	Tolerance
DOWN_A	0.4378	43.344	0.0001	0.5903
DOWN_P	0.0848	5.156	0.0020	0.4389
DOWN_L	0.0652	3.882	0.0102	0.8386
DOWN_B	0.0995	6.154	0.0005	1.0000
RIGHT_P	0.0401	2.325	0.0767	0.2551
RIGHT_L	0.1299	8.308	0.0001	0.5372
RIGHT_B	0.0899	5.502	0.0013	0.4726

Variable DOWN_A will be entered

The following variable(s) have been entered:
            DOWN_A   RIGHT_A

Multivariate Statistics

Wilks' Lambda = 0.40270650   F(6, 334) =   32.054   Prob > F = 0.0001
Pillai's Trace =   0.695126   F(6, 336) =   29.832   Prob > F = 0.0001

Average Squared Canonical Correlation = 0.23170874

Stepwise Selection: Step 3

Statistics for Removal, DF = 3, 167

Variable	Partial R**2	F	Prob > F
DOWN_A	0.4378	43.344	0.0001
RIGHT_A	0.4811	51.614	0.0001

No variables can be removed

FIG. 7.1
Pages 23, 24, and 27–29 of computer printout, Example 7.2

On the top of page 28, the procedure looks for the variable to be selected next. In checking, it uses all of the variables previously selected as covariates. The variable DOWN_B is selected because it is significant at the 0.40 level. However, at the bottom of page 28, we can see that this variable is not significant at the 0.15 level, so the variable is removed from the candidate list. When no other variables are significant at the 0.40 level and the variable removed is the one that was just entered, the process stops.

```
        Ex. 7.2 - SIZE MEASUREMENTS ON WHEAT KERNELS              27

              STEPWISE DISCRIMINANT ANALYSIS

Stepwise Selection:  Step 5

                  Statistics for Entry, DF = 3, 164

                   Partial
     Variable       R**2          F        Prob > F     Tolerance

     DOWN_L        0.0135       0.751       0.5234        0.1931
     DOWN_B        0.0266       1.496       0.2177        0.2204
     RIGHT_P       0.0157       0.872       0.4569        0.1073
     RIGHT_B       0.0695       4.083       0.0079        0.1439


              Variable RIGHT_B will be entered

          The following variable(s) have been entered:
          DOWN_A   DOWN_P   RIGHT_A  RIGHT_L  RIGHT_B

                 Multivariate Statistics

     Wilks' Lambda  = 0.28389128    F( 15, 453) =   17.459
                                    Prob > F = 0.0001
        Pillai's Trace =   0.888455    F( 15, 498) =   13.969
                                    Prob > F = 0.0001

     Average Squared Canonical Correlation = 0.29615176

---------------------------------------------------------------------------


Stepwise Selection:  Step 6

                 Statistics for Removal,  DF = 3, 164

                   Partial
      Variable      R**2          F        Prob > F

      DOWN_A       0.4666       47.828      0.0001
      DOWN_P       0.1256        7.853      0.0001
      RIGHT_A      0.2898       22.302      0.0001
      RIGHT_L      0.1046        6.386      0.0004
      RIGHT_B      0.0695        4.083      0.0079


              No variables can be removed
```

F I G. 7.7

Pages 23, 24, and 27–29 of computer printout, Example 7.2

On page 29, the steps in the stepwise process are summarized. (Note that it is possible to avoid printing the previous pages and have only this summary page printed.) The final subset of variables selected by this procedure consists of DOWN_A, DOWN_P, RIGHT_A, RIGHT_L, and RIGHT_B. Each of these variables is significant at a level less than 0.01 when tested by

Ex. 7.2 - SIZE MEASUREMENTS ON WHEAT KERNELS   28

STEPWISE DISCRIMINANT ANALYSIS

Stepwise Selection: Step 6

Statistics for Entry, DF = 3, 163

Variable	Partial R**2	F	Prob > F	Tolerance
DOWN_L	0.0135	0.745	0.5267	0.1432
DOWN_B	0.0292	1.635	0.1834	0.1437
RIGHT_P	0.0152	0.836	0.4758	0.1061

Variable DOWN_B will be entered

The following variable(s) have been entered:
DOWN_A   DOWN_P   DOWN_B   RIGHT_A   RIGHT_L   RIGHT_B

Multivariate Statistics

Wilks' Lambda   = 0.27559977   F(18, 462) =   14.800
                                 Prob > F = 0.0001
Pillai's Trace =   0.912381   F(18, 495) =   12.019
                                 Prob > F = 0.0001

Average Squared Canonical Correlation = 0.30412701

Stepwise Selection: Step 7

Statistics for Removal, DF = 3, 163

Variable	Partial R**2	F	Prob > F
DOWN_A	0.4309	41.147	0.0001
DOWN_P	0.1260	7.831	0.0001
DOWN_B	0.0292	1.635	0.1834
RIGHT_A	0.2888	22.065	0.0001
RIGHT_L	0.1036	6.280	0.0005
RIGHT_B	0.0720	4.212	0.0067

Variable DOWN_B will be removed

No further steps are possible

F I G. 7.7
Pages 23, 24, and 27–29 of computer printout, Example 7.2

using all of the other variables in this subset as covariates (see the bottom of page 27). Unfortunately, the ASCC for these variables is only 0.2962, which is not too close to 1. So these variables are not likely to discriminate accurately between all four groups of wheat kernels. This should not be surprising if you recall that when using all of the variables, the ability to dis-

Ex. 7.2 - SIZE MEASUREMENTS ON WHEAT KERNELS   29

STEPWISE DISCRIMINANT ANALYSIS

Stepwise Selection: Summary

Step	Variable Entered	Removed	Number In	Partial R**2	F Statistic	Prob > F
1	RIGHT_A		1	0.2837	22.183	0.0001
2	DOWN_A		2	0.4378	43.344	0.0001
3	DOWN_P		3	0.1606	10.586	0.0001
4	RIGHT_L		4	0.0975	5.939	0.0007
5	RIGHT_B		5	0.0695	4.083	0.0079
6	DOWN_B		6	0.0292	1.635	0.1834
7		DOWN_B	5	0.0292	1.635	0.1834

Step	Variable Entered	Removed	Number In	Wilks' Lambda	Prob < Lambda	Average Squared Canonical Correlation	Prob > ASCC
1	RIGHT_A		1	0.71627052	0.0001	0.09457649	0.0001
2	DOWN_A		2	0.40270650	0.0001	0.23170874	0.0001
3	DOWN_P		3	0.33803754	0.0001	0.26066796	0.0001
4	RIGHT_L		4	0.30509260	0.0001	0.28592638	0.0001
5	RIGHT_B		5	0.28389128	0.0001	0.29615176	0.0001
6	DOWN_B		6	0.27559977	0.0001	0.30412701	0.0001
7		DOWN_B	5	0.00000000	0.0	0.00000000	0.0

F I G. **7.7**

Pages 23, 24, and 27–29 of computer printout, Example 7.2

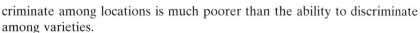

criminate among locations is much poorer than the ability to discriminate among varieties.

In the exercises at the end of this chapter you are asked to perform a discriminant analysis with the variables selected by the stepwise procedure in the preceding example.

7.6
Canonical Discriminant Functions

The idea of canonical discriminant analysis was first introduced by Fisher, and many authors refer to the method as Fisher's *between-within method.* Canonical discriminant analysis creates new variables by taking special linear combinations of the original variables. The canonical variables are created so that they contain all of the useful information in a set of original variables. In some sense, they are similar to principal components and factors. However, they are not computed in the same way.

In a few cases, a researcher may be able to interpret the canonical variables, which increases their usefulness. One advantage the canonical functions have, regardless of whether they are interpretable, is that they often allow a researcher to visualize the actual distances between the populations under investigation in a reduced dimensional space.

The canonical functions allow researchers to develop simple rules that can be used when they do not have easy access to computing facilities. For example, in emergency situations, the anesthesiologist would not want to have to locate a computer prior to making a determination as to the safeness of an anesthetic for a particular patient, and one or two simple canonical functions may allow the anesthesiologist to make accurate predictions.

The First Canonical Function

Suppose a researcher has a random sample of size n_i from Π_i, which is assumed to be $N_p(\boldsymbol{\mu}_i, \boldsymbol{\Sigma})$, for $i = 1, 2, \ldots, m$. Note that this development assumes that there are m populations under consideration and that the populations have equal variance–covariance matrices. Let

$$\mathbf{B} = \sum_{i=1}^{m} n_i(\hat{\boldsymbol{\mu}}_i - \hat{\boldsymbol{\mu}}_.)(\hat{\boldsymbol{\mu}}_i - \hat{\boldsymbol{\mu}}_.)'$$

where

$$\hat{\boldsymbol{\mu}}_. = \frac{1}{n_.} \sum_{i=1}^{m} n_i \hat{\boldsymbol{\mu}}_i \qquad \text{and} \qquad n_. = \sum_{i=1}^{m} n_i$$

and let

$$\mathbf{W} = \sum_{i=1}^{m} \sum_{r=1}^{n_i} (\mathbf{x}_{ri} - \hat{\boldsymbol{\mu}}_i)(\mathbf{x}_{ri} - \hat{\boldsymbol{\mu}}_i)'$$

$\mathbf{B}$ is called a "between sample mean" sum of squares and cross-products matrix, and $\mathbf{W}$ is called a "within sample" sum of squares and cross-products matrix. These are multivariate generalizations of the between and within sums of squares in a one-way analysis of variance.

We can show that

$$\max_{\mathbf{b} \neq 0} \frac{\mathbf{b}'\mathbf{B}\mathbf{b}}{\mathbf{b}'(\mathbf{B} + \mathbf{W})\mathbf{b}}$$

is the largest eigenvalue of $(\mathbf{B} + \mathbf{W})^{-1}\mathbf{B}$. This largest eigenvalue is denoted by ℓ_1, and this maximum is attained when $\mathbf{b} = \mathbf{b}_1$ where $\mathbf{b}_1$ is an eigenvector of $(\mathbf{B} + \mathbf{W})^{-1}\mathbf{B}$ corresponding to the eigenvalue ℓ_1.

The linear combination $y_1 = \mathbf{b}_1'\mathbf{x}$ can be regarded as the single linear discriminant function that provides the maximum separation between the m mean vectors.

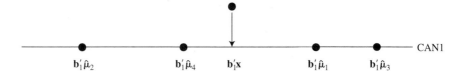

If there are just two populations, $\mathbf{b}_1'\mathbf{x}$ is equivalent to the linear discriminant function defined in Section 7.1. In the case of two populations, a single discriminant function is adequate. However, when $m > 2$, a single function is generally not adequate unless all population means should happen to lie on a straight line within the p-dimensional sample space.

Should all means lie on a straight line, effective discrimination of $\mathbf{x}$ can be based on $\mathbf{b}_1'\mathbf{x}$, $\mathbf{b}_1'\hat{\boldsymbol{\mu}}_1$, $\mathbf{b}_1'\hat{\boldsymbol{\mu}}_2$, ..., and $\mathbf{b}_1'\hat{\boldsymbol{\mu}}_m$, the first canonical function evaluated at $\mathbf{x}$ and at each of the m sample mean vectors, respectively. Then we compute $d_i = |\mathbf{b}_1'\mathbf{x} - \mathbf{b}_1'\hat{\boldsymbol{\mu}}_i|$ for $i = 1, 2, \ldots, m$, and $\mathbf{x}$ is assigned to the population that gives the smallest value for d_i.

Geometrically, if all of the group means lie on a straight line, then we can place a coordinate system on that line. The first canonical discriminant function, evaluated at the group means $\mathbf{b}_1'\hat{\boldsymbol{\mu}}_1$, $\mathbf{b}_1'\hat{\boldsymbol{\mu}}_2$, ..., and $\mathbf{b}_1'\hat{\boldsymbol{\mu}}_m$, gives the relative locations of the means with respect to one another on this line. The new observation vector, $\mathbf{x}$, probably will not lie on this line, but the sample mean that it lies closest to will be the same one it is closest to if we project the point perpendicularly onto this line. The location of the projected point onto the line is given by $\mathbf{b}_1'\mathbf{x}$. Figure 7.8 illustrates the preceding discussion where the means of four populations are plotted on the horizontal axis and the projection of the new point to be classified to the axis is shown by the arrow.

A Second Canonical Function

Some might feel that a researcher would be really fortunate if the sample mean vectors actually fell onto a single line. But if the mean vectors do not fall on one line, what would be the next best thing? The next best thing would be to have all of the mean vectors fall onto a plane.

When the mean vectors lie on a plane within the p-dimensional sample space, we need two mutually orthogonal canonical discriminant functions,

$$y_1 = \mathbf{b}_1'\mathbf{x} \qquad \text{and} \qquad y_2 = \mathbf{b}_2'\mathbf{x}$$

to discriminate effectively. The best two functions are defined by $\mathbf{b}_1$ and $\mathbf{b}_2$, the eigenvectors of $(\mathbf{B} + \mathbf{W})^{-1}\mathbf{B}$ corresponding to its two largest eigenvalues, ℓ_1 and ℓ_2. Given the two canonical functions

$$y_1 = \mathbf{b}_1'\mathbf{x} \qquad \text{and} \qquad y_2 = \mathbf{b}_2'\mathbf{x}$$

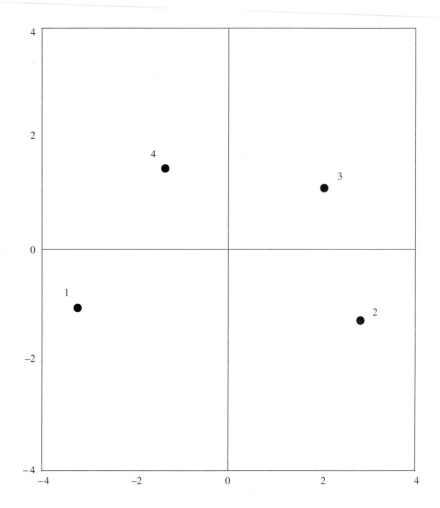

F I G. **7.9**
Canonical dis-
criminant plot in
two dimensions

we can calculate

$$d_i^2 = [(\mathbf{b}_1'\mathbf{x} - \mathbf{b}_1'\hat{\boldsymbol{\mu}}_i)^2 + (\mathbf{b}_2'\mathbf{x} - \mathbf{b}_2'\hat{\boldsymbol{\mu}}_i)^2] \qquad \text{for } i = 1, 2, \ldots, m$$

and assign $\mathbf{x}$ to the population that gives the minimum value for d_i^2.

Note　Defining vectors $\mathbf{b}_1$ and $\mathbf{b}_2$ can also be obtained from the eigenvectors of $\mathbf{W}^{-1}\mathbf{B}$ corresponding to its two largest eigenvalues.

　　　Geometrically, if the sample mean vectors fall onto a plane, we need a coordinate system in this plane. The canonical functions allow us to place a coordinate system in this plane. The relative locations of the sample mean vectors in the canonical space are given by the pairs $(\mathbf{b}_1'\hat{\boldsymbol{\mu}}_i, \mathbf{b}_2'\hat{\boldsymbol{\mu}}_i)$, for $i = 1, 2, \ldots, m$. The projection of a new $\mathbf{x}$ onto this plane has coordinates $(\mathbf{b}_1'\mathbf{x}, \mathbf{b}_2'\mathbf{x})$. Then $\mathbf{x}$ is assigned to the group to whose mean it lies closest to in the

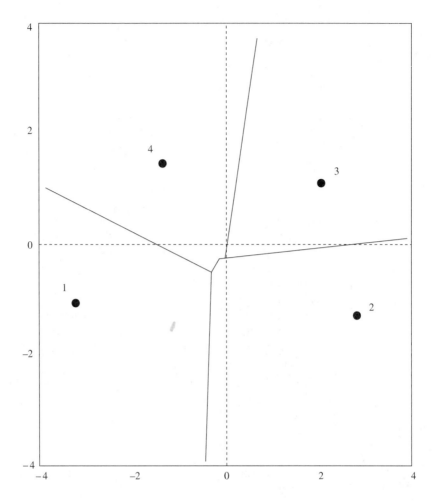

F I G. **7.10**
Canonical discriminant plot in two dimensions with territorial map

canonical space. Figure 7.9 illustrates the location of four means in a two-dimensional canonical space.

Remark   When the sample means lie on a plane and there are just two canonical discriminant functions, the canonical space can be partitioned into regions by constructing perpendicular bisectors between all pairs of means. Then **x** can be assigned to a group according to which partition the projection of **x** falls into. Figure 7.10 illustrates the four regions obtained for the four points in Figure 7.9.

If the sample means do not appear to fall on a line or onto a plane, then the next best thing would be for them to fall into a three-dimensional subspace of the *p*-dimensional sample space. This situation would require three canonical discriminant functions, and simplifications are not quite as easy. Obviously, things continue to get more complicated as the required number of canonical discriminant functions increases and, as a result, the advantages of using the canonical functions decrease.

Determining the Dimensionality of the Canonical Space

When using canonical discriminant functions it is important to determine the dimensionality of the space in which the sample mean vectors lie. Let the dimensionality of this space be denoted by s. The dimensionality of the canonical space is bounded above by the minimum of p and $m - 1$. That is, $s \leq \min(p, m - 1)$.

Subjectively, we can estimate the dimensionality of the canonical space by looking at the eigenvalues of $(\mathbf{B} + \mathbf{W})^{-1}\mathbf{B}$ (or $\mathbf{W}^{-1}\mathbf{B}$) and using methods similar to those that were recommended for principal components analysis. We can construct SCREE plots of the eigenvalues or consider what proportion of the total variability is being accounted for by each canonical function and select enough to account for a large proportion of the total variability.

When working with normally distributed data, an objective hypothesis testing approach is available, based on the eigenvalues of $(\mathbf{B} + \mathbf{W})^{-1}\mathbf{B}$ (or $\mathbf{W}^{-1}\mathbf{B}$), that can be used to estimate the dimensionality of the canonical space. Such tests are given by many statistical computing packages and are discussed in the following examples.

E X A M P L E *7.2* (continued)

A SAS procedure that produces the information required for a canonical discriminant analysis is `CANDISC`. The data in Example 7.2 were analyzed by using those variables selected by the stepwise procedure in the previous section and the SAS commands:

```
PROC CANDISC DATA=FINAL OUT=CANSCRS NCAN=2;
 CLASSES GRP;
 VAR DOWN_A DOWN_P RIGHT_A RIGHT_L RIGHT_B;
 RUN;

PROC PRINT DATA=CANSCRS;
 VAR LOC VARIETY GRP CAN1 CAN2;
 RUN;
```

The results can be found on pages 30–37 of the computer output results (a few of the pages are shown in Figure 7.11; the others are available by executing the file called `EX7_2.SAS` on the enclosed disk).

In the middle of page 32, look for the eigenvalues of $\mathbf{W}^{-1}\mathbf{B}$ and likelihood ratio test statistics for determining the dimensionality of the canonical

```
Ex. 7.2 - SIZE MEASUREMENTS ON WHEAT KERNELS                    32

                    CANONICAL DISCRIMINANT ANALYSIS

                        Adjusted       Approx      Squared
            Canonical   Canonical     Standard    Canonical
            Correlation Correlation     Error     Correlation

    1       0.777977    0.768984      0.030187    0.605248
    2       0.523995    0.511377      0.055475    0.274570
    3       0.092932    0.032644      0.075811    0.008636

                  Eigenvalues of INV(E)*H
                    = CanRsq/(1-CanRsq)

          Eigenvalue    Difference    Proportion    Cumulative

    1       1.5332        1.1547        0.7984        0.7984
    2       0.3785        0.3698        0.1971        0.9955
    3       0.0087          .           0.0045        1.0000

    Test of H0: The canonical correlations in the current row
                and all that follow are zero

            Likelihood
              Ratio      Approx F      Num DF      Den DF     Pr > F
```

	Likelihood Ratio	Approx F	Num DF	Den DF	Pr > F	
1	0.28389128	17.4592	15	453.1332	0.0001	} Determines
2	0.71916449	7.3918	8	330	0.0001	} the number
3	0.99136357	0.4820	3	166	0.6952	of canonical functions.

FIG. 7.11

Pages 32–34 of computer print-out, Example 7.2

space. Note that two canonical functions are statistically significant and that these two account for 99.55% of the total variability. Thus, for these data, the four group means seem to fall into a two-dimensional subspace of the five-dimensional sample space of the variables being used in this analysis.

The vectors corresponding to the Pooled Within Canonical Structure printed at the top of page 33 are not useful. These are eigenvectors corresponding to the two largest eigenvalues of **W**.

The interesting vectors are the two sets of vectors in the middle of page 33. Those corresponding to the Standardized Canonical Coefficients can be used to define the first two canonical functions in terms of standardized data. Those corresponding to the Raw Canonical Coefficients can be used to define the first two canonical functions from raw (unstandardized) data values. Thus values for the first (raw) canonical function would be computed by

```
CAN1 = −0.920·DOWN_A+0.120·DOWN_P+0.962·RIGHT_A−
0.107·RIGHT_L−0.229·RIGHT_B
```

while values for the second (raw) canonical variable would be computed by

```
CAN2 = 0.271·DOWN_A−0.016·DOWN_P−0.015·RIGHT_A+
0.186·RIGHT_L−0.019·RIGHT_B.
```

CANONICAL DISCRIMINANT ANALYSIS

Pooled Within Canonical Structure

	CAN1	CAN2
DOWN_A	-0.186076	0.788588
DOWN_P	0.162418	0.769543
RIGHT_A	0.371794	0.697189
RIGHT_L	0.182725	0.938213
RIGHT_B	0.135727	0.226097

Standardized Canonical Coefficients

	CAN1	CAN2
DOWN_A	-2.040404410	0.601758422
DOWN_P	1.022667219	-0.139467394
RIGHT_A	2.134294658	-0.033827551
RIGHT_L	-0.502220415	0.877830962
RIGHT_B	-0.636991804	-0.052110464
	$\hat{b}_1^*$	$\hat{b}_2^*$

Raw Canonical Coefficients

	CAN1	CAN2
DOWN_A	-.9195280752	0.2711882807
DOWN_P	0.1204288044	-.0164236139
RIGHT_A	0.9622840313	-.0152517425
RIGHT_L	-.1066536116	0.1864198262
RIGHT_B	-.2290268149	-.0187360239
	$\hat{b}_1$	$\hat{b}_2$

Class Means on Canonical Variables

GRP	CAN1	CAN2
1	1.090728597	-0.889750618
2	-1.130733366	-0.498726912
3	1.313272529	0.606990369
4	-1.284469095	0.392713453

F I G. 7.11
Pages 32–34 of
computer print-
out, Example 7.2

On the bottom of page 33, the procedure gives the locations of the four
sample means in terms of values of the standardized canonical functions.
The relative locations of the four group means in terms of these standardized canonical functions are plotted in Figure 7.12.

The canonical functions are interesting in that the first canonical function seems to be trying to discriminate between varieties (recall that groups
1 and 3 are Arkan, and groups 2 and 4 are Arthur), while the second canonical function is trying to discriminate between locations (recall that
groups 1 and 2 are from one location, and groups 3 and 4 are from another

Ex. 7.2 - SIZE MEASUREMENTS ON WHEAT KERNELS   34

OBS	LOC	VARIETY	GRP	CAN1	CAN2
1	MAS0	ARKAN	1	1.76610	-0.30122
2	MAS0	ARKAN	1	1.05251	0.25084
3	MAS0	ARKAN	1	1.75775	-0.05993
4	MAS0	ARKAN	1	0.81550	-0.87226
5	MAS0	ARKAN	1	1.03811	-2.26940
37	MAS0	ARTHUR	2	-0.33406	-1.52181
38	MAS0	ARTHUR	2	0.78483	-0.27099
39	MAS0	ARTHUR	2	-0.12251	-1.38702
40	MAS0	ARTHUR	2	-1.07502	0.77872
41	MAS0	ARTHUR	2	-0.59469	0.51447

Canonical Scores

FIG. 7.11
Pages 32–34 of computer print-out, Example 7.2

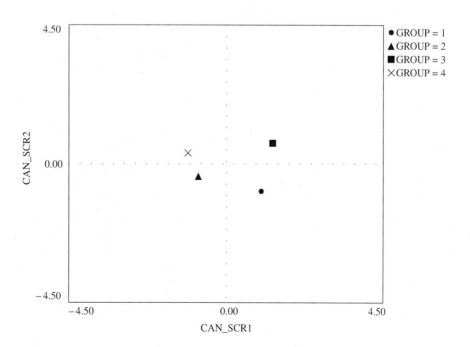

FIG. 7.12
Plots of wheat group means in their canonical space

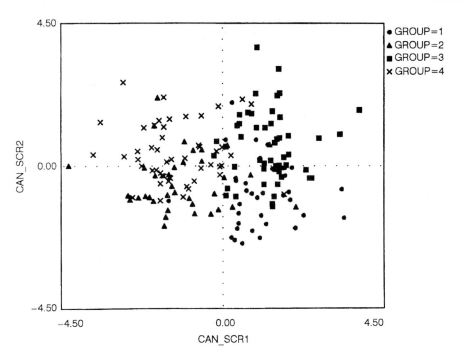

F I G. 7.13

Plot of wheat data in the canonical space

location). Also note that the distance between means with respect to the CAN1 axis is much greater than the distance between means with respect to the CAN2 axis. This is consistent with the earlier observation that it was easier to discriminate correctly among varieties than among locations.

Pages 34 through 37 (a portion of page 34 is shown in Figure 7.11) give the values (scores) of the projections of all of the kernels in the sample data onto the two-dimensional canonical space in standardized units. A plot of the values of these two canonical functions is shown in Figures 7.13 and 7.14. Note that the kernels in groups 1 and 3 tend to be located to the right, and those in groups 2 and 4 tend to be to the left. Thus, the Arkan variety tends to be located to the right, while the Arthur variety tends to be located to the left, and there is a fair amount of overlap between the two varieties. Similarly, groups 1 and 2 (location 1) tend to be located toward the bottom while groups 3 and 4 (location 2) tend to be located toward the top. There is even more overlap among the groups with respect to locations than there is with respect to varieties. The plots make it clear as to why it is not possible to discriminate between groups as accurately as we would like. There is a lot of overlap among the populations under study. If the researcher wants to improve the accuracy of the discrimination process, he or she might need to look for new response variables that might be better discriminators than the present variables.

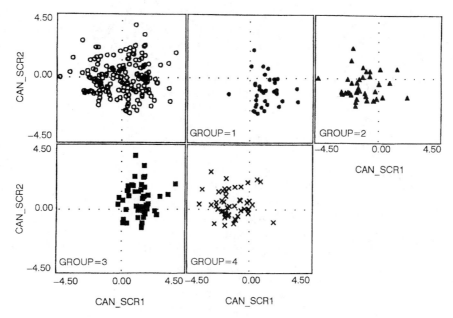

FIG. 7.14
Plot of wheat
data in the canoni-
cal space

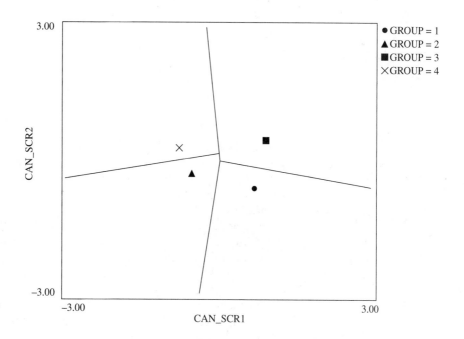

FIG. 7.15
Territorial map
for group discrimi-
nation

Figure 7.15 shows how the canonical space would be subdivided into four regions. A new observation would be classified into a population corresponding to the region into which its two canonical scores would fall. This region is obtained by constructing perpendicular bisectors of lines connecting all pairs of group means and then erasing the portions of these bisectors that are not informative.

E X A M P L E *7.3*

This example illustrates the use of SPSS's DISCRIMINANT procedure. The data used are known as Fisher's iris data. These data consist of 50 samples from each of three varieties of iris plants. The variables measured are sepal length (SL), sepal width (SW), petal length (PL), and petal width (PW). The data can be found in the file named EX7_3.SSX on the enclosed disk. These data were analyzed with the following SPSS commands:

```
TITLE 'Ex. 7.3 Discriminant Analysis on Fisher''s Iris Data'
DISCRIMINANT GROUPS=IRIS(1,3)
               /VARIABLES=SL SW PL PW
               /ANALYSIS=SL SW PL PW
               /METHOD=DIRECT
               /PRIORS=EQUAL
STATISTICS 1 2 7 10 11 13 14 15
BEGIN DATA
  1 5.1 3.5 1.4 0.2
  1 4.9 3.0 1.4 0.2
  .   .    .    .    .
  .   .    .    .    .
  .   .    .    .    .
  3 5.9 3.0 5.1 1.8
END DATA
```

The results from these commands can be viewed by executing the file EX7_3.SSX on the enclosed disk. Portions of the results are shown on the computer printout pages of Figure 7.16.

Page 2 shows the means and standard deviations of the variables for each of the varieties.

In the middle of page 3, we see the eigenvalues of $\mathbf{W}^{-1}\mathbf{B}$ (they are 32.1919 and 0.2854), as well as statistical tests for determining the dimensionality of the canonical space. For this example, both eigenvalues are significant ($P = 0.0000$). However, the first accounts for 99.12% of the total variability, so the second is likely to be of little practical importance even though it is statistically significant. Consequently, the means for these three varieties come very close to lying on a straight line within the four-dimensional sample space of the measured variables.

Ex. 7.3 Discriminant Analysis on Fisher's Iris Data   Page   2

- - - - - - - -   D I S C R I M I N A N T   A N A L Y S I S   - - - - - - - -

GROUP MEANS

IRIS	SL	SW	PL	PW
1	5.00600	3.42800	1.46200	0.24600
2	5.93600	2.77000	4.26000	1.32600
3	6.58800	2.97400	5.55200	2.02600
TOTAL	5.84333	3.05733	3.75800	1.19933

GROUP STANDARD DEVIATIONS

IRIS	SL	SW	PL	PW
1	0.35249	0.37906	0.17366	0.10539
2	0.51617	0.31380	0.46991	0.19775
3	0.63588	0.32250	0.55189	0.27465
TOTAL	0.82807	0.43587	1.76530	0.76224

F I G. 7.16

Pages 2–15 of computer printout, Example 7.3

Ex. 7.3 Discriminant Analysis on Fisher's Iris Data   Page   3

- - - - - - - -   D I S C R I M I N A N T   A N A L Y S I S   - - - - - - - -

CANONICAL DISCRIMINANT FUNCTIONS

FCN	EIGENVALUE	PCT OF VARIANCE	CUM PCT	CANONICAL CORR	AFTER FCN	WILKS' LAMBDA	CHISQUARE	DF	SIG
					: 0	0.0234	546.115	8	0.0000
1*	32.1919	99.12	99.12	0.9848	: 1	0.7780	36.530	3	0.0000
2*	0.2854	0.88	100.00	0.4712	:				

   * MARKS THE   2 CANONICAL DISCRIMINANT FUNCTIONS REMAINING IN THE ANALYSIS

STANDARDIZED CANONICAL DISCRIMINANT FUNCTION COEFFICIENTS

	FUNC 1	FUNC 2
SL	-0.42695	0.01241
SW	-0.52124	0.73526
PL	0.94726	-0.40104
PW	0.57516	0.58104
	$\hat{b}_1^*$	$\hat{b}_2^*$

F I G. 7.16

Pages 2–15 of computer printout, Example 7.3

UNSTANDARDIZED CANONICAL DISCRIMINANT FUNCTION COEFFICIENTS

	FUNC 1	FUNC 2
SL	-0.8293776	0.2410215E-01
SW	-1.534473	2.164521
PL	2.201212	-0.9319212
PW	2.810460	2.839188
(CONSTANT)	-2.105106	-6.661473

$$\hat{b}_1 \qquad \hat{b}_2$$

CANONICAL DISCRIMINANT FUNCTIONS EVALUATED AT GROUP MEANS (GROUP CENTROIDS)

GROUP	FUNC 1	FUNC 2
1	-7.60760	0.21513
2	1.82505	-0.72790
3	5.78255	0.51277

F I G. 7.16
Pages 2–15 of computer printout, Example 7.3

TEST OF EQUALITY OF GROUP COVARIANCE MATRICES USING BOX'S M

THE RANKS AND NATURAL LOGARITHMS OF DETERMINANTS PRINTED ARE THOSE

OF THE GROUP COVARIANCE MATRICES.

GROUP LABEL	RANK	LOG DETERMINANT
1 IS	4	-13.067360
2 IC	4	-10.874325
3 IV	4	-8.927058
POOLED WITHIN-GROUPS COVARIANCE MATRIX	4	-9.958539

BOX'S M	APPROXIMATE F	DEGREES OF FREEDOM	SIGNIFICANCE
146.66	7.0453	20, 77566.8	0.0000

Tests $\Sigma_1 = \Sigma_2 = \Sigma_3$

SYMBOLS USED IN TERRITORIAL MAP

SYMBOL	GROUP	LABEL
------	-----	--------------------
1	1	IS
2	2	IC
3	3	IV
*		GROUP CENTROIDS

F I G. 7.16
Pages 2–15 of computer printout, Example 7.3

```
Ex. 7.3 Discriminant Analysis on Fisher's Iris Data           Page   6

                  TERRITORIAL MAP  * INDICATES A GROUP CENTROID

                    CANONICAL DISCRIMINANT FUNCTION 1

         -12.0      -8.0      -4.0        .0       4.0       8.0      12.0
            +---------+---------+---------+---------+---------+---------+
  C   12.0  +                        12 223                            +
  A         I                        12  233                           I
  N         I                        112 223                           I
  O         I                        122  233                          I
  N         I                        12   223                          I
  I         I                        12    233                         I
  C    8.0  +         +         +     12 + 223       +         +        +
  A         I                        12    233                         I
  L         I                        112   223                         I
            I                        122    233                        I
  D         I                        12    2233                        I
  I         I                        12     223                        I
  S    4.0  +         +         +     12  +  233 +         +            +
  C         I                        12     223                        I
  R         I                        112    233                        I
  I         I                        122    223                        I
  M         I                        12      233                       I
  I         I                        12      223    * IV               I
  N    .0   +       +*  IS     +     12   +  233          +            +
  A         I                        12      223                       I
  N         I                        112  *  233                       I
  T         I                        122  IC 223                       I
            I                        12       233                      I
  F         I                        12       223                      I
  U   -4.0  +         +         +  12 +       + 233      +             +
  N         I                        12        223                     I
  C         I                        112        233                    I
  T         I                        122        223                    I
  I         I                        12         233                    I
  O         I                        12         223                    I
  N   -8.0  +         +         +12      +      +  233      +          +
            I                        12           223                  I
  2         I                        112          233                  I
            I                        122         2233                  I
            I                        12           223                  I
            I                        12            233                 I
       -12.0 +                       12            223                 +
            +---------+---------+---------+---------+---------+---------+
         -12.0      -8.0      -4.0        .0       4.0       8.0      12.0
```

F I G. **7.16**

Pages 2–15 of computer printout, Example 7.3

Ex. 7.3 Discriminant Analysis on Fisher's Iris Data   Pages 7-13

CASE SEQNUM	MIS VAL	SEL	ACTUAL GROUP	HIGHEST PROBABILITY GROUP P(D/G) P(G/D)	2ND HIGHEST GROUP P(G/D)	DISCRIM SCORES
1			1	1 0.8987 1.0000	2 0.0000	-8.0618 0.3004
2			1	1 0.5398 1.0000	2 0.0000	-7.1287 -0.7867
3			1	1 0.8848 1.0000	2 0.0000	-7.4898 -0.2654
51			2	2 0.7026 0.9999	3 0.0001	1.4593 0.0285
52			2	2 0.4794 0.9993	3 0.0007	1.7977 0.4844
70			2	2 0.5098 1.0000	3 0.0000	1.0904 -1.6266
71			2 **	3 0.1026 0.7468	2 0.2532	3.7159 1.0445
84			2 **	3 0.1656 0.8566	2 0.1434	4.4985 -0.8827
85			2	2 0.4065 0.9636	3 0.0364	2.9340
101			3	3 0.0321 1.0000	2 0.0000	7.8395 2.1397
102			3	3 0.8284 0.9989	2 0.0011	5.5075 -0.0358
134			3 **	2 0.1349 0.7294	3 0.2706	3.8152 -0.9430
150			3	3 0.5376 0.9825	2 0.0175	4.6832

Posterior probabilities (handwritten annotation pointing to P(D/G) P(G/D) columns)

Canonical scores (handwritten annotation pointing to DISCRIM SCORES column)

SYMBOLS USED IN PLOTS

SYMBOL	GROUP	LABEL
1	1	IS
2	2	IC
3	3	IV

FIG. 7.16
Pages 2–15 of computer printout, Example 7.3

Ex. 7.3 Discriminant Analysis on Fisher's Iris Data   Page   14
          ALL-GROUPS SCATTERPLOT - * INDICATES A GROUP CENTROID
                    CANONICAL DISCRIMINANT FUNCTION 1

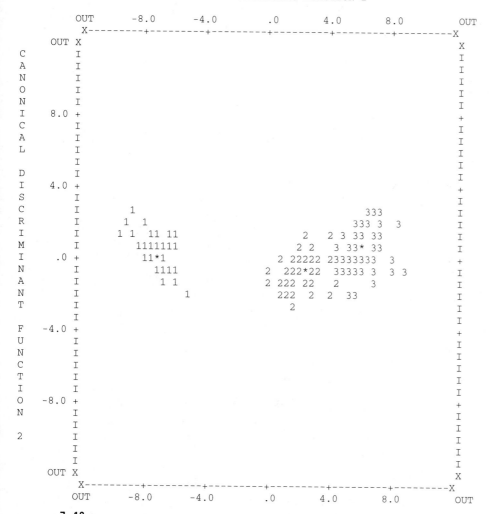

F I G. 7.16
Pages 2–15 of computer printout, Example 7.3

```
Ex. 7.3 Discriminant Analysis on Fisher's Iris Data          Page  15

CLASSIFICATION RESULTS -

                        NO. OF   PREDICTED GROUP MEMBERSHIP
ACTUAL GROUP            CASES       1          2            3
-------------------     ------   --------   --------    --------
GROUP    1                50        50          0            0
  IS                               100.0%      0.0%         0.0%
GROUP    2                50         0         48            2
  IC                                 0.0%      96.0%        4.0%
GROUP    3                50         0          1           49
  IV                                 0.0%      2.0%         98.0%

PERCENT OF ''GROUPED'' CASES CORRECTLY CLASSIFIED:  98.00%
```

F I G. 7.16
Pages 2–15 of computer printout, Example 7.3

The bottom of page 3 lists vectors that define standardized canonical functions—these could be used on data that has been standardized to determine the projections of data points onto the canonical space. Page 4 lists vectors that define unstandardized canonical functions. For example, we could compute unstandardized canonical scores (the projections of the observations to the canonical space) via

$$CAN1 = -0.829 \cdot SL - 1.534 \cdot SW + 2.201 \cdot PL + 2.810 \cdot PW - 2.105$$

and

$$CAN2 = 0.024 \cdot SL + 2.165 \cdot SW - 0.932 \cdot PL + 2.839 \cdot PW - 6.661$$

The locations of the three variety means in the unstandardized canonical space are shown at the bottom of page 4. These variety means are plotted on page 6; their locations are given by the asterisks (*) on the plot. Note that the asterisks come very close to falling onto a straight line.

SPSS also locates the perpendicular bisectors between the variety means on this plot. SPSS calls this plot a *territorial map*. These bisectors divide the canonical space into three distinct regions. If the projection of a new data point falls into one of the regions, then the new point is closest to the mean (*) in that region which determines the variety to which the observation would be classified.

Page 5 shows a test for homogeneity of the variance–covariance matrices for the three varieties. This test is significant and the hypothesis of equal variance–covariance matrices would be rejected. Readers might recall that SAS's DISCRIM procedure creates a quadratic discriminant rule when the variance–covariance matrices are determined to be unequal. This option is not available in SPSS. As noted earlier, the linear discriminant functions often work quite well even though the variance–covariance matrices are unequal. If the probabilities of correct classification are high enough to satisfy the user, then the user should not be too concerned that he or she is using a linear discriminant rule rather than a quadratic rule. On the other hand, if the user is not satisfied, then a quadratic rule should be considered. Unfortunately, SPSS users will need to switch to a different statistical computing package.

Portions of pages 7–13 are shown on one page of Figure 7.16. These pages provide a listing of how each iris plant in the data set would be classified by the discriminant rule developed. The column labeled ACTUAL GROUP identifies the variety from which the observation came; the column labeled HIGHEST GROUP shows the variety to which the observation would be assigned by the discriminant rule. The first column labeled P(G/D) is the posterior probability for the group to which the observation is assigned. This posterior probability is 0.9825 for CASE 150. While SAS gives posterior probabilities for all possible groups, SPSS gives posterior probabilities for only the best group and the second best group. The posterior probabilities for the second most likely group are shown in the second column that is labeled P(G/D). The column labeled P(D/G) can be ignored.

The last column, labeled DISCRIM SCORES, on pages 7–13 gives the locations of the projections of the observations to the canonical space. A plot of these projections for all the observations in the data set is shown on page 14. The projections of the means of the three iris groups are also located by the asterisks on this plot. This plot is interesting in that we can see that IS (variety 1) is quite distinct from IC and IV (varieties 2 and 3, respectively). Also there is very little overlap between varieties 2 and 3. So discrimination should be quite good for this iris data.

Page 15 gives a summary of the classification results by the resubstitution method. Examination of this summary shows that all 50 of the IS variety plants are reclassified into IS, 48 of the 50 IC plants are reclassified into IC, and 49 of the 50 IV plants are reclassified into IV. Overall, 98% of the observations are correctly classified. These classifications are made by applying the rule to the data used to build the rule, and so they may overestimate the actual probabilities of correct classification. SPSS does not have an option for cross-validation of the data as was available in SAS. SPSS does allow us to create a holdout data set by using its SELECT option.

Although variable selection methods are not illustrated here, SPSS DISCRIM does allow us to select variables through the use of several different stepwise procedures.

Discriminant Analysis with Categorical Predictor Variables

Each of the discriminant procedures discussed in the previous sections was developed under an assumption that the data vectors have multivariate normal distributions. Quite often these rules are applied to nonnormal data. One advantage that researchers have with discriminant rules is that they can see how well they work by using them. If the rules work well in cases when the data are nonnormal, then there is no reason to be too concerned with the fact that the data are nonnormal. In fact, I have used these types of rules on categorical variables by introducing dummy variables and using these dummy variables in discriminant programs.

To illustrate, suppose a researcher wishes to use RACE as a variable for discrimination and suppose that RACE takes on values BLK, HISP, WHT, and ASIAN. To use RACE (with four categories) in a discriminant program, the researcher must define three dummy variables (one less than the number of race categories). These dummy variables, denoted by DUM1, DUM2, and DUM3, could be defined as follows:

DUM1 = 1 if RACE='BLK', otherwise DUM1 = 0,

DUM2 = 1 if RACE='HISP', otherwise DUM2 = 0, and

DUM3 = 1 if RACE='WHT', otherwise DUM3 = 0.

Then the variables DUM1–DUM3 would be included in the variable set when using a statistical computing package. Note that a fourth dummy variable for the ASIAN category is not needed because if the first three dummy variables all have values equal to 0, then RACE must be equal to ASIAN. If we were to include a fourth dummy variable for the last race, the discriminant programs would not work because the sample variance–covariance matrix would not be invertible.

Because dummy variables only take on two values, they obviously are not distributed normally. And, while there is no reason to expect that a discriminant rule based on normal assumptions would work well for these kinds of variables, there is no reason not to use such a rule if it works well. In a manner similar to that described earlier, additional categorical variables could be included in the discriminating set of variables as well.

When using dummy variables to replace categorical variables, you must be careful when using stepwise procedures. Generally, you should not eliminate any single dummy variable corresponding to a categorical variable unless you can remove the whole set of dummy variables that correspond to that discrete variable.

Another technique for developing discriminant rules is based on logistic regression. This technique does not require that the discriminating variables be multivariate normal. This technique should be given serious consideration when it is known that some of the variables are nonnormal. In particular, logistic regression methods should be considered when one or more of the discriminating variables is categorical. Logistic regression methods are discussed in Chapter 8.

7.7
Nearest Neighbor Discriminant Analysis

Nearest neighbor discriminant analysis is a nonparametric discrimination procedure. It does not depend on the data being distributed normally. It does depend on the Mahalanobis distances between pairs of observation vectors.

The basic idea of nearest neighbor discriminant analysis is as follows: For a new observation that is to be classified, first find the observation in the calibration data set that is closest to the new observation (i.e., its Mahalanobis distance is smallest). Then assign the new observation to the group from which the observation's nearest neighbor comes.

If there should be a tie (i.e., if the distances between the new observation and two or more other observations are identical), then the procedure looks for its next nearest neighbor unless the tied observations are from the same group. If the tied observations are from the same group, the new observation is assigned to that group. If the tied observations are not from the same group and the next nearest neighbor matches one of these groups, then the new observation is assigned to that group. If there is no match, then the procedure looks for the next nearest neighbor, etc.

A variation of this process is to look at the k nearest neighbors of a new observation, and assign each new observation to the group to which a majority of its k nearest neighbors belongs. For example, suppose $k = 5$, and suppose in the set of the five nearest neighbors of a new observation, three are from group 1 and two are from one or two other groups. Since the majority of the five nearest neighbors are from group 1, the new observation would be assigned to group 1.

A nice variation of the nearest neighbor procedure is available in SAS's DISCRIM procedure by adding a METHOD=NPAR option and a K=k option. This procedure is illustrated for the wheat kernel data introduced in Example 7.2.

E X A M P L E 7 . 4

The nearest neighbor procedure with $k = 1$ and then with $k = 5$ is applied to the wheat kernel data by using the following SAS commands:

```
TITLE 'Ex. 7.4- NEAREST NEIGHBOR ANALYSIS ON WHEAT KERNELS';

PROC DISCRIM DATA=FINAL METHOD=NPAR K=1 LIST CROSSLIST;
 CLASSES GRP;
 VAR DOWN_A--DOWN_B RIGHT_A--RIGHT_B;
 RUN;
```

```
PROC DISCRIM DATA=FINAL METHOD=NPAR K=5 LIST CROSSLIST;
 CLASSES GRP;
 VAR DOWN_A--DOWN_B RIGHT_A--RIGHT_B;
 RUN;
```

A portion of the results of these commands can be found on the pages of the computer printout shown in Figure 7.17. All of the output can be seen by executing the file named EX7_4.SAS on the enclosed disk.

Pages 2–10 (not shown in Figure 7.17) provide a listing of how each kernel would be classified by looking at its nearest neighbor. Since each kernel is obviously its own nearest neighbor, these results are neither interesting nor useful. Each kernel is classified by itself. Pages 11 and 12 summarize the results, but since there are no errors, none of these pages is shown.

Pages 13–21 are more interesting and useful. A portion of these pages is shown in Figure 7.17. These pages show how each kernel would be classified if we were using the cross-validation procedure. Here this means that a kernel's nearest neighbor is selected from all other kernels, excluding the kernel being examined. Those kernels that would be classified incorrectly are flagged with an asterisk (*). Page 22 summarizes the results of the cross-validation procedure. An examination of these pages reveals that 52.78% of group 1 kernels are classified correctly, 44.44% of group 2, 62% of group 3, and 74% of group 4. If we consider only the cases where variety is being classified and not location (recall that groups 1 and 3 are Arkan, and groups 2 and 4 are Arthur), then group 1 kernels are classified into the correct variety 88.89% of the time, group 2 kernels are classified correctly 77.78% of the time, group 3 kernels are classified correctly 90% of the time, and group 4 kernels are classified correctly 86% of the time. These results compare favorably with those obtained in Example 7.2.

Pages 24–46 show the results obtained when we consider the five nearest neighbors of each kernel. For a given kernel, the program finds its five nearest neighbors, and then computes a posterior probability for each of the four possible classification groups. It assigns the kernel to the group that produces the highest posterior probability.

If all samples from the groups were of the same size, then a new kernel would be classified into the group from which a majority of its nearest neighbors came. For example, suppose in the set of a kernel's five nearest neighbors, three were from group 1, one was from group 2, and one was from group 3. Then the posterior probabilities of the four classification groups are 0.6, 0.2, 0.2, and 0.0, respectively. The observation would be classified into group 1.

Next suppose the samples from the different populations are unequal. Suppose groups 1 and 2 contain 36 observations each and that groups 3 and 4 contain 50 observations each. Then the posterior probability for the ith group is computed by $p_i/(p_1 + p_2 + p_3 + p_4)$ where p_i is equal to the proportion of the observations in the ith group that falls into the kernel's

```
Ex. 7.4- NEAREST NEIGHBOR ANALYSIS ON WHEAT KERNELS      13-21

                    DISCRIMINANT ANALYSIS

      Classification Results for Calibration Data: WORK.FINAL
          Cross-validation Results using Nearest Neighbor

Squared Distance Function:

 2              -1
D (X,Y) = (X-Y)' COV  (X-Y)


Posterior Probability of Membership in each GRP:

 m (X) = Proportion of obs in group k in nearest neighbor of X
  k

Pr(j|X) = m (X) PRIOR  / SUM ( m (X) PRIOR  )
           j        j    k     k        k
```

			Posterior Probability of Membership in GRP:		
Obs	From	Classified			
	GRP	into GRP	1	2	3
			4		
1	1	1	1.0000	0.0000	0.0000
			0.0000		
2	1	3 *	0.0000	0.0000	1.0000
			0.0000		
3	1	3 *	0.0000	0.0000	1.0000
			0.0000		
37	2	1 *	1.0000	0.0000	0.0000
			0.0000		
38	2	1 *	1.0000	0.0000	0.0000
			0.0000		
39	2	1 *	1.0000	0.0000	0.0000
			0.0000		
73	3	3	0.0000	0.0000	1.0000
			0.0000		
74	3	3	0.0000	0.0000	1.0000
			0.0000		
75	3	4 *	0.0000	0.0000	0.0000
			1.0000		
123	4	4	0.0000	0.0000	0.0000
			1.0000		
124	4	4	0.0000	0.0000	0.0000
			1.0000		
125	4	4	0.0000	0.0000	0.0000
			1.0000		

```
                    * Misclassified observation
```

F I G. **7.17**

Pages 13–22, 25–34, and 36–45 of computer printout, Example 7.4

Ex. 7.4- NEAREST NEIGHBOR ANALYSIS ON WHEAT KERNELS   22

DISCRIMINANT ANALYSIS

Classification Summary for Calibration Data: WORK.FINAL

Cross-validation Summary using Nearest Neighbor

Squared Distance Function:

$$D^2(X,Y) = (X-Y)'\ COV^{-1}\ (X-Y)$$

Posterior Probability of Membership in each GRP:

$m_k(X)$ = Proportion of obs in group k in nearest neighbor of X

$$Pr(j|X) = m_j(X)\ PRIOR_j\ /\ SUM_k\ (\ m_k(X)\ PRIOR_k\ )$$

Number of Observations and Percent Classified into GRP:

From GRP	1	2	3	4	Total
1	19	2	13	2	36
	52.78	5.56	36.11	5.56	100.00
2	6	16	2	12	36
	16.67	44.44	5.56	33.33	100.00
3	14	2	31	3	50
	28.00	4.00	62.00	6.00	100.00
4	4	6	3	37	50
	8.00	12.00	6.00	74.00	100.00
Total	43	26	49	54	172
Percent	25.00	15.12	28.49	31.40	100.00
Priors	0.2500	0.2500	0.2500	0.2500	

F I G. 7.17
Pages 13–22, 25–34, and 36–45 of computer printout, Example 7.4

```
     Ex. 7.4- NEAREST NEIGHBOR ANALYSIS ON WHEAT KERNELS        25-33

                     DISCRIMINANT ANALYSIS

        Classification Results for Calibration Data: WORK.FINAL
            Resubstitution Results using 5 Nearest Neighbors
```

Squared Distance Function:

$$D^2(X,Y) = (X-Y)' \, COV^{-1} \, (X-Y)$$

Posterior Probability of Membership in each GRP:

$m_k(X)$ = Proportion of obs in group k in 5 nearest neighbors of X

$$Pr(j|X) = m_j(X) \, PRIOR_j \; / \; SUM_k \, (\, m_k(X) \, PRIOR_k \,)$$

Obs	From GRP	Classified into GRP	1 / 4	2	3
1	1	1	0.6757 / 0.0000	0.0000	0.3243
2	1	1	0.4808 / 0.1731	0.0000	0.3462
3	1	1	0.4505 / 0.0000	0.2252	0.3243
37	2	OTHER #	0.4237 / 0.1525	0.4237	0.0000
38	2	1 *	0.4505 / 0.0000	0.2252	0.3243
39	2	1 *	0.6000 / 0.0000	0.4000	0.0000
73	3	3	0.0000 / 0.0000	0.2577	0.7423
74	3	3	0.4808 / 0.0000	0.0000	0.5192
75	3	3	0.0000 / 0.4000	0.0000	0.6000
123	4	4	0.0000 / 0.5192	0.4808	0.0000
124	4	4	0.0000 / 0.7423	0.2577	0.0000
125	4	4	0.0000 / 0.7423	0.2577	0.0000

Posterior Probability of Membership in GRP:

    * Misclassified observation   # Tie for largest probability

FIG. 7.17

Pages 13–22, 25–34, and 36–45 of computer printout, Example 7.4

Ex. 7.4- NEAREST NEIGHBOR ANALYSIS ON WHEAT KERNELS   34

DISCRIMINANT ANALYSIS

Classification Summary for Calibration Data: WORK.FINAL

Resubstitution Summary using 5 Nearest Neighbors

Squared Distance Function:

$$D^2(X,Y) = (X-Y)' COV^{-1} (X-Y)$$

Posterior Probability of Membership in each GRP:

$m_k(X)$ = Proportion of obs in group k in 5 nearest neighbors of X

$$Pr(j|X) = m_j(X) PRIOR_j / SUM_k (m_k(X) PRIOR_k)$$

Number of Observations and Percent Classified into GRP:

From GRP	1	2	3	4	OTHER	Total
1	26	1	8	0	1	36
	72.22	2.78	22.22	0.00	2.78	100.00
2	4	15	2	14	1	36
	11.11	41.67	5.56	38.89	2.78	100.00
3	8	1	39	0	2	50
	16.00	2.00	78.00	0.00	4.00	100.00
4	3	6	1	37	3	50
	6.00	12.00	2.00	74.00	6.00	100.00
Total	41	23	50	51	7	172
Percent	23.84	13.37	29.07	29.65	4.07	100.00
Priors	0.2500	0.2500	0.2500			

F I G. **7.17**

Pages 13–22, 25–34, and 36–45 of computer printout, Example 7.4

```
        Ex. 7.4- NEAREST NEIGHBOR ANALYSIS ON WHEAT KERNELS       36-44

                    DISCRIMINANT ANALYSIS

        Classification Results for Calibration Data: WORK.FINAL
          Cross-validation Results using 5 Nearest Neighbors

Squared Distance Function:

  2                 -1
D (X,Y) = (X-Y)' COV  (X-Y)


Posterior Probability of Membership in each GRP:

m (X) = Proportion of obs in group k in 5 nearest neighbors of X
 k

Pr(j|X) = m (X) PRIOR  / SUM ( m (X) PRIOR  )
          j       j    k    k          k
```

			Posterior Probability of Membership in GRP:		
Obs	From GRP	Classified into GRP	1 / 4	2	3
1	1	3 *	0.4808 / 0.0000	0.0000	0.5192
2	1	3 *	0.2577 / 0.1856	0.0000	0.5567
3	1	3 *	0.2404 / 0.0000	0.2404	0.5192
37	2	1 *	0.6356 / 0.1525	0.2119	0.0000
38	2	3 *	0.4808 / 0.0000	0.0000	0.5192
39	2	1 *	0.6000 / 0.0000	0.4000	0.0000
73	3	3	0.0000 / 0.0000	0.2577	0.7423
74	3	3	0.4808 / 0.0000	0.0000	0.5192
75	3	3	0.0000 / 0.4000	0.0000	0.6000
123	4	4	0.0000 / 0.5192	0.4808	0.0000
124	4	4	0.0000 / 0.7423	0.2577	0.0000
125	4	4	0.0000 / 0.7423	0.2577	0.0000

```
    * Misclassified observation     # Tie for largest probability
```

F I G. 7.17

Pages 13–22, 25–34, and 36–45 of computer printout, Example 7.4

282   *Chapter 7: Discriminant Analysis*

Ex. 7.4- NEAREST NEIGHBOR ANALYSIS ON WHEAT KERNELS   45

                      DISCRIMINANT ANALYSIS

        Classification Summary for Calibration Data: WORK.FINAL

          Cross-validation Summary using 5 Nearest Neighbors

Squared Distance Function:

   2   -1
  D (X,Y) = (X-Y)' COV   (X-Y)

Posterior Probability of Membership in each GRP:

  m (X) = Proportion of obs in group k in 5 nearest neighbors of X
   k

  Pr(j|X) = m (X) PRIOR / SUM (m (X) PRIOR)
            j   j   k   k   k

        Number of Observations and Percent Classified into GRP:

From GRP   1   2   3   4   OTHER   Total

    1   20   1   14   0   1   36
              55.56   2.78   38.89   0.00   2.78   100.00

    2   5   9   2   18   2   36
              13.89   25.00   5.56   50.00   5.56   100.00

    3   12   3   34   0   1   50
              24.00   6.00   68.00   0.00   2.00   100.00

    4   4   10   3   31   2   50
               8.00   20.00   6.00   62.00   4.00   100.00

Total   41   23   53   49   6   172
Percent   23.84   13.37   30.81   28.49   3.49   100.00

Priors   0.2500   0.2500   0.2500
```

**FIG. 7.17**
Pages 13–22, 25–34, and 36–45 of computer printout, Example 7.4

nearest neighbors' set. For the situation described in the previous paragraph, $p_1 = 3/36 = 0.0833$, $p_2 = 1/36 = 0.0278$, $p_3 = 1/50 = 0.02$, and $p_4 = 0/50 = 0$. Then $p_1 + p_2 + p_3 + p_4 = 0.1311$, and the posterior probabilities for the four groups are $0.0833/0.1311 = 0.6354$, $0.0278/0.1311 = 0.2121$, $0.02/0.1311 = 0.1526$, and $0/0.1311 = 0$, respectively. Since the posterior probability for the first group is larger than the rest, the observation would be assigned to the first group.
```

7.8
Classification Trees

Another approach to discriminant analysis has been suggested. This method may be worth considering if you have a large number of discrete variables. This procedure is proposed by Breiman, Friedman, Olshen and Stone (1984) in their book titled *Classification and Regression Trees.* This procedure, available in software called CART, develops a series of questions that can be answered yes or no. The answers to these questions take the researcher through the branches of a tree until the researcher reaches a location on the tree at which a classification can be made. At each node on the tree, the procedure finds the best question to ask at that particular stage to help the researcher make the best decision. At the end of each branch a decision is made as to where a particular observation should be classified.

Summary

This chapter introduces the basic ideas behind discriminant analysis where we are attempting to classify individuals or experimental units into two or more uniquely defined populations. The primary methods used in this chapter were based on an assumption that the predictor variables have a multivariate normal distribution. Fortunately, in many cases these methods also work very well for cases where the predictor variables do not have a multivariate normal distribution. Stepwise methods for choosing those variables from a predictor set of variables that are likely to be the best discriminators were discussed. Canonical discriminant functions were also defined. These are most useful for creating visual displays of the original data projected down to a canonical space.

Exercises

1   Consider Example 7.1 and its SAS analysis. Answer each of the following questions:

   a   Which turkeys in this data set were misclassified by the discriminant rule when the rule was applied to the calibration data?

   b   What are the posterior probabilities for both DOMESTIC and WILD classifications for those turkeys that were misclassified in part a?

   c   How many turkeys of each type are misclassified by using the cross-validation method? What is the estimated probability for classifying a

WILD turkey incorrectly? What is the estimated probability for classifying a DOMESTIC turkey incorrectly?

d Determine the value of each of the classification functions for turkeys whose IDs are B710 and L674.

e What is the Mahalanobis distance between the sample means of the DOMESTIC and WILD turkeys?

2 Consider the TURKEY data of Example 7.1.

a Determine a subset of the bone measurement variables that should be good candidates for discrimination.

b Evaluate the effectiveness of the variables selected in part a. What are the chances for correct classifications using these variables?

c Suppose it is four times as costly to declare an innocent person guilty than it is to declare a guilty person innocent. Using this information and assuming equal prior probabilities, develop a discriminant rule for these data using the variables selected in part a. (*Note:* Declaring an innocent person guilty is equivalent to claiming that a wild turkey is domestic.)

d Write a program that will classify turkeys according to the following rule: Choose the DOMESTIC classification if $\mathbf{b}'\mathbf{x} > -25$, and choose WILD otherwise, where $\mathbf{b}'\mathbf{x}$ is the linear discriminant function for these data. That is, $\mathbf{x}$ is the vector that contains all of the variables being used as discriminators. Use this rule to classify turkeys. How do these classifications compare with those given on page 9 of the computer printout (see Figure 7.1)?

3 Consider the data in Example 7.2.

a Do a discriminant analysis using the variables selected by the stepwise analysis. That is, the variables that should be used are RIGHT_A, DOWN_A, DOWN_P, RIGHT_L, and RIGHT_B.

b Do another discriminant analysis using only the first three variables selected by the stepwise procedure.

c Compare the results of parts a and b with each other and with those obtained when using all eight variables. Which of these sets of variables does the better job of discriminating among the four groups? Why do you believe this is the best set?

4 Consider the canonical discriminant analysis of the data in Example 7.2 reported on pages 32–34 of the computer output (see Figure 7.11). Use the first two raw canonical functions to compute the locations of the four group means in the corresponding canonical space. If a computer is available, compute the locations of the projections of each of the original wheat kernels onto this canonical space. In addition, plot the first two canonical scores against one another in such a way that the group from which the kernel comes is identified.

Finally, draw in lines that will partition the discriminant space into regions—so that kernels can be classified into groups according to the region in which their canonical discriminant scores fall.

5   Consider the iris data discussed in Example 7.3.

   a   Create a data set that contains the first two canonical discriminant scores for each data point. These scores should be computed using the UNSTANDARDIZED CANONICAL DISCRIMINANT FUNCTION COEFFICIENTS.

   b   Perform a discriminant analysis of the iris data using only the first of the two canonical discriminant scores created in part a. How well does this one variable discriminate?

   c   Repeat part a using both of the canonical discriminant scores. How well do these two variables discriminate? Note that the results obtained when using both discriminant scores should be exactly the same as those obtained when using all of the original variables. Is this true in your analysis?

6   The enclosed disk contains a file labeled ORANGE.DAT. This file contains data that were collected from orange juice samples from several different countries. The countries are identified only by labels: BEL, LSP, NSP, TME, VME. On each sample of orange juice, a scientist measured the amounts of several different chemical elements: boron (B), barium (BA), calcium (CA), potassium (K), magnesium (MG), manganese (MN), phosphorous (P), rubidium (RB), and zinc (ZN).

   a   Determine a subset of these chemical elements that is likely to contain the best predictors of those countries from which orange juice comes.

   b   Determine how well the subset of variables selected in part a discriminates.

   c   Perform a canonical discriminant analysis using the elements boron, barium, calcium, potassium, manganese, and phosphorous.

   d   How many of the canonical functions are statistically significant? How many are likely to be of practical importance?

   e   Create a plot of the means of the countries in a two-dimensional canonical space and draw lines that partition the canonical space into regions so that orange juice samples can be classified into countries according to the region in which their canonical discriminant scores fall.

Logistic Regression Methods 8

If we know that predictor variables have a multivariate normal distribution, then the discriminant rules described in Sections 7.1 through 7.5 are known to be the best at discriminating. Nearest neighbor discriminant analysis can be used if the data are not multivariate normal provided that the predictor variables are still continuous. Logistic regression methods can be considered in situations where the predictor variables are not normally distributed and in situations where some or all of the predictor variables are discrete or categorical.

Logistic regression is similar to multiple regression; the primary difference is that the dependent variable in logistic regression is usually binary (i.e., it takes on only two possible values), whereas the dependent variable in multiple regression is continuous. As an example consider the turkey data introduced in Example 7.1. We might define a dependent variable y to be equal to 1 if a turkey is wild and equal to 0 if a turkey is domestic. Logistic regression tries to model the probability that y is equal to 1 given observed values of predictor variables. To use the logistic regression model for discrimination purposes, you simply compute an estimate of the probability that y is equal to 1 given the values of the predictor variables for a selected turkey. If this estimated probability is greater than 0.5, the turkey is classified as a wild turkey. If the estimated probability is less than 0.5, the turkey is classified as a domestic turkey.

8.1
Logistic Regression Model

The notation used here for logistic regression is similar to that used in the excellent book by Hosmer and Lemeshow (1989). Let $\mathbf{x}$ be a data vector for a randomly selected experimental unit and let y be the value of a binary outcome variable so that $y = 1$ if $\mathbf{x}$ comes from population 1 and $y = 0$ if $\mathbf{x}$ comes from population 2. Let $p(y = 1 \mid \mathbf{x})$ equal the probability that $y = 1$ given the observed data vector $\mathbf{x}$. The form of the logistic regression model is

$$p(y = 1 \mid \mathbf{x}) = \frac{\exp(\beta_0 + \beta_1' \mathbf{x})}{1 + \exp(\beta_0 + \beta_1' \mathbf{x})} \tag{8.1}$$

8.2
The Logit Transformation

When we study logistic regression it is customary to consider the logit transformation, a transformation performed on $p(y = 1 \mid \mathbf{x})$. The logit transformation is the log of the odds that $y = 1$ versus $y = 0$ and defined by

$$g(\mathbf{x}) = \log\{p(y = 1 \mid \mathbf{x})/[1 - p(y = 1 \mid \mathbf{x})]\} \tag{8.2}$$

Note that $g(\mathbf{x}) = \beta_0 + \beta' \mathbf{x}$. The logit transformation has many properties similar to those possessed by regression models. Foremost among these is that the logit transformation is linear in the parameters of the model. Furthermore, many of the things that we are accustomed to doing in regression modeling have counterparts in logistic regression. In particular, variable selection methods such as backward and stepwise can be used in both types of models.

Model Fitting

The logistic regression model is fit via the use of maximum likelihood methods, whereas regression models are usually fit via least squares. Maximum likelihood methods are beyond the scope of this text, but fortunately computer programs exist that will produce maximum likelihood analyses of logistic regression models. Those readers who want more information about the theory of logistic regression models should see Hosmer and Lemeshow (1989). The following example is used to illustrate many of the fundamental concepts of logistic regression.

E X A M P L E   **8.1**

A department store wants to develop a discriminant rule to determine whether local college students should be given credit for future purchases. During the preceding two years the store collected information from students that were given credit. Some of the variables collected included the students' gender (SEX), college major (MAJOR), age (AGE), grade point average (GPT), and the hours worked per week (HRS). Based on each student's past credit history at the store, each student was classified according to whether the student was a good credit risk or a bad credit risk (RISK). A random sample of 170 students from the store's database was selected for use in developing a discriminant rule. The data are given in the file labeled EX8_5.DAT on the enclosed disk. A portion of the data is shown in the following set of computer commands and on page 1 of the computer printout reproduced as Figure 8.1. Note that the data used in this example are not real, so you should not infer anything about college students from these data. The computer program used to simulate the data in this example is given in EX8_1.GEN on the enclosed computer disk.

An examination of page 1 of the computer output verifies that SEX and MAJOR are discrete variables and that AGE and GPT are continuous variables. The variable HRS could probably be considered either discrete or continuous. Hours worked per week could be considered discrete since it actually takes on only the values 0, 5, 10, and 20. Here HRS is taken as a continuous variable since the values that HRS takes on are quantitatively ordered. The possible values for SEX are male (MALE) and female (FEMALE). The possible values for MAJOR are science (SCI), humanities (HUM), social science (SOC), and business (BUS).

EX. 8.1 - DISCRIMINANT ANALYSIS USING LOGISTIC REGRESSION   1

FIG. 8.1

Pages 1, 5, 7, and
11 of computer
printout, Exam-
ple 8.1

OBS	ID	SEX	MAJOR	AGE	GPT	HRS	RISK
1	1	FEMALE	SCI	25	4.0	5	GOOD
2	2	MALE	HUM	28	3.3	5	BAD
3	3	FEMALE	SOC	25	3.3	0	BAD
4	4	FEMALE	BUS	24	2.2	20	GOOD
5	5	MALE	BUS	23	2.9	5	BAD

EX. 8.1 - DISCRIMINANT ANALYSIS USING LOGISTIC REGRESSION   5

The LOGISTIC Procedure

Data Set: WORK.CREDIT
Response Variable: RISK
Response Levels: 2
Number of Observations: 170
Link Function: Logit

Response Profile

Ordered Value	RISK	Count
1	BAD	87
2	GOOD	83

Criteria for Assessing Model Fit

Criterion	Intercept Only	Intercept and Covariates	Chi-Square for Covariates
AIC	237.576	61.490	.
SC	240.712	86.576	.
-2 LOG L	235.576	45.490	190.086 with 7 DF (p=0.0001)
Score	.	.	117.343 with 7 DF (p=0.0001)

Analysis of Maximum Likelihood Estimates

Variable	DF	Parameter Estimate	Standard Error	Wald Chi-Square	Pr > Chi-Square	Standardized Estimate	Odds Ratio
INTERCPT	1	33.2162	12.5345	7.0224	0.0080	.	999.000
NSEX	1	-0.6175	0.9579	0.4155	0.5192	-0.169004	0.539
DUM1	1	0.8180	1.3276	0.3796	0.5378	0.170129	2.266
DUM2	1	5.4875	1.8668	8.6412	0.0033	1.499330	241.653
DUM3	1	5.0109	1.7642	8.0674	0.0045	1.154355	150.043
GPT	1	-3.9438	1.3683	8.3079	0.0039	-1.055750	0.019
AGE	1	-0.5519	0.3593	2.3597	0.1245	-0.354071	0.576
HRS	1	-1.2542	0.3155	15.8076	0.0001	-4.924725	0.285

FIG. 8.1

Pages 1, 5, 7, and 11 of computer printout, Example 8.1

OBS	ID	SEX	MAJOR	AGE	GPT	HRS	RISK	NSEX	DUMMAJ1	DUMMAJ2	DUMMAJ3	_LEVEL_	PHAT	PREDICT	PBAD	PGOOD
1	1	FEMALE	SCI	25	4.0	5	GOOD	0	1	0	0	BAD	0.14071	GOOD	0.14071	0.85929
2	2	MALE	HUM	28	3.3	5	BAD	1	0	0	1	BAD	0.94640	BAD	0.94640	0.05360
3	3	FEMALE	SOC	25	3.3	0	BAD	0	0	1	0	BAD	0.99999	BAD	0.99999	0.00001
4	4	FEMALE	BUS	24	2.2	20	GOOD	0	0	0	0	BAD	0.00000	GOOD	0.00000	1.00000
5	5	MALE	BUS	23	2.9	5	BAD	1	0	0	0	BAD	0.89999	BAD	0.89999	0.10001
6	6	MALE	BUS	23	3.4	10	GOOD	1	0	0	0	BAD	0.00236	GOOD	0.00236	0.99764
7	7	FEMALE	HUM	25	3.3	10	GOOD	0	0	0	1	BAD	0.24473	GOOD	0.24473	0.75527
8	8	MALE	SCI	23	2.8	10	GOOD	1	1	0	0	BAD	0.05408	GOOD	0.05408	0.94592
9	9	FEMALE	SOC	25	3.4	0	BAD	0	0	1	0	BAD	0.99999	BAD	0.99999	0.00001
10	10	MALE	SOC	24	2.9	15	GOOD	1	0	1	0	BAD	0.00445	GOOD	0.00445	0.99555
11	11	FEMALE	SCI	23	3.7	20	GOOD	0	1	0	0	BAD	0.00000	GOOD	0.00000	1.00000
12	12	FEMALE	BUS	25	2.3	10	BAD	0	0	0	0	BAD	0.10029	BAD	0.10029	0.89971
13	13	MALE	BUS	22	3.3	5	GOOD	1	0	0	0	BAD	0.76342	BAD	0.76342	0.23658
14	14	FEMALE	BUS	22	3.3	0	BAD	0	0	0	0	BAD	0.99968	BAD	0.99968	0.00032
15	15	FEMALE	SOC	22	2.8	0	BAD	0	0	1	0	BAD	1.00000	BAD	1.00000	0.00000
16	16	MALE	SOC	21	3.0	0	BAD	1	0	1	0	BAD	1.00000	BAD	1.00000	0.00000
.	.	.	.	.	.	.	.	.	.	.	.	.	.	.	.	.
.	.	.	.	.	.	.	.	.	.	.	.	.	.	.	.	.
166	166	MALE	HUM	24	3.8	0	BAD	1	0	0	1	BAD	0.99992	BAD	0.99992	0.00008
167	167	MALE	SOC	21	2.9	20	GOOD	1	0	1	0	BAD	0.00004	GOOD	0.00004	0.99996
168	168	FEMALE	SOC	25	3.0	20	GOOD	0	0	1	0	BAD	0.00001	GOOD	0.00001	0.99999
169	169	FEMALE	HUM	23	2.5	5	BAD	0	0	0	1	BAD	0.99992	BAD	0.99992	0.00008
170	170	FEMALE	SOC	23	3.0	20	GOOD	0	0	1	0	BAD	0.00002	GOOD	0.00002	0.99998

FIG. 8.1

Pages 1, 5, 7, and 11 of computer printout, Example 8.1

TABLE OF RISK BY PREDICT

RISK   PREDICT

Frequency ~~Percent~~ Row Pct ~~Col Pct~~	BAD	GOOD	Total
BAD	79 ~~46.47~~ 90.80 ~~94.05~~	8 ~~4.71~~ 9.20 ~~9.30~~	87 ~~51.18~~
GOOD	5 ~~2.94~~ 6.02 ~~5.95~~	78 ~~45.88~~ 93.98 ~~90.70~~	83 ~~48.82~~
Total	84 ~~49.41~~	86 ~~50.59~~	170 ~~100.00~~

FIG. 8.1

Pages 1, 5, 7, and 11 of computer printout, Example 8.1

The following SAS commands are used to read the data into a SAS data set and to print the data:

```
TITLE 'Ex. 8.1 - DISCRIMINANT ANALYSIS USING LOGISTIC
REGRESSION';
DATA CREDIT;
 INPUT ID SEX $ MAJOR $ AGE GPT HRS RISK $;
CARDS;
              1      FEMALE     SCI      25     4.0      5     GOOD
              2      MALE       HUM      28     3.3      5     BAD
              3      FEMALE     SOC      25     3.3      0     BAD
              4      FEMALE     BUS      24     2.2     20     GOOD
              5      MALE       BUS      23     2.9      5     BAD
              6      MALE       BUS      23     3.4     10     GOOD
              7      FEMALE     HUM      25     3.3     10     GOOD
              .        .         .        .      .       .      .
              .        .         .        .      .       .      .
              .        .         .        .      .       .      .
            170      FEMALE     SOC      23     3.0     20     GOOD
PROC PRINT;
 RUN;
```

A portion of the data produced by the PRINT procedure is shown on page 1 of the computer output (see Figure 8.1).

We need to create dummy variables to use the discrete variables in logistic discriminant rules. The following SAS commands are used to create these dummy variables. Since SEX takes on two values, one dummy variable must be created for SEX. This dummy variable is named NSEX. Since MAJOR takes on four values, three dummy variables must be created for MAJOR. These are denoted by DUM1–DUM3. Note that DUM1 has the value 1 if a student's major is in the sciences and the value 0 otherwise. Similarly, DUM2 has the value 1 if a student's major is in the social sciences and the value 0 otherwise. Finally, DUM3 has the value 1 if a student's major is in the humanities and the value 0 otherwise. If the values of DUM1–DUM3 are all equal to 0, then a student's major is known to be business.

```
DATA CREDIT; SET CREDIT;
 IF SEX='MALE' THEN NSEX= 1; ELSE NSEX=0;
 IF MAJOR = 'SCI' THEN DUM1=1; ELSE DUM1=0;
 IF MAJOR = 'SOC' THEN DUM2=1; ELSE DUM2=0;
 IF MAJOR= 'HUM' THEN DUM3=1; ELSE DUM3=0;
```

Now the data are ready to be analyzed with the LOGISTIC procedure in SAS. Recommended SAS commands for performing logistic discriminant analysis are given below. The model statement that is used in these commands should be self-explanatory because the dependent variable is on the left-hand side of the equal sign and all of the predictor variables are listed on the right-hand side. The OUTPUT statement is used to produce an output data set that has been named PDICTS. In addition to containing those variables that were in the original data set, this data set contains a variable called PHAT. This variable gives the estimated probability that a student belongs to the first population (the BAD risk group in this example).

```
PROC LOGISTIC DATA=CREDIT;
 MODEL RISK = NSEX DUM1 DUM2 DUM3 GPT AGE HRS;
 OUTPUT OUT=PDICTS PREDICTED = PHAT;
 RUN;
```

A portion of the results of these commands is shown on page 5 of the computer printout of Figure 8.1. Near the top of page 5 under the label Response Profile we can see that this student data set contains 87 students that were known to be bad credit risks; this population is identified by SAS as population 1. Eighty-three students were good credit risks, and this population is identified by SAS as population 2. (SAS orders the populations according to the alphabetical order of the values of the dependent variable.) In the middle of page 5 under the label Chi-Square for Covariates we can see the value of a chi-square test statistic of 190.086 with 7 degrees of freedom. This has a corresponding p value equal to 0.0001. This chi-square test indicates that the predictor variables being used (called *covariates* by SAS) are statistically significant predictors of credit risk. This test considers all predictor variables simultaneously.

The bottom of page 5 lists the parameter estimates of the parameters in the logit function. For this example, the logit function $g(\mathbf{x})$ is estimated by

$$\hat{g} = 33.22 - 0.62\,\text{NSEX} + 0.82\,\text{DUM1} + 5.49\,\text{DUM2} + 5.01\,\text{DUM3}$$
$$- 3.94\,\text{GPT} - 0.55\,\text{AGE} - 1.25\,\text{HRS}$$

Looking at each parameter estimate's corresponding p values, we see that the variable that corresponds to sex, NSEX, is not statistically significant and the variable that corresponds to age, AGE, is also not statistically significant. Thus we should be able to develop a simpler model by removing these two variables from the predictor set. The nonsignificance of these variables indicates that credit risk does not depend on the sex of the student and that credit risk does not depend on the age of the student.

Note also that DUM1 is not statistically significant. Recall that DUM1, DUM2, and DUM3, the three dummy variables representing majors, had values equal to 1 if a student's major was science, social science, or humanities, respectively, and had values of 0 otherwise. All three of these dummy variables have a value equal to 0 if and only if a student's major is business. Thus each dummy variable can be interpreted as to whether there is a difference in credit risk between students for which the major is identified to have a value equal to 1 by the dummy variable and students that are business majors. For the data being analyzed, the nonsignificance of DUM1 indicates that the probability that a science major is a bad credit risk is essentially the same as the probability that a business major is a bad credit risk, provided that all remaining variables (SEX, GPT, AGE, and HRS) have identical values for any two students being compared. In other words, if we condition on SEX, GPT, AGE, and HRS, then business majors and science majors are equally bad and/or equally good credit risks. Remember that these data are not real.

DUM2 is statistically significant and has an estimated value of 5.49. Since the estimated value is positive and since DUM2 has a value of 1 if a student's major is in the social sciences, we can conclude that for this manufactured data, social science majors tend to be poorer credit risks than business majors. (Remember that population 1 is the population of students that have been identified as bad credit risks.) A similar statement can be made about students whose major is in the humanities since DUM2 is positive and statistically significant. You might also note that the estimated coefficients for DUM2 and DUM3 are nearly equal in size, 5.4875 for DUM2 and 5.0109 for DUM3 (see bottom of page 5 of Figure 8.1 computer printout), and you might guess that social science majors and humanities majors are equally bad and/or equally good credit risks. You should note that the near equality of the parameters corresponding to DUM2 and DUM3 needs to be verified by a statistical test before actually concluding that social science majors and humanities majors are equally bad and/or equally good credit risks.

Summarizing the discussion in the preceding two paragraphs and assuming that a comparison of social science majors and humanities majors is not significant, we could say that science majors and business majors have equal risks and that social science majors and humanities majors also have equal risks, but social science majors and humanities majors are poorer credit risks than science majors or business majors. This statement assumes that all remaining variables in the predictor set have equal values for students being compared.

The statistical significance of the parameter estimate of -3.94 corresponding to grade point average (GPT) and the fact that the sign of this estimate is negative imply that the probability that a student is a bad credit risk decreases as the student's grade point average increases. Again this interpretation assumes that two students being compared according to grade point average are of the same sex, have the same major, are of the same age, and work the same number of hours per week. That is, when comparing one student to another with respect to grade point average, all of their other variables must be identical for both students.

The statistical significance of the parameter estimate of -1.25 corresponding to hours worked per week (HRS) and the fact that the sign of this estimate is negative imply that the probability that a student is a bad credit risk decreases as the student's hours worked per week increases provided that all remaining predictor variables for students being compared have identical values.

Information that is not yet available that would be interesting and helpful is an answer to questions such as these two: "How much of a greater risk is a social science major than a business major?" and "How much less risk is associated with a student with a 3.0 grade point average versus one with a 2.0 grade point average?"

To answer the first question, note that the estimated odds that a social science major falls into the bad credit risk group rather than the good credit risk group is equal to $e^{\hat{g}_{soc}}$ where

$$\hat{g}_{soc} = 33.22 - 0.62 \text{ NSEX} + 5.49 - 3.94 \text{ GPT} - 0.55 \text{ AGE} - 1.25 \text{ HRS}$$

and the estimated odds that a business major falls into the bad credit risk rather than the good credit risk group is equal to $e^{\hat{g}_{bus}}$ where

$$\hat{g}_{bus} = 33.22 - 0.62 \text{ NSEX} - 3.94 \text{ GPT} - 0.55 \text{ AGE} - 1.25 \text{ HRS}$$

The ratio of these two probabilities, called the *odds ratio,* is equal to $e^{5.49} = 242.3$. This can be interpreted to mean that a social science major is 242.3 times more likely to be a bad credit risk than a business major (assuming that all the other variables for two such students have identical values).

The last column near the bottom of page 5 of the computer printout (see Figure 8.1) gives the odds ratios for each of the predictor variables in the model. The difference between the 242.3 calculated earlier and the 241.653 printed on the computer printout is due to round-off error. In this case, the computer printout gives the more accurate estimate of the odds ratio.

From the computer printout we can also see that the odds ratio for DUM3 is predicted to be 150.043; this means that a humanities major is 150 times more likely to be a bad credit risk than a business major for these made-up data.

We can compare the relative risks of social science majors to humanities majors by computing the ratio of their respective odds ratios. This is true because these respective ratios have common denominators, so that when we take the ratio of the two, their denominators cancel. To illustrate, a social science major is estimated to be $241.653/150.043 = 1.6$ times more likely to be a bad credit risk than a humanities major. Note, however, that there may not be a statistically significant difference between these two groups of students.

The odds ratios for continuous variables usually need some adjustment to be meaningful. Whether they are interesting or not depends on the units in which the continuous predictor variables are measured.

First let's consider grade point average for two students. Assume that the two students are identical on all other variables. The odds ratio for grade point average from the computer printout is 0.019. This means that an increase in grade point average of one unit, say, from 2.5 to 3.5 or from 2.0 to 3.0, reduces the risk of being a bad credit risk by a factor of 0.019. Equivalently, we could say that a 3.0 grade point average student is 0.019 times as likely to be a bad credit risk as a 2.0 grade point average student. Alternatively, we could also say that a 2.0 grade point average student is $1/0.019 = 52.6$ times more likely to be a bad credit risk than a 3.0 grade point average student.

How can we compare the relative risks of a 2.0 grade point average student and a 2.5 grade point average student? The difference in grade point averages is equal to 0.5 units. To compute the odds ratio to compare the relative risks of a 2.5 student compared to a 2.0 student one can compute $\exp[(2.5 - 2.0)(-3.9438)] = \exp(-1.9719) = 0.1392$. Then we can say that a 2.5 grade point average student is 0.1392 times less likely to be a bad credit risk than a 2.0 student. Equivalently, a 2.0 student is $1/0.1392 = 7.2$ times more likely to be a bad credit risk than a 2.5 student.

Next consider the hours that students work per week. It makes very little sense to compare two students who differ in hours worked per week by only 1 hour. So the odds ratio for HRS is really not useful. It might be of some interest to compare the relative risks of offering credit to students who work 15 hours per week versus those who work 10 hours per week. The difference in hours worked per week for these two kinds of students is equal to 5 hours. So the odds ratio to compare students who work 10 hours per week to those who work 5 hours per week is $\exp[(10 - 5)(-1.2542)] = \exp(-6.2710) = 0.0019$. So that a student who works 10 hours per week is 0.0019 times less likely to be a bad credit risk than one who works only 5 hours per week. Equivalently, a student who works 5 hours per week is about 526 times more likely to be a bad credit risk than a student that works 10 hours per week.

Caution　The reader should remember that the data being analyzed in this example are not real data, but simulated data with majors randomly assigned to the various risk groups. So the reader should infer nothing about real college students from these data and the analyses.

Next recall that a data set named PDICTS was created by the LOGISTIC procedure. Recall also that this data set contains a new variable named PHAT. This variable gives the predicted probability that a student with the given values for the predictor variable would fall into the BAD credit risk group. The first set of commands that follows is used to create a predicted group for each of the students in the data set. With this set of commands, a new variable called PREDICT is created that identifies the population to which a student would be classified using the logistic discriminant rule created by the LOGISTIC procedure. The second group of commands is used to create a summary classification matrix of all classifications produced by the preceding commands.

```
DATA PDICTS; SET PDICTS;
 IF PHAT>.5 THEN PREDICT='BAD '; ELSE PREDICT='GOOD';
 PBAD=PHAT; PGOOD=1-PHAT;
 RUN;

PROC PRINT;
 RUN;

PROC FREQ;
 TABLES RISK*PREDICT;
 RUN;
```

All of the results of these commands can be obtained by executing the file labeled EX8_5.SAS on the enclosed disk and looking at pages 7–11. A portion of the data set that was created by the LOGISTIC procedure and modified by the preceding commands is printed on page 7 of the computer printout of Figure 8.1. The variable labeled PREDICT gives the group to which a student would be classified by the logistic discriminant rule. The variables labeled PBAD and PGOOD are the estimated probabilities that a student would fall into each of the bad and good groups, respectively.

Page 11 contains the classification summary matrix for the 170 students in this data set. The labels on the left show the groups from which these students came, and the labels on the top show the groups to which the students are classified. One can see that 79 of 87 (90.8%) of the students in the BAD risk group would be classified correctly by the logistic discriminant rule, and that 78 of the 83 (94.0%) of the students in the GOOD risk group would be classified correctly.

8.3
Variable Selection Methods

In the previous example, we noted that age and sex were not statistically significant variables. Hence, we should be able to develop a simpler model by removing these variables from the model. However, we should refrain from removing both variables at the same time because it could be that sex is not important as long as age is in the model and that age is not important as long as sex is in the model. From the initial fitted model we cannot really tell whether both variables can be removed simultaneously.

Stepwise logistic regression procedures exist that can be used to help a researcher identify an appropriate subset of candidate variables to use as predictors, similar to the stepwise procedures discussed in Section 7.5. It is possible to select variables in a backward elimination procedure as well as in a stepwise procedure. The following example illustrates variable selection in the SAS LOGISTIC procedure using the backward elimination procedure. The

SELECTION=BACKWARD option on the MODEL statement tells the computer to choose variables by using a backward elimination procedure. The INCLUDE=3 option tells the computer to always include the first three variables in the model statement; these are the three dummy variables in the following example. Dummy variables should generally only be removed if none of the variables in a group of dummy variables is significant. In the first run we saw that DUM2 and DUM3 were both significant. So all three dummy variables are included in the model even though DUM1 was not significant.

E X A M P L E **8 . 1 (continued)**

The data in Example 8.1 were reanalyzed using the following SAS commands:

```
PROC LOGISTIC DATA=CREDIT;
 MODEL RISK = DUM1 DUM2 DUM3 NSEX GPT AGE HRS/
  SELECTION=BACKWARD INCLUDE=3;
 OUTPUT OUT=PDICTS PREDICTED = PHAT;
 RUN;

DATA; SET PDICTS;
IF PHAT>.5 THEN PREDICT='BAD '; ELSE PREDICT='GOOD';

IF PHAT=. THEN DELETE;

 PBAD=PHAT; PGOOD=1-PHAT;

PROC PRINT;
RUN;

PROC FREQ;
 TABLES RISK*PREDICT;
RUN;
```

A portion of the results from these commands can be found on pages 12–19 of the computer printout. Some of these pages are shown in Figure 8.2. All of the results can be obtained by executing the file labeled EX8_5.SAS on the enclosed floppy disk. From the bottom of page 12, we can see that the first variable removed from the set of predictor variables is NSEX. In the middle of page 13, we can see that the second variable removed from the set of predictor variables is AGE. On the bottom of page 13, we can see the significance probabilities corresponding to each variable when the variable was removed from the set of possible predictors.

The final model includes the variables DUM1–DUM3, GPT, and HRS as shown on page 14 of the computer printout. The estimates of the parame-

EX. 8.1 - DISCRIMINANT ANALYSIS USING LOGISTIC REGRESSION   12

The LOGISTIC Procedure

Data Set: WORK.CREDIT
Response Variable: RISK
Response Levels: 2
Number of Observations: 170
Link Function: Logit

Response Profile

	Ordered Value	RISK	Count
	1	BAD	87
	2	GOOD	83

Backward Elimination Procedure

The following variables will be included in each model:

INTERCPT DUM1 DUM2 DUM3

Step 0. The following variables were entered:

INTERCPT DUM1 DUM2 DUM3 NSEX   GPT
AGE   HRS

Criteria for Assessing Model Fit

F I G. **8**.2

Pages 12–15 and
19 of computer
printout, Example 8.1

Criterion	Intercept Only	Intercept and Covariates	Chi-Square for Covariates
AIC	237.576	61.490	.
SC	240.712	86.576	.
-2 LOG L	235.576	45.490	190.086 with 7 DF (p=0.0001)
Score	.	.	117.343 with 7 DF (p=0.0001)

Step 1. Variable NSEX is removed:

ters of the final logit model are given under the column labeled `Parameter Estimate`. In the exercises at the end of this chapter you will be asked to interpret the results shown on page 14. Page 19 summarizes classifications obtained from the final model. We can see that 90.8% of those in the BAD risk group are classified correctly and that 95.2% of those in the GOOD risk group are classified correctly. The columns labeled PBAD and

```
      EX. 8.1 - DISCRIMINANT ANALYSIS USING LOGISTIC REGRESSION        13

                        The LOGISTIC Procedure

                    Criteria for Assessing Model Fit

                                Intercept
                    Intercept      and
     Criterion        Only      Covariates    Chi-Square for Covariates

     AIC             237.576      59.923             .
     SC              240.712      81.873             .
     -2 LOG L        235.576      45.923       189.653 with 6 DF (p=0.0001)
     Score              .            .         117.107 with 6 DF (p=0.0001)


          Residual Chi-Square = 0.4218 with 1 DF (p=0.5161)


Step  2. Variable AGE is removed:


                    Criteria for Assessing Model Fit

                                Intercept
                    Intercept      and
     Criterion        Only      Covariates    Chi-Square for Covariates

     AIC             237.576      60.199             .
     SC              240.712      79.014             .
     -2 LOG L        235.576      48.199       187.376 with 5 DF (p=0.0001)
     Score              .            .         116.657 with 5 DF (p=0.0001)


          Residual Chi-Square = 2.6516 with 2 DF (p=0.2656)


NOTE: No (additional) variables met the 0.05 significance level for
      removal from the model.


             Summary of Backward Elimination Procedure

                  Variable   Number     Wald        Pr >
          Step    Removed      In    Chi-Square  Chi-Square

            1      NSEX         6       0.4155      0.5192
            2      AGE          5       2.1615      0.1415
```

F I G. 8.2

Pages 12–15 and 19 of computer printout, Example 8.1

PGOOD on page 15 give the posterior probabilities for the BAD and GOOD risk groups, respectively. Some of the predictions for individual students in this data set are also shown on page 15 and a final summary classification matrix is shown on page 19.

EX. 8.1 - DISCRIMINANT ANALYSIS USING LOGISTIC REGRESSION   14

The LOGISTIC Procedure

Analysis of Maximum Likelihood Estimates

Variable	DF	Parameter Estimate	Standard Error	Wald Chi-Square	Pr > Chi-Square	Standardized Estimate	Odds Ratio
INTERCPT	1	16.7720	4.4494	14.2087	0.0002	.	999.000
DUM1	1	0.5459	1.2559	0.1889	0.6638	0.113538	1.726
DUM2	1	4.4005	1.4711	8.9478	0.0028	1.202324	81.489
DUM3	1	4.4676	1.5715	8.0818	0.0045	1.029189	87.146
GPT	1	-3.1824	1.0344	9.4646	0.0021	-0.851920	0.041
HRS	1	-1.1109	0.2568	18.7141	0.0001	-4.362078	0.329

F I G. 8.2
Pages 12–15 and 19 of computer printout, Example 8.1

EX. 8.1 - DISCRIMINANT ANALYSIS USING LOGISTIC REGRESSION   15

								D	D	D							
								U	U	U	_			P			
								M	M	M	L			R			
			M					_	_	_	E			E			P
			A				R	N	M	M	M	V	P	D	P		G
O		S	J	A	G	H	I	S	A	A	A	E	H	I	B		O
B	I	E	O	G	P	R	S	E	J	J	J	L	A	C	A		O
S	D	X	R	E	T	S	K	X	1	2	3	_	T	T	D		D
1	1	FEMALE	SCI	25	4.0	5	GOOD	0	1	0	0	BAD	0.27560	GOOD	0.27560		0.72440
2	2	MALE	HUM	28	3.3	5	BAD	1	0	0	1	BAD	0.99442	BAD	0.99442		0.00558
3	3	FEMALE	SOC	25	3.3	0	BAD	0	0	1	0	BAD	0.99998	BAD	0.99998		0.00002
4	4	FEMALE	BUS	24	2.2	20	GOOD	0	0	0	0	BAD	0.00000	GOOD	0.00000		1.00000
5	5	MALE	BUS	23	2.9	5	BAD	1	0	0	0	BAD	0.87957	BAD	0.87957		0.12043
6	6	MALE	BUS	23	3.4	10	GOOD	1	0	0	0	BAD	0.00572	GOOD	0.00572		0.99428
7	7	FEMALE	HUM	25	3.3	10	GOOD	0	0	0	1	BAD	0.40815	GOOD	0.40815		0.59185
8	8	MALE	SCI	23	2.8	10	GOOD	1	1	0	0	BAD	0.06285	GOOD	0.06285		0.93715
9	9	FEMALE	SOC	25	3.4	0	BAD	0	0	1	0	BAD	0.99997	BAD	0.99997		0.00003
10	10	MALE	SOC	24	2.9	15	GOOD	1	0	1	0	BAD	0.00883	GOOD	0.00883		0.99117
11	11	FEMALE	SCI	23	3.7	20	GOOD	0	1	0	0	BAD	0.00000	GOOD	0.00000		1.00000
12	12	FEMALE	BUS	25	2.3	10	BAD	0	0	0	0	BAD	0.16019	GOOD	0.16019		0.83981
13	13	MALE	BUS	22	3.3	5	GOOD	1	0	0	0	BAD	0.67160	BAD	0.67160		0.32840
14	14	FEMALE	BUS	22	3.3	0	BAD	0	0	0	0	BAD	0.99811	BAD	0.99811		0.00189
15	15	FEMALE	SOC	22	2.8	0	BAD	0	0	1	0	BAD	1.00000	BAD	1.00000		0.00000
16	16	MALE	SOC	21	3.0	0	BAD	1	0	1	0	BAD	0.99999	BAD	0.99999		0.00001

F I G. 8.2
Pages 12–15 and 19 of computer printout, Example 8.1

```
                    TABLE OF RISK BY PREDICT


           RISK        PREDICT


           Frequency|
           Percent  |
           Row Pct  |
           Col Pct  |BAD      |GOOD     |  Total
           ---------+---------+---------+
           BAD      |     79  |      8  |     87
                    |  46.47  |   4.71  |  51.18
                    |  90.80  |   9.20  |
                    |  95.18  |   9.20  |
           ---------+---------+---------+
           GOOD     |      4  |     79  |     83
                    |   2.35  |  46.47  |  48.82
                    |   4.82  |  95.18  |
                    |   4.82  |  90.80  |
           ---------+---------+---------+
           Total          83        87      170
                       48.82     51.18   100.00
```

F I G. 8.2

Pages 12–15 and 19 of computer printout, Example 8.1

8.4
Logistic Discriminant Analysis (More Than Two Populations)

A Kansas State University researcher wanted to see if it would be possible to discriminate between different groups of turkeys using measurements that could be taken on various bones of the turkeys. The turkeys consisted of big-breasted turkeys produced by a well-known turkey farm, wild turkeys, and various types of hybrids and domestic turkeys raised by individual farmers. The turkeys were initially placed into one of three groups: big-breasted turkeys (BB), wild turkeys (W), and all other types (OT).

This example is used to discuss extensions of logistic discriminant analysis to the case where there are more than two groups.

Logistic Regression Models

The simplest way to generalize to the case for three classification groups is to let $\mathbf{x}$ be a data vector for a randomly selected experimental unit as before, and let $y = 0$ if $\mathbf{x}$ comes from population 1, let $y = 1$ if $\mathbf{x}$ comes from population 2, and let $y = 2$ if $\mathbf{x}$ comes from population 3. Next let the logit transformation for comparing $y = 1$ to $y = 0$ be defined by $g_1(\mathbf{x}) = \beta_{01} + \beta_1'\mathbf{x}$, and let the logit transformation for comparing $y = 2$ to $y = 0$ be defined by $g_2(\mathbf{x}) = \beta_{02}$

$+ \beta_2'\mathbf{x}$. Then the probability that $y = 0$ given $\mathbf{x}$ is

$$p(y = 0 \mid \mathbf{x}) = \frac{1}{1 + \exp(\beta_{01} + \beta_1'\mathbf{x}) + \exp(\beta_{02} + \beta_2'\mathbf{x})}$$

the probability that $y = 1$ given $\mathbf{x}$ is

$$p(y = 1 \mid \mathbf{x}) = \frac{\exp(\beta_{01} + \beta_1'\mathbf{x})}{1 + \exp(\beta_{01} + \beta_1'\mathbf{x}) + \exp(\beta_{02} + \beta_2'\mathbf{x})}$$

and the probability that $y = 2$ given $\mathbf{x}$ is

$$p(y = 2 \mid \mathbf{x}) = \frac{\exp(\beta_{02} + \beta_2'\mathbf{x})}{1 + \exp(\beta_{01} + \beta_1'\mathbf{x}) + \exp(\beta_{02} + \beta_2'\mathbf{x})}$$

Model Fitting

The logistic regression model defined by the preceding logits must be fit via the use of maximum likelihood methods that are beyond the scope of this text. Unfortunately, computer programs that produce estimates of the parameters of the given logistic regression models are not easily available. See Hosmer and Lemeshow (1989) for more information. Begg and Gray (1984) proposed an alternative method for fitting these models that seems to provide an adequate fit to the data. Their procedure can be accomplished by a simple modification of SAS's LOGISTIC procedure.

Begg and Gray proposed approximating the parameters in the first logit by using standard logistic regression to model the $p(y = 1 \mid \mathbf{x})$ assuming only the two outcomes $y = 0$ and $y = 1$. Then they estimated the parameters in the second logit by using standard logistic regression to model the $p(y = 2 \mid \mathbf{x})$ assuming only the two outcomes $y = 0$ and $y = 2$. Then the estimates of the probabilities for each of the three groups are calculated by using the three formulas in the preceding section, replacing each of the parameters with their respective estimates. The following example is used to illustrate the fundamentals of using logistic regression when there are more than two groups.

E X A M P L E   8 . 2

As stated earlier, a researcher wanted to see if it would be possible to discriminate among three different groups of turkeys using bone measurements. Turkeys consisted of big-breasted turkeys produced by a well-known turkey farm (BB), wild turkeys (W), and various other types of hybrids and domestic turkeys raised by individual farmers (OT). Bone measurements on a large number of turkeys is given in the file labeled EX8_6.DAT on the enclosed floppy disk. The data were set up for analysis by means of the following commands. Only the male turkeys are being analyzed here. The CHK variable is created so that any turkeys that have missing values for any of the bones identified by the CHK variable can be deleted from the data set. This is done primarily to reduce the amount of printout because the LOGISTIC procedure would have eliminated these turkeys anyway. The IF statement just prior to the CARDS statement is used to label all turkeys that are neither wild or big breasted as other types.

```
OPTIONS LINESIZE=75 PAGESIZE=54 NODATE PAGENO=1;
TITLE 'EX8_2 - Three Group Disc Anal using Logistic
Regression';

DATA TURKEY;
  INPUT ID 1-4 TYPE $ 6-7 SEX $9 AGE 11-12
    (HUM RAD ULN CAR FEM TIB TIN MAX MIN MMS BC STL STB KEL
GB ANL HN TK WBL)
    (7*4.0 3*4.1 7*4.0 4.1 4.0);
  IF SEX='M';

  CHK=HUM+FEM+TIB+TIN+MAX+MIN+MMS+BC+STL; IF CHK=. THEN
DELETE;
  DROP RAD ULN CAR KEL STB GB ANL HN TK WBL CHK;
  IF TYPE ^= 'BB' AND TYPE ^= 'W' THEN TYPE= 'OT';
CARDS;
2237 BB M  5 145        140 219 146 192 135 162  28 135  85 110    74 55 165
2229 BB M  5 153        146 232 148 201 143 171  28 161 101 134 130  87 72 174
2228 BB M  5 154        132 208 133 193 141 174  30 166  98 146 112  98 69 160
 .    .  .  .   .         .   .   .   .   .   .    .   .    .   .   .    .  .   .
 .    .  .  .   .         .   .   .   .   .   .    .   .    .   .   .    .  .   .
 .    .  .  .   .         .   .   .   .   .   .    .   .    .   .   .    .  .   .
1020  D M  S 146 132 141  82 133 207 141 190 144 172  28 127                113
2122 GF M  9 142 124 141    130 211 135 149  98 135  22 196  76 135   104 53  85 124
2085  W M  6 143 129 140    129 221 140 112  86 102  20 148  62 108  74 93 45  78 120
;

PROC PRINT;
  RUN;
```

A portion of the results from the print procedure can be found on page 1 of the computer printout shown in Figure 8.3. All of the results can be obtained by executing the file labeled EX8_2.SAS on the enclosed disk.

Next the data must be used to create two data sets so that, within each set, the dependent variable is a binary variable. In this analysis, the data set labeled TURKEY2 contains the BB turkeys and the OT turkeys, and the data set labeled TURKEY3 contains the BB turkeys and the W turkeys. Next the logistic procedure is run on both data sets. The DESCENDING option was selected in each run so that in TURKEY2 the logit corresponds to the probabil-

EX8_2 - Three Group Disc Anal using Logistic Regression   1

OBS	ID	TYPE	SEX	AGE	HUM	FEM	TIB	TIN	MAX	MIN	MMS	BC	STL
1	2237	BB	M	5	145	140	219	146	19.2	13.5	16.2	28	135
2	2229	BB	M	5	153	146	232	148	20.1	14.3	17.1	28	161
3	2228	BB	M	5	154	132	208	133	19.3	14.1	17.4	30	166
4	2232	BB	M	5	154	141	228	146	20.6	14.5	17.5	30	155
5	2240	BB	M	5	154	136	217	140	20.7	13.7	16.1	31	142
6	2244	BB	M	5	154	146	227	144	19.1	14.0	16.1	30	156
7	2231	BB	M	5	155	143	222	146	22.0	16.3	18.2	29	165
8	2233	BB	M	5	155	144	231	151	19.6	14.9	17.3	30	142
9	2236	BB	M	5	155	148	227	140	20.4	14.7	17.8	28	153
10	2239	BB	M	5	155	141	215	146	20.3	14.9	17.5	28	150
11	2245	BB	M	5	156	138	231	145	18.4	14.3	14.2	30	154
12	2247	BB	M	5	156	148	236	148	19.7	14.2	16.3	31	154
13	2230	BB	M	5	157	142	220	141	18.0	13.4	16.8	30	162
14	2235	BB	M	5	157	151	229	143	18.8	14.2	16.8	30	149
15	2241	BB	M	5	157	146	225	139	21.1	16.1	17.7	29	147
16	2243	BB	M	5	157	141	224	144	18.9	14.3	16.8	29	164
17	2242	BB	M	5	158	146	230	144	20.4	15.4	17.1	29	160
18	2234	BB	M	5	160	149	231	149	20.1	14.3	16.0	29	148
19	2238	BB	M	5	160	150	230	135	19.3	14.8	17.6	29	158
20	2246	BB	M	5	165	150	235	152	18.9	13.7	15.8	31	158
21	1020	OT	M	.	146	133	207	141	19.0	14.4	17.2	28	127
22	1021	OT	M	.	152	139	214	145	19.0	14.0	16.8	29	134
23	1023	OT	M	.	147	133	198	132	18.1	12.5	16.1	26	134
24	1024	OT	M	.	145	129	193	128	20.0	13.6	17.7	26	124
25	1025	OT	M	.	144	130	200	135	18.5	13.8	16.0	27	132
26	1035	OT	M	.	150	137	212	142	19.1	14.2	17.5	27	128
27	1052	OT	M	.	148	134	206	141	18.9	13.5	16.6	26	125
28	2219	OT	M	12	140	132	203	132	13.9	10.5	12.5	22	212
29	2122	OT	M	9	142	130	211	135	14.9	9.8	13.5	22	196
30	2249	OT	M	9	143	131	211	135	14.8	10.1	12.7	22	198
31	2330	OT	M	9	154	142	229	149	14.7	10.3	12.8	24	205
32	2327	OT	M	9	157	144	239	155	13.7	9.6	12.2	24	196
33	2326	OT	M	9	164	148	242	155	14.4	10.5	13.1	25	226
34	2334	OT	M	24	153	145	233	147	14.1	10.6	12.7	25	226
35	2040	OT	M	.	147	136	222	146	14.9	9.5	13.1	21	218
36	2085	W	M	6	143	129	221	140	11.2	8.6	10.2	21	148
37	2084	W	M	6	148	129	220	142	11.8	8.8	10.6	21	143
38	2135	W	M	7	154	140	232	153	14.3	9.9	12.4	22	207
39	2130	W	M	9	154	143	238	154	13.6	10.2	12.4	22	218
40	2113	W	M	10	150	131	217	142	12.8	9.3	11.3	19	189
41	2053	W	M	10	153	138	222	150	14.1	9.3	12.5	22	208
42	2127	W	M	10	163	149	241	157	13.9	10.2	12.5	22	223
43	2055	W	M	10	165	151	248	162	14.0	10.1	12.2	22	214
44	2052	W	M	11	153	142	230	148	14.3	10.4	12.5	23	219
45	2099	W	M	11	153	141	230	151	13.8	10.1	12.8	23	218
46	2101	W	M	12	147	135	232	153	13.2	9.2	12.5	21	229
47	2150	W	M	12	148	137	231	148	12.4	9.3	11.5	21	220
48	2096	W	M	12	150	137	225	144	11.8	9.0	10.8	22	209

F I G. **8.3**

Pages 1, 3–5, 8–13, and 16 of computer printout, Example 8.2

```
EX8_2 - Three Group Disc Anal using Logistic Regression          3
                         ANALYSIS 1

                        The LOGISTIC Procedure

Data Set: WORK.TURKEY2
Response Variable: TYPE
Response Levels: 2
Number of Observations: 46
Link Function: Logit


                          Response Profile

                 Ordered
                   Value   TYPE      Count

                       1   OT          26
                       2   BB          20


              Analysis of Maximum Likelihood Estimates

           Parameter  Standard    Wald        Pr >    Standardized    Odds
Variable DF Estimate    Error   Chi-Square Chi-Square   Estimate      Ratio

INTERCPT 1  32.1633   13.0391     6.0845     0.0136           .      999.000
HUM      1  -0.3663    0.1209     9.1802     0.0024    -1.264205      0.693
TIB      1   0.0111    0.0652     0.0292     0.8642     0.077964      1.011
TIN      1   0.1497    0.1005     2.2186     0.1364     0.625778      1.161
```

F I G. **8.3**

Pages 1, 3–5, 8–13, and 16 of computer printout, Example 8.2

ity that a turkey is OT rather than BB, which would be the default, and so that in TURKEY3 the logit corresponds to the probability that a turkey is W rather than BB. To simplify things a bit more, only the three variables HUM, TIB, and TIN are used in this analysis. The SAS commands that produce these analyses are given here:

```
PROC SORT DATA=TURKEY OUT=TURKEY; BY ID;


DATA TURKEY2; SET TURKEY; IF TYPE='W' THEN DELETE;
DATA TURKEY3; SET TURKEY; IF TYPE='OT' THEN DELETE;


PROC LOGISTIC DATA=TURKEY2 DESCENDING;
 MODEL TYPE = HUM TIB TIN;
RUN;


PROC  LOGISTIC DATA=TURKEY3 DESCENDING;
 MODEL TYPE =HUM TIB TIN;
RUN;
```

```
        EX8_2 - Three Group Disc Anal using Logistic Regression              4
                               ANALYSIS 1

                          The LOGISTIC Procedure

Data Set: WORK.TURKEY3
Response Variable: TYPE
Response Levels: 2
Number of Observations: 74
Link Function: Logit


                            Response Profile

                    Ordered
                      Value   TYPE    Count

                        1     W         54
                        2     BB        20



                Analysis of Maximum Likelihood Estimates

                  Parameter  Standard    Wald       Pr >     Standardized    Odds
    Variable DF   Estimate    Error   Chi-Square Chi-Square    Estimate      Ratio

    INTERCPT 1     6.1005    12.9052    0.2235     0.6364            .       446.100
    HUM      1    -0.5759     0.1631   12.4634     0.0004     -1.641465       0.562
    TIB      1     0.1831     0.1034    3.1335     0.0767      0.812629       1.201
    TIN      1     0.2827     0.1217    5.3937     0.0202      1.070879       1.327
```

F I G. 8.3

Pages 1, 3–5, 8–13, and 16 of computer printout, Example 8.2

A portion of the results from these two runs is shown on pages 3 and 4 of the Figure 8.3 computer printout. From these two runs, we must identify the parameters in the two logit functions. We can see that the estimate of the first logit is

G1 = 32.1633−.3663*HUM+.0111*TIB+.1497*TIN

and the estimate of the second logit is

G2 = 6.1005−.5759*HUM+.1831*TIB+.2827*TIN

Using these estimates of the two logit functions, we can create a new data set that contains estimated probabilities for each of the three groups. This can be accomplished by using the following SAS commands. In these commands, G1 and G2 are the estimates of the two logit functions obtained from the preceding analyses. P1, P2, and P3 are the estimated probabilities of membership for the three groups BB, OT, and W, respectively. The PRINT procedure produces a listing of predicted classifications for each of the turkeys in the data set, and the FREQ procedure produces an overall

EX8_2 - Three Group Disc Anal using Logistic Regression   5
                              ANALYSIS 1

OBS	ID	TYPE	P1	P2	P3	PREDICT
1	1012	W	0.29327	0.35503	0.35170	OT
2	1013	W	0.04423	0.11496	0.84081	W
3	1013	W	0.04423	0.11496	0.84081	W
4	1020	OT	0.10122	0.81743	0.08135	OT
5	1021	OT	0.32816	0.57889	0.09295	OT
6	1023	OT	0.43031	0.56675	0.00294	OT
7	1024	OT	0.41201	0.58684	0.00115	OT
8	1025	OT	0.13402	0.84863	0.01735	OT
9	1033	W	0.17927	0.11254	0.70819	W
10	1034	W	0.04136	0.10802	0.85061	W
11	1035	OT	0.28114	0.64404	0.07481	OT
12	1049	W	0.07151	0.07147	0.85702	W
13	1051	W	0.00864	0.06736	0.92400	W
14	1052	OT	0.19800	0.76012	0.04188	OT
15	2002	OT	0.01299	0.26324	0.72377	W
16	2003	OT	0.02319	0.15830	0.81851	W
17	2004	OT	0.36692	0.11912	0.51396	W
24	2028	W	0.01073	0.10054	0.88873	W
25	2031	W	0.00224	0.03234	0.96542	W
26	2032	W	0.06132	0.09199	0.84668	W
27	2035	W	0.00720	0.25702	0.73579	W
28	2040	OT	0.02276	0.31822	0.65902	W
29	2048	W	0.00261	0.04252	0.95487	W
30	2052	W	0.09756	0.22329	0.67916	W
31	2053	W	0.15021	0.42438	0.42540	W
32	2054	W	0.19522	0.10669	0.69809	W
33	2055	W	0.09019	0.02528	0.88454	W
34	2084	W	0.10151	0.52872	0.36976	OT
62	2228	BB	0.88292	0.11619	0.00089	BB
63	2229	BB	0.07474	0.17489	0.75037	W
64	2230	BB	0.84649	0.14047	0.01304	BB
65	2231	BB	0.49138	0.36665	0.14197	BB
66	2232	BB	0.27085	0.31158	0.41757	W
67	2233	BB	0.11219	0.19554	0.69227	W
68	2234	BB	0.71233	0.14741	0.14025	BB
69	2235	BB	0.72032	0.17820	0.10148	BB
70	2236	BB	0.68794	0.22101	0.09105	BB
71	2237	BB	0.01219	0.34300	0.64480	W
72	2238	BB	0.97248	0.02447	0.00305	BB
73	2239	BB	0.56479	0.38992	0.04529	BB
74	2240	BB	0.68850	0.28552	0.02597	BB
75	2241	BB	0.86814	0.11288	0.01898	BB
76	2242	BB	0.75322	0.15172	0.09506	BB
77	2243	BB	0.74258	0.20185	0.05556	BB
78	2244	BB	0.38870	0.32780	0.28350	BB
79	2245	BB	0.46985	0.23125	0.29890	BB
80	2246	BB	0.90252	0.04900	0.04848	BB

F I G. 8.3

Pages 1, 3–5, 8–13, and 16 of computer printout, Example 8.2

EX8_2 - Three Group Disc Anal using Logistic Regression   8
ANALYSIS 1

TABLE OF TYPE BY PREDICT

TYPE   PREDICT

```
Frequency|
Percent  |
Row Pct  |
Col Pct  |BB      |OT      |W       | Total
---------+--------+--------+--------+
BB       |     15 |      0 |      5 |     20
         |  15.00 |   0.00 |   5.00 |  20.00
         |  75.00 |   0.00 |  25.00 |
         |  88.24 |   0.00 |   7.58 |
---------+--------+--------+--------+
OT       |      2 |     13 |     11 |     26
         |   2.00 |  13.00 |  11.00 |  26.00
         |   7.69 |  50.00 |  42.31 |
         |  11.76 |  76.47 |  16.67 |
---------+--------+--------+--------+
W        |      0 |      4 |     50 |     54
         |   0.00 |   4.00 |  50.00 |  54.00
         |   0.00 |   7.41 |  92.59 |
         |   0.00 |  23.53 |  75.76 |
---------+--------+--------+--------+
Total          17       17       66      100
            17.00    17.00    66.00   100.00
```

FIG. 8.3

Pages 1, 3–5, 8–13, and 16 of computer printout, Example 8.2

classification matrix for the three groups of turkeys. A portion of the results from these commands can be found on pages 5–8 of the computer printout (pages 5 and 8 are shown in Figure 8.3). All of the results can be seen by executing the file labeled EX8_2.SAS on the enclosed floppy disk.

```
DATA FINAL; SET TURKEY;
  G1 = 32.1633-.3663*HUM+.0111*TIB+.1497*TIN;
  G2 =  6.1005-.5759*HUM+.1831*TIB+.2827*TIN;
  EG1=EXP(G1); EG2=EXP(G2);
  SUM= 1+EG1+EG2;
  P1=1/SUM; P2=EG1/SUM; P3=EG2/SUM;

IF P1>P2 AND P1>P3 THEN PREDICT='BB';
IF P2>P1 AND P2>P3 THEN PREDICT='OT';
IF P3>P1 AND P3>P2 THEN PREDICT='W ';
RUN;

PROC PRINT DATA=FINAL;
  VAR ID TYPE P1 P2 P3 PREDICT;
  RUN;

PROC FREQ;
  TABLES TYPE*PREDICT;
```

```
          EX8_2 - Three Group Disc Anal using Logistic Regression        9
                         ANALYSIS 2

                      The LOGISTIC Procedure

Data Set: WORK.TURKEY
Response Variable: TYPE
Response Levels: 3
Number of Observations: 100
Link Function: Logit


                        Response Profile

                   Ordered
                    Value   TYPE      Count

                       1    BB          20
                       2    OT          26
                       3    W           54



                 Backward Elimination Procedure


Step  0. The following variables were entered:

     INTERCP1  INTERCP2  HUM       FEM       TIB       TIN
     MAX       MIN       MMS       BC        STL


         Score Test for the Proportional Odds Assumption

            Chi-Square = 9.5528 with 9 DF (p=0.3879)


                 Criteria for Assessing Model Fit

                              Intercept
                   Intercept     and
     Criterion       Only     Covariates   Chi-Square for Covariates

     AIC            204.973     92.506             .
     SC             210.184    121.162             .
     -2 LOG L       200.973     70.506      130.468 with 9 DF (p=0.0001)
     Score             .           .        74.185 with 9 DF (p=0.0001)
```

F I G. 8.3

Pages 1, 3–5, 8–13, and 16 of computer printout, Example 8.2

The LOGISTIC Procedure

Step 1. Variable STL is removed:

Score Test for the Proportional Odds Assumption

Chi-Square = 8.5491 with 8 DF (p=0.3817)

Criteria for Assessing Model Fit

Criterion	Intercept Only	Intercept and Covariates	Chi-Square for Covariates
AIC	204.973	90.642	.
SC	210.184	116.693	.
-2 LOG L	200.973	70.642	130.332 with 8 DF (p=0.0001)
Score	.	.	74.176 with 8 DF (p=0.0001)

Residual Chi-Square = 0.1376 with 1 DF (p=0.7107)

Step 2. Variable MMS is removed:

Score Test for the Proportional Odds Assumption

Chi-Square = 8.3903 with 7 DF (p=0.2994)

Criteria for Assessing Model Fit

Criterion	Intercept Only	Intercept and Covariates	Chi-Square for Covariates
AIC	204.973	89.025	.
SC	210.184	112.472	.
-2 LOG L	200.973	71.025	129.948 with 7 DF (p=0.0001)
Score	.	.	74.021 with 7 DF (p=0.0001)

Residual Chi-Square = 0.4883 with 2 DF (p=0.7834)

F I G. 8.3

Pages 1, 3–5, 8–13, and 16 of computer printout, Example 8.2

```
        EX8_2 - Three Group Disc Anal using Logistic Regression        11
                              ANALYSIS 2

                          The LOGISTIC Procedure

Step  3. Variable FEM is removed:


                 Score Test for the Proportional Odds Assumption

                   Chi-Square = 7.3658 with 6 DF (p=0.2883)


                        Criteria for Assessing Model Fit

                               Intercept
                   Intercept     and
    Criterion        Only      Covariates    Chi-Square for Covariates

    AIC             204.973     88.173              .
    SC              210.184    109.014              .
    -2 LOG L        200.973     72.173      128.800 with 6 DF (p=0.0001)
    Score              .           .         73.895 with 6 DF (p=0.0001)


            Residual Chi-Square = 1.6398 with 3 DF (p=0.6504)


NOTE: No (additional) variables met the 0.1 significance level for removal
      from the model.


                   Summary of Backward Elimination Procedure

                   Variable    Number      Wald         Pr >
            Step   Removed        In     Chi-Square   Chi-Square

              1    STL           8        0.1364       0.7119
              2    MMS           7        0.3160       0.5740
              3    FEM           6        1.1154       0.2909
```

F I G. **8.3**

Pages 1, 3–5, 8–13, and 16 of computer printout, Example 8.2

The LOGISTIC Procedure

Analysis of Maximum Likelihood Estimates

Variable	DF	Parameter Estimate	Standard Error	Wald Chi-Square	Pr > Chi-Square	Standardized Estimate	Odds Ratio
INTERCP1	1	-17.7641	9.5975	3.4259	0.0642	.	0.000
INTERCP2	1	-10.3502	8.8049	1.3818	0.2398	.	0.000
HUM	1	-0.2673	0.1295	4.2631	0.0389	-0.843013	0.765
TIB	1	0.2134	0.0860	6.1549	0.0131	1.311201	1.238
TIN	1	-0.2568	0.0850	9.1237	0.0025	-1.115858	0.774
MAX	1	1.5546	0.5975	6.7698	0.0093	2.330622	4.733
MIN	1	-1.6238	0.8765	3.4321	0.0639	-1.852686	0.197
BC	1	1.5020	0.4049	13.7613	0.0002	2.696444	4.491

Association of Predicted Probabilities and Observed Responses

Concordant = 78.2%	Somers' D = 0.777	
Discordant = 0.5%	Gamma = 0.986	
Tied = 21.3%	Tau-a = 0.471	
(3004 pairs)	c = 0.888	

F I G. 8.3

Pages 1, 3–5, 8–13, and 16 of computer printout, Example 8.2

```
EX8_2 - Three Group Disc Anal using Logistic Regression          13
                    ANALYSIS 2
```

OBS	ID	TYPE	P1	P2	P3	PREDICT
1	1012	W	0.00040	0.39579	0.60382	W
2	1013	W	0.00043	0.41502	0.58456	W
3	1013	W	0.00043	0.41502	0.58456	W
4	1020	OT	0.34896	0.64992	0.00112	OT
5	1021	OT	0.59638	0.40321	0.00041	BB
6	1023	OT	0.13967	0.85663	0.00370	OT
7	1024	OT	0.46118	0.53811	0.00070	OT
8	1025	OT	0.20639	0.79129	0.00231	OT
9	1033	W	0.00015	0.19440	0.80545	W
10	1034	W	0.00003	0.04460	0.95537	W
11	1035	OT	0.12959	0.86638	0.00403	OT
12	1049	W	0.00001	0.01232	0.98767	W
13	1051	W	0.00048	0.44066	0.55886	W
14	1052	OT	0.04437	0.94281	0.01282	OT
15	2002	OT	0.00004	0.06697	0.93298	W
16	2003	OT	0.00028	0.31661	0.68311	W
24	2028	W	0.00012	0.16742	0.83246	W
25	2031	W	0.00000	0.00470	0.99530	W
26	2032	W	0.00016	0.20762	0.79222	W
27	2035	W	0.00022	0.26399	0.73580	W
28	2040	OT	0.00037	0.37921	0.62042	W
29	2048	W	0.00000	0.00290	0.99710	W
30	2052	W	0.00045	0.42707	0.57248	W
31	2053	W	0.00005	0.07309	0.92686	W
32	2054	W	0.00019	0.24201	0.75779	W
33	2055	W	0.00001	0.00870	0.99130	W
34	2084	W	0.00001	0.02102	0.97897	W
62	2228	BB	0.96962	0.03036	0.00002	BB
63	2229	BB	0.94856	0.05141	0.00003	BB
64	2230	BB	0.90747	0.09247	0.00006	BB
65	2231	BB	0.87735	0.12256	0.00008	BB
66	2232	BB	0.99687	0.00313	0.00000	BB
67	2233	BB	0.93392	0.06604	0.00004	BB
68	2234	BB	0.88845	0.11147	0.00008	BB
69	2235	BB	0.97428	0.02570	0.00002	BB
70	2236	BB	0.96024	0.03974	0.00002	BB
71	2237	BB	0.93659	0.06337	0.00004	BB
72	2238	BB	0.96775	0.03223	0.00002	BB
73	2239	BB	0.19820	0.79936	0.00243	OT
74	2240	BB	0.99963	0.00037	0.00000	BB
75	2241	BB	0.94249	0.05748	0.00004	BB
76	2242	BB	0.91377	0.08618	0.00006	BB
77	2243	BB	0.69038	0.30935	0.00027	BB
78	2244	BB	0.98948	0.01051	0.00001	BB
79	2245	BB	0.95395	0.04602	0.00003	BB
80	2246	BB	0.94952	0.05045	0.00003	BB

F I G. 8.3

Pages 1, 3–5, 8–13, and 16 of computer printout, Example 8.2

TABLE OF TYPE BY PREDICT

TYPE   PREDICT

```
Frequency|
Percent  |
Row Pct  |
Col Pct  |BB      |OT      |W       |  Total
---------+--------+--------+--------+
BB       |    19  |     1  |     0  |     20
         | 19.00  |  1.00  |  0.00  |  20.00
         | 95.00  |  5.00  |  0.00  |
         | 95.00  |  5.00  |  0.00  |
---------+--------+--------+--------+
OT       |     1  |    15  |    10  |     26
         |  1.00  | 15.00  | 10.00  |  26.00
         |  3.85  | 57.69  | 38.46  |
         |  5.00  | 75.00  | 16.67  |
---------+--------+--------+--------+
W        |     0  |     4  |    50  |     54
         |  0.00  |  4.00  | 50.00  |  54.00
         |  0.00  |  7.41  | 92.59  |
         |  0.00  | 20.00  | 83.33  |
---------+--------+--------+--------+
Total         20       20       60      100
            20.00    20.00    60.00   100.00
```

F I G. 8.3

Pages 1, 3–5, 8–13, and 16 of computer printout, Example 8.2

From the summary classification matrix, we can see that 15 of 20 (75%) of the big-breasted turkeys are classified correctly, 13 of 26 (50%) of the other types are classified correctly, and 50 of 54 (92.6%) of the wild birds are correctly classified.

Another SAS LOGISTIC Analysis

The SAS LOGISTIC procedure will automatically analyze data sets when we have three groups. In this case we let $y = 1$ if the observation comes from population 1, $y = 2$ if the observation comes from population 2, and $y = 3$ if the observation comes from population 3. If we let $u(\mathbf{x}) = \beta_{01} + \beta'\mathbf{x}$ and $v(\mathbf{x}) = \beta_{02} + \beta'\mathbf{x}$, then the SAS LOGISTIC procedure models the probability that $y = 1$ by

$$P(y = 1) = \frac{\exp(\beta_{01} + \beta'\mathbf{x})}{1 + \exp(\beta_{01} + \beta'\mathbf{x})}$$

and the probability that y is less than or equal 2 by

$$P(y \leq 2) = \frac{\exp(\beta_{02} + \beta'\mathbf{x})}{1 + \exp(\beta_{02} + \beta'\mathbf{x})}$$

Then it can be shown that

$$P(y = 2) = P(y \leq 2) - P(y = 1) \quad \text{and} \quad P(y = 3) = \frac{1}{1 + \exp(\beta_{02} + \beta'\mathbf{x})}$$

The SAS analysis commands for reanalyzing the turkey data using the above models are given later in this section. In this analysis, the variables that are going to be used as discriminators are selected by the backward elimination procedure.

The LOGISTIC procedure can create an output data set by using the OUTPUT option. This data set contains two rows of data for each experimental unit. The first row identified by _LEVEL_='BB' contains a new variable PHAT, which gives the estimated probability that the experimental unit is in the BB group. The second row identified by _LEVEL_='OT' also contains the variable PHAT, but in this case it gives the estimated probability that the experimental unit is either in the BB group or the OT group. In the data set labeled ONE, created by the following SAS commands, P1 is the estimated probability that the experimental unit is in the BB group. In the data set labeled TWO, P2 is the estimated probability that the experimental unit is in either the BB or OT group. These two data sets are merged into a third data set labeled THREE. In this data set P1 is the estimated probability that an experimental unit is in the BB group, P2 is the estimated probability that an experimental unit is in the OT group, and P3 is the estimated probability that an experimental unit is in the W group.

A portion of the results from these SAS commands can be found on pages 9–16 of the computer printout. Some of these pages are shown in Figure 8.3. All of the results can be obtained by executing the file labeled EX8_2.SAS on the enclosed floppy disk.

```
TITLE2 'ANALYSIS 2';
PROC LOGISTIC DATA=TURKEY;
 MODEL TYPE =HUM--STL/SELECTION=BACKWARD SLSTAY=.10;
OUTPUT OUT=PDICTS  PREDICTED=PHAT;

DATA ONE;
 SET PDICTS;
 IF _LEVEL_='BB' THEN P1=PHAT;

 IF _LEVEL_='BB';

DATA TWO;
 SET PDICTS;
 IF _LEVEL_='OT' THEN P2=PHAT;
 IF _LEVEL_='OT';
```

```
DATA THREE; DROP _LEVEL_;
 MERGE ONE TWO;
 P2=P2-P1;
 P3=1-P1-P2;
RUN;

DATA FINAL; SET THREE;
 IF P1>P2 AND P1>P3 THEN PREDICT='BB';
 IF P2>P1 AND P2>P3 THEN PREDICT='OT';
 IF P3>P1 AND P3>P2 THEN PREDICT='W ';

PROC PRINT; VAR ID TYPE P1 P2 P3 PREDICT;

PROC FREQ;
 TABLES TYPE*PREDICT;
```

Exercises

1 Consider the student credit risk problem described in Example 8.1.

a Explain how you would attempt to determine if there is a statistically significant difference between the credit risks of social science majors and humanities majors.

b Carry out the process you described in part a to see if there is a statistically significant difference between the credit risks of social science majors and humanities majors.

c What is the odds ratio for comparing social science majors to science majors?

d What is the odds ratio for comparing 3.5 grade point average students to 1.5 grade point average students?

e What is the odds ratio for comparing students who do not work to students who work 20 hours per week?

f The estimated coefficient for NSEX in the logit model was -0.6175. If this were statistically significant, who would be the better credit risk—males or females?

g Analyze the data in this example using a traditional discriminant analysis procedure. How do the results from the two analyses compare with one another?

h Consider the results from the LOGISTIC procedure after removing the variables NSEX and AGE from the set of possible variables. Give the estimated logit function for the remaining variables.

i Using the estimated logit function from part h, what is the estimated probability that a 20-year-old female science major who works 20 hours per week and has a 3.0 grade point average will be a good credit risk?

j Using the estimated logit function from part h, who would be the bigger credit risk: business majors, science majors, social science majors, or humanities majors? Explain your answer.

k Using the analysis that produced the logit function in part h, what is the odds ratio for comparing humanities majors to science majors?

l Give new answers to parts c–e using the logit function given in part h.

2 Reanalyze the turkey data used in Example 7.1 by means of a logistic regression approach.

3 Reanalyze the turkey data used in Example 8.2 using a traditional discriminant analysis approach. Compare your results to the results given by the two logistic regression approaches described in the text. Write a short report describing what you have learned from comparing these three analyses of these data.

4 Consider the iris data described in Example 7.3. Perform a discriminant analysis on these data using a logistic regression approach. You may use either of the two approaches described in the text for cases where there are more than two populations under study.

Cluster Analysis

9

Suppose a marketing professional has data that have been collected on consumers. Variables measured might include such things as age, education level, income level, marital status, employment status, number of children less than 5 years old, number of children between 6 and 13, and number of children 14 or older. This marketing professional might want to use this information to partition consumers into subgroups, called *clusters,* so that consumers who fall into distinct subgroups have similar characteristics with respect to the measured variables.

If it is possible to partition consumers into subgroups, then the marketing professional might be able to study the buying habits of the consumers in each of the distinct subgroups. This information might be extremely valuable when making decisions about how to best advertise to consumers within these different subgroups. In other words, this information could allow for more efficient use of the resources a company has devoted to advertising.

In more general terms, suppose a researcher has collected data on a large number of experimental units. The basic question posed for a cluster analysis is whether it is possible to devise a classification or grouping scheme that will allow partitioning of the experimental units into classes or groups, called clusters, so that the units within a class or group are similar to one another while those in distinct classes or groups are not similar to those in the other groups.

Cluster analysis involves techniques that produce classifications from data that are initially unclassified, and must not be confused with discriminant analysis, in which you initially know how many distinct groups exist and you have data that are known to come from each of these distinct groups.

Many disciplines have their own terminology for clustering and cluster analysis. Some other names that exist include topology, grouping, classification, and numerical taxonomy.

9.1
Measures of Similarity and Dissimilarity

To perform a cluster analysis, we must first be able to measure the similarity or dissimilarity between two individual observations and then later the similarity or dissimilarity between two clusters of observations.

Ruler Distance

One simple measure of dissimilarity is standard Euclidean distance, also called *ruler distance.* This is the distance between two observations if we were able to plot the two observations in the p-dimensional sample space and measure the distance between them using a ruler. The formula that calculates Euclidean distance is given later by Eq. (9.1).

Standardized Ruler Distance

Another possibility for measuring the distance between a pair of points is to first standardize all of the variables and then compute the standard Euclidean distance between points using their standardized Z scores. For most situations this is probably the best choice for measuring dissimilarities. The formula that calculates standardized Euclidean distance is given later in Eq. (9.2).

A Mahalanobis Distance

A third possibility is to compute a Mahalanobis-type distance between points. This would require estimates of within cluster variance–covariance matrices, after which these matrices would be pooled across clusters. However, estimating a pooled variance–covariance matrix is impossible, because to do so, a researcher would need to know where the clusters are, and this is precisely why he or she is performing a cluster analysis in the first place! Thus, we cannot really compute a Mahalanobis distance until we have grouped the points into initial clusters. Even then the usefulness of this measure of distance would depend on how well the clusters are identified. If the clusters are well identified, then we probably no longer need the Mahalanobis distance measure. If the clusters are not well defined, then the Mahalanobis distance measure would not be well defined either. The formula that calculates a Mahalanobis-type distance is given in the next section in Eq. (9.3).

Dissimilarity Measures

Thus, the three choices of dissimilarity between the two points $\mathbf{x}_r$ and $\mathbf{x}_s$ discussed above are:

1 ruler distance where

$$d_{rs} = [(\mathbf{x}_r - \mathbf{x}_s)'(\mathbf{x}_r - \mathbf{x}_s)]^{1/2} \tag{9.1}$$

2 standardized ruler distance where

$$d_{rs} = [(\mathbf{z}_r - \mathbf{z}_s)'(\mathbf{z}_r - \mathbf{z}_s)]^{1/2} \tag{9.2}$$

and where $\mathbf{z}_r$ is the vector of Z scores corresponding to the rth experimental unit, and

3 the Mahalanobis distance where

$$d_{rs} = [(\mathbf{x}_r - \mathbf{x}_s)'\mathbf{\Sigma}^{-1}(\mathbf{x}_r - \mathbf{x}_s)]^{1/2} \tag{9.3}$$

and where $\mathbf{\Sigma}$ is replaced by some reasonable estimate of $\mathbf{\Sigma}$.

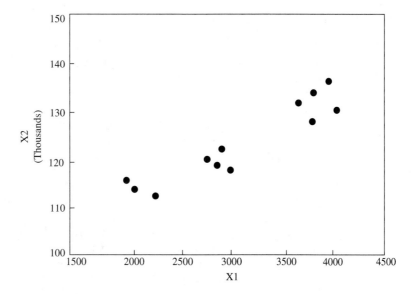

F I G. 9.1
Scatter plot show-
ing three clusters

9.2
Graphical Aids in Clustering

There are many different clustering algorithms, and often, different algorithms applied to the same set of data produce clusters that are substantially different. This is unfortunate, but it happens because the choice of a clustering algorithm imposes a structure on the sample. Furthermore, clustering methods often detect clusters that may not really exist.

These undesirable characteristics of clustering algorithms suggest that researchers need ways to verify and/or evaluate the results of clustering programs. Many of the graphical methods discussed in Chapter 3 are extremely valuable for verifying, evaluating, and fine-tuning the results of clustering programs.

Scatter Plots

When $p = 2$, that is, when only two variables are being measured on each experimental unit, perhaps the best and safest way to identify clusters of points is to plot the data and then select the clusters visually. For example, researchers faced with a plot similar to the one shown in Figure 9.1, would likely agree that there are three clusters of points in these data.

Using Principal Components

When $p > 2$, a principal components analysis can be performed to see if the data actually fall within a reduced dimensional space. If it is possible to get the effective dimensionality down to two, that is, if the data lie on a plane within the p-dimensional sample space, then the first two principal component scores corresponding to each experimental unit in the data set could be plotted, after which clusters could be selected visually as described earlier.

If more than two principal components are required, it is still probably safer and easier to apply clustering programs to the first few principal component scores rather than applying clustering programs to raw data values.

Caution   When applying clustering programs to principal component scores and when plotting principal component scores to locate the clusters visually, it is extremely important that standardized principal component scores not be used. Both cluster analysis and principal component analysis may be applied to standardized data when necessary, but the principal component scores must not be standardized. The reason for this is that standardized principal component scores do not realistically illustrate the actual distances between pairs of points.

Andrews' Plots

Andrews' plots, discussed in Section 3.2, are very useful for identifying clusters and for validating results from clustering programs. Observations that fall into the same cluster should produce Andrews' curves that are similar to one another, while those that fall into different clusters should produce curves that are not similar.

Other Methods

Other graphical methods that are likely to be useful when doing cluster analysis are three-dimensional plots, blob plots or bubble plots, Chernoff faces, and star or sun-ray plots. However, Chernoff faces and sun-ray plots lose their simplicity as the sample size gets large and, consequently, their usefulness decreases.

9.3
Clustering Methods

There are two basic ways to search for clusters. The two basic methods are distinguished by being either hierarchical or nonhierarchical in nature.

Nonhierarchical Clustering Methods

One way to search for clusters is to select initially a set of cluster seed points and then build clusters around each of the seeds. This is accomplished by assigning every point in the data set to its closest cluster seed, using one of the dissimilarity measures to measure the distances between each of the points and the cluster seeds. After this step, clusters that are too large may be split, and clusters that are real close to one another may be combined. This type of clustering is classified as nonhierarchical clustering, and while it is a very reasonable approach, it has three major disadvantages.

One disadvantage of nonhierarchical clustering methods is that the approach requires us to guess initially at the number of clusters that are going to exist. Another is that the approach is greatly influenced by the choice of the initial cluster seeds. Furthermore, if we let the computing package choose the seeds, their selection often depends on the order in which the data are read into the computer. So two researchers could perform a cluster analysis on the same set of data and produce entirely different clusters. Finally, the procedure is very often not feasible computationally because there are just too many possible choices, for not only the number of clusters, but for locations of cluster seeds as well.

Several nonhierarchical clustering possibilities are available in the SAS `FASTCLUS` procedure. An example using the `FASTCLUS` procedure is shown later in Example 9.4.

Hierarchical Clustering

Other methods for selecting clusters are classified as hierarchical cluster analysis methods. With these methods the observed data points are grouped into clusters in a nested sequence of clusterings. The most efficient of the hierarchical clustering methods are those that are known as *single-link clustering methods*.

Nearest Neighbor Method

One example of a single-link clustering method is one that is known as the *nearest neighbor method*. The nearest neighbor method utilizes the following steps:

1 Start with N clusters where each cluster contains exactly one data point.

2 Unite the two closest points according to one of the three distance measures selected.

3 Define the dissimilarity between this new cluster and any other point as the minimum distance between the two points in the cluster and this point.

4 Continue combining clusters that are closest to one another, so that at each stage the number of clusters is reduced by one, and the dissimilarity between any two clusters is always defined to be the distance between their two closest members.

Thus the nearest neighbor method starts with N clusters, where each cluster contains one observation, and continues to combine points and clusters until all observations are within one cluster. Clearly, the appropriate number of clusters is somewhere between the beginning of this process and its end. Some methods for deciding when to stop this process and/or to determine the number of clusters that exist are described later in this section.

E X A M P L E 9.1

To illustrate how the nearest neighbor method works, consider a sample containing six points, and suppose that the distances between the points are given by the following dissimilarity matrix:

	1	2	3	4	5	6
1		0.31	0.23	0.32	0.26	0.25
2			0.34	0.21	0.36	0.28
3				0.31	0.04	0.07
4					0.31	0.28
5						0.09
6						

The initial clustering is denoted by C_0 and has every point in a cluster by itself. Thus the initial clustering is

$$C_0 = \{[1], [2], [3], [4], [5], [6]\}$$

Searching through the dissimilarity matrix, we can see that the two points closest to one another are points 3 and 5. So the first step of the clustering process would produce the cluster:

$$C_1 = \{[1], [2], [3,5], [4], [6]\}$$

Then we must compute a new distance matrix between the clusters in C_1. The nearest neighbor method takes the distance between [1] and [3,5] to be the minimum of 0.23 and 0.26. So the distance between [1] and [3,5] is 0.23. In a similar way the distances between all other clusters can be determined. A new distance matrix for the clustering defined by C_1 is

	1	2	[3,5]	4	6
1		0.31	0.23	0.32	0.25
2			0.34	0.21	0.28
[3,5]				0.31	0.07
4					0.28
6					

Here the two closest clusters are [6] and [3,5] and, hence, these two clusters would be combined. The results of the second step produce this clustering:

$$C_2 = \{[1], [2], [3,5,6], [4]\}$$

Then a new distance matrix must be computed. The advantage of the single-link methods is that the new distance matrix can be computed from the distance matrix in the previous step. Thus we do not need to return to the original distance matrix when using single-link methods. The distance matrix for the clustering defined by C_2 is

	1	2	[3,5,6]	4
1		0.31	0.23	0.32
2			0.28	0.21
[3,5,6]				0.28
4				

This distance matrix produces the following clustering:

$$C_3 = \{[1], [2,4], [3,5,6]\}$$

The new distance matrix for the clustering defined by C_3 is

	1	[2,4]	[3,5,6]
1		0.31	0.23
[2,4]			0.28
[3,5,6]			

which produces the following clustering:

$$C_4 = \{[1,3,5,6], [2,4]\}$$

The distance matrix for this clustering is

	[1,3,5,6]	[2,4]
[1,3,5,6]		0.28
[2,4]		

and the final clustering is

$$C_5 = \{[1,2,3,4,5,6]\}$$

A Hierarchical Tree Diagram

One way to help decide when to stop the clustering process is to construct a hierarchical tree diagram. A hierarchical tree diagram contains branches connecting data points and shows the order in which the points are assigned

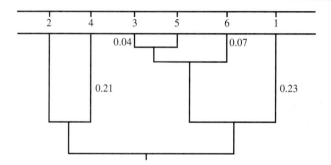

F I G. **9.2**
Hierarchical tree
diagram

to clusters. The lengths of its branches are proportional to the distances between points and clusters when points and clusters are combined. For the preceding example, points 3 and 5 were the first two points combined and they were 0.04 units apart when they were assigned to the same cluster; thus, a branch connecting these two points would be 0.04 units long. In the second step, point 6 was combined with the cluster containing points 3 and 5. The distance between this point and the cluster containing points 3 and 5, as calculated by the nearest neighbor method, was equal to 0.07 units. Hence, the branch that connects point 6 and the cluster containing points 3 and 5 would have a length equal to 0.07 units. In the third step, points 2 and 4 were placed in a common cluster. They were 0.21 units apart, so the length of the branch connecting points 2 and 4 would be 0.21 units in length. This process continues until one has all points combined into a single cluster. A hierarchical tree diagram for Example 9.1 is shown in Figure 9.2.

Other Hierarchical Clustering Methods

Other hierarchical clustering methods include (1) the furthest neighbor method where the distance between clusters is defined to be the distance between their two furthest members, (2) the centroid method where the distance between clusters is defined to be the distance between the cluster means, (3) the average method where the distance between clusters is defined as the average of all the dissimilarities between all possible pairs of points such that one of each pair is in each cluster, and (4) Ward's minimum variance method where the distance between two clusters is defined to be the square of the distance between the cluster means divided by the sum of the reciprocals of the number of points within each cluster. The preceding methods as well as other hierarchical clustering methods are available in the SAS CLUSTER procedure. Milligan (1980) gives comparisons between 15 different clustering procedures.

Comparisons of Clustering Methods

The nearest neighbor method tends to maximize the "connectedness" of a pair of clusters and has the tendency to create fewer clusters than the furthest neighbor method. The furthest neighbor method tends to minimize the intracluster distances at each step and, as a result, tends to find compact clusters. Most other methods fall somewhere between these two extremes.

Accurate clustering is not a simple task, and it is difficult to make general recommendations. It is always advisable to try more than one method. If several methods give similar results, then one can believe that natural clusterings actually exist.

Verification of Clustering Methods

As mentioned earlier, in all cases involving clustering, it is mandatory that researchers verify and/or adjust or fine-tune the results of clustering programs by looking at multivariate plots of the data being clustered. Some of the types of plots that are very useful are (1) scatter plots when there are only two variables being used to define the clusters; (2) scatter plots of the first two principal component scores especially when the first two principal components account for most of the variability in the data being clustered; (3) three-dimensional scatter plots, blob plots, and/or bubble plots when there are only three variables being used to define the clusters; (4) three-dimensional scatter plots, blob plots, and/or bubble plots of the first three principal component scores especially when the first three principal components account for most of the variability in the data being clustered; (5) sun-ray and/or star plots using either the original data or the important principal components; (6) plots of data values or principal component scores using Chernoff faces; and (7) Andrews' plots for each experimental unit using either the original data if the variables can be ordered by their relative importance or by using principal component scores.

How Many Clusters?

This section considers some ways that can help us determine the actual number of clusters in a data set. One way is to use a hierarchical tree diagram as previously described. Many statistical software programs produce hierarchical tree diagrams (also sometimes called *icicle plots*). An example of an ideal tree diagram is shown in Figure 9.3. If a researcher obtained a diagram similar to the one in Figure 9.3, the researcher would be quite confident that the data fall into three distinct clusters. The hierarchical tree diagram shown in Figure 9.2 does not reveal any particular cluster structure in the data used to produce this diagram.

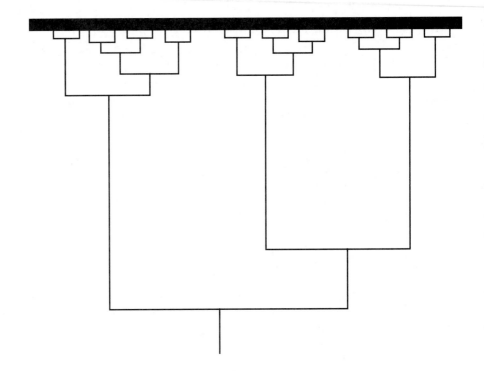

F I G. 9.3
Tree diagram
showing three
clusters

Beale's F-Type Statistic

Another approach was suggested by Beale. Suppose there are two possible clusterings with the first consisting of c_1 clusters and the second consisting of c_2 clusters and suppose that $c_2 < c_1$. That is, the second clustering contains fewer clusters than the first. Let W_1 and W_2 be the corresponding sums of squares of within cluster distances computed from the cluster means. That is, suppose there are n_r data points in the rth cluster, $r = 1, 2, \ldots, c_1$. If $\mathbf{x}_{rq}$ represents the qth observation vector in the rth cluster and if $\bar{x}_r$ represents the mean of the observations in the rth cluster, then

$$W_1 = \sum_{r=1}^{c_1} \sum_{q=1}^{n_r} (\mathbf{x}_{rq} - \bar{\mathbf{x}}_r)'(\mathbf{x}_{rq} - \bar{\mathbf{x}}_r)$$

In a similar manner, W_2 can be defined. If W_1 and W_2 are nearly the same size, then the clustering consisting of the fewer number of clusters is just as good as that consisting of the larger number of clusters and, for simplicity, we would select the clustering consisting of the fewer number of clusters. However, if W_1 is much smaller than W_2, then we could say that the first clustering is an improvement over the second, and we would select the clustering with the larger number of clusters.

To determine if the first clustering is better than the second, Beale suggested computing a pseudo F-type statistic by

$$F^* = \frac{(W_2 - W_1)}{W_1} \cdot \frac{(N - c_1)k_1}{(N - c_2)k_2 - (N - c_1)k_1} \quad \text{where} \quad k_1 = c_1^{-2/p} \quad \text{and} \quad k_2 = c_2^{-2/p}$$

If F^* is greater than an F critical point with $k_2(N - c_2) - k_1(N - c_1)$ degrees of freedom for the numerator and $k_1(N - c_1)$ for the denominator, then we would choose the first clustering (the one with more clusters) over the second (the one with fewer clusters). This particular method is not directly available in most statistical computing packages, but the SAS FASTCLUS procedure produces information that can be used to calculate Beale's statistic. This is illustrated later in Example 9.4.

A Pseudo Hotelling's T^2 Test

In Chapter 10, we see that Hotelling's T^2 statistic can be used to compare the means of two multivariate normal populations. A variation of this can be used when we are trying to decide whether two clusters should be combined into a single cluster in that one can compute a pseudo Hotelling's T^2 statistic. This statistic is used to compare the means of two clusters. Basically, if the means of two clusters are not significantly different, then the clusters could be combined; but if the means are significantly different, then the clusters should not be combined. This type of statistic is available to users of the SAS CLUSTER procedure.

The Cubic Clustering Criterion

Sarle (1983) introduced a cubic clustering criterion (CCC) that can be plotted against the number of clusters for various choices for the number of clusters. Peaks on this plot that have CCC > 3 are supposed to correspond to an appropriate number of clusters. The cubic clustering criterion is also available to users of the SAS CLUSTER procedure.

E X A M P L E 9.2

To illustrate clustering, let's use data collected on frozen pizzas found in supermarkets. The data can be found in the file labeled EX9_2.SAS on the enclosed floppy disk. This file contains data on more than 350 samples of frozen pizza. Each individual pizza was pureed and thoroughly mixed after which a sample of the mixture was taken for nutrient analysis. The variables measured on each sample included the percent moisture of the pizza sample (MOIS), the amount of protein per 100 g in the sample (PROT), the amount of fat per 100 g in the sample (FAT), the amount of ash per 100 g in the sample (ASH), the amount of sodium per 100 g in the sample (SODIUM), the amount of carbohydrates per 100 g in the sample (CARB), and the calories per gram in the sample (CAL).

This file actually contains too much data to illustrate in this book, so approximately 20% of the data points were randomly selected from the total data set to illustrate the process of performing a cluster analysis. The total data set can be found in the file labeled EX9_2.SAS where it has the name PIZZA. The sample being analyzed here was randomly selected from the total data set by using the SAS commands:

```
TITLE 'Ex. 9.2 - Nutrients in Pizza Samples';

DATA PIZZA2; SET PIZZA; RAN=UNIFORM(1115823); IF RAN < .20;

PROC PRINT DATA=PIZZA2;
  VAR ID MOIS PROT FAT ASH SODIUM CARB CAL;
  RUN;
```

A portion of the data set created from these commands can be found on page 1 of the computer printout shown in Figure 9.4. All of the data can be seen by executing the file labeled EX9_2.SAS on the enclosed floppy disk.

Values were missing on some of the variables for some of the pizzas in these data, and the pizzas with missing data are eliminated by computing packages before performing a cluster analysis. However, the tree diagrams produced by SAS are affected (not seriously) by including those pizzas with missing data. To produce clearer plots, pizzas that did not have complete data were removed by using the following commands:

```
DATA PIZZA3; SET PIZZA2;
  CHK=MOIS+PROT+FAT+ASH+SODIUM+CARB+CAL;
  IF CHK=. THEN DELETE; DROP CHK;
```

Thus, the data actually being analyzed in this example are the data found in the SAS data set labeled PIZZA3. These data were analyzed by the CLUSTER procedure in SAS using the AVERAGE method. This method calculates the distance between clusters by taking the average of all pairwise differences between the points within each cluster.

The data were also standardized before performing the cluster analysis by including a STANDARD option. Thus Z scores will be used by the CLUSTER procedure. This is a necessity for these data because the variables being analyzed have extremely different standard deviations. For example, the standard deviation of the protein measurements is 6.877, while that for the sodium measurements is 0.6556. Failure to standardize would mean that protein would be much more important than sodium when it comes to identifying clusters. However, from a health standpoint, the sodium content of a pizza may be more important than its protein content.

Ex. 9.2 - Nutrients in Pizza Samples   1

OBS	ID	MOIS	PROT	FAT	ASH	SODIUM	CARB	CAL
1	14025	28.35	19.99	45.78	5.08	1.63	0.80	4.95
2	14164	28.70	20.00	45.12	4.93	1.56	1.25	4.91
3	14154	30.91	19.65	42.45	4.81	1.65	2.81	4.72
4	24082	31.02	19.05	42.29	5.27	1.71	2.37	4.66
5	24138	29.62	21.10	43.37	5.05	1.69	0.86	4.78
6	14047	49.99	13.35	29.20	3.52	1.05	3.94	3.32
7	14074	50.72	12.93	29.88	3.60	1.03	2.87	3.32
8	14149	54.96	14.26	22.99	3.19	0.90	4.60	2.82
9	14113	54.12	14.06	24.95	3.14	0.82	3.73	2.96
10	24118	51.75	13.18	28.38	3.04	0.86	3.65	3.23
21	24057	47.70	22.23	21.33	3.98	.	4.76	3.00
22	24153	48.81	22.43	18.68	4.10	0.72	5.98	2.82
23	24123	46.28	21.51	25.44	4.58	0.60	2.19	3.24
24	24043	52.19	26.00	16.64	4.17	0.61	1.00	2.58
25	14012	85.70	1.98	.	1.85	0.38	10.47	0.50
26	14048	85.42	1.93	.	1.82	0.36	10.83	0.51
27	14167	85.95	1.42	.	1.71	0.36	10.92	0.49
28	34001	85.99	1.68	.	1.67	0.32	10.66	0.49
29	24105	85.12	1.69	.	1.94	0.44	11.25	0.52
30	24116	85.17	1.42	.	1.68	0.37	11.73	0.53
41	14118	29.79	8.17	14.35	1.49	0.46	46.20	3.46
42	14143	29.54	7.79	15.08	1.41	0.45	46.18	3.52
43	14008	.	8.29	16.69	1.46	0.69	46.64	3.70
44	24035	27.65	7.78	17.30	1.29	0.40	46.25	3.72
45	24049	28.33	7.82	17.96	1.41	0.45	44.48	3.71
46	24103	30.15	8.06	12.23	1.50	0.47	48.06	3.35
47	34034	28.36	7.62	19.29	1.45	0.47	43.28	3.77
48	14153	28.83	8.26	20.10	1.37	0.42	41.44	3.80
49	14077	33.09	7.87	12.07	1.37	0.44	45.60	3.23
50	14166	27.56	8.25	14.65	1.45	0.46	48.09	3.57
61	34022	54.28	10.75	13.87	2.13	0.46	18.97	2.40
62	34021	54.54	10.40	13.22	2.10	0.47	19.74	2.40
63	34014	56.77	10.37	12.17	2.02	.	18.67	2.26
64	34026	54.17	10.13	13.25	2.07	0.46	20.38	2.41
65	14043	47.84	10.16	14.56	2.27	0.54	25.17	2.72
66	24073	46.13	10.84	13.99	2.38	0.64	26.66	2.76
67	24071	46.22	11.26	15.93	2.47	0.63	24.12	2.85
68	24056	43.45	10.81	19.49	2.51	0.68	23.74	3.14
69	24070	45.21	9.39	16.23	2.14	0.55	27.03	2.92
70	24069	43.15	11.79	18.46	2.43	0.67	24.17	3.10
71	34039	44.55	11.01	16.03	2.43	0.64	25.98	2.92

F I G. 9.4

Pages 1, 3–14, 16, and 19–21 of computer printout, Example 9.2

Furthermore, calories are measured in per gram units while protein, fat, ash, sodium, and carbohydrates are measured in per 100 gram units. These variables are not measured in the same units that would also require that variables be standardized.

Additional options used in the CLUSTER procedure included the S option, which computes the mean and standard deviation of each of the measured variables; the CCC option, which computes the cubic clustering

Average Linkage Cluster Analysis

Simple Statistics

	Mean	Std Dev	Skewness	Kurtosis	Bimodality
MOIS	41.7064	10.0873	-0.2015	-1.6365	0.6782
PROT	14.0175	6.8770	0.8156	-0.8531	0.7184
FAT	20.1141	8.5428	1.9515	3.1636	0.7591
ASH	2.6927	1.2603	0.6394	-0.8473	0.6063
SODIUM	0.6566	0.3553	2.0013	3.2515	0.7793
CARB	21.4871	17.7911	0.2014	-1.5932	0.6596
CAL	3.2304	0.6429	1.0879	1.1031	0.5109

Eigenvalues of the Correlation Matrix

	Eigenvalue	Difference	Proportion	Cumulative
1	3.90916	1.38582	0.558451	0.55845
2	2.52334	2.08087	0.360477	0.91893
3	0.44247	0.34408	0.063210	0.98214
4	0.09838	0.07181	0.014055	0.99619
5	0.02658	0.02652	0.003797	0.99999
6	0.00006	0.00005	0.000009	1.00000
7	0.00001	.	0.000001	1.00000

The data have been standardized to mean 0 and variance 1
Root-Mean-Square Total-Sample Standard Deviation =   1
Root-Mean-Square Distance Between Observations   = 3.741657

F I G. **9.4**

Pages 1, 3–14, 16, and 19–21 of computer printout, Example 9.2

criterion for many choices as to the number of clusters in the data; and the PSEUDO option, which computes a pseudo T^2 test statistic that helps us decide whether pairs of clusters should be combined.

The final option used was the OUTTREE=TREE option. This option creates a data set named TREE, which contains information calculated by the CLUSTER procedure. This data set contains values of CCC for different numbers of clusters. These pairs can be plotted allowing us to look for peaks where values of the CCC are greater than 3. There is also information in this output data set that can be used by SAS's TREE procedure to construct a hierarchical tree diagram.

To summarize, a cluster analysis of these data was performed by using the following SAS commands:

```
PROC CLUSTER DATA=PIZZA3 S STANDARD METHOD=AVERAGE
  CCC PSEUDO OUTTREE=TREE;
  VAR MOIS PROT FAT ASH SODIUM CARB CAL;
  ID ID;
  RUN;
```

Ex. 9.2 - Nutrients in Pizza Samples   4

Average Linkage Cluster Analysis

NCL	-Clusters Joined-		FREQ	SPRSQ	RSQ	ERSQ	CCC	PSF	PST2	Norm RMS Dist	Tie
55	34021	34026	2	0.0000	1.00	.	.	2461	.	0.0203	
54	24107	34022	2	0.0000	1.00	.	.	1552	.	0.0304	
53	14072	24030	2	0.0000	1.00	.	.	1186	.	0.0366	
52	CL54	CL55	4	0.0000	1.00	.	.	864	3.5	0.0386	
51	24049	24033	2	0.0000	1.00	.	.	832	.	0.0402	
50	14118	14143	2	0.0000	1.00	.	.	805	.	0.0418	
49	CL52	14067	5	0.0000	1.00	.	.	761	1.8	0.0426	
48	34037	34034	2	0.0000	1.00	.	.	758	.	0.0427	
47	14047	14074	2	0.0000	1.00	.	.	757	.	0.0432	
46	14099	14122	2	0.0000	1.00	.	.	728	.	0.0507	
45	14100	24100	2	0.0000	1.00	.	.	703	.	0.0524	
44	24056	24069	2	0.0001	1.00	.	.	674	.	0.0567	
43	24071	34039	2	0.0001	1.00	.	.	640	.	0.0618	
42	CL48	CL51	4	0.0001	.999	.	.	577	3.4	0.0619	
41	14025	14164	2	0.0001	.999	.	.	555	.	0.0680	
40	14166	24119	2	0.0001	.999	.	.	531	.	0.0730	
39	CL49	24111	6	0.0002	.999	.	.	482	5.9	0.0756	
38	CL42	24035	5	0.0002	.999	.	.	437	3.3	0.0868	
37	CL39	14157	7	0.0002	.999	.	.	398	4.0	0.0892	
36	24073	CL43	3	0.0002	.998	.	.	377	2.8	0.0942	
35	CL50	CL40	4	0.0003	.998	.	.	347	4.1	0.0949	
34	14154	24138	2	0.0002	.998	.	.	339	.	0.1007	
33	14106	34010	2	0.0002	.998	.	.	333	.	0.1020	
32	CL45	34007	3	0.0002	.998	.	.	323	4.8	0.1032	
31	CL38	14153	6	0.0003	.997	.	.	312	3.0	0.1043	
30	24103	14077	2	0.0002	.997	.	.	312	.	0.1062	
29	14149	14113	2	0.0002	.997	.	.	312	.	0.1077	
28	CL53	24153	3	0.0003	.997	.	.	307	11.9	0.1106	
27	CL31	24041	7	0.0003	.996	.	.	303	2.4	0.1124	
26	CL34	24082	3	0.0003	.996	.	.	303	1.4	0.1163	
25	14151	24043	2	0.0003	.996	.	.	305	.	0.1225	
24	CL46	CL32	5	0.0006	.995	.	.	289	5.1	0.1288	
23	14043	CL36	4	0.0004	.995	.	.	288	3.0	0.1288	
22	CL23	24070	5	0.0004	.994	.	.	291	1.7	0.1300	
21	CL33	24058	3	0.0004	.994	.	.	293	2.1	0.1374	
20	14107	24123	2	0.0003	.994	.	.	300	.	0.1382	
19	CL21	14014	4	0.0006	.993	.	.	297	2.1	0.1653	
18	CL27	CL35	11	0.0020	.991	.	.	251	13.4	0.1660	
17	CL41	CL26	5	0.0010	.990	.	.	246	5.5	0.1667	
16	CL47	24118	3	0.0008	.989	.	.	247	24.7	0.1872	
15	CL19	34003	5	0.0009	.988	.	.	249	2.4	0.2014	
14	CL22	CL44	7	0.0020	.986	.	.	235	9.2	0.2099	
13	CL24	14126	6	0.0014	.985	.	.	235	6.2	0.2284	
12	CL28	CL20	5	0.0020	.983	.	.	231	9.1	0.2318	
11	CL13	CL30	8	0.0025	.981	.894	17.8	227	5.8	0.2426	
10	CL15	CL25	7	0.0027	.978	.883	17.8	226	5.5	0.2579	

F I G. 9.4

Pages 1, 3–14, 16, and 19–21 of computer printout, Example 9.2

Ex. 9.2 - Nutrients in Pizza Samples   5

Average Linkage Cluster Analysis

									Norm	T
									RMS	i
NCL	-Clusters Joined-		FREQ	SPRSQ	RSQ	ERSQ	CCC	PSF	PST2	Dist	e
9	CL11	CL18	19	0.0084	.969	.869	15.8	187	17.2	0.2743	
8	CL16	CL29	5	0.0033	.966	.853	16.2	196	9.1	0.2942	
7	CL12	14140	6	0.0027	.963	.834	14.1	215	4.1	0.3230	
6	CL37	CL14	14	0.0165	.947	.809	12.3	178	56.7	0.3790	
5	CL10	CL7	13	0.0191	.928	.776	11.5	164	20.1	0.4551	
4	CL8	CL5	18	0.0446	.883	.726	9.36	131	21.1	0.6532	
3	CL9	CL6	33	0.1497	.733	.645	3.64	72.9	126	0.7578	
2	CL4	CL3	51	0.3298	.404	.437	-.81	36.5	61.0	1.0267	
1	CL17	CL2	56	0.4036	.000	.000	0.00	.	36.5	1.6634	

F I G. 9.4
Pages 1, 3–14, 16, and 19–21 of computer printout, Example 9.2

Clustering Order

The results of the preceding SAS commands are shown on pages 3–5 of the computer printout of Figure 9.4. On the top of page 3 we find the means and standard deviations for each variable as well as measures of skewness, kurtosis, and bimodality for each variable.

On the bottom of page 3, we find the eigenvalues of the correlation matrix. It is important to note that there are two eigenvalues that are greater than one and these two account for 91.9% of the total variability in the measured variables. This implies that the measured variables nearly fall within a two-dimensional subspace (a plane) of the seven-dimensional sample space, and this implies that a plot of the first two principal component scores should be extremely useful for clustering these data. Such a plot should help us verify whether the clusters obtained by the CLUSTER procedure are reasonable or not. The plot should also help us determine whether there are certain "adjustments" or "fine-tunings" that should be made to the results of the CLUSTER procedure as well as providing guidance as to how such adjustments might be made.

Fifty-six pizzas are being clustered, so initially there will be 56 clusters with each cluster containing exactly one pizza. On the very top of page 4 (see Figure 9.4), we can see that the two pizzas that are closest to one another are pizzas 34021 and 34026. After these pizzas are placed in the same cluster, there are 55 clusters, as shown in the first column on the left of page 4. The FREQ column shows the number of pizzas in this cluster. This first cluster is also labeled as CL55, which will be important later.

In the second line, we can see that pizzas 24107 and 34022 are placed in the same cluster; there are now 54 clusters, and this cluster is labeled CL54. CL54 also contains two pizzas. In the third line, pizzas 14072 and 24030 are placed in the same cluster, and this cluster is labeled CL53.

Ex. 9.2 - Nutrients in Pizza Samples   6

Plot of _CCC_*_NCL_.   Symbol is value of _NCL_.

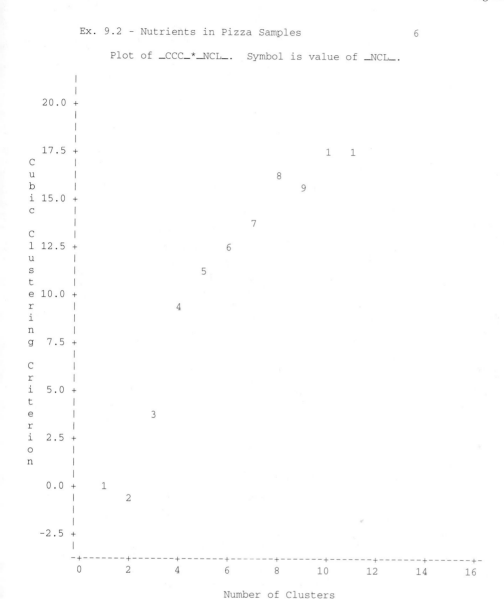

NOTE: 5 obs had missing values.   95 obs out of range.

F I G. 9.4

Pages 1, 3–14, 16, and 19–21 of computer printout, Example 9.2

```
                    Ex. 9.2 - Nutrients in Pizza Samples                7
                      Average Linkage Cluster Analysis
                                    ID

           1 1 1 2 2 1 1 2 1 1 1 3 2 1 3 1 2 1 2 2 1 2 1 2 1 1 1 1 2 3 1 2 1 3 3
           4 4 4 4 4 4 4 4 4 4 4 4 4 4 4 4 4 4 4 4 4 4 4 4 4 4 4 4 4 4 4 4 4 4 4
           0 1 1 1 0 0 0 1 1 1 1 0 0 0 0 1 0 0 0 1 1 1 1 0 1 1 1 0 1 1 0 1 1 0 0
           2 6 5 3 8 4 7 1 4 1 0 1 5 1 0 5 4 7 3 5 0 2 4 9 2 0 0 0 2 0 7 3 3
           5 4 4 8 2 7 4 8 9 3 6 0 8 4 3 1 3 2 0 3 7 3 0 9 2 0 0 7 6 3 7 7 4
    1.75 +
         |
         |XXXXXXXXX XXXXXXXXXXXXXXXXXXXXXXXXXXXXXXXXXXXXXXXXXXXXXXXXXXXXXXXXXXXXXXX
         |XXXXXXXXX XXXXXXXXXXXXXXXXXXXXXXXXXXXXXXXXXXXXXXXXXXXXXXXXXXXXXXXXXXXXXXX
         |XXXXXXXXX XXXXXXXXXXXXXXXXXXXXXXXXXXXXXXXXXXXXXXXXXXXXXXXXXXXXXXXXXXXXXXX
A        |XXXXXXXXX XXXXXXXXXXXXXXXXXXXXXXXXXXXXXXXXXXXXXXXXXXXXXXXXXXXXXXXXXXXXXXX
v    1.5 +XXXXXXXXX XXXXXXXXXXXXXXXXXXXXXXXXXXXXXXXXXXXXXXXXXXXXXXXXXXXXXXXXXXXXXXX
e        |XXXXXXXXX XXXXXXXXXXXXXXXXXXXXXXXXXXXXXXXXXXXXXXXXXXXXXXXXXXXXXXXXXXXXXXX
r        |XXXXXXXXX XXXXXXXXXXXXXXXXXXXXXXXXXXXXXXXXXXXXXXXXXXXXXXXXXXXXXXXXXXXXXXX
a        |XXXXXXXXX XXXXXXXXXXXXXXXXXXXXXXXXXXXXXXXXXXXXXXXXXXXXXXXXXXXXXXXXXXXXXXX
g        |XXXXXXXXX XXXXXXXXXXXXXXXXXXXXXXXXXXXXXXXXXXXXXXXXXXXXXXXXXXXXXXXXXXXXXXX
e        |XXXXXXXXX XXXXXXXXXXXXXXXXXXXXXXXXXXXXXXXXXXXXXXXXXXXXXXXXXXXXXXXXXXXXXXX
    1.25 +XXXXXXXXX XXXXXXXXXXXXXXXXXXXXXXXXXXXXXXXXXXXXXXXXXXXXXXXXXXXXXXXXXXXXXXX
D        |XXXXXXXXX XXXXXXXXXXXXXXXXXXXXXXXXXXXXXXXXXXXXXXXXXXXXXXXXXXXXXXXXXXXXXXX
i        |XXXXXXXXX XXXXXXXXXXXXXXXXXXXXXXXXXXXXXXXXXXXXXXXXXXXXXXXXXXXXXXXXXXXXXXX
s        |XXXXXXXXX XXXXXXXXXXXXXXXXXXXXXXXXXXXXXXXXXXXXXXXXXXXXXXXXXXXXXXXXXXXXXXX
t        |XXXXXXXXX XXXXXXXXXXXXXXXXXXXXXXXXXXXXXXXXXXXXXXXXXXXXXXXXXXXXXXXXXXXXXXX
a        |XXXXXXXXX XXXXXXXXXXXXXXXXXXXXXXXXXXXXXXXXXXXXXXXXXXXXXXXXXXXXXXXXXXXXXXX
n      1 +XXXXXXXXX XXXXXXXXXXXXXXXXXXXXXXXXXXXXX XXXXXXXXXXXXXXXXXXXXXXXXXXXX
c        |XXXXXXXXX XXXXXXXXXXXXXXXXXXXXXXXXXXXXX XXXXXXXXXXXXXXXXXXXXXXXXXXXX
e        |XXXXXXXXX XXXXXXXXXXXXXXXXXXXXXXXXXXXXX XXXXXXXXXXXXXXXXXXXXXXXXXXXX
         |XXXXXXXXX XXXXXXXXXXXXXXXXXXXXXXXXXXXXX XXXXXXXXXXXXXXXXXXXXXXXXXXXX
B        |XXXXXXXXX XXXXXXXXXXXXXXXXXXXXXXXXXXXXX XXXXXXXXXXXXXXXXXXXXXXXXXXXX
e        |XXXXXXXXX XXXXXXXXXXXXXXXXXXXXXXXXXXXXX XXXXXXXXXXXXXXXXXXXXXXXXXXXX
t   0.75 +XXXXXXXXX XXXXXXXXXXXXXXXXXXXXXXXXXXXXX XXXXXXXXXXXXXXXXXXXXXXXXXXXX
w        |XXXXXXXXX XXXXXXXXXXXXXXXXXXXXXXXXXXXXX XXXXXXXXXXXXXXXXXXXXXXXXXXXX
e        |XXXXXXXXX XXXXXXXXXXXXXXXXXXXXXXXXXXXXX XXXXXXXXXXXXXXXXXXXXXXXXXXXX
e        |XXXXXXXXX XXXXXXXXX XXXXXXXXXXXXXXXXXXXXXXXXXX XXXXXXXXXXXXXXXXXXXXXXXXXXXX
n        |XXXXXXXXX XXXXXXXXX XXXXXXXXXXXXXXXXXXXXXXXXXX XXXXXXXXXXXXXXXXXXXXXXXXXXXX
         |XXXXXXXXX XXXXXXXXX XXXXXXXXXXXXXXXXXXXXXXXXXX XXXXXXXXXXXXXXXXXXXXXXXXXXXX
C   0.5  +XXXXXXXXX XXXXXXXXX XXXXXXXXXXXXXXXXXXXXXXXXXX XXXXXXXXXXXXXXXXXXXXXXXXXXXX
l        |XXXXXXXXX XXXXXXXXX XXXXXXXXXXXXXXXXXXXXXXXXXX XXXXXXXXXXXXXXXXXXXXXXXXXXXX
u        |XXXXXXXXX XXXXXXXXX XXXXXXXXXXXXX XXXXXXXXXXX XXXXXXXXXXXXXXXXXXXXXXXXXXXX
s        |XXXXXXXXX XXXXXXXXX XXXXXXXXXXXXX XXXXXXXXXXX XXXXXXXXXXXXXXXXXXXXXXXXXXXX
t        |XXXXXXXXX XXXXXXXXX XXXXXXXXXXXXX XXXXXXXXXXX XXXXXXXXXXXXXXXXXXXXXXXXXXXX
e        |XXXXXXXXX XXXXXXXXX XXXXXXXXXXXXX XXXXXXXXX . XXXXXXXXXXXXXXXXXXXXXXXXXXXX
r   0.25 +XXXXXXXXX XXXXX XXX XXXXXXXXXXXXX XXXXXXXXX . XXXXXXXXXXXX XXX XXX
s        |XXXXXXXXX XXXXX XXX XXXXXXXX XXX XXX XXXXX XXX . XXXXXXXXXXX XXX XXX
         |XXXXXXXXX XXXXX XXX XXXXXXX . XXX XXXXX XXX . XXXXXXXXX . XXX XXX
         |XXX XXXXX XXX . XXX XXXXX . . XXX XXXXX XXX . XXXXXXXXX . XXX XXX
         |XXX XXX . XXX . . . XXX . . . . XXX . . . . . XXX XXXXX . . . XXX
         |. . . . . XXX . . . . . . . . . XXX . . . . XXX XXX . . . . XXX
       0 +. . . . . . . . . . . . . . . . . . . . . . . . . . . . . . . . .
```

F I G. **9.4**

Pages 1, 3–14, 16, and 19–21 of computer printout, Example 9.2

Ex. 9.2 - Nutrients in Pizza Samples
Average Linkage Cluster Analysis   8

ID

```
2 2 2 1 2 1 1 1 2 2 3 3 3 1 2 1 1 2 2 3 2 2 2
4 4 4 4 4 4 4 4 4 4 4 4 4 4 4 4 4 4 4 4 4 4 4
0 0 0 1 0 1 1 1 1 1 0 0 0 0 1 1 0 0 0 0 0 0 0
4 3 3 5 4 1 4 6 1 0 2 2 2 6 1 5 4 7 7 3 7 5 6
9 3 5 3 1 8 3 6 9 7 2 1 6 7 1 7 3 3 1 9 0 6 9
```

```
XXXXXXXXXXXXXXXXXXXXXXXXXXXXXXXXXXXXXXXXXXXXXXX
XXXXXXXXXXXXXXXXXXXXXXXXXXXXXXXXXXXXXXXXXXXXXXX
XXXXXXXXXXXXXXXXXXXXXXXXXXXXXXXXXXXXXXXXXXXXXXX
XXXXXXXXXXXXXXXXXXXXXXXXXXXXXXXXXXXXXXXXXXXXXXX
XXXXXXXXXXXXXXXXXXXXXXXXXXXXXXXXXXXXXXXXXXXXXXX
XXXXXXXXXXXXXXXXXXXXXXXXXXXXXXXXXXXXXXXXXXXXXXX
XXXXXXXXXXXXXXXXXXXXXXXXXXXXXXXXXXXXXXXXXXXXXXX
XXXXXXXXXXXXXXXXXXXXXXXXXXXXXXXXXXXXXXXXXXXXXXX
XXXXXXXXXXXXXXXXXXXXXXXXXXXXXXXXXXXXXXXXXXXXXXX
XXXXXXXXXXXXXXXXXXXXXXXXXXXXXXXXXXXXXXXXXXXXXXX
XXXXXXXXXXXXXXXXXXXXXXXXXXXXXXXXXXXXXXXXXXXXXXX
XXXXXXXXXXXXXXXXXXXXXXXXXXXXXXXXXXXXXXXXXXXXXXX
XXXXXXXXXXXXXXXXXXXXXXXXXXXXXXXXXXXXXXXXXXXXXXX
XXXXXXXXXXXXXXXXXXXXXXXXXXXXXXXXXXXXXXXXXXXXXXX
XXXXXXXXXXXXXXXXXXXXXXXXXXXXXXXXXXXXXXXXXXXXXXX
XXXXXXXXXXXXXXXXXXXXXXXXXXXXXXXXXXXXXXXXXXXXXXX
XXXXXXXXXXXXXXXXXXXXXXXXXXXXXXXXXXXXXXXXXXXXXXX
XXXXXXXXXXXXXXXXXXXXXXXXXXXXXXXXXXXXXXXXXXXXXXX
XXXXXXXXXXXXXXXXXXXXXXXXXXXXXXXXXXXXXXXXXXXXXXX
XXXXXXXXXXXXXXXXXXXXXXXXXXXXXXXXXXXXXXXXXXXXXXX
XXXXXXXXXXXXXXXXXXXXXXXXXXXXXXXXXXXXXXXXXXXXXXX
XXXXXXXXXXXXXXXXXXXXXXXXXXXXXXXXXXXXXXXXXXXXXXX
XXXXXXXXXXXXXXXXXXXXXXXXXXXXXXXXXXXXXXXXXXXXXXX
XXXXXXXXXXXXXXXX XXXXXXXXXXXXXXXXXXXXXXXXXXXXXXX
XXXXXXXXXXXXXXXX XXXXXXXXXXXXXXXXXXXXXXXXXXXXXXX
XXXXXXXXXXXXXXXX XXXXXXXXXXXXXXXXXXXXXXXXXXXXXXX
XXXXXXXXXXXXXXXX XXXXXXXXXXXXXXXXXXXXXXXXXXXXXXX
XXXXXXXXXXXXXXXX XXXXXXXXXXXXXXXXXXXXXXXXXXXXXXX
XXXXXXXXXXXXXXXX XXXXXXXXXXXXXXXXXXXXXXXXXXXXXXX
XXXXXXXXXXXXXXXX XXXXXXXXXXXXXXXXXXXXXXXXXXXXXXX
XXXXXXXXXXXXXXXX XXXXXXXXXXXXXXXXXXXXXXXXXXXXXXX
XXXXXXXXXXXXXXXX XXXXXXXXXXXXX XXXXXXXXXXXXX
XXXXXXXXXXXXXXXX XXXXXXXXXXXXX XXXXXXXXXXXXX
XXXXXXXXXXXXXXXX XXXXXXXXXXXXX XXXXXXXXXXXXX
XXXXXXXXXXXXXXXX XXXXXXXXXXXXX XXXXXXXXXXXXX
XXXXXXXXXXXXXXXX XXXXXXXXXXXXX XXXXXXXXX XXX
XXXXXXXXXX XXXXXX XXXXXXXXXXXXX XXXXXXXXX XXX
XXXXXX . . XXXXXXX XXXXXXXXXXXXX . XXXXX . XXX
XXXX . . . XXX . . XXXXXXXXX . . . . XXX . XXX
. . . . . . . . . . . XXX . . . . . . . . .
```

F I G. 9.4
Pages 1, 3–14, 16, and 19–21 of computer printout, Example 9.2

Ex. 9.2 - Nutrients in Pizza Samples   9

OBS	ID	CLUSTER
1	34021	1
2	34026	1
3	24107	1
4	34022	1
5	14067	1
6	24111	1
7	14157	1
8	14072	2
9	24030	2
10	24153	2
11	14107	2
12	24123	2
13	24049	3
14	24033	3
15	14118	3
16	14143	3
17	34037	3
18	34034	3
19	14166	3
20	24119	3
21	24035	3
22	14153	3
23	24041	3
24	14047	4
25	14074	4
26	24118	4
27	14099	5
28	14122	5
29	14100	5
30	24100	5
31	34007	5
32	24103	5
33	14077	5
34	14126	5
35	24056	6
36	24069	6
37	24071	6
38	34039	6
39	24073	6
40	14043	6
41	24070	6
42	14025	7
43	14164	7
44	14154	7

F I G. **9.4**

Pages 1, 3–14, 16, and 19–21 of computer printout, Example 9.2

```
Ex. 9.2 - Nutrients in Pizza Samples                    10

        OBS      ID       CLUSTER

        45     24138        7
        46     24082        7
        47     14106        8
        48     34010        8
        49     14151        8
        50     24043        8
        51     24058        8
        52     14014        8
        53     34003        8
        54     14149        9
        55     14113        9
        56     14140       10
```

F I G. **9.4**

Pages 1, 3–14, 16, and 19–21 of computer printout, Example 9.2

In the fourth line, clusters CL54 and CL55 are combined into one cluster that is labeled cluster CL52. CL52 contains four pizzas, the two that were in CL54 plus the two that were in CL55. In a similar manner, we can continue to follow the sequential steps that are executed in the clustering process.

Estimating the Number of Clusters

Next we must estimate the number of clusters for these data. Many of the statistics computed on pages 4 and 5 (see Figure 9.4) can help to determine an appropriate number of clusters, but this book concentrates on only two of these, the cubic clustering criterion (CCC) and Hotelling's pseudo T^2 statistic (PST2). Note that the pseudo F statistic (PSF) printed by the CLUSTER procedure is not Beale's pseudo F statistic.

In using both the CCC and PST2 statistics, it is best to start at the bottom of page 5 and work upward, since the objective is usually to have as few clusters as might be reasonable. A plot of the CCC values against the number of clusters (NCL) is shown on page 6 of Figure 9.4. This plot was obtained by using the following SAS commands:

```
PROC PLOT DATA=TREE;
  PLOT _CCC_*_NCL_=_NCL_/HAXIS=0 TO 16 BY 2;
  RUN;
```

On the plot on page 6, the symbol plotted is equal to the number of clusters except for the two 1s in the upper right corner—these correspond to 10 and 11 clusters, respectively—but only the first digit of each number is plotted. If we were to connect these points sequentially with lines, a small peak occurs when the number of clusters is equal to 8. There may be another peak beginning to occur at around 10 or 11 clusters, but we cannot

Ex. 9.2 - Nutrients in Pizza Samples   11

Principal Component Analysis

56 Observations
7 Variables

Simple Statistics

	MOIS	PROT	FAT	ASH
Mean	41.70642857	14.01750000	20.11410714	2.692678571
StD	10.08730569	6.87700391	8.54278064	1.260323899

	SODIUM	CARB	CAL
Mean	0.6566071429	21.48714286	3.230357143
StD	0.3552887521	17.79114240	0.642888268

Correlation Matrix

	MOIS	PROT	FAT	ASH	SODIUM	CARB	CAL
MOIS	1.0000	0.3906	-.2425	0.2701	-.1356	-.6217	-.8123
PROT	0.3906	1.0000	0.3761	0.8144	0.3482	-.8458	-.0583
FAT	-.2425	0.3761	1.0000	0.7248	0.9314	-.5378	0.7615
ASH	0.2701	0.8144	0.7248	1.0000	0.7708	-.8860	0.2346
SODIUM	-.1356	0.3482	0.9314	0.7708	1.0000	-.5580	0.6449
CARB	-.6217	-.8458	-.5378	-.8860	-.5580	1.0000	0.1024
CAL	-.8123	-.0583	0.7615	0.2346	0.6449	0.1024	1.0000

Eigenvalues of the Correlation Matrix

	Eigenvalue	Difference	Proportion	Cumulative
PRIN1	3.90916	1.38582	0.558451	0.55845
PRIN2	2.52334	2.08087	0.360477	0.91893
PRIN3	0.44247	0.34408	0.063210	0.98214
PRIN4	0.09838	0.07181	0.014055	0.99619
PRIN5	0.02658	0.02652	0.003797	0.99999
PRIN6	0.00006	0.00005	0.000009	1.00000
PRIN7	0.00001	.	0.000001	1.00000

F I G. **9.4**

Pages 1, 3–14, 16, and 19–21 of computer printout, Example 9.2

tell this for sure because the program does not compute the value of CCC when the number of clusters is greater than 20% of the number of data points. Thus an examination of the CCC values leads one to believe that the appropriate number of clusters is somewhere around 8 clusters.

The values of the pseudo T^2 statistic, PST2, printed in each row can be used to help determine whether the two clusters combined in that row

Ex. 9.2 - Nutrients in Pizza Samples   12

Principal Component Analysis

Eigenvectors

	PRIN1	PRIN2	PRIN3	PRIN4
MOIS	0.072468	-.591440	0.457949	-.201367
PROT	0.376602	-.287747	-.724057	-.064028
FAT	0.439164	0.281253	0.210489	-.512520
ASH	0.486412	-.108591	-.093574	0.552362
SODIUM	0.440202	0.227130	0.432383	0.449484
CARB	-.431373	0.319985	-.074430	0.345541
CAL	0.208799	0.567914	-.143065	-.257169

	PRIN5	PRIN6	PRIN7
MOIS	-.005564	0.557763	0.289051
PROT	0.400693	0.073813	0.284625
FAT	-.193821	-.374282	0.493351
ASH	-.656636	0.071263	0.038761
SODIUM	0.604676	-.001702	-.002599
CARB	-.006053	0.197706	0.739908
CAL	-.070763	0.706538	-.207532

F I G. 9.4
Pages 1, 3–14, 16, and 19–21 of computer printout, Example 9.2

should have been combined. If PST2 is large, the two clusters should not be combined; but if PST2 is small, then the two clusters can safely be combined. Unfortunately, the designation of values of PST2 as "large" and others as "small" is somewhat relative to the data being analyzed. Usually, it is best to use these values according to how they compare to one another. It is also advisable to ignore the PST2 values for levels of clustering greater than 20% of the number of data points, which is $0.20 \cdot 56 = 11.2$ for these data. When the program reduces 7 clusters to 6 clusters, the value of the pseudo T^2 statistic is 56.7, which can be considered large when it is compared to many of the other values of PST2 for larger numbers of clusters. When one goes from 10 clusters to 9 clusters, the pseudo T^2 value is 17.2, which is "kind of" large. The PST2 values that precede this value tend to be much smaller—at least until we get to the line where we go from 17 clusters to 16. This value should be ignored because 17 clusters are probably too many to be useful, at least for partitioning data that contain only 56 points, because 17 clusters would tend to give an average of about three pizzas per cluster.

Summarizing the information in the preceding two paragraphs, we conclude that the number of clusters in these data is likely to be somewhere between 7 and 10, with the cubic clustering criteria suggesting 8 clusters

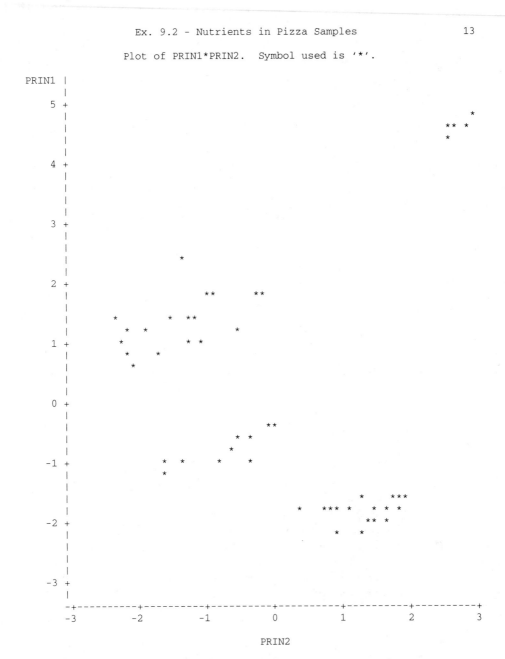

Ex. 9.2 - Nutrients in Pizza Samples   13

Plot of PRIN1*PRIN2.   Symbol used is '*'.

NOTE: 6 obs hidden.

F I G. 9.4

Pages 1, 3–14, 16, and 19–21 of computer printout, Example 9.2

Ex. 9.2 - Nutrients in Pizza Samples   14

Plot of PRIN1*PRIN2.   Symbol is value of CLUSTER.

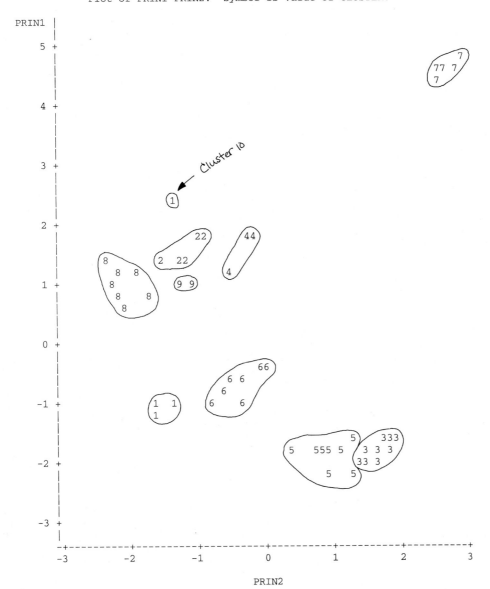

NOTE: 6 obs hidden.

F I G. **9.4**
Pages 1, 3–14, 16, and 19–21 of computer printout, Example 9.2

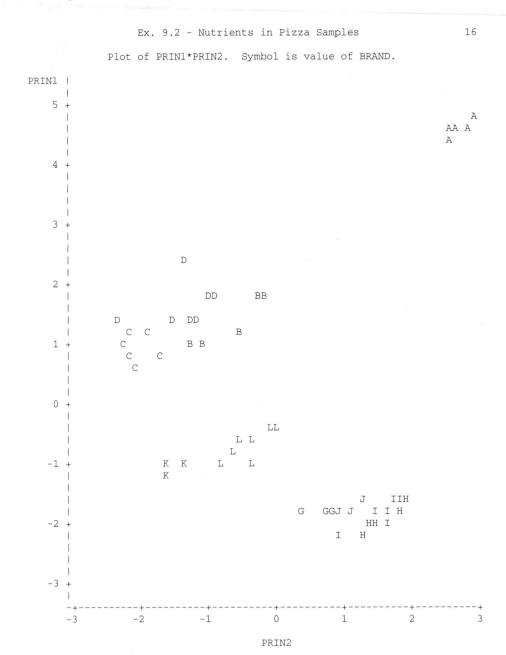

Ex. 9.2 - Nutrients in Pizza Samples   16

Plot of PRIN1*PRIN2. Symbol is value of BRAND.

NOTE: 6 obs hidden.

FIG. 9.4

Pages 1, 3–14, 16, and 19–21 of computer printout, Example 9.2

Ex. 9.2 - Nutrients in Pizza Samples   19

OBS	ID	CLUSTER
1	34021	1
2	34026	1
3	24107	1
4	34022	1
5	14067	1
6	24111	1
7	14157	1
8	14072	2
9	24030	2
10	24153	2
11	14107	2
12	24123	2
13	14140	2
14	24049	3
15	24033	3
16	14118	3
17	14143	3
18	34037	3
19	34034	3
20	14099	3
21	14122	3
22	14100	3
23	24100	3
24	14166	3
25	24119	3
26	24035	3
27	34007	3
28	14153	3
29	24103	3
30	14077	3
31	24041	3
32	14126	3

F I G. 9.4

Pages 1, 3–14, 16, and 19–21 of computer printout, Example 9.2

and the pseudo T^2 statistic suggesting 7 or 10 clusters. Some of the plotting methods discussed earlier will now be used to help make a final decision.

A hierarchical tree diagram was printed by SAS's TREE procedure by using the following commands:

```
PROC TREE DATA=TREE OUT=TREEOUT NCLUSTERS=10;
 COPY MOIS PROT FAT ASH SODIUM CARB CAL;
 ID ID;
 RUN;
```

The OUT=TREEOUT and NCLUSTERS=10 options in the preceding command list cause a new data set, named TREEOUT, to be created; this data set contains a new variable that is associated with each pizza. For a given pizza the value of this new variable, named CLUSTER, is equal to the clus-

Ex. 9.2 - Nutrients in Pizza Samples   20

OBS	ID	CLUSTER
33	14047	4
34	14074	4
35	14149	4
36	14113	4
37	24118	4
38	24056	5
39	24069	5
40	24071	5
41	34039	5
42	24073	5
43	14043	5
44	24070	5
45	14025	6
46	14164	6
47	14154	6
48	24138	6
49	24082	6
50	14106	7
51	34010	7
52	14151	7
53	24043	7
54	24058	7
55	14014	7
56	34003	7

F I G. **9.4**

Pages 1, 3–14, 16, and 19–21 of computer printout, Example 9.2

ter to which the pizza has been assigned. With NCLUSTERS=10, this new variable takes on integer values between 1 and 10. To see the cluster to which each pizza is assigned, we can print a portion of the TREEOUT data set by using the following commands:

```
PROC SORT DATA=TREEOUT; BY CLUSTER;
PROC PRINT DATA=TREEOUT;
 VARIABLES ID CLUSTER;
 RUN;
```

The results of the preceding commands are shown on pages 9 and 10 of the computer printout of Figure 9.4. The hierarchical tree diagram printed by the TREE procedure is shown on pages 7 and 8 of the computer printout. To be most useful, the plot on page 8 should be attached to the right of the plot on page 7. An examination of this diagram reveals the existence of columns corresponding to each of the 56 pizzas as identified by the ID variable at the top of each column. The 54 single dots on the bottom line of this plot indicate that the corresponding pizzas are in clusters by themselves. The three X's (XXX) in the last row on page 8 indicate that pizzas 34021 and 34026 have been placed into the same cluster. Summarizing, the

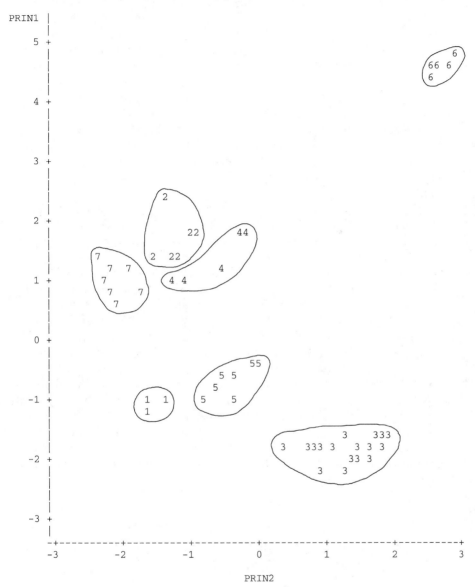

Ex. 9.2 - Nutrients in Pizza Samples   21

Plot of PRIN1*PRIN2. Symbol is value of CLUSTER.

NOTE: 6 obs hidden.

FIG. 9.4

Pages 1, 3–14, 16, and 19–21 of computer printout, Example 9.2

last row on pages 7 and 8 illustrates the clustering for 55 clusters where 54 of the clusters contain single pizzas and 1 cluster contains 2 pizzas.

For each of the other rows or levels of clustering, a single dot occurs in a column if the pizza represented by that column remains in a cluster all by itself, and pizzas that are connected by XXXXX's belong to the same cluster. Gaps in the XXX XXXX's indicate pizzas that are in different clusters. The number on the left side of the plot gives the "average distance between clusters."

Hierarchical plots are useful because they make it easy to compare different possibilities for reasonable numbers of clusters.

For this example, an appropriate number of clusters was determined to be somewhere between 7 and 10 clusters. Find the row in the hierarchical tree diagram on pages 7 and 8 that corresponds to 10 clusters. This row happens to be the row that is labeled by the 0.25 on the left side of the plot on page 7.

If you look at the row labeled by 0.25, you can see that the first 5 pizzas (columns) are in a single cluster (this is actually cluster 7 on page 9 of the printout); the next 3 pizzas are in a second cluster (this is cluster 4 on page 9); the next 2 pizzas are in a third cluster (this is cluster 9 on page 10); the next 7 pizzas are in a fourth cluster (this is cluster 8 on pages 9 and 10); the next 5 are in a fifth cluster (this is cluster 2 on page 9); the next pizza, represented by the single dot, is in a cluster all by itself (this is cluster 10 on page 9). In a similar manner you can identify the other four clusters on this line. The last four clusters (from left to right) contain 8, 11, 7, and 7 pizzas, respectively. These correspond to clusters 5, 3, 1, and 6, respectively, on pages 9 and 10.

If you move up one line on the tree diagram, you can see that there are now 8 clusters instead of 10. The 10 clusters were reduced to 8 by combining clusters 4 and 9 and by combining clusters 5 and 3. If you move up one more line, you can see that cluster 10, pizza 14140, is combined with cluster 2. In a similar manner you can easily see the consequences of combining clusters by moving up in the diagram and/or splitting clusters by moving down in the diagram.

Principal Components Plots

Next a principal components analysis is performed on these data for the purpose of creating plots of their principal component scores. Careful examinations of these plots should help "fine-tune" the results of the clustering process. The SAS commands used to create the principal component plots follow:

```
PROC PRINCOMP DATA=PIZZA3 OUT=SCORES;
 VAR MOIS PROT FAT ASH SODIUM CARB CAL;
 RUN;
PROC SORT DATA=TREEOUT; BY ID;
PROC SORT DATA=SCORES; BY ID;
```

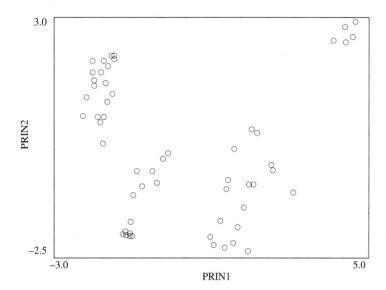

FIG. 9.5

Plot of the first two PC scores for the pizza data

```
DATA COMB; MERGE SCORES TREEOUT; BY ID;

PROC PLOT DATA=COMB;
 PLOT PRIN1*PRIN2='*';
 PLOT PRIN1*PRIN2=CLUSTER;
 PLOT PRIN1*PRIN2=PRIN3/CONTOUR=10;
 PLOT PRIN1*PRIN2=BRAND;
 RUN;
```

Pages 11 and 12 of the Figure 9.4 computer printout show the results of the principal component analysis. As noted earlier, the first two principal components account for 91.9% of the total variability in the Z scores for these variables, and the first three principal components account for 98.2% of the total variability. Obviously, the data take up no more than three dimensions and the data nearly fall into a two-dimensional subspace of the sample space. An examination of the eigenvectors corresponding to the first three principal components do not reveal anything interesting and these principal components do not seem to be interpretable, but for examining clusters, it is not necessary for them to be interpretable.

Page 13 of the computer printout shows a plot of the first two principal component (PC) scores as printed by SAS's PLOT procedure. Figure 9.5 shows another plot of the first two PC scores. This second plot was created so that both of the principal component axes are scaled equally. It is important to use the same scale on each axis when using the plots visually to make decisions about clusters. When the axes have different scales, perceived distances between points can be affected. Plots that are created automatically by many plotting programs usually have different scales on each

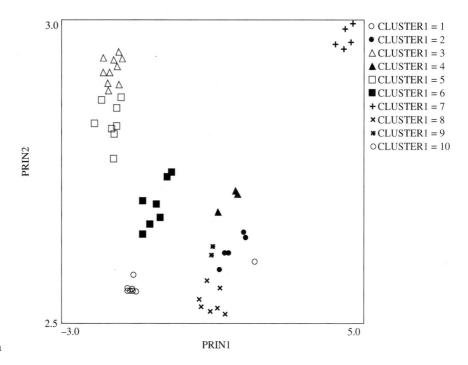

F I G. **9.6**

Scatter plot of
the first two PCs
for the pizza data

axis. You must be careful using such plots when clustering points. For example, the plots created by the PLOT procedure on pages 13–16 (see Figure 9.4) have slightly different scales on the two axes.

Now suppose you are asked the same questions that are being asked of the CLUSTER procedure. These questions are:

How many clusters are there?

Which pizzas belong to which cluster?

On the basis of what you see in the plots in Figure 9.5 and/or on page 13 of the computer printout, what are your answers to the preceding questions? Do you believe your answers are obvious? Do you think every other researcher will agree with your answers?

Clearly these are not easy questions to answer, yet many researchers place a lot of faith in statistical packages and their package's ability to answer these questions correctly. The plots on page 14 of the printout and in Figure 9.6 show the assignment of the pizzas to clusters according to SAS's CLUSTER procedure. The symbol plotted identifies the cluster to which the pizza was assigned by the CLUSTER procedure, except that the single 1 above the 2s on page 14 identifies cluster 10 rather than cluster 1 (only the first digit of 10 is plotted by SAS).

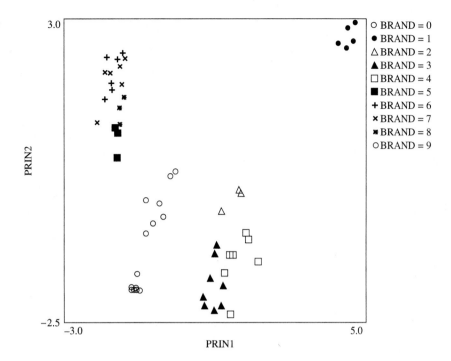

F I G. **9.7**

Scatter plots of
the first two PCs
for the pizza data

These plots illustrate results for 10 clusters. Do these clusters agree with
the ones you selected in Figure 9.5? Do they agree in the number of clusters selected? Where do they agree with your answers and where do they
disagree?

At this time, we should point out that these pizzas were actually samples from 12 different brands of pizza. The brands have been labeled A, B,
C, . . . , L in the original data file. Brands E and F do not occur in this analysis because no fat measurements were available for those two brands.
Thus the SAS data set `PIZZA3` contains data from 10 brands of pizzas. On
page 16 of the printout and in Figure 9.7 you can find additional plots of
the first two principal component scores. In these plots, the plot symbol
identifies the true brand of each pizza. Obviously, the brands should represent the true clusters.

It is interesting to compare the plots on pages 14 and 16 of the printout
with each other to see where the clusters obtained by the `CLUSTER` procedure and the true brands agree and disagree. Clusters 1, 6, and 7 agree exactly with brands K, L, and A, respectively. Clusters 3 and 5 correspond to
brands G, H, I, and J. By looking at the locations of these four brands on
page 16, it is unrealistic to expect a clustering algorithm to separate these
pizzas into the correct brands. Clusters 2, 4, 8, 9 and 10 contain pizzas from
brands B, C and D. Again it seems unrealistic to expect a clustering algorithm to separate these pizzas into the correct brands.

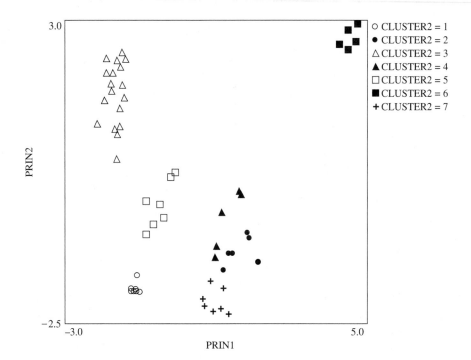

F I G. **9.8**

Plot of the first two PC scores showing seven clusters

A scatter plot of the first two principal component scores showing the assignment of pizzas to clusters when there are seven clusters selected is shown in Figure 9.8. How do these compare to the other plots?

Additional plots of the pizza data are shown in Figures 9.9, 9.10, and 9.11. Each of these plots was created by using all of the original variables after they were standardized. Figure 9.9 shows Andrews' curves for each of the pizzas, Figure 9.10 shows a star plot for each of the pizza samples, and Figure 9.11 shows Chernoff faces for each of the pizza samples. The pizzas plotted in Figures 9.10 and 9.11 are plotted in the order in which they occurred in the original data. In Figures 9.10 and 9.11, pizzas 1–5 were brand A, 6–10 were brand B, 11–16 were brand C, 17–23 were brand D, 24–26 were brand G, 27–33 were brand H, 34–39 were brand I, 40–42 were brand J, 43–49 were brand K, and 50–56 were brand L. Note the similarities and dissimilarities among the pizzas in these plots. How do the locations of the true brands compare to the clusters you selected visually earlier? Where do they agree and where do they disagree?

In summary, it is comforting to see that pizzas that were quite different actually did get placed in different clusters by the CLUSTER procedure. And while the pizzas that were close together did not always get placed in the correct cluster, the mistakes in assignment are not likely to be too serious because some of the brands involved are obviously quite similar to one another.

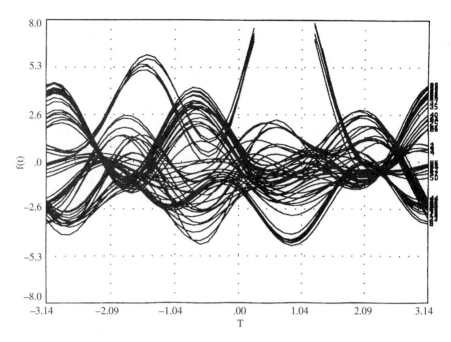

F I G. 9.9
Andrews' plots of
pizza data using
the original vari-
ables

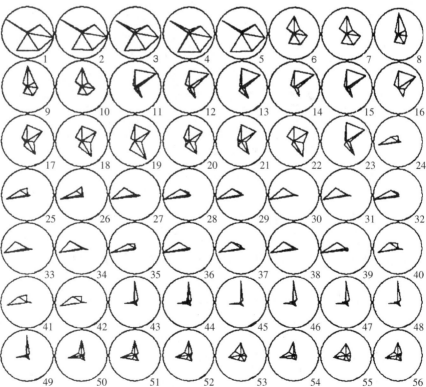

F I G. 9.10
Star plots of
pizza data

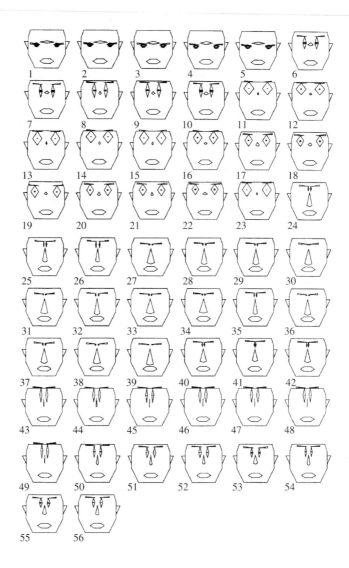

Next, how would you "fine-tune" the results of the clustering procedure?

As I look at the plot of principal component scores on page 14 of Figure 9.4, I would be inclined to combine clusters 2, 9, and 10 into one cluster and move one of the points in cluster 5 to cluster 3. This gives 8 clusters. How does this compare to what you would do?

If the computer program were asked to select eight clusters, which eight would it select? From the tree diagram, we can see that clusters 4 and 9 would be combined and that clusters 5 and 3 would be combined into one cluster. If the computer were asked to choose seven clusters, then the single point in cluster 10 would be combined with cluster 2.

If we want to reproduce the plots for a computer clustering resulting in seven clusters, we can use the following SAS commands:

```
PROC TREE DATA=TREE OUT=TREEOUT NCLUSTERS=7;
 COPY MOIS PROT FAT ASH SODIUM CARB CAL;
 ID ID;
 RUN;

PROC SORT DATA=TREEOUT; BY CLUSTER;
PROC PRINT DATA=TREEOUT;
 VARIABLES ID CLUSTER;
 RUN;

PROC SORT DATA=TREEOUT; BY ID;
PROC SORT DATA=SCORES; BY ID;

DATA COMB; MERGE SCORES TREEOUT; BY ID;

PROC PLOT DATA=COMB;
 PLOT PRIN1*PRIN2=CLUSTER;
 RUN;
```

The assignments of points to clusters are shown on pages 19 and 20 of Figure 9.4, and a new plot of principal component scores is shown on page 21.

Clustering with SPSS

SPSS's CLUSTER procedure is a hierarchical procedure, but it does not provide nearly as much information as SAS's CLUSTER procedure. We do get a listing of the order in which samples are clustered and three kinds of tree diagrams are available, but there are no pseudo test statistics and no cubic clustering criteria available. As a consequence, determining an appropriate number of clusters must generally be based on the tree diagrams and/or the distances between clusters when they are combined. Of course, one can and should look at scatter plots of principal component scores, Andrews' plots, Chernoff faces, sun-ray plots, etc. These plots can be used to make visual assessments as to the uniqueness of the clusters formed as well as to the number of clusters formed. The CLUSTER procedure in SPSS is illustrated in Example 9.3.

E X A M P L E 9.3

This example uses SPSS's CLUSTER procedure to perform a cluster analysis on data collected from orange juice samples from five different countries. A researcher measured the amounts of nine different chemical elements present in orange juice. The chemicals measured were boron (B), barium (BA), calcium (CA), potassium (K), magnesium (MG), manganese (MN), phosphorous (P), rubidium (RB), and zinc (ZN). One question of interest is whether samples from the same country will form clusters that are different from clusters formed from other countries.

All of the commands used to perform a cluster analysis on the orange juice data are available in the file labeled EX9_3.SSX on the enclosed disk. They are repeated here. The first set of commands is used to read the data into a file for execution of program statements.

```
SET WIDTH = 80
TITLE "EX. 9.3 - CLUSTER ANALYSIS ON ORANGE JUICE SAMPLES"

DATA LIST LIST
     /ID 1-5 (A) COUNTRY 10-12 (A) B BA CA K MG MN P RB ZN

PRINT
      /ID COUNTRY B BA CA K MG MN P RB ZN
BEGIN DATA
2939 BEL 0.298 0.237 109.1 1802 102.3 0.453 122.2 2.98 0.323
2940 BEL 0.288 0.236 108.4 1797 102.3 0.455 121.9 2.99 0.328
2962 BEL 0.421 0.271  43.4 1634  85.1 0.160 124.9 7.26 0.277
2984 BEL 0.434 0.117  53.4 1504  84.5 0.265 135.5 2.35 0.397
4395 BEL 0.578 0.145  82.9 1486  94.5 0.330 148.6 1.98 0.356
4396 BEL 0.575 0.144  81.8 1464  93.8 0.327 147.6 1.96 0.365
4427 BEL 0.811 0.196 266.2 1761 113.0 0.360 194.5 5.28 0.477
4428 BEL 0.794 0.193 270.4 1731 113.2 0.353 192.9 5.23 0.439
 159 LSP 0.181 0.092  31.2 1233  64.3 0.253  52.4 1.89 0.306
 193 LSP 0.733 0.034  34.7 1640  69.4 0.132 103.4 1.37 0.244
 228 LSP 0.514 0.026  28.9 1753  61.9 0.119 104.1 1.44 0.232
 261 LSP 0.504 0.040  36.7 1759  73.2 0.133 124.3 2.99 0.194
 339 LSP 0.744 0.040  43.3 1814  76.8 0.149 125.3 1.45 0.291
 340 LSP 0.771 0.045  43.3 1793  77.2 0.151 125.4 1.45 0.305
 367 LSP 0.453 0.059  42.0 1819  81.5 0.172 121.1 3.11 0.279
 368 LSP 0.445 0.054  40.3 1779  81.1 0.168 120.0 3.11 0.265
 476 LSP 0.127 0.077  39.0 1570  98.1 0.136  80.4 1.18 0.276
1017 LSP 0.431 0.084  44.0 1564  86.1 0.200  95.6 1.11 0.268
1018 LSP 0.433 0.084  44.4 1585  87.3 0.202  96.6 1.10 0.273
1045 LSP 0.609 0.065  39.7 1553  81.9 0.126  89.7 2.14 0.247
1046 LSP 0.604 0.068  38.6 1518  80.9 0.125  88.9 2.12 0.231
1128 LSP 0.410 0.073  48.1 1713 119.4 0.176 119.5 1.82 0.479
1129 LSP 0.404 0.073  45.3 1693 119.4 0.172 117.7 1.85 0.464
1162 LSP 0.564 0.048  34.5 1128  76.6 0.114  95.6 0.83 0.174
1163 LSP 0.554 0.051  33.4 1130  76.6 0.112  94.8 0.80 0.170
 126 NSP 0.517 0.020  42.9 1593  84.0 0.131 137.0 1.11 0.304
2852 TME 1.013 0.050 102.6 1555  75.0 0.135 132.1 0.22 0.270
```

```
2859 TME 1.017 0.050  95.0 1559 74.3 0.130 132.5 0.26 0.261
2891 TME 1.491 0.032 111.5 1462 70.5 0.157 115.1 0.59 0.225
2898 TME 1.500 0.031 109.7 1456 70.5 0.156 114.3 0.60 0.220
4267 TME 0.795 0.089 389.5 1305 74.6 0.217 126.9 0.26 0.411
4268 TME 0.805 0.087 399.7 1327 76.2 0.217 127.4 0.23 0.393
2766 VME 0.440 0.709  64.7 1549 93.6 0.162 157.5 2.74 0.355
2767 VME 0.417 0.646  79.8 1420 84.7 0.148 144.3 2.48 1.045
2773 VME 0.402 0.633  77.0 1387 82.2 0.150 143.7 2.48 1.050
4300 VME 0.754 0.472  76.4 1507 91.6 0.542 170.6 4.12 0.324
4301 VME 0.758 0.471  75.5 1484 90.7 0.544 171.9 4.08 0.337
END DATA
```

The next set of commands is used to standardize the measured chemical element concentrations so that all measured variables are in comparable units.

```
CONDESCRIPTIVE B TO ZN
STATISTICS 1 5
OPTION 3
```

The next set of commands is used to ask SPSS to perform a cluster analysis on these data using Ward's method with distances measured by squaring the Euclidean distance (ruler distance) between points and clusters. Other clustering methods that exist in SPSS include the nearest neighbor method, the furthest neighbor method, the average method, and the centroid method. SPSS also allows users to measure distances between points and clusters by several different distance measures; however, only two can be recommended. These two are SEUCLID (squared Euclidean distance) and EUCLID (standard Euclidean distance). For many methods of clustering, these two distance measures will produce identical results.

```
CLUSTER ZB TO ZZN
 /METHOD = WARD
 /MEASURE = SEUCLID
 /ID = ID
 /PRINT SCHEDULE CLUSTER (3,8)
 /PLOT VICICLE DENDROGRAM
```

The next set of commands is used to perform a principal components analysis on the data and to compute principal component scores for each orange juice sample. The first three (standardized) principal component scores have the variable names (PCSCR1, PCSCR2, and PCSCR3, respectively).

```
FACTOR VARIABLES = ZB TO ZZN
  /EXTRACTION=PC
  /ROTATE=NOROTATE
  /SAVE REG (ALL PCSCR)
```

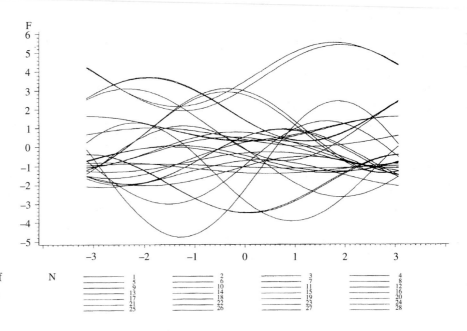

F I G. 9.12

Andrews' plots of
orange juice data
using three PCs

The next set of commands is used to compute, print, and plot un-standardized principal component scores from the standardized principal component scores automatically produced by SPSS. To compute unstandardized principal component scores, we must multiply each of the standardized principal component scores by the square root of its corresponding eigenvalue. The first three unstandardized principal component scores are denoted by UPCSCR1, UPCSCR2, and UPCSCR3, respectively.

```
COMPUTE UPCSCR1=PCSCR1*SQRT(3.2101)
COMPUTE UPCSCR2=PCSCR2*SQRT(1.6857)
COMPUTE UPCSCR3=PCSCR3*SQRT(1.4608)

LIST VARIABLES = ID UPCSCR1 UPCSCR2 UPCSCR3

PLOT /SYMBOLS = NUMERIC
     /FORMAT = CONTOUR(9)
     /PLOT UPCSCR1 WITH UPCSCR2 BY UPCSCR3
```

Andrews' plots of the orange juice data using the first three unstandardized principal component scores are shown in Figure 9.12, and a three-dimensional plot of the first three unstandardized principal component scores is shown in Figure 9.13.

Some of the results from the preceding commands can be found on the pages of computer output shown in Figure 9.14. The total output can be seen by executing the file labeled EX9_3.SSX on the enclosed floppy disk.

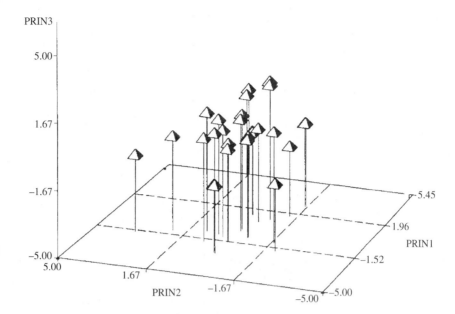

FIG. 9.13

Three-dimensional plot of the first three unstandardized PCs

Page 9 of the computer output shows the order in which samples are combined. The first two data points combined are those listed in `Stage 1`—they are cases 1 and 2 and their IDs are 2939 and 2940, respectively. Note that the IDs and the case numbers are given in the first two columns of the computer output shown on page 10. The column labeled `Coefficient` gives the squared Euclidean distance between the clusters that are being combined.

The two data points combined in stage 2 are cases 24 and 25—their IDs are 1162 and 1163, respectively. An examination of page 9 reveals that in the first 14 stages, only pairs of points are combined into clusters. Note also that each cluster is named by the minimum case number of all of the cases in the cluster. Thus the names of the 14 clusters formed during the first 14 stages are 1, 24, 29, 27, 5, 18, 22, 36, 13, 20, 31, 15, 34, and 7, respectively. At stage 15, one can see that case 12 is combined with the cluster named cluster 15, which contains cases 15 and 16. This new cluster is now named Cluster 12 since 12 is the minimum of the case numbers 12, 15, and 16.

To help clarify the points made in the previous paragraph, consider the cluster formed at stage 23. This cluster was formed by combining clusters 10 and 12. As a result, this new cluster contains cases 10, 11, and 13 (points that were placed in cluster 10 at stage 16) and cases 12, 15, and 16 (points that were placed in cluster 12 at stage 15). This cluster is now named cluster 10 since 10 is smaller than 12.

Page 10 shows the assignment of cases to clusters for situations where there are eight clusters, seven clusters, . . . , three clusters. Pages 11 and 12

EX. 9.3 - CLUSTER ANALYSIS ON ORANGE JUICE SAMPLES   Page   9

* * * * * * H I E R A R C H I C A L   C L U S T E R   A N A L Y S I S * * * * *

Agglomeration Schedule using Ward Method

	Clusters Combined			Stage Cluster 1st Appears		
Stage	Cluster 1	Cluster 2	Coefficient	Cluster 1	Cluster 2	Next Stage
1	1	2	.001537	0	0	26
2	24	25	.003281	0	0	24
3	29	30	.005245	0	0	22
4	27	28	.012612	0	0	22
5	5	6	.023031	0	0	17
6	18	19	.033968	0	0	18
7	22	23	.046208	0	0	28
8	36	37	.059781	0	0	29
9	13	14	.073961	0	0	19
10	20	21	.098521	0	0	18
11	31	32	.123393	0	0	31
12	15	16	.151821	0	0	15
13	34	35	.187389	0	0	34
14	7	8	.227502	0	0	30
15	12	15	.690105	0	12	23
16	10	11	1.293336	0	0	20
17	4	5	2.212379	0	5	26
18	18	20	3.655457	6	10	21
19	13	26	5.265358	9	0	20
20	10	13	7.499083	16	19	23
21	17	18	9.910087	0	18	27
22	27	29	13.414657	4	3	31
23	10	12	17.008163	20	15	27
24	9	24	21.770691	0	2	32
25	3	33	29.758347	0	0	29
26	1	4	38.134644	1	17	28
27	10	17	47.361084	23	21	32
28	1	22	57.209152	26	7	33
29	3	36	70.173172	25	8	30
30	3	7	88.566650	29	14	33
31	27	31	108.067719	22	11	35
32	9	10	129.939651	24	27	35
33	1	3	153.189819	28	30	34
34	1	34	192.416107	33	13	36
35	9	27	232.115753	32	31	36
36	1	9	323.998779	34	35	0

F I G. 9.14
Pages 9–13, 16, 19, 22, and 25 of computer printout, Example 9.3

EX. 9.3 - CLUSTER ANALYSIS ON ORANGE JUICE SAMPLES   Page   10

* * * * * * H I E R A R C H I C A L   C L U S T E R   A N A L Y S I S * * * * * *

Cluster Membership of Cases using Ward Method

		Number of Clusters					
Label	Case	8	7	6	5	4	3
2939	1	1	1	1	1	1	1
2940	2	1	1	1	1	1	1
2962	3	2	2	2	2	1	1
2984	4	1	1	1	1	1	1
4395	5	1	1	1	1	1	1
4396	6	1	1	1	1	1	1
4427	7	3	2	2	2	1	1
4428	8	3	2	2	2	1	1
159	9	4	3	3	3	2	2
193	10	5	4	4	3	2	2
228	11	5	4	4	3	2	2
261	12	5	4	4	3	2	2
339	13	5	4	4	3	2	2
340	14	5	4	4	3	2	2
367	15	5	4	4	3	2	2
368	16	5	4	4	3	2	2
476	17	5	4	4	3	2	2
1017	18	5	4	4	3	2	2
1018	19	5	4	4	3	2	2
1045	20	5	4	4	3	2	2
1046	21	5	4	4	3	2	2
1128	22	1	1	1	1	1	1
1129	23	1	1	1	1	1	1
1162	24	4	3	3	3	2	2
1163	25	4	3	3	3	2	2
126	26	5	4	4	3	2	2
2852	27	6	5	5	4	3	3
2859	28	6	5	5	4	3	3
2891	29	6	5	5	4	3	3
2898	30	6	5	5	4	3	3
4267	31	7	6	5	4	3	3
4268	32	7	6	5	4	3	3
2766	33	2	2	2	2	1	1
2767	34	8	7	6	5	4	1
2773	35	8	7	6	5	4	1
4300	36	2	2	2	2	1	1
4301	37	2	2	2	2	1	1

F I G. 9.14

Pages 9–13, 16, 19, 22, and 25 of computer printout, Example 9.3

```
        EX. 9.3 - CLUSTER ANALYSIS ON ORANGE JUICE SAMPLES          Page   11

 * * * * * H I E R A R C H I C A L   C L U S T E R   A N A L Y S I S * * * * * *

 Vertical Icicle Plot using Ward Method

    (Down) Number of Clusters   (Across) Case Label and number

        4  4  2  2  2  2  1  1  1  1  4  3  3  2  1  3  3  2  1  1  1  1  2  2
        2  2  8  8  8  8  0  0  0  0  7  6  6  6  2  4  3  2  9  1  1  5  7  7
        6  6  9  9  5  5  4  4  1  1  6  8  7  1  6  0  9  8  3  6  6  9  7  6
        8  7  8  1  9  2  6  5  8  7                          3  2        3  7

        3  3  3  2  2  2  2  2  1  1  1  1  1  1  2  1  1  1  1  2  2        3  3
        2  1  0  9  8  7  1  0  9  8  7  6  5  2  6  4  3  1  0  5  4  9     5  4
     1 +XXXXXXXXXXXXXXXXXXXXXXXXXXXXXXXXXXXXXXXXXXXXXXXXXXXXXXXXXXXXXXXXXXXXXXX
     2 +XXXXXXXXXXXXXXXXXXXXXXXXXXXXXXXXXXXXXXXXXXXXXXXXXXXXXXXXXXXXXXXXXX XXXX
     3 +XXXXXXXXXXXXXXX XXXXXXXXXXXXXXXXXXXXXXXXXXXXXXXXXXXXXXXXXXXXXXXXXX XXXX
     4 +XXXXXXXXXXXXXXX XXXXXXXXXXXXXXXXXXXXXXXXXXXXXXXXXXXXXXXXXXXXXXXXXX XXXX
     5 +XXXXXXXXXXXXXXX XXXXXXXXXXXXXXXXXXXXXXXXXXXXXXXXXXXXXXXXXXXXXXXXXX XXXX
     6 +XXXXXXXXXXXXXXX XXXXXXXXXXXXXXXXXXXXXXXXXXXXXXXXXXXXXX XXXXXXX XXXX
     7 +XXXX XXXXXXXXXX XXXXXXXXXXXXXXXXXXXXXXXXXXXXXXXXXXXXXX XXXXXXX XXXX
     8 +XXXX XXXXXXXXXX XXXXXXXXXXXXXXXXXXXXXXXXXXXXXXXXXXXXXX XXXXXXX XXXX
     9 +XXXX XXXXXXXXXX XXXXXXXXXXXXXXXXXXXXXXXXXXXXXXXXXXXXXX XXXXXXX XXXX
    10 +XXXX XXXXXXXXXX XXXXXXXXXXXXXXXXXXXXXXXXXXXXXXXXXXXXXX XXXXXXX XXXX
    11 +XXXX XXXXXXXXXX XXXXXXXXXXXXXX XXXXXXXXXXXXXXXXXXXXXX XXXXXXX XXXX
    12 +XXXX XXXXXXXXXX XXXXXXXXXXXXXX XXXXXXXXXXXXXXXXXXXXXX XXXXXXX XXXX
    13 +XXXX XXXXXXXXXX XXXXXXXXXXXXXX XXXXXXXXXXXXXXXXXXXXXX XXXXXXX XXXX
    14 +XXXX XXXXXXXXXX XXXXXXXXXXXXXX XXXXXXXXXXXXXXXXXXXXXX XXXX  X  XXXX
    15 +XXXX XXXXXXXXXX XXXXXXXXXXXXXX XXXXXXX XXXXXXXXXXXXXX XXXX  X  XXXX
    16 +XXXX XXXX  XXXX XXXXXXXXXXXXXX XXXXXXX XXXXXXXXXXXXXX XXXX  X  XXXX
    17 +XXXX XXXX  XXXX XXXXXXXXXX  X  XXXXXXX XXXXXXXXXXXXXX XXXX  X  XXXX
    18 +XXXX XXXX  XXXX XXXXXXXXXX  X  XXXXXXX XXXXXXX  XXXX XXXX  X  XXXX
    19 +XXXX XXXX  XXXX XXXXXXXXXX  X  XXXXXXX  X  XXXX XXXX XXXX  X  XXXX
    20 +XXXX XXXX  XXXX XXXX  XXXX  X  XXXXXXX  X  XXXX XXXX XXXX  X  XXXX
    21 +XXXX XXXX  XXXX XXXX  XXXX  X  XXXXXXX  X  XXXX XXXX XXXX  X  XXXX
    22 +XXXX XXXX  XXXX XXXX  XXXX  X  XXXXXXX  X  XXXX  X  X XXXX  X  XXXX
    23 +XXXX XXXX  XXXX XXXX  XXXX  X  XXXX  X  X  XXXX  X  X XXXX  X  XXXX
    24 +XXXX XXXX  XXXX XXXX  XXXX  X  XXXX  X  X  XXXX  X  X XXXX  X  XXXX
    25 +XXXX XXXX  XXXX XXXX  XXXX  X  XXXX  X  X  XXXX  X  X XXXX  X  X  X
    26 +XXXX XXXX  XXXX XXXX  XXXX  X  X  X  X  X  XXXX  X  X XXXX  X  X  X
    27 +X  X XXXX  XXXX XXXX  XXXX  X  X  X  X  X  XXXX  X  X XXXX  X  X  X
    28 +X  X XXXX  XXXX  X  X XXXX  X  X  X  X  X  XXXX  X  X XXXX  X  X  X
    29 +X  X XXXX  XXXX  X  X XXXX  X  X  X  X  X  X  X  X  X XXXX  X  X  X
    30 +X  X XXXX  XXXX  X  X XXXX  X  X  X  X  X  X  X  X  X XXXX  X  X  X
    31 +X  X XXXX  XXXX  X  X XXXX  X  X  X  X  X  X  X  X  X XXXX  X  X  X
    32 +X  X XXXX  XXXX  X  X  X  X  X  X  X  X  X  X  X  X  X XXXX  X  X  X
    33 +X  X XXXX  XXXX  X  X  X  X  X  X  X  X  X  X  X  X  X XXXX  X  X  X
    34 +X  X XXXX  X  X  X  X  X  X  X  X  X  X  X  X  X  X  X XXXX  X  X  X
    35 +X  X  X  X  X  X  X  X  X  X  X  X  X  X  X  X  X  X XXXX  X  X  X
    36 +X  X  X  X  X  X  X  X  X  X  X  X  X  X  X  X  X  X  X  X  X  X  X
```

F I G. 9.14

Pages 9–13, 16, 19, 22, and 25 of computer printout, Example 9.3

EX. 9.3 - CLUSTER ANALYSIS ON ORANGE JUICE SAMPLES   Page 12

* * * * * * H I E R A R C H I C A L   C L U S T E R   A N A L Y S I S * * * *   *

Vertical Icicle Plot using Ward Method (CONT.)

(Down) Number of Clusters (Across) Case Label and number

```
          4  4  4  4  2  2  1  1  4  4  2  2  2
          4  4  3  3  7  9  1  1  3  3  9  9  9
          2  2  0  0  6  6  2  2  9  9  8  4  3
          8  7  1  0  6  2  9  8  6  5  4  0  9
                3  3  3        2  2
          8  7  7  6  3  3  3  2  6  5  4  2  1
   1 +XXXXXXXXXXXXXXXXXXXXXXXXXXXXXXXXXXXXXXXXX
   2 +XXXXXXXXXXXXXXXXXXXXXXXXXXXXXXXXXXXXXXXXX
   3 +XXXXXXXXXXXXXXXXXXXXXXXXXXXXXXXXXXXXXXXXX
   4 +XXXXXXXXXXXXXXXXXXXXXXXXXXXXXXXXXXXXXXXXX
   5 +XXXXXXXXXXXXXXXX  XXXXXXXXXXXXXXXXXXXXXXX
   6 +XXXXXXXXXXXXXXXX  XXXXXXXXXXXXXXXXXXXXXXX
   7 +XXXXXXXXXXXXXXXX  XXXXXXXXXXXXXXXXXXXXXXX
   8 +XXXX  XXXXXXXXXX  XXXXXXXXXXXXXXXXXXXXXXX
   9 +XXXX  XXXX  XXXX  XXXXXXXXXXXXXXXXXXXXXXX
  10 +XXXX  XXXX  XXXX  XXXX  XXXXXXXXXXXXXX
  11 +XXXX  XXXX  XXXX  XXXX  XXXXXXXXXXXXXX
  12 +XXXX  XXXX  XXXX  XXXX   XXXXXXX   XXXX
  13 +XXXX  XXXX  X  X  XXXX   XXXXXXX   XXXX
  14 +XXXX  XXXX  X  X  XXXX   XXXXXXX   XXXX
  15 +XXXX  XXXX  X  X  XXXX   XXXXXXX   XXXX
  16 +XXXX  XXXX  X  X  XXXX   XXXXXXX   XXXX
  17 +XXXX  XXXX  X  X  XXXX   XXXXXXX   XXXX
  18 +XXXX  XXXX  X  X  XXXX   XXXXXXX   XXXX
  19 +XXXX  XXXX  X  X  XXXX   XXXXXXX   XXXX
  20 +XXXX  XXXX  X  X  XXXX   XXXXXXX   XXXX
  21 +XXXX  XXXX  X  X  XXXX   XXXX  X   XXXX
  22 +XXXX  XXXX  X  X  XXXX   XXXX  X   XXXX
  23 +XXXX  XXXX  X  X  XXXX   XXXX  X   XXXX
  24 +X  X  XXXX  X  X  XXXX   XXXX  X   XXXX
  25 +X  X  XXXX  X  X  XXXX   XXXX  X   XXXX
  26 +X  X  XXXX  X  X  XXXX   XXXX  X   XXXX
  27 +X  X  XXXX  X  X  XXXX   XXXX  X   XXXX
  28 +X  X  XXXX  X  X  XXXX   XXXX  X   XXXX
  29 +X  X  XXXX  X  X  XXXX   XXXX  X   XXXX
  30 +X  X  X  X  X  X  XXXX   XXXX  X   XXXX
  31 +X  X  X  X  X  X  X  X   XXXX  X   XXXX
  32 +X  X  X  X  X  X  X  X   XXXX  X   XXXX
  33 +X  X  X  X  X  X  X  X  X  X  X   XXXX
  34 +X  X  X  X  X  X  X  X  X  X  X   XXXX
  35 +X  X  X  X  X  X  X  X  X  X  X   XXXX
  36 +X  X  X  X  X  X  X  X  X  X  X   XXXX
```

F I G. 9.14

Pages 9–13, 16, 19, 22, and 25 of computer printout, Example 9.3

```
      EX. 9.3 - CLUSTER ANALYSIS ON ORANGE JUICE SAMPLES          Page   13

*  *  *  *  *  * H I E R A R C H I C A L   C L U S T E R    A N A L Y S I S * * * * *
*


Dendrogram using Ward Method

                   Rescaled Distance Cluster Combine

  C A S E    0         5        10        15        20        25
 Label  Seq  +---------+---------+---------+---------+---------+

  2939    1    -+---+
  2940    2    -+    +---+
  4395    5    -+   I   I
  4396    6    -+---+   +-------------+
  2984    4    -+       I             I
  1128   22    -+-------+             +-----+
  1129   23    -+                 I   I
  4427    7    -+-----------+      I   I
  4428    8    -+           +---------+   I
  4300   36    -+---------+ I           +-------------------+
  4301   37    -+         +-+       I             I
  2962    3    -----+-----+       I             I
  2766   33    -----+               I             I
  2767   34    -+-------------------------+     I
  2773   35    -+                           I
  4267   31    -+---------------+           I
  4268   32    -+             +-----------------+   I
  2891   29    -+-+           I               I   I
  2898   30    -+ +-------------+             I   I
  2852   27    -+-+                         I   I
  2859   28    -+                       +-------------+
  1162   24    -+-+                       I
  1163   25    -+ +---------------+       I
  159     9    ---+           I       I
  1017   18    -+             I       I
  1018   19    -+             +-------------+
  1045   20    -+-----+       I
  1046   21    -+    I       I
  476    17    -+    +-------------+
  367    15    -+    I
  368    16    -+-+  I
  261    12    -+ +---+
  193    10    -+ I
  228    11    -+-+
  339    13    -+
  340    14    -+
  126    26    -+
```

F I G. 9.14
Pages 9–13, 16, 19, 22, and 25 of computer printout, Example 9.3

```
EX. 9.3 - CLUSTER ANALYSIS ON ORANGE JUICE SAMPLES          Page  16

- - - - - - - - - - -   F A C T O R   A N A L Y S I S   - - - - - - - - - - -

ANALYSIS NUMBER  1  LISTWISE DELETION OF CASES WITH MISSING VALUES

EXTRACTION  1  FOR ANALYSIS  1, PRINCIPAL-COMPONENTS ANALYSIS (PC)

INITIAL STATISTICS:

VARIABLE      COMMUNALITY  *  FACTOR   EIGENVALUE   PCT OF VAR   CUM PCT
                           *
ZB              1.00000    *    1       3.21010       35.7        35.7
ZBA             1.00000    *    2       1.68569       18.7        54.4
ZCA             1.00000    *    3       1.46084       16.2        70.6
ZK              1.00000    *    4        .78452        8.7        79.3
ZMG             1.00000    *    5        .75756        8.4        87.8
ZMN             1.00000    *    6        .43799        4.9        92.6
ZP              1.00000    *    7        .33658        3.7        96.4
ZRB             1.00000    *    8        .18082        2.0        98.4
ZZN             1.00000    *    9        .14590        1.6       100.0
     PC EXTRACTED    3 FACTORS.
```

FIG. 9.14

Pages 9–13, 16, 19, 22, and 25 of computer printout, Example 9.3

give a vertical icicle plot and page 13 shows a hierarchical tree diagram constructed by SPSS.

A portion of the results of a principal components analysis are shown on page 16. From this page we can see that there are three eigenvalues greater than 1 and that these three account for 70.6% of the total variability in the standardized data values.

A listing of the first three (unstandardized) principal component scores for each of the cases is shown on page 19. The column labeled CLUSMEM4 shows where each of the cases should be assigned if we were to select four clusters.

Page 22 shows a scatter plot of the first three (unstandardized) principal component scores where vertical and horizontal axes correspond to the first and second principal components, respectively, and the value of the symbol plotted is proportional to the value of the third principal component. Page 25 gives a scatter plot of the first two principal component scores with the plot symbol equal to the cluster to which the case has been assigned.

An examination of the plot on page 25 makes us feel good about the assignment of cases to clusters 2 and 3, but there might be some concern about the assignment of cases to clusters 1 and 4. It seems strange that the two cases assigned to cluster 4 are located in the middle of the cases assigned to cluster 1. I would be inclined to reassign the four cases labeled by the 1s near the top of the plot to cluster 4. What would you do?

ID	UPCSCR1	UPCSCR2	UPCSCR3	CLUSMEM4	COUNTRY
2939	1.90	-1.18	.89	1	BEL
2940	1.91	-1.20	.85	1	BEL
2962	1.35	-1.61	.22	1	BEL
2984	.29	-.33	-.28	1	BEL
4395	.85	.25	.17	1	BEL
4396	.79	.29	.07	1	BEL
4427	3.82	1.14	1.77	1	BEL
4428	3.68	1.16	1.77	1	BEL
159	-2.08	-.99	-1.74	2	LSP
193	-1.75	-.27	.35	2	LSP
228	-1.83	-.87	.38	2	LSP
261	-.76	-1.05	.88	2	LSP
339	-.88	-.38	.93	2	LSP
340	-.86	-.28	.86	2	LSP
367	-.16	-1.32	.85	2	LSP
368	-.28	-1.29	.77	2	LSP
476	-1.04	-1.85	-.49	2	LSP
1017	-1.06	-.85	-.12	2	LSP
1018	-.98	-.89	-.05	2	LSP
1045	-1.38	-.73	.02	2	LSP
1046	-1.48	-.68	-.06	2	LSP
1128	.81	-1.37	.21	1	LSP
1129	.73	-1.40	.17	1	LSP
1162	-2.36	.24	-1.09	2	LSP
1163	-2.39	.21	-1.11	2	LSP
126	-.79	-.35	.19	2	NSP
2852	-1.45	1.28	.51	3	TME
2859	-1.49	1.23	.53	3	TME
2891	-2.02	2.28	.88	3	TME
2898	-2.06	2.28	.88	3	TME
4267	-.60	3.22	-.37	3	TME
4268	-.55	3.24	-.21	3	TME
2766	1.89	-.24	-1.42	1	VME
2767	2.25	.45	-3.84	4	VME
2773	2.13	.49	-3.96	4	VME
4300	2.94	.64	.35	1	VME
4301	2.92	.73	.25	1	VME

NUMBER OF CASES READ =   37   NUMBER OF CASES LISTED =   37

F I G. 9.14
Pages 9–13, 16, 19, 22, and 25 of computer printout, Example 9.3

EX. 9.3 - CLUSTER ANALYSIS ON ORANGE JUICE SAMPLES   Page 22

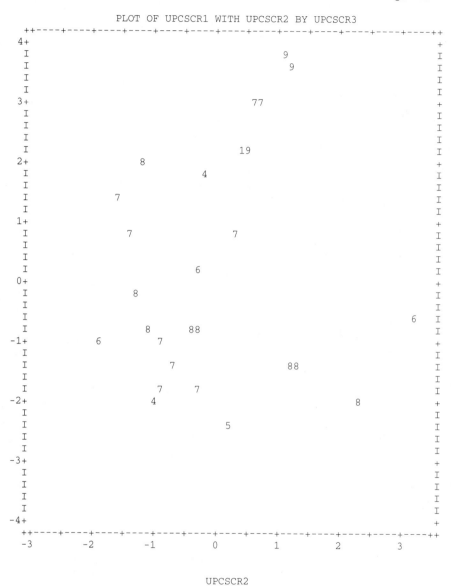

```
              PLOT OF UPCSCR1 WITH UPCSCR2 BY UPCSCR3
      ++----+----+----+----+----+----+----+----+----+----+----+----++
    4+                                                              +
     I                                        9                     I
     I                                        9                     I
     I                                                              I
     I                                                              I
    3+                                    77                        +
     I                                                              I
     I                                                              I
     I                                                              I
     I                                  19                          I
    2+              8                                               +
     I                       4                                      I
     I                                                              I
     I          7                                                   I
     I                                                              I
    1+                                                              +
     I              7              7                                I
     I                                                              I
     I                                                              I
     I                        6                                     I
    0+                                                              +
     I          8                                                   I
     I                                                          6   I
     I              8     88                                        I
   -1+      6          7                                            +
     I                                                              I
     I                  7                    88                     I
     I              7     7                                         I
   -2+          4                                  8               +
     I                                                              I
     I                      5                                       I
     I                                                              I
   -3+                                                              +
     I                                                              I
     I                                                              I
     I                                                              I
   -4+                                                              +
      ++----+----+----+----+----+----+----+----+----+----+----+----++
      -3        -2        -1         0         1         2         3
```

                              UPCSCR2

    37 cases plotted. The bounds of UPCSCR3 are defined as follows:

1: -3.958- -3.322 2: -3.322- -2.686 3: -2.686- -2.049
4: -2.049- -1.413 5: -1.413-   -.777 6:   -.777-   -.141
7:   -.141-   .495 8:   .495-   1.131 9:   1.131-   1.767

F I G. 9.14

Pages 9–13, 16, 19, 22, and 25 of computer printout, Example 9.3

EX. 9.3 - CLUSTER ANALYSIS ON ORANGE JUICE SAMPLES   Page 25

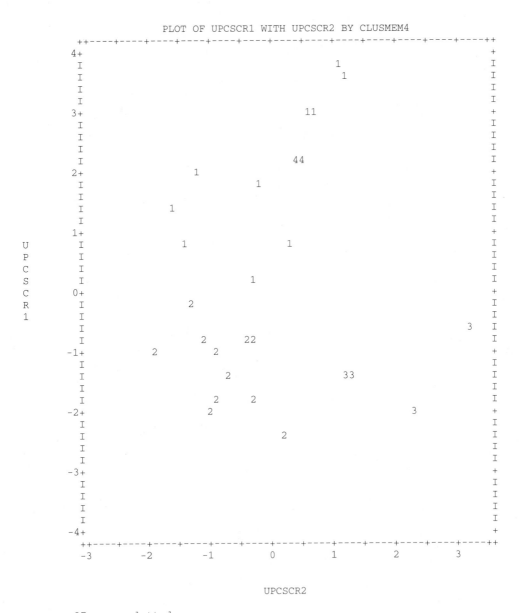

PLOT OF UPCSCR1 WITH UPCSCR2 BY CLUSMEM4

UPCSCR2

37 cases plotted.

F I G. **9.14**

Pages 9–13, 16, 19, 22, and 25 of computer printout, Example 9.3

SAS's FASTCLUS Procedure

SAS® states that "the FASTCLUS procedure is designed for disjoint clustering of very large data sets and can find good clusters with only two or three passes over the data."

The FASTCLUS procedure offers two basic options for determining clusters. In one option, the user must specify the maximum number of clusters to be created. Because a user generally will not know exactly how many clusters there should be, the user will generally want to run the FASTCLUS procedure several times by specifying a different value for the maximum number of clusters each time. Another option requires the user to estimate the "size" of the clusters to be formed. This is done by specifying the minimum "radius" of the clusters. Again a user may not generally know a good radius to choose, so the user will generally want to run the FASTCLUS procedure several times by specifying a different radius each time.

The FASTCLUS procedure operates in four steps. First, observations called *cluster seeds* are selected. Second, temporary clusters are formed by assigning each observation to the cluster with the nearest seed. If the DRIFT option is selected, the cluster seed is changed each time an observation is assigned to a cluster. In this case the cluster seed is taken to be the mean of the observations in the cluster. Third, after all observations are assigned, the initially selected cluster seeds are changed to the cluster means, and the process starts all over again. These latter two steps are repeated until the changes in the cluster seeds become small or zero. Four, final clusters are formed by assigning each observation to the cluster with the nearest seed.

One unfortunate drawback to this procedure is that the clusters formed are influenced by the order in which the data are read into the computer. That is, two or more researchers could take the same set of data and find entirely different clusters using this procedure. Because of this, it is recommended that the procedure be performed several times on a given data set by reordering the data according to some random procedure in each run. In FASTCLUS this is easy to do by using a RANDOM option that has been built into the procedure.

E X A M P L E 9.4

This example illustrates nonhierarchical clustering. The data used are Fisher's iris data. The data consist of measurements of petal length (PL), petal width (PW), stamen length (SL), and stamen width (SW) on each of 150 iris plants. The data set contains 50 plants of each of three varieties denoted by IS, IC, and IV. This analysis ignores knowledge of the varieties except to compare results from the cluster procedures to the actual varieties from which the plants are known to have come.

The SAS statements used in this example can be found on the enclosed disk in the file named EX9_4.SAS. The SAS command statements are described next.

The following statements are used to read the data into the computer and print the data:

```
TITLE 'EX. 9.4 - Cluster Analyses of Fisher''s Iris Data';

DATA IRIS;
 INPUT (IRIS SL SW PL PW) (@27 $3. 4*8.0);
 IF IRIS='IS' THEN VARIETY='S';
 IF IRIS='IC' THEN VARIETY='C';
 IF IRIS='IV' THEN VARIETY='V';
CARDS;
                      1     IS    5.1    3.5    1.4    0.2
                      2     IS    4.9    3.0    1.4    0.2
                      .     .     .      .      .      .
                      .     .     .      .      .      .
                      .     .     .      .      .      .
                     50     IS    5.0    3.3    1.4    0.2
                     51     IC    7.0    3.2    4.7    1.4
                     52     IC    6.4    3.2    4.5    1.5
                      .     .     .      .      .      .
                      .     .     .      .      .      .
                      .     .     .      .      .      .
                    100     IC    5.7    2.8    4.1    1.3
                    101     IV    6.3    3.3    6.0    2.5
                    102     IV    5.8    2.7    5.1    1.9
                      .     .     .      .      .      .
                      .     .     .      .      .      .
                      .     .     .      .      .      .
                    150     IV    5.9    3.0    5.1    1.8
;
PROC PRINT;
 TITLE 2 'A Print of the Data';
 RUN;
```

The following statements standardize the raw data and create Z scores for each of the iris plants. The FASTCLUS procedure does not have an option that automatically standardizes the data. In these data, petal length and petal width have much less variability than stamen length and stamen width. Standardizing the data is necessary, if we want all of the measured variables to contribute equally to the formation of the clusters.

```
PROC STANDARD MEAN=0 STD=1;
 VAR SL SW PL PW;
 RUN;
```

FIG. 9.15

Pages 5–22 of
computer print-
out, Example 9.4

```
Ex. 9.4 - Cluster Analyses of Fisher's Iris Data        5
                    A Print of the Data

          NAME          MEAN            STD           N

          SL       5.8433333333    0.828066128       150
          SW       3.0573333333    0.4358662849      150
          PL          3.758        1.7652982333      150
          PW       1.1993333333    0.762237669       150
```

The following commands perform a principal components analysis on the standardized data and create a scatter plot of the first two principal component scores.

```
PROC PRINCOMP OUT=SCRS;
 VAR SL SW PL PW;
 TITLE2 'A PCA on the Iris Data';
 RUN;

PROC PLOT;
 PLOT PRIN2*PRIN1='*'/VAXIS=-4 TO 4 BY 2 HAXIS=-4 TO 4 BY 2
   VPOS=35 HPOS=60;
 TITLE2 ' A plot of the first two principal component
scores';
 RUN;
```

A portion of the results of the preceding SAS commands can be found on the computer printout pages shown in Figure 9.15. All of the results can be seen by executing the file named EX9_4.SAS on the enclosed disk.

The first four pages of the computer output give a print of the data being analyzed and are not reproduced here. Page 5 was created by the STANDARD procedure and gives the means and standard deviations of each of the measured variables in the data set. Page 6 contains some of the results from the principal components analysis. You can see that the first two eigenvalues account for almost 96% of the total variability in the measured variables. Thus for all practical purposes, the standardized data essentially fall onto a plane within the four-dimensional sample space.

A plot of the first two principal component scores is shown on page 7. Take some time to study this plot. If you were asked to define clusters for these data based on the plot, how many clusters do you think there are? Where do you think they lie? With a light pencil, you might want to identify clusters to compare your classifications with those that will be obtained by the cluster procedures. Suppose you were told that there are three clusters; where would your clusters lie?

The following commands are used to create three clusters from the iris data by inputting the data in the same order that it occurs in the data set, letting the FASTCLUS procedure select the initial cluster seeds, and passing

```
        Ex. 9.4 - Cluster Analyses of Fisher's Iris Data          6
                    A PCA on the Iris Data

              Principal Component Analysis

  150 Observations
    4 Variables

                    Correlation Matrix

              SL          SW          PL          PW

    SL     1.0000      -.1176      0.8718      0.8179
    SW     -.1176      1.0000      -.4284      -.3661
    PL     0.8718      -.4284      1.0000      0.9629
    PW     0.8179      -.3661      0.9629      1.0000


     Eigenvalues of the Correlation Matrix

           Eigenvalue    Difference    Proportion    Cumulative

  PRIN1      2.91850       2.00447      0.729624      0.72962
  PRIN2      0.91403       0.76727      0.228508      0.95813
  PRIN3      0.14676       0.12604      0.036689      0.99482
  PRIN4      0.02071          .         0.005179      1.00000
```

F I G. **9.15**
Pages 5–22 of computer printout, Example 9.4

through the data once. The OUT=CLUSTER option creates a new data set named CLUSTER. This data set contains two new variables, one named CLUSTER, which shows the cluster to which a data point has been assigned, and one named DISTANCE, which shows the distance each data point is from its cluster mean. These distances can be squared and summed to find a portion of the information needed to perform Beale's pseudo F test.

```
PROC FASTCLUS OUT=CLUSTER MAXCLUSTERS=3;
 VAR SL SW PL PW;
 TITLE2 'Obtaining three clusters with the data in the order
     given';
 RUN;
```

The following plot commands create a scatter plot that shows how the computer formed clusters:

```
PROC PLOT;
 PLOT PRIN2*PRIN1=CLUSTER/VAXIS=-4 TO 4 BY 2 HAXIS=-4 TO 4
BY    2 VPOS=35 HPOS=60;
 TITLE3 'A plot of the three clusters obtained';
 RUN;
```

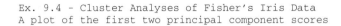

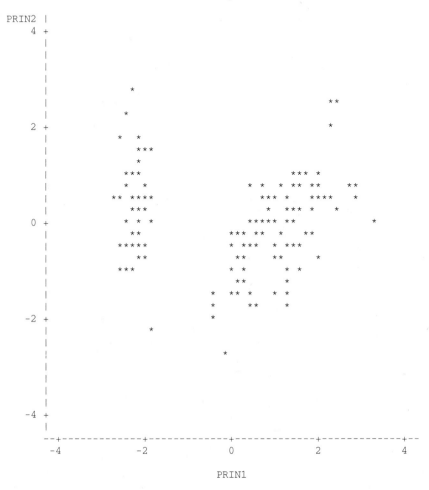

NOTE: 33 obs hidden.

F I G. **9.15**

Pages 5–22 of computer printout, Example 9.4

The following commands create Beale's sum of squares of within cluster distances computed from the cluster means for the clustering obtained by the procedure:

```
PROC MEANS USS; VAR DISTANCE;
TITLE3 'Beale''s intracluster residual sum of squares';
 RUN;
```

Ex. 9.4 - Cluster Analyses of Fisher's Iris Data   8
Obtaining three clusters with the data in the order given

FASTCLUS Procedure

Replace=FULL   Radius=0   Maxclusters=3   Maxiter=1

Initial Seeds

Cluster	SL	SW	PL	PW
1	2.24217	-1.04925	1.77987	1.44399
2	-1.62225	-1.73754	-1.39240	-1.17986
3	-0.17309	3.08046	-1.27910	-1.04867

Cluster Summary

Cluster	Frequency	RMS Std Deviation	Maximum Distance from Seed to Observation	Nearest Cluster	Centroid Distance
1	75	0.5644	2.6275	2	2.3679
2	36	0.6514	1.9424	3	2.3387
3	39	0.4032	1.8118	2	2.3387

Cluster Means

Cluster	SL	SW	PL	PW
1	0.81555	-0.21719	0.83499	0.83176
2	-0.72659	-0.80070	-0.33970	-0.39635
3	-0.89767	1.15679	-1.29218	-1.23368

FIG. 9.15

Pages 5–22 of computer printout, Example 9.4

The results from the preceding commands are shown on pages 8–11 of the computer output of Figure 9.15. Near the top of page 8, we find the initial cluster seed, and on the bottom of page 8, we find the final cluster means. On page 9, we see the standard deviations of the variables within each of the clusters.

The portion of the output on page 8 labeled `Cluster Summary` gives some information about the clusters formed. The `Frequency` column shows the number of data points in each cluster, and the column labeled `RMS Std Deviation` gives the square root of the average of the observed variances of the four response variables for each cluster. This number gives the user some information about the relative sizes of the radii of the clusters formed, and it may help a researcher decide whether clusters should be split. The number is not quite as useful in those cases where the response variables being used in the clustering process are correlated with

```
         Ex. 9.4 - Cluster Analyses of Fisher's Iris Data              9
        Obtaining three clusters with the data in the order given

                      Cluster Standard Deviations

    Cluster          SL            SW            PL            PW
    --------------------------------------------------------------
       1          0.650843      0.698312      0.371716      0.474168
       2          0.575461      0.683626      0.693178      0.646894
       3          0.388858      0.683469      0.101884      0.146579
```

```
PROC FASTCLUS OUT=CLUSTER MAXCLUSTERS=3 RANDOM=2342901 DRIFT MAXITER=3;
 VAR SL SW PL PW;
 TITLE2 'Obtaining three clusters with the data in a random
     order';
 TITLE3;
 RUN;

PROC PLOT;
 PLOT PRIN2*PRIN1=CLUSTER/VAXIS=-4 TO 4 BY 2 HAXIS=-4 TO 4 BY
     2 VPOS=35 HPOS=60;
 TITLE3 'A plot of the three clusters obtained';
 RUN;

PROC MEANS USS; VAR DISTANCE;
 TITLE3 'Beale''s intracluster residual sum of squares';
 RUN;
```

F I G. 9.15

Pages 5–22 of computer printout, Example 9.4

one another as it would be in those cases where the response variables are uncorrelated with one another. For this example, the clusters seem to be similar in size (the amount of space they take up) even though they have different numbers of data points in them.

The fourth column shows the distance between the cluster mean and the point in the cluster that is furthest from the cluster mean, the column labeled Nearest Cluster shows the cluster whose mean is closest to that of the given cluster, and the last column gives the distance between these two cluster means. These may help users determine whether certain clusters should be combined.

Page 10 shows a scatter plot of the first two principal component scores for these data. The symbol plotted is equal to the cluster to which the plotted data point was assigned. How do these clusters compare with the ones you were asked to identify from the scatter plot on page 7? Where do they agree and where do they disagree? Are you happy with the results of the clustering program? Do you agree with the results?

The value of Beale's sum of squares of within cluster distances is shown on page 11. This will be used to compare these results to those of another clustering performed later in this section.

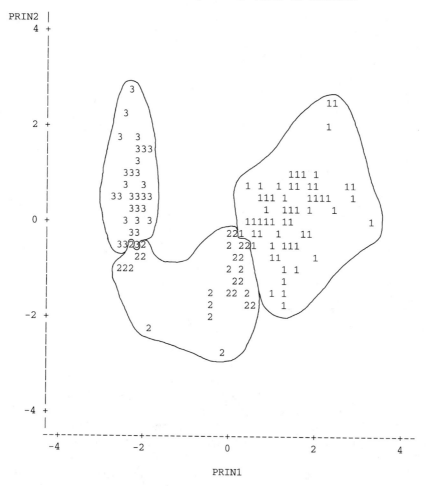

Ex. 9.4 - Cluster Analyses of Fisher's Iris Data   10
Obtaining three clusters with the data in the order given
A plot of the three clusters obtained

Plot of PRIN2*PRIN1.   Symbol is value of CLUSTER.

NOTE: 33 obs hidden.

F I G. **9.15**

Pages 5–22 of computer printout, Example 9.4

We now produce another clustering by adding options RAN-DOM=2342901, DRIFT, and MAXITER=3. Three clusters are requested, but the order of the data is randomized prior to performing the clustering. In addition, the options request that the data be passed through the cluster-ing cycle three times rather than once as was the case in the previous clus-

```
       Ex. 9.4 - Cluster Analyses of Fisher's Iris Data          11
    Obtaining three clusters with the data in the order given
           Beale's intracluster residual sum of squares


    Analysis Variable : DISTANCE Distance to Cluster Seed



                   N Obs              USS
               --------------------------
                   150       185.5382728
               --------------------------
```

F I G. 9.15
Pages 5–22 of computer printout, Example 9.4

```
          Ex. 9.4 - Cluster Analyses of Fisher's Iris Data        12
       Obtaining three clusters with the data in a random order


                         FASTCLUS Procedure

   Replace=RANDOM  Drift  Radius=0  Maxclusters=3   Maxiter=3  Converge=0.02


                           Initial Seeds

     Cluster          SL            SW            PL            PW
     ----------------------------------------------------------------
        1        -0.17309       3.08046      -1.27910      -1.04867
        2        -1.13920      -1.27868       0.42033       0.65684
        3         0.55149      -1.27868       0.64692       0.39445


            Minimum Distance Between Seeds = 2.027692
```

F I G. 9.15
Pages 5–22 of computer printout, Example 9.4

tering attempt. Once again, the clusters obtained are illustrated with the PLOT procedure and Beale's intracluster residual sum of squares is computed by the MEANS procedure.

A portion of the results from the previous analysis are shown on pages 12–16 of the Figure 9.15 computer output. Page 12 lists the initial cluster seeds. Notice the difference between these seeds and those that were on page 8 of the initial run. On the top of page 13, we can see how far the cluster seeds moved in each of the three iterations. The remaining results on page 12 as well as those on page 13 are similar to those obtained in the previous run except that clusters formed are more equal in size both with respect to the numbers within each cluster and with respect to their root mean square standard deviations. A plot of the principal component scores showing the clusters to which the data points were assigned is shown on page 15 and Beale's intracluster distance sum of squares is shown on page

```
        Ex. 9.4 - Cluster Analyses of Fisher's Iris Data        13
    Obtaining three clusters with the data in a random order

        Iteration    Change in Cluster Seeds
                             1            2            3
        ----------------------------------------------------
               1     0.046749     0.212161     0.119462
               2            0     0.173718     0.102476
               3            0     0.154844     0.107505


            Iteration Limit Exceeded


                    Cluster Summary
```

Cluster	Frequency	RMS Std Deviation	Maximum Distance from Seed to Observation	Nearest Cluster	Centroid Distance
1	49	0.4571	2.3375	2	2.9519
2	44	0.4854	2.5832	3	1.8703
3	57	0.5162	2.4058	2	1.8703

```
                    Cluster Means
```

Cluster	SL	SW	PL	PW
1	-0.99872	0.90323	-1.29876	-1.25215
2	-0.19780	-1.00232	0.25296	0.17381
3	1.01123	-0.00274	0.92121	0.94224

F I G. **9.15**
Pages 5–22 of computer printout, Example 9.4

```
        Ex. 9.4 - Cluster Analyses of Fisher's Iris Data        14
    Obtaining three clusters with the data in a random order

                Cluster Standard Deviations
```

Cluster	SL	SW	PL	PW
1	0.420760	0.793551	0.098491	0.139309
2	0.497974	0.594951	0.406469	0.418492
3	0.611717	0.588162	0.368177	0.458587

F I G. **9.15**
Pages 5–22 of computer printout, Example 9.4

16. How do these clusters compare with the ones you were asked to identify from the scatter plot on page 7? How do they compare with those on page 10. Where do they agree and where do they disagree? Are you happier with the results of this clustering than with the previous one? Do you agree with these results?

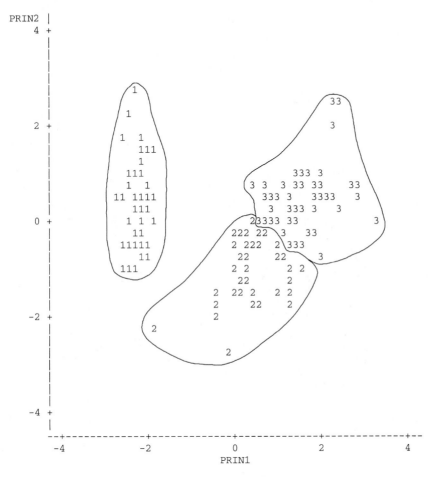

Ex. 9.4 - Cluster Analyses of Fisher's Iris Data   15
Obtaining three clusters with the data in a random order
A plot of the three clusters obtained

Plot of PRIN2*PRIN1. Symbol is value of CLUSTER.

NOTE: 33 obs hidden.

FIG. 9.15
Pages 5–22 of computer printout, Example 9.4

The following commands were used to produce a third clustering of the iris data. This time an option was used to restrict the size (area) of the clusters by using the RADIUS=2 option rather than specifying the number of clusters as was done in the previous two runs. Once again a plot of the principal component scores was created showing the assignment of points to clusters and Beale's intracluster sum of squared distances was computed.

```
        Ex. 9.4 - Cluster Analyses of Fisher's Iris Data              16
     Obtaining three clusters with the data in a random order
            Beale's intracluster residual sum of squares

       Analysis Variable : DISTANCE Distance to Cluster Seed


                N Obs              USS
                ------------------------
                 150    141.2351169
                ------------------------
```

F I G. 9.15
Pages 5–22 of computer printout, Example 9.4

```
          Ex. 9.4 - Cluster Analyses of Fisher's Iris Data            17
       A cluster analysis by specifying the maximum cluster radii

                        FASTCLUS Procedure

   Replace=FULL  Drift  Radius=2  Maxclusters=100   Maxiter=3  Converge=0.02

                        Initial Seeds

       Cluster        SL            SW           PL           PW
       ------------------------------------------------------------
          1        -0.89767       1.01560     -1.33575     -1.31105
          2        -0.17309       3.08046     -1.27910     -1.04867
          3        -1.62225      -1.73754     -1.39240     -1.17986
          4         1.39683       0.32732      0.53362      0.26326
          5        -0.41462      -1.73754      0.13709      0.13207
          6         2.24217       1.70389      1.66657      1.31280
          7         2.24217      -1.04925      1.77987      1.44399

            Minimum Distance Between Seeds = 1.369874
```

F I G. 9.15
Pages 5–22 of computer printout, Example 9.4

```
PROC FASTCLUS OUT=CLUSTER RADIUS=2 DRIFT MAXITER=3;
VAR SL SW PL PW;
TITLE2 'A cluster analysis by specifying the maximum
cluster   radii';
TITLE3;
RUN;
```

```
PROC PLOT;
PLOT PRIN2*PRIN1=CLUSTER/VAXIS=-4 TO 4 BY 2 HAXIS=-4 TO 4
BY   2 VPOS=35 HPOS=60;
TITLE3 'A plot of the clusters obtained';
RUN;
```

Iteration	Change in Cluster Seeds			
	1	2	3	4
		5	6	7
1	0	0.08626	0	0.065958
		0	0.065561	0.188199
2	0	0	0	0.063156
		0	0	0.135556
3	0	0	0	0.06489
		0.041438	0	0.082341

Iteration Limit Exceeded

Cluster Summary

Cluster	Frequency	RMS Std Deviation	Maximum Distance from Seed to Observation	Nearest Cluster	Centroid Distance
1	36	0.3056	1.0656	2	1.5418
2	13	0.3007	1.2076	1	1.5418
3	1	.	0	1	2.3139
4	47	0.3861	1.2584	7	1.3053
5	25	0.4048	1.3864	4	1.4014
6	3	0.3153	0.6915	7	1.7617
7	25	0.4151	1.5937	4	1.3053

F I G. 9.15

Pages 5–22 of computer printout, Example 9.4

Cluster Means

Cluster	SL	SW	PL	PW
1	-1.15262	0.52488	-1.31057	-1.27097
2	-0.57254	1.95096	-1.26603	-1.20004
3	-1.62225	-1.73754	-1.39240	-1.17986
4	0.43843	-0.33656	0.59871	0.53681
5	-0.33733	-1.29703	0.13935	0.05860
6	2.12141	1.55093	1.49663	1.35653
7	1.28090	0.04283	1.15675	1.27082

Cluster Standard Deviations

Cluster	SL	SW	PL	PW
1	0.346148	0.475445	0.096177	0.136049
2	0.301443	0.490833	0.101213	0.140138
3	.	.	.	.
4	0.454935	0.456942	0.245214	0.346920
5	0.458371	0.512675	0.281572	0.321534
6	0.435418	0.264921	0.169943	0.330161
7	0.564797	0.460193	0.267206	0.295111

F I G. 9.15

Pages 5–22 of computer printout, Example 9.4

Ex. 9.4 - Cluster Analyses of Fisher's Iris Data   20
A cluster analysis by specifying the maximum cluster radii
A plot of the clusters obtained

Plot of PRIN2*PRIN1. Symbol is value of CLUSTER.

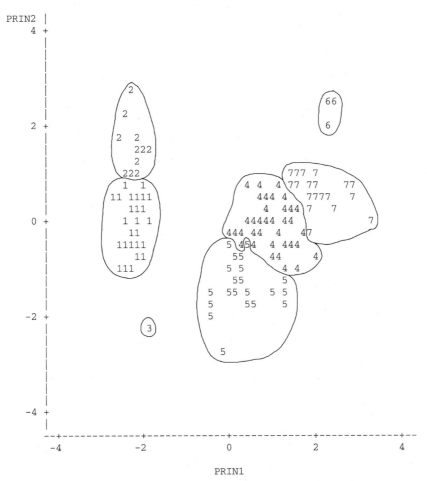

NOTE: 33 obs hidden.

F I G. 9.15

Pages 5–22 of computer printout, Example 9.4

```
       Ex. 9.4 - Cluster Analyses of Fisher's Iris Data          21
     A cluster analysis by specifying the maximum cluster radii
          Beale's intracluster residual sum of squares

      Analysis Variable : DISTANCE Distance to Cluster Seed


                  N Obs              USS
              -------------------------
                  150      77.9691928
              -------------------------
```

F I G. 9.15
Pages 5–22 of computer printout, Example 9.4

```
PROC MEANS USS; VAR DISTANCE;
 TITLE3 'Beale''s intracluster residual sum of squares';
 RUN;
```

A portion of the results of the preceding commands is shown on pages 17–21 of the computer printout. An examination of these results shows that seven clusters were formed. A plot of the principal component scores showing cluster assignments is shown on page 20. How do these clusters compare with the ones you were asked to identify from the scatter plot on page 7? How do they compare with those on pages 10 and 15? Where do they agree and where do they disagree? Are you happier with the results of this clustering program than with the previous ones? Do you agree with the results?

Next Beale's pseudo F statistic is computed to see if the seven-cluster solution is better than the previous three-cluster solution. From page 21 of the computer printout $W_1 = 77.969$ with $c_1 = 7$. From page 16, $W_2 = 141.235$ with $c_2 = 3$. Thus, Beale's pseudo F is

$$F^* = \frac{(W_2 - W_1)}{W_1} \cdot \frac{(N - c_1)k_1}{(N - c_2)k_2 - (N - c_1)k_1}$$

$$= \frac{141.235 - 77.969}{77.969} \cdot \frac{(150 - 7) \cdot 0.3780}{(150 - 3) \cdot 0.5774 - (150 - 7) \cdot 0.3780} = 1.4230$$

since $k_1 = c_1^{-2/p} = 7^{-2/4} = 0.3780$ and $k_2 = c_2^{-2/p} = 3^{-2/4} = 0.5774$. This would be compared to a critical point from the F distribution with numerator degrees of freedom equal to

$$k_2(N - c_2) - k_1(N - c_1) = (0.5774) \cdot 147 - (0.3780) \cdot 143 = 30.8$$

and denominator degrees of freedom equal to

$$k_1(N - c_1) = 54.1$$

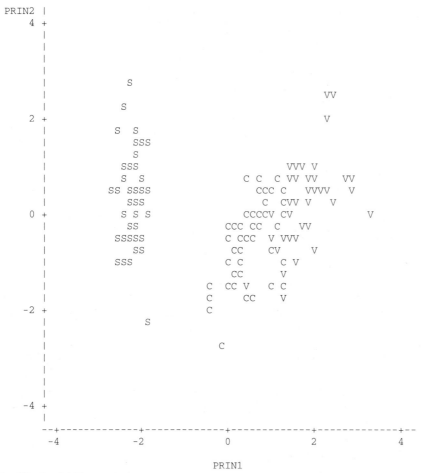

```
        Ex. 9.4 - Cluster Analyses of Fisher's Iris Data          22
             A cluster analysis by specifying the maximum cluster radii
                    A plot of the clusters obtained

          Plot of PRIN2*PRIN1.  Symbol is value of VARIETY.
```

NOTE: 33 obs hidden.

FIG. **9.15**

Pages 5–22 of computer printout, Example 9.4

The observed significance level of this test statistic is $P = 0.1277$, and we would conclude that the seven-cluster solution is not significantly better than the three-cluster solution.

The following commands produce a scatter plot of the principal component scores showing the three varieties from which these data actually come. The plot is shown on page 22 of the computer printout. How do the actual varieties compare with the clusters obtained from each of the three clusterings performed? Obviously, the second clustering comes closest to reality for these data.

```
PROC PLOT;
 PLOT PRIN2*PRIN1=VARIETY/VAXIS=-4 TO 4 BY 2
  HAXIS=-4 TO 4 BY 2 VPOS=35 HPOS=60;
 TITLE3 'A plot of the clusters obtained';
 RUN;
```

9.4
Multidimensional Scaling

Multidimensional scaling (MDS) is a mathematical technique that allows us to map the distances between points in a high dimensional space into a lower dimensional space. It is most useful when we can map distances from a high dimensional space into a two-dimensional space. In this case the data points can be plotted in the two-dimensional space, and we can examine the plot to see which points tend to fall close to one another. Consequently, multidimensional scaling can be used as another technique to use when we want to cluster observations into groups. An excellent account of many of the issues involved in multidimensional scaling is given by Young and Hamer (1987).

As an example, suppose that a producer of underarm deodorants wants to know who the closest competitors might be. This knowledge could be helpful in determining how to spend the company's advertising dollars wisely. The company might consider surveying a random sample of consumers of underarm deodorants and ask each consumer to rate each of the products on a number of criteria. For instance, consumers could evaluate deodorants as to fragrance, ability to control wetness, ability to stop odor, ease of application, and lack of leaving residue on clothes. You might compute means for each of the brands of deodorants for each of these criteria and then calculate the distances between all pairs of brand means. Multidimensional scaling might be used to create a plot that attempts to show the relative locations of the brands with respect to one another in a two-dimensional space.

To apply multidimensional scaling to a set of data points (possibly a set of group means), we must first calculate the distances between all pairs of points. One of the most reasonable distance measures to use is to standardize the data, and then use the standardized ruler distance formula as described in Section 9.1.

In the next section, mathematical formulas are given that reduce distances between points to a two-dimensional space. Note that these arguments can be extended to cases where we are attempting to reduce a p-dimensional data set to q dimensions where q is any number less than p. Because of a desire for simplicity, only the case where $q = 2$ is considered here.

Suppose D_{rs} represents the actual distance between the rth point and the sth point in the p-dimensional sample space for $r = 1, 2, \ldots, N$ and $s = 1, 2, \ldots, N$ where N is the total number of data points. If one lets $\mathbf{z}_r$ and $\mathbf{z}_s$ be vectors of Z scores corresponding to the rth and sth data points, respectively, then the standardized ruler distance between these two data points is given by

$$D_{rs} = [(\mathbf{z}_r - \mathbf{z}_s)'(\mathbf{z}_r - \mathbf{z}_s)]^{1/2}$$

These distances can be ordered from smallest to largest. Let $D_{r_1 s_1}$ be the distance between the two closest points, $D_{r_2 s_2}$ be the distance between the next two closest points, $\ldots$, and $D_{r_{N(N-1)/2} s_{N(N-1)/2}}$ be the distance between the two farthest points. Note that the number of distinct pairs of points is equal to $N(N-1)/2$ so there are $N(N-1)/2$ possible pairwise distances.

Next consider plotting N points in a two-dimensional space and let the distance between the rth and sth points in this two-dimensional space be denoted by d_{rs}. In this two-dimensional space, let $d_{r_1 s_1}$ be distance between the two closest points, $d_{r_2 s_2}$ be the distance between the next two closest points, $\ldots$, and $d_{r_{N(N-1)/2} s_{N(N-1)/2}}$ be the distance between the two farthest points. Multidimensional scaling tries to locate N points in a two-dimensional space so that the distances between the pairs of points in this space match as closely as possible with the true ordered distances between the observed points, namely,

$$D_{r_1 s_1} < D_{r_2 s_2} < \cdots < D_{r_{N(N-1)/2} s_{N(N-1)/2}}$$

To assess the quality of the fit, it is customary to plot the actual differences between pairs of points against their modeled distances. If the plot of these pairs of distances reveals a monotonically increasing trend, then we can deduce that the two-dimensional plot accurately illustrates the closeness of the pairs of points.

MDS programs attempt to locate the observed data points in a reduced dimensional space so that

$$E = \frac{\left[\sum_{r=1}^{N} \sum_{s=1}^{r-1} (D_{rs} - d_{rs})^2 / D_{rs} \right]}{\left[\sum_{r=1}^{N} \sum_{s=1}^{r-1} D_{rs} \right]}$$

is minimized where D_{rs} is the distance between the rth and sth observations and d_{rs} is the distance between those same two points in the reduced space.

E X A M P L E 9.5

The file labeled CARS.DAT on the enclosed disk contains information about a large number of late model cars. The variables that are contained in this data set are make and model (MAKE and MAKE_MOD), base price (PRICE), engine type (ET), engine horsepower (HP), time to reach a speed of 60 mph (TIM1), elapsed time for one-quarter mile (TIM2), top speed (TS), braking distance from 60 mph (BRAK1), braking distance from 80 mph (BRAK2), steady speed cornering grip (SP), speed through a slalom course (SS), and fuel consumption in miles per gallon (MPG). From this set of data, a subset of cars was selected so that there was one car selected from each of 35 manufacturers. The data utilized in this example can be found in the file labeled EX9_5.SAS. Generally, the car selected from each manufacturer was the manufacturer's most expensive model. These data were analyzed with the MDS procedure in SAS using the following commands:

```
OPTIONS LINESIZE=75 PAGESIZE=54 NODATE PAGENO=1;
TITLE 'EX9_5 - An Example using Multidimensional Scaling';
DATA CARS;
 INPUT #1 MAKE $ 1 - 9 MAKE_MOD $ 1 - 24
      #2 PRICE ET $ HP TIM1 TIM2 TS BRAK1 BRAK2 SP SS MPG;
CARDS;
Acura NSX
 68600 V-6   270    5.8  14.0 168  120   200    .87 62.3 18.0
Alfa   Romeo 164 Quad
 36690 V-6   230    7.6  15.8 150  150   268   0.77 59.3 17.5
Audi Cabriolet
 38950 V-6   172    9.7  17.1 130  140   254   0.82 57.6 22.6
Audi S4
 43750 I-5t  227    6.5  15.0 130  147   246   0.83 60.0 16.0
Bentley Turbo R
 174900 V-8t 315    7.1  15.5 128  147   253   0.73 56.8 11.5
BMW 850CSi
 98500 V-12  372    5.9  14.4 155  135   220   0.89 62.0 13.0
Bugatti EB110
 35000 V-12t 611    4.4  12.5 207  112   209   0.99 64.3 16.0
Buick Riviera
 27632 V-6s  225    7.9  16.0 114  139   254   0.75 57.9 19.0
Cadillac Eldorado Touring Coupe
 41535 V-8   300    7.2  15.3 150  138   249   0.78 61.7 16.0
Chevrolet Corvette ZR1
 65318 V-8   375    5.6  13.9 178  142   256   0.91 63.6 19.0
Chrysler LHS
 28254 V-6   214    9.1  16.8 112  153   275   0.73 55.0 19.0
Dinan M5 (BMW)
 60700 I-6   382    5.6  14.1 160  122   216   0.87 63.1 14.0
Dodge Viper RT/10
 50000 V-10  400    4.8  13.1 160  156   261   0.96 62.7 14.0
Ford Mustang Cobra
 23535 V-8   240    6.9  15.3 140  130   236   0.87 61.1 18.0
 .
 .
 .
```

```
PROC PRINT;
TITLE2 'PERFORMANCE DATA ON CARS';
RUN;

/* The following two commands are used to eliminate any cars
   that have any of the variables HP, TIM1, ..., MPG missing.
*/

DATA CARS2;
 CHK=HP+TIM1+TIM2+TS+BRAK1+BRAK2+SP+SS+MPG;
 IF CHK=. THEN DELETE; DROP CHK;

PROC PRINT DATA=CARS2;
RUN;

DATA CARS3; SET CARS2; DROP MAKE MAKE_MOD PRICE ET;
DATA CARS4; SET CARS2; KEEP MAKE;

PROC STANDARD DATA=CARS2 MEAN=0 STD=1 OUT=CARS2;
 VAR HP--MPG;
 RUN;

/* The next group of commands is used to compute the
pairwise distances H between all pairs of cars. */

DATA ONE; SET CARS2;
 I = _N_;
 KEEP HP--MPG I;

DATA ORIG; SET ONE;
 DO J=1 TO 34;
 OUTPUT; END;

DATA DUP; SET ORIG;
II=J;  JJ=I;  I=II;  J=JJ;  DROP JJ II;
Y1=HP; Y2=TIM1; Y3=TIM2; Y4=TS; Y5=BRAK1; Y6=BRAK2; Y7=SP;
Y8=SS; Y9=MPG;
NN = _N_;
DROP HP--MPG;

PROC SORT DATA=DUP; BY I J;

DATA COMB; MERGE ORIG DUP; BY I J;
DROP I J;

*IF _N_ = NN THEN DELETE;

H = SQRT((HP-Y1)**2 + (TIM1-Y2)**2 +(TIM2-Y3)**2 +
(TS-Y4)**2 +(BRAK1-Y5)**2 +(BRAK2-Y6)**2 +(SP-Y7)**2
+(SS-Y8)**2 +(MPG-Y9)**2);
KEEP H;
RUN;
```

```
/* The following commands are used to create a square
matrix of the distances computed above for use by the SAS
MDS procedure. */

PROC IML; RESET NOLOG;
 USE  COMB;
 READ ALL INTO DIST;
 USE CARS4;
 READ ALL VAR _CHAR_ INTO CARS;
 PRINT CARS;
 N=SQRT(NROW(DIST));
DD=SHAPE(DIST,N,N);
*PRINT 'DISTANCE MATRIX = ',DD;
CREATE DIST_DAT FROM DD[COLNAME=CARS];
APPEND FROM DD;
QUIT;

/* The following commands create the final input for the SAS
MDS procedure merging the distances with the car names. */

DATA FINAL;
MERGE CARS4 DIST_DAT;
RUN;

PROC PRINT;
TITLE3 'DISTANCES BETWEEN CAR MAKES';
RUN;

PROC MDS DATA=FINAL OUT=OUT OUTRES=RES SHAPE=SQUARE; ID
MAKE;
TITLE3 'MDS Analyses and Plots';
RUN;

PROC PRINT DATA=OUT;
 RUN;
PROC PLOT DATA=OUT; PLOT DIM2 * DIM1 $ MAKE; RUN;
```

A portion of the results of the preceding commands can be found on the computer printout pages shown in Figure 9.16. All of the results can be obtained by executing the file labeled EX9_5.SAS on the enclosed floppy disk. Pages 1 and 2 contain a printout of the data being analyzed. Pages 6–10 give a printout of the distances between the cars (only page 6 is shown). Page 11 contains the output produced automatically by the MDS procedure. The most important thing to note on this page is that the convergence criteria were satisfied. Page 12 shows the coordinates of the locations of the cars in a two-dimensional space, and, finally, the primary reason for conducting the multidimensional scaling analysis is to produce the plot shown on page 13. This is a plot of the coordinates shown on page

EX9_5 - An Example using Multidimensional Scaling

EX8_5 PERFORMANCE DATA ON CARS

OBS	MAKE	MAKE_MOD	PRICE	ET	HP	TIM1	TIM2	TS	BRAK1	BRAK2	SP	SS	MPG
1	Acura NSX	Acura NSX	68600	V-6	270	5.8	14.0	168	120	200	0.87	62.3	18.0
2	Alfa Rom	Alfa Romeo 164 Quad	36690	V-6	230	7.6	15.8	150	150	268	0.77	59.3	17.5
3	Audi S4	Audi S4	43750	I-5t	227	6.5	15.0	130	147	246	0.83	60.0	16.0
4	Bentley T	Bentley Turbo R	174900	V-8t	315	7.1	15.5	128	147	253	0.73	56.8	11.5
5	BMW 850CS	BMW 850CSi	98500	V-12	372	5.9	14.4	155	135	220	0.89	62.0	13.0
6	Bugatti E	Bugatti EB110	35000	V-12t	611	4.4	12.5	207	112	209	0.99	64.3	16.0
7	Buick Riv	Buick Riviera	27632	V-6s	225	7.9	16.0	114	139	254	0.75	57.9	19.0
8	Cadillac	Cadillac Eldorado Tourin	41535	V-8	300	7.2	15.3	150	138	249	0.78	61.7	16.0
9	Chevrolet	Chevrolet Corvette ZR1	65318	V-8	375	5.6	13.9	178	142	256	0.91	63.6	19.0
10	Chrysler	Chrysler LHS	28254	V-6	214	9.1	16.8	112	153	275	0.73	55.0	19.0
11	Dinan M5	Dinan M5 (BMW)	60700	I-6	382	5.6	14.1	160	122	216	0.87	63.1	14.0
12	Dodge Vip	Dodge Viper RT/10	50000	V-10	400	4.8	13.1	160	156	261	0.96	62.7	14.0
13	Ford Must	Ford Mustang Cobra	23535	V-8	240	6.9	15.3	140	130	236	0.87	61.1	18.0
14	Honda Pre	Honda Prelude VTEC	24500	I-4	190	7.1	15.4	139	137	242	0.84	61.7	24.0
15	Infiniti	Infiniti Q45	44000	V-8	278	7.5	15.6	150	159	272	0.81	59.5	18.3
16	Jaguar XJ	Jaguar XJR-S Coupe	73000	V-12	318	7.0	15.2	167	137	225	0.86	59.3	12.0
17	Lincoln M	Lincoln Mark VIII	38800	V-8	280	7.6	15.8	130	137	248	0.77	59.9	17.0
18	Lotus Esp	Lotus Esprit Turbo	67345	I-4t	264	5.3	13.7	167	121	225	0.86	60.6	16.0
19	Mazda Mil	Mazda Millenia S	31400	V-6	210	8.0	16.1	142	151	269	0.80	59.8	21.1
20	Mercedes	Mercedes 600SEC	132000	V-12	389	6.6	14.8	155	125	240	0.75	54.6	13.5
21	Mercury V	Mercury Villager	21798	V-6	151	13.4	19.4	115	178	293	0.76	55.4	19.0
22	Mitsubish	Mitsubishi 3000GT VR-4	40900	V-6t	320	5.7	14.2	159	122	218	0.86	63.7	16.3
23	Nissan 30	Nissan 300ZX Turbo	39500	V-6t	300	6.0	14.4	155	124	279	0.88	63.0	18.7
24	Olsmobile	Olsmobile Aurora	31995	V-8	250	8.6	16.5	135	131	244	0.81	57.3	18.0
25	Plymouth	Plymouth Grand Voyager L	22883	V-6	162	10.7	17.8	112	161	285	0.68	51.4	15.0
26	Pontiac F	Pontiac Firebird Formula	24279	V-8	275	6.7	15.1	150	123	279	0.85	60.4	18.0
27	Porsche 9	Porsche 911 Turbo 3.6	99000	F-6t	355	4.5	12.9	174	126	223	0.91	61.6	15.0
28	Saab 9000	Saab 9000 Aero	38235	I-4t	225	7.7	15.8	140	122	221	0.84	60.6	20.0
29	Saleen Mu	Saleen Mustang S-351 (Fo	33500	V-8	370	5.9	14.3	170	125	226	0.97	64.1	15.0
30	Subaru SV	Subaru SVX	25000	F-6	230	7.3	13.7	143	134	229	0.85	60.2	20.0
31	Toyota Su	Toyota Supra Turbo	43900	I-6t	320	5.3	13.7	155	122	208	0.98	65.4	18.9
32	Vector W8	Vector W8 TwinTurbo	448000	V-8t	625	4.2	12.0	218	145	250	0.97	60.6	13.5
33	Volkswage	Volkswagen Corrado SLC	21840	V-6	178	6.9	15.5	139	154	257	0.86	61.0	23.0
34	Volvo 850	Volvo 850 Turbo Sportswa	30985	I-5t	222	7.4	15.5	155	130	245	0.80	61.2	23.5

FIG. 9.16

Pages 1, 6, 11, 12, and 13 of computer printout, Example 9.5

```
        EX9_5 - An Example using Multidimensional Scaling          6
                    PERFORMANCE DATA ON CARS
                   DISTANCES BETWEEN CAR MAKES
```

OBS	MAKE	ACURA_NS	ALFA__RO	AUDI_S4	BENTLEY	BMW_850C	BUGATTI
1	Acura NSX	0.00000	4.23485	3.40093	4.86142	2.38311	4.2833
2	Alfa Rom	4.23485	0.00000	1.77467	2.59301	3.72810	7.1009
3	Audi S4	3.40093	1.77467	0.00000	2.42033	2.65289	6.4896
4	Bentley T	4.86142	2.59301	2.42033	0.00000	3.53591	7.3232
5	BMW 850CS	2.38311	3.72810	2.65289	3.53591	0.00000	4.2496
6	Bugatti E	4.28335	7.10095	6.48957	7.32320	4.24963	0.0000
7	Buick Riv	4.45774	1.94103	2.09744	2.77585	4.26736	7.7392

OBS	BUICK_RI	CADILLAC	CHEVROLE	CHRYSLER	DINAN_M5	DODGE_VI	FORD_MUS
1	4.45774	3.07570	3.05244	6.05175	1.85211	4.19465	2.34424
2	1.94103	1.65555	3.49230	2.52556	4.18307	4.16190	2.46162
3	2.09744	1.57724	3.24919	3.32892	3.23409	3.40641	1.60459
4	2.77585	2.52725	4.79344	3.25109	4.14456	4.51979	3.49268
5	4.26736	2.41731	2.76137	5.59385	1.10233	2.67268	2.43658
6	7.73921	5.86035	4.21100	9.06394	3.70762	4.73063	5.69898
7	0.00000	2.42477	4.56169	1.83731	4.57991	5.21367	2.50593

OBS	HONDA_PR	INFINITI	JAGUAR_X	LINCOLN	LOTUS_ES	MAZDA_MI	MERCEDES
1	3.43353	4.46577	2.90278	3.57466	1.39536	4.48674	3.96931
2	2.87242	0.97749	3.14888	1.56753	3.61590	1.32974	3.36609
3	2.81014	1.97072	2.52154	1.44330	2.79671	2.32006	3.24746
4	4.80062	2.99213	2.84092	2.27184	4.04173	3.68707	2.36212
5	4.25848	3.76869	1.51153	3.09351	1.96339	4.39422	3.24401
6	6.78167	6.95329	5.13725	6.65290	4.41958	7.47299	5.98927
7	2.70957	2.44667	3.88828	1.32773	4.04721	1.93146	3.48186

OBS	MERCURY	MITSUBIS	NISSAN_3	OLSMOBIL	PLYMOUTH	PONTIAC	PORSCHE	SAAB_900
1	8.6689	1.22118	3.37958	3.80845	7.7806	3.55740	2.03000	2.42400
2	5.0714	3.82078	3.01237	2.07902	4.1900	2.31956	4.46047	3.02966
3	6.0718	2.92819	2.87095	2.22174	5.0333	2.43292	3.54646	2.55805
4	5.9478	4.28323	4.36387	2.89701	4.0587	3.61360	4.67451	4.09432
5	8.0650	1.63045	3.24098	3.66619	7.0601	3.33022	1.79366	3.27282
6	11.4603	4.11738	5.28628	6.96994	10.6836	5.90958	3.31718	6.10607
7	5.0436	4.16519	3.67652	1.52157	3.8907	2.83451	5.08203	2.57929

OBS	SALEEN_M	SUBARU_S	TOYOTA_S	VECTOR_W	VOLKSWAG	VOLVO_85
1	2.26996	2.48366	1.96732	5.4333	4.16061	3.22145
2	4.47300	2.40114	5.06753	6.4508	2.39142	2.66376
3	3.61227	1.88920	4.01264	6.1455	2.50875	2.98947
4	4.95062	3.73270	5.82588	6.5523	4.53257	4.61929
5	1.69391	2.99290	2.76788	4.4375	4.39228	4.15837
6	3.35502	6.02815	3.83289	3.2197	7.21533	6.46568
7	5.10927	2.30370	5.30333	7.5795	2.73253	2.71955

F I G. **9.16**

Pages 1, 6, 11, 12, and 13 of computer printout, Example 9.5

Multidimensional Scaling: Data=WORK.FINAL
Shape=SQUARE Cond=MATRIX Level=ORDINAL Coef=IDENTITY Dim=2 Formula=1 Fit=1

Mconverge=0.01 Gconverge=0.01 Maxiter=100 Over=2 Ridge=0.0001

| | | Badness-of-Fit | Change in | Convergence Measures | |
Iteration	Type	Criterion	Criterion	Monotone	Gradient
0	Initial	0.145929	.	.	.
1	Monotone	0.115136	0.030793	0.083736	0.506053
2	Gau-New	0.098257	0.016879	.	.
3	Monotone	0.095073	0.003184	0.024235	0.198999
4	Gau-New	0.094357	0.000716	.	.
5	Monotone	0.091164	0.003194	0.023609	0.075566
6	Gau-New	0.091061	0.000102	.	.
7	Monotone	0.090862	0.000200	0.005960	0.047352
8	Gau-New	0.090755	0.000107	.	0.008989

Convergence criteria are satisfied.

F I G. 9.16

Pages 1, 6, 11, 12, and 13 of computer printout, Example 9.5

```
        EX9_5 - An Example using Multidimensional Scaling          12
                    PERFORMANCE DATA ON CARS
                    MDS Analyses and Plots
```

OBS	_DIMENS_	_MATRIX_	_TYPE_	MAKE	_NAME_	DIM1	DIM2
2	2	.	CONFIG	Acura NSX	ACURA_NS	-0.97296	-0.66553
3	2	.	CONFIG	Alfa Rom	ALFA__RO	0.81257	0.16103
4	2	.	CONFIG	Audi S4	AUDI_S4	0.31963	0.16961
5	2	.	CONFIG	Bentley T	BENTLEY	0.72261	1.09592
6	2	.	CONFIG	BMW 850CS	BMW_850C	-0.81755	0.33629
7	2	.	CONFIG	Bugatti E	BUGATTI	-2.86683	-0.00838
8	2	.	CONFIG	Buick Riv	BUICK_RI	1.12860	-0.09334
9	2	.	CONFIG	Cadillac	CADILLAC	0.10908	0.15486
10	2	.	CONFIG	Chevrolet	CHEVROLE	-0.83263	0.00758
11	2	.	CONFIG	Chrysler	CHRYSLER	1.86550	0.26183
12	2	.	CONFIG	Dinan M5	DINAN_M5	-1.08821	0.04042
13	2	.	CONFIG	Dodge Vip	DODGE_VI	-1.14881	0.97148
14	2	.	CONFIG	Ford Must	FORD_MUS	-0.02511	-0.31027
15	2	.	CONFIG	Honda Pre	HONDA_PR	0.51202	-0.97174
16	2	.	CONFIG	Infiniti	INFINITI	0.80907	0.25932
17	2	.	CONFIG	Jaguar XJ	JAGUAR_X	-0.40495	0.63661
18	2	.	CONFIG	Lincoln M	LINCOLN	0.54741	0.05515
19	2	.	CONFIG	Lotus Esp	LOTUS_ES	-0.75473	-0.22960
20	2	.	CONFIG	Mazda Mil	MAZDA_MI	1.00853	-0.39738
21	2	.	CONFIG	Mercedes	MERCEDES	0.09602	1.19978
22	2	.	CONFIG	Mercury V	MERCURY	3.46176	0.11197
23	2	.	CONFIG	Mitsubish	MITSUBIS	-0.85610	-0.34186
24	2	.	CONFIG	Nissan 30	NISSAN_3	-0.33942	-0.63629
25	2	.	CONFIG	Olsmobile	OLSMOBIL	0.80291	-0.10018
26	2	.	CONFIG	Plymouth	PLYMOUTH	2.76251	0.93502
27	2	.	CONFIG	Pontiac F	PONTIAC	0.12408	-0.15382
28	2	.	CONFIG	Porsche 9	PORSCHE	-1.33617	0.08043
29	2	.	CONFIG	Saab 9000	SAAB_900	0.13896	-0.72272
30	2	.	CONFIG	Saleen Mu	SALEEN_M	-1.30745	-0.16914
31	2	.	CONFIG	Subaru SV	SUBARU_S	0.18411	-0.47729
32	2	.	CONFIG	Toyota Su	TOYOTA_S	-1.38096	-0.82512
33	2	.	CONFIG	Vector W8	VECTOR_W	-2.50589	1.35025
34	2	.	CONFIG	Volkswage	VOLKSWAG	0.81341	-0.82575
35	2	.	CONFIG	Volvo 850	VOLVO_85	0.41900	-0.89915

F I G. 9.16

Pages 1, 6, 11, 12, and 13 of computer printout, Example 9.5

```
EX9_5 - An Example using Multidimensional Scaling
             PERFORMANCE DATA ON CARS
              MDS Analyses and Plots

     Plot of DIM2*DIM1$MAKE.   Symbol points to label.

D 1.5 +
i
m       |                > Vector W8
e 1.0 +
n       |                              > Mercedes
s       |                  > Dodge Vip         > Bentley T
i 0.5 +
o       |                          > Jaguar XJ
n       |
  0.0 +                     > BMW 850CS
2       |                              Cadillac    > Infiniti    >Chrysler
        |        > Bugatti E   Dinan M5<                ^ > Alfa  Rom
        |                   > Porsche 9          Audi S4 > Lincoln M
        |                       > Chevrolet   Olsmobile <    > Buick Riv
        |                          > Saleen Mu      > Pontiac F
        |                            > Lotus Esp
        |                       > Mitsubish > Ford Must
        |                                      > Subaru SV    > Mazda Mil
-0.5 +
        |         Acura NSX <       > Nissan 30
        |                                > Saab 9000
        |            > Toyota Su                  > Volkswage
        |                                    > Volvo 850
-1.0 +                                          > Honda Pre
        --+----------+----------+----------+----------+----------+----------+--
         -3         -2         -1          0          1          2          3          4
                                      Dimension 1
```

> Plymouth

>Mercury V

NOTE: 1 obs had missing values.

F I G. 9.16
Pages 1, 6, 11, 12, and 13 of computer printout, Example 9.5

12 along with a label that identifies each point. The > symbols give the location of the car and the symbol is plotted so that it points to the car being plotted at this location. From this plot, we can see that the Volvo 850 Turbo Sportswagon, the Honda Prelude VTEC, and the Volkswagen Corrado SLC are very close to one another. Likewise the Mercedes-Benz 600SEC and the Bentley Turbo R are very close to one another as are the Buick Riviera and the Oldsmobile Aurora. Occasionally, we may be able to identify the two dimensions in some meaningful way, but that does not seem possible here. Perhaps someone who is a real expert at identifying and understanding cars would be able to identify the dimensions.

Exercises

1 Use the distance matrix shown in Example 9.1 for this exercise.

    a Create the sequence of clusterings that would result if we were to use the furthest neighbor method.

    b Construct a hierarchical tree diagram for the clusterings in part a.

    c Repeat parts a and b using the average method.

2 Consider the 3-D plot of the first three PC scores for the U.S. Navy Bachelor Officers' Quarters shown in Figure 5.11. How many clusters do there appear to be? Which sites belong to each of the clusters?

3 Consider Example 9.2 and its computer output in Figure 9.4.

    a Identify the pizzas that would fall into each cluster for the case when there are five clusters.

    b What proportion of the total variability in the measured variables is accounted for by the first two principal components?

    c Make a copy of the plot shown in Figure 9.5. On this copy show the locations of the five clusters obtained in part a by drawing, if possible, circles about the clusters of points.

    d Consider the cluster history shown on page 4 of the computer printout and the row labeled NCL=18 where CL27 and CL35 are combined. Identify the pizzas that are in CL27 and CL35.

4 Consider Example 9.3 and its computer printout in Figure 9.14.

    a Consider the cluster formed at stage=23. Identify the orange juice samples that are in each of the clusters formed at this stage.

    b Identify the orange juice samples that would be in each cluster for the case when three clusters are formed.

    c Make a copy of the plot shown in Figure 9.13. On this copy show the locations of the three clusters from part b by drawing, if possible, circles about the clusters of points.

5 Perform a cluster analysis on another random sample of the pizza data by selecting a different random sample than the one used in Example 9.2. Write a short report summarizing what you have learned about this sample.

6 Perform a principal components analysis on the data used in Exercise 5 and plot the first two (unstandardized) principal component scores. Identify the clusters on this plot. Do you believe that the clusters identified by the clustering program are reasonable? How might you fine-tune the results of the clustering program?

7 Use at least two of the graphical techniques discussed in Chapter 3 to verify and validate the clusters found in Exercise 5.

8 Perform a cluster analysis on the data you provided for Exercise 1 of Chapter 1. Write a short report summarizing what you have learned about your data from the cluster analysis.

9 Consider the data on Big 8 football teams described in Exercise 7 of Chapter 1. Use multidimensional scaling to create a two-dimensional map that shows how Big 8 football teams compare with each other. Interpret your map and write a short report describing what you have learned.

10 Consider the data described in Appendix C.

    a Perform a cluster analysis on these data using only the responses to the Kansas Marital Satisfaction Scale. Validate, if possible, the results of your cluster analysis using some plotting techniques. Write a short report summarizing your results.

    b Perform a cluster analysis on these data using only the responses to the Kansas Family Life Satisfaction Scale. Validate, if possible, the results of your cluster analysis using some plotting techniques. Write a short report summarizing your results.

    c Perform a cluster analysis on these data using only the responses to the Family Adaptability Cohesion Evaluation Scale. Validate, if possible, the results of your cluster analysis using principal components analysis and some plotting techniques. Write a short report summarizing your results.

11 Consider the data described in Appendix B. Perform a cluster analysis on these data using only the responses to the job security questions (questions 34–53). Validate, if possible, the results of your cluster analysis using some plotting techniques. Write a short report summarizing your results.

12 Perform cluster analyses on the U.S. Navy Bachelor Officers' Quarters described in Example 5.2. Use the centroid method, Ward's method, and a third method of your choice. Write a report describing what you have learned from these analyses. Do the results of these analyses agree with your answers to Exercise 2 in this problem set?

Mean Vectors and
Variance–Covariance Matrices

10

Consider the survey of 304 married adults described in Appendix C. Researchers might propose several questions of interest when analyzing these data. For example, a researcher might want to know if there are differences between the males and females in this data set. In particular, consider individuals' responses to the Family Adaptability Cohesion Evaluation Scales (FACES). A researcher might wish to know if the relationships between responses are the same for both males and females. An answer to this question could be obtained by comparing the variance–covariance matrices of the two groups of individuals. A second question might concern whether the averages of the male responses to the questions in FACES are equal to the averages of the female responses to these questions. While we could compare males to females on a question-by-question basis, it is more appropriate to consider responses to all of the questions simultaneously. When we are analyzing many variables one at a time, the probability of finding significant differences by chance alone is very high. The only way to make sure that real differences actually exist is to analyze all response variables simultaneously. This chapter is concerned with multivariate methods that can be used to provide answers to these kinds of questions as well as some other types of questions.

This chapter considers multivariate generalizations of many univariate inference procedures that are usually taught in a first course in statistics. Multivariate generalizations of the one-sample t test, two-sample unpaired t test, and the two-sample paired t test are given, along with their corresponding confidence interval counterparts. Multivariate generalizations of chi-square tests on sample variances and F tests for comparing two sample variances are also given. Also included is a multivariate generalization of Bartlett's test for testing the equality of variances for more than two populations.

Most statistical computing packages do not contain routines for computing the test statistics discussed in this chapter, so most users will need to write their own programs using a matrix language software such as SAS-IML, GAUSS, MATLAB, or S+. In this text, SAS-IML is used.

10.1
Inference Procedures for Variance–Covariance Matrices

Procedures for testing hypotheses about variance–covariance matrices are given in this section.

Suppose $x_1, x_2, \ldots, x_N$ is a random sample from a univariate normal distribution, N (μ, σ^2). Let σ_0^2 be a user-specified positive constant. In a first course in statistics, students usually learn how to test $H_0 : \sigma^2 = \sigma_0^2$ versus $H_1 : \sigma^2 \neq \sigma_0^2$ using a chi-square test. The test requires us to compute

$$U = \frac{(N-1)\hat{\sigma}^2}{\sigma_0^2}$$

where

$$\hat{\sigma}^2 = \frac{1}{N-1} \sum_{i=1}^{N} (x_i - \overline{x})^2$$

and then H_0 is rejected if $U > \chi^2_{\alpha/2,N-1}$ or if $U < \chi^2_{1-\alpha/2,N-1}$ where $\chi^2_{\gamma,v}$ is the upper $\gamma \cdot 100\%$ critical point of the chi-square distribution with v degrees of freedom.

A Test for a Specific Variance–Covariance Matrix

Next consider a multivariate generalization of the preceding situation. Suppose $\mathbf{x}_1, \mathbf{x}_2, \ldots, \mathbf{x}_N$ is a random sample of vectors from a multivariate normal distribution, $N_p(\boldsymbol{\mu}, \boldsymbol{\Sigma})$. Consider testing $H_0 : \boldsymbol{\Sigma} = \boldsymbol{\Sigma}_0$ versus $H_a : \boldsymbol{\Sigma} \neq \boldsymbol{\Sigma}_0$ where $\boldsymbol{\Sigma}_0$ is a known (user-specified) variance–covariance matrix. Let $\mathbf{W} = (N - 1)\hat{\boldsymbol{\Sigma}}$, then calculate $L = -2 \cdot \log(\Lambda^*)$ where

$$\Lambda^* = (e/v)^{pv/2} \cdot |\boldsymbol{\Sigma}_0^{-1}\mathbf{W}|^{v/2} \cdot \exp[-(1/2)\,\text{tr}(\boldsymbol{\Sigma}_0^{-1}\mathbf{W})] \tag{10.1}$$

and $v = N - 1$. A formula for calculating an approximate p value for this test is given in Srivastava and Carter (1983, p. 325). Currently, no statistical software is available to compute this test statistic. The following example shows how SAS-IML can be used to compute the preceding test statistic as well as its approximate significance level using the results in Srivastava and Carter.

E X A M P L E 10.1

Consider the following data consisting of three responses on each of six experimental units:

ID	x_1	x_2	x_3
1	1	3	5
2	2	3	5
3	1	4	6
4	1	4	4
5	3	4	7
6	2	5	6

Now consider testing

$$H_0 : \boldsymbol{\Sigma} = \begin{bmatrix} 2 & 1 & 0 \\ 1 & 2 & 0 \\ 0 & 0 & 1 \end{bmatrix}$$

versus the alternative that H_0 is not true. The following SAS-IML commands, also given in the file named `EX10_1.IML` on the enclosed disk, produce the test statistic given in Eq. (10.1) and its approximate significance level:

```
OPTIONS LINESIZE=75 PAGESIZE=54 NODATE PAGENO=1;
TITLE 'Ex. 10.1 - A test on SIGMA, the variance-covariance
     matrix.';
DATA DAT;
 INPUT ID X1-X3;
 CARDS;
 1 1 3 5
 2 2 3 5
 3 1 4 6
 4 1 4 4
 5 3 4 7
 6 2 5 6
DATA XX; SET DAT; DROP ID;

PROC IML;
 RESET NOLOG;
 USE XX;   READ ALL INTO X;
 PRINT "The Data Matrix is" X;
 N=NROW(X);   P=NCOL(X);
 XBAR = X(|+,|)`/N;
 PRINT, "XBAR = " XBAR;
 SUMSQ=X`*X-(XBAR*XBAR`)#N;
 S=SUMSQ/(N-1);
 PRINT , "The Variance-Covariance Matrix is " S;
 NU=N-1;
 W=NU#S;
 SIG0 = {2 1 0, 1 2 0, 0 0 1};
 Q=INV(SIG0)*W;
 LAM_STAR = (EXP(1)/NU)##(P#NU/2)#
               (DET(Q))##(NU/2)#EXP(-.5#TRACE(Q));
 PRINT, "LAM_STAR = " LAM_STAR;
 L = -2#LOG(LAM_STAR);
 PRINT, "L = " L;
 B2 = P#(2#P#P+3#P-1)/24;
 B3 = -P#(P-1)#(P+1)#(P+2)/32;
 F=P#(P+1)/2;
 A1 = 1-PROBCHI(L,F);
 A2 = 1-PROBCHI(L,F+2);
 A3 = 1-PROBCHI(L,F+4);
 ALPHA = A1+(1/NU)#B2#(A2-A1)
        +(1/(6#NU#NU))#((3#B2#B2-4#B3)#A3-6#B2#B2#A2
        +(3#B2#B2+4#B3)#A1);
 PRINT, "ALPHA = ", ALPHA;
QUIT;
```

```
        Ex. 10.1 - A test on SIGMA, the variance-covariance matrix.        1

                                              X
                 The Data Matrix is           1          3          5
                                              2          3          5
                                              1          4          6
                                              1          4          4
                                              3          4          7
                                              2          5          6

                                           XBAR
                         XBAR =      1.6666667
                                     3.8333333
                                          5.5

                                                   S
        The Variance-Covariance Matrix is   0.6666667 0.1333333        0.6
                                            0.1333333 0.5666667        0.3
                                                  0.6       0.3        1.1

                                         LAM_STAR
                     LAM_STAR =      0.0162956

                                              L
                       L =           8.2337203

                                                 ALPHA
                            ALPHA = 0.3842768

        Exiting IML.
```

F I G. 10.1

Page 1 of computer printout, Example 10.1

The results of the preceding commands are shown in Figure 10.1. The top of page 1 shows a listing of the data matrix. Following the data matrix are values for $\bar{\mathbf{x}}$ and $\hat{\boldsymbol{\Sigma}}$. Given next are Λ^*, L, and its approximate significance level. H_0 would be rejected for small values of ALPHA; in this example, ALPHA=0.3843, and we cannot reject H_0.

A Test for Sphericity

Next consider testing $H_0: \boldsymbol{\Sigma} = \sigma^2 \mathbf{I}$ versus $H_a: \boldsymbol{\Sigma} \neq \sigma^2 \mathbf{I}$. This provides a test of the hypothesis that all measured variables are independently distributed (uncorrelated) and that they have the same variance, σ^2, which is assumed to be unknown. To understand why such a test may be important, suppose that white rats are given a specified dose of a drug. Then 2 hours later, suppose each rat is sacrificed, and the amounts of this drug deposited in several of the rat's organs are measured. Perhaps the amount of drug that occurs in each rat's heart, lung, kidney, and liver is measured. One question that should be of some interest is whether the amounts of the drug found in the various organs are distributed independently from organ to organ, and whether the amounts in these locations all have the same variance. This question can be answered by testing $H_0: \mathbf{\Sigma} = \sigma^2 \mathbf{I}$.

To test $H_0: \mathbf{\Sigma} = \sigma^2 \mathbf{I}$ versus $H_a: \mathbf{\Sigma} \neq \sigma^2 \mathbf{I}$, we calculate

$$\Lambda = |\mathbf{W}| / [(1/p)\, \mathrm{tr}(\mathbf{W})]^p \tag{10.2}$$

A formula for approximating a p value for this test statistic is given by Srivastava and Carter (1983, p. 327).

Currently, no statistical software is available for computing this test statistic. The following example, which uses the data from Example 10.1 and the p-value formulas in Srivastava and Carter, shows how SAS-IML can be used to compute the preceding test statistic and its approximate significance level.

E X A M P L E 10.2

Consider the data in Example 10.1, and test $H_0: \mathbf{\Sigma} = \sigma^2 \mathbf{I}$ versus $H_a: \mathbf{\Sigma} \neq \sigma^2 \mathbf{I}$. The following SAS-IML program produces the test of H_0. This program, named `EX10_2.IML`, can be found on the enclosed disk.

```
OPTIONS LINESIZE=75 PAGESIZE=54 NODATE PAGENO=1;
TITLE  'Ex. 10.2 - A test for sphericity of the
variance-covariance matrix.';
DATA DAT;
 INPUT ID X1-X3;
 CARDS;
 1 1 3 5
 2 2 3 5
 3 1 4 6
 4 1 4 4
 5 3 4 7
 6 2 5 6
DATA XX; SET DAT; DROP ID;
PROC IML;
 RESET NOLOG;
 USE XX;   READ ALL INTO X;
 PRINT, "The Data Matrix is" X;
N=NROW(X);   P=NCOL(X);
 XBAR = X(|+,|)`/N;
PRINT, "XBAR = " XBAR;
SUMSQ=X`*X-(XBAR*XBAR`)#N;
S=SUMSQ/(N-1);
PRINT , "The Variance-Covariance Matrix is " S;
NU=N-1;
```

```
                    Ex. 10.2 - A test for sphericity of the variance-covariance matrix.   1

                                              X
                    The Data Matrix is        1         3         5
                                              2         3         5
                                              1         4         6
                                              1         4         4
                                              3         4         7
                                              2         5         6

                                             XBAR
                             XBAR =    1.6666667
                                       3.8333333
                                             5.5

                                                            S
            The Variance-Covariance Matrix is    0.6666667 0.1333333       0.6
                                                 0.1333333 0.5666667       0.3
                                                       0.6       0.3       1.1

                                            LAMDA
                          LAMDA =    0.3825656

                                              L
                        L =         3.5765164

                                        ALPHA =

                                        ALPHA
                                     0.654943
        Exiting IML.
```

FIG. 10.2

Page 1 of computer printout, Example 10.2

```
W=NU#S;
LAMDA = DET(W)/((1/P)#TRACE(W))##P;
PRINT, "LAMDA = " LAMDA;
M = NU - (2#P#P+P+2)/6/P;
L = -M#LOG(LAMDA);
PRINT, "L = " L;
A = (P+1)#(P-1)#(P+2)#(2#P#P#P+6#P#P+3#P+2)/288/P/P;
F=P#(P+1)/2-1;
A1 = 1-PROBCHI(L,F);
A3 = 1-PROBCHI(L,F+4);
ALPHA = A1+(A/M/M)#(A3-A1);
PRINT, "ALPHA = ", ALPHA;
QUIT;
```

The results of the preceding commands are shown in Figure 10.2. First, we see the data matrix, $\bar{\mathbf{x}}$ and $\hat{\boldsymbol{\Sigma}}$. Next, we see Λ, L, and its corresponding significance level. We reject H_0 for small values of ALPHA and, for this example ALPHA=0.655 and H_0 cannot be rejected.

A Test for Compound Symmetry

A variance–covariance matrix is said to possess compound symmetry if it can be written in the form

$$
\Sigma = \sigma^2
\begin{bmatrix}
1 & \rho & \cdots & \rho \\
\rho & 1 & \cdots & \rho \\
\vdots & \vdots & \ddots & \vdots \\
\rho & \rho & \cdots & 1
\end{bmatrix}
$$

for some σ^2 and ρ. This special form requires that all variables have the same variance and that all pairs of variables be equally correlated with one another. Such a test is likely to be more interesting for the experimental situation involving rats described in the previous section than the test for sphericity, because we would not really expect the amounts of drug in the various organs of rats to be uncorrelated with one another. We might expect them to be equally variable and equally correlated.

This test is also of interest in repeated measures experiments. Suppose a drug is given to a large number of rats. Then suppose that the pulse rate of each rat is measured at different points in time, say, 5, 10, 15, 30, and 60 minutes after injecting the drug into the rat. A question of interest is whether the pulse rates have the same variance at all times of measurements and whether the correlations between time measurements are all equal. The answers to these questions require a test for compound symmetry of the variance–covariance matrix of time measurements.

To test for compound symmetry, we compute

$$
\Lambda = \frac{|\hat{\Sigma}|}{(s^2)^p (1 - r)^{p-1} [1 + (p - 1)r]}
$$

where

$$
s^2 = \left(\frac{1}{p}\right) \sum_{i=1}^{p} \hat{\sigma}_{ii}
\quad \text{and} \quad
r = \left[\frac{2}{p(p - 1)s^2}\right] \sum_{i<j}^{p} \hat{\sigma}_{ij}
$$

Compound symmetry is rejected for large N if $Q > \chi^2_{\alpha,f}$ where

$$
Q = -\frac{N - 1 - p(p + 1)^2 (2p - 3)}{6(p - 1)(p^2 + p - 4)} \log \Lambda
\quad \text{and} \quad
f = [p(p + 1) - 4]/2
$$

$$(10.3)$$

E X A M P L E 10.3

The following SAS-IML program uses the data in Example 10.1 to illustrate a test for compound symmetry. Furthermore, the program provides an approximate significance level for the test. The program is in the file labeled EX10_3.IML on the enclosed disk.

```
OPTIONS LINESIZE=75 PAGESIZE=54 NODATE PAGENO=1;
TITLE  'Ex. 10.3 - A test for compound symmetry of a var-cov
matrix.';
DATA DAT;
 INPUT ID X1-X3;
 CARDS;
 1 1 3 5
 2 2 3 5
 3 1 4 6
 4 1 4 4
 5 3 4 7
 6 2 5 6
DATA XX; SET DAT; DROP ID;


PROC IML;
 RESET NOLOG;
 USE XX;   READ ALL INTO X;
 PRINT, "The Data Matrix is" X;
 N=NROW(X);   P=NCOL(X);
 XBAR = X(|+,|)`/N;
PRINT, "XBAR = " XBAR;

SUMSQ=X` *X-(XBAR*XBAR)`)#N;
S=SUMSQ/(N-1);
PRINT, "The Variance-Covariance Matrix is " S;
S2 = (1/P)#TRACE(S);
R = (2/(P#(P-1)#S2))#(S(|+,+|)-TRACE(S))/2;
LAMDA = DET(S)/((S2##P)#((1-R)##(P-1))#(1+(P-1)#R));
Q= -(N-1-(P#(P+1)#(P+1)#(2#P-3))/
     (6#(P-1)#(P#P+P-4)))#LOG(LAMDA);
F=(P#(P+1)-4)/2;
ALPHA = 1 - PROBCHI(Q,F);
PRINT, "S2 = " S2, "R = " R, "LAMDA = " LAMDA, "Q = " Q;
PRINT, "DEGREES OF FREEDOM = " F;
PRINT, "ALPHA = " ALPHA;
QUIT;
```

The results from the preceding program statements are shown in Figure 10.3. As before, we first look at the data matrix, $\bar{x}$ and $\hat{\Sigma}$. Next we find s^2, r, Λ, Q, its degree of freedom, and its corresponding observed significance level. We reject for small values of ALPHA and, for this example, compound symmetry cannot be rejected.

```
Ex. 10.3 - A test for compound symmetry of a var-cov matrix.          1

                                      X
        The Data Matrix is            1          3          5
                                      2          3          5
                                      1          4          6
                                      1          4          4
                                      3          4          7
                                      2          5          6

                                   XBAR
                   XBAR =     1.6666667
                              3.8333333
                                    5.5

                                   S
  The Variance-Covariance Matrix is  0.6666667 0.1333333        0.6
                                     0.1333333 0.5666667        0.3
                                           0.6       0.3        1.1
                                  S.2
               S2 =     0.7777778

                                  R
               R =      0.4428571

                                LAMDA
            LAMDA =     0.6535772

                                  Q
               Q =      1.4885312

                                             F
           DEGREES OF FREEDOM =              4

                                ALPHA
            ALPHA =     0.8286711
Exiting IML.
```

F I G. **10.3**
Page 1 of computer printout, Example 10.3

A Test for the Huynh–Feldt Conditions

In repeated measures experiments, discussed in more detail in Section 10.3, we often need to know whether the variance–covariance matrix of the repeated measures satisfies conditions that are known as the Huynh–Feldt (H-F) conditions. A variance–covariance matrix Σ is said to satisfy the H-F conditions if $\Sigma = \eta \mathbf{I} + \boldsymbol{\gamma} \mathbf{j}' + \mathbf{j} \boldsymbol{\gamma}'$ for some η and some vector $\boldsymbol{\gamma}$ where $\mathbf{j}$ is a $p \times 1$ vector of ones. When repeated measures satisfy the H-F conditions, experimental data can be analyzed using analysis of variance methods similar to those used for analyzing split-plot experiments. See Milliken and Johnson (1992) for additional details.

Let $\mathbf{P}$ be a $p \times (p-1)$ matrix whose columns are orthogonal to $\mathbf{j}$ (recall that **j** is a vector of ones) and to each other such that $\mathbf{P}'\mathbf{P} = \mathbf{I}_{p-1}$. In other words, the columns of **P** are orthogonal contrasts of length 1. We can show that $\boldsymbol{\Sigma}$ satisfies the H-F conditions if and only if $\boldsymbol{\Sigma}^* = \mathbf{P}'\boldsymbol{\Sigma}\mathbf{P} = \eta\mathbf{I}$. That is, $\boldsymbol{\Sigma}$ satisfies the H-F conditions if and only if $\boldsymbol{\Sigma}^*$ possesses the sphericity property, and this can be tested by using the test given in Eq. (10.2).

Remark   Repeated measures whose variance–covariance matrix satisfies the conditions for compound symmetry also satisfy the H-F conditions.

An example illustrating the H-F conditions is delayed until Example 10.5 in Section 10.2 where a repeated measures experiment is discussed and analyzed.

A Test for Independence

Let

$$
\boldsymbol{\Sigma} = \begin{bmatrix}
\sigma_{11} & \sigma_{12} & \cdots & \sigma_{1k} \\
\sigma_{21} & \sigma_{22} & \cdots & \sigma_{2k} \\
\vdots & \vdots & \ddots & \vdots \\
\sigma_{k1} & \sigma_{k2} & \cdots & \sigma_{kk}
\end{bmatrix}
$$

and consider testing $H_0 : \sigma_{ij} = 0$ for all $i \neq j$ versus $H_a : \sigma_{ij} \neq 0$ for some $i \neq j$. This hypothesis is of interest in those cases where we want to know whether the variables have independent distributions, but do not care whether the variables have equal variances. Earlier, in Chapters 5 and 6 this test was recommended as one to perform before doing a principal components analysis (PCA) or a factor analysis when a researcher believes the data to be analyzed are multivariate normal.

To test the above hypothesis, calculate

$$
V = |\hat{\boldsymbol{\Sigma}}|/(\hat{\sigma}_{11}\hat{\sigma}_{22} \cdots \hat{\sigma}_{pp}) \qquad \text{or equivalently}
$$
$$
V = |\mathbf{R}|
$$

This test and its corresponding p value can be obtained by using the SAS FACTOR procedure when one specifies the ML option for solving the factor analysis equations and looks at the test labeled:

```
Test of HO: No common factors.
    vs HA: At least one common factor.
```

See page 10 of the computer printout for Example 6.2 (see Figure 6.8) for an illustration of this test.

A Test for Independence of Subsets of Variables

The test in the previous section can be generalized even more. Let

$$
\Sigma = \begin{bmatrix} \Sigma_{11} & \Sigma_{12} & \cdots & \Sigma_{1k} \\ \Sigma_{21} & \Sigma_{22} & \cdots & \Sigma_{2k} \\ \vdots & \vdots & \ddots & \vdots \\ \Sigma_{k1} & \Sigma_{k2} & \cdots & \Sigma_{kk} \end{bmatrix}
$$

where Σ_{ij} is a $p_i \times p_j$ matrix for $i = 1, 2, \ldots, k$, $j = 1, 2, \ldots, k$, and where $\sum_{i=1}^{k} p_i = p$. Consider testing $H_0 : \Sigma_{ij} = 0$ for all $i \neq j$. This test is interesting when the measured variables are partitioned into subgroups, and a researcher wants to know if variables in different subgroups are independent. That is, the variables within a subgroup are allowed to be correlated with one another, but those in different subgroups are expected to be independent. To test this hypothesis, calculate

$$
V = |\mathbf{R}| / (|\mathbf{R}_{11}| \cdot |\mathbf{R}_{22}| \cdots |\mathbf{R}_{kk}|)
$$

where the sample correlation matrix is partitioned in a manner similar to the partitioning on Σ. That is,

$$
\mathbf{R} = \begin{bmatrix} \mathbf{R}_{11} & \mathbf{R}_{12} & \cdots & \mathbf{R}_{1k} \\ \mathbf{R}_{21} & \mathbf{R}_{22} & \cdots & \mathbf{R}_{2k} \\ \vdots & \vdots & \ddots & \vdots \\ \mathbf{R}_{k1} & \mathbf{R}_{k2} & \cdots & \mathbf{R}_{kkk} \end{bmatrix}
$$

where $\mathbf{R}_{ij}$ is a $p_i \times p_j$ matrix.

The p value for the preceding test can be approximated using the formulas given next. Recall that the submatrix $\mathbf{R}_{ii}$ is a dimension $p_i \times p_i$ for $i = 1, 2, \ldots, k$. Let

$$
f = \frac{1}{2}\left[p^2 - \sum_{i=1}^{k} p_i^2 \right]
$$

$$
\rho = 1 - \frac{2\left[\left(p^3 - \sum_{i=1}^{k} p_i^3\right) + 9\left(p^2 - \sum_{i=1}^{k} p_i^2\right) \right]}{6N\left(p^2 - \sum_{i=1}^{k} p_i^2\right)}
$$

$$
a = \rho N, \quad \text{and}
$$

$$
\gamma_2 = \frac{1}{48}\left(p^4 - \sum_{i=1}^{k} p_i^4\right) - \frac{5}{96}\left(p^2 - \sum_{i=1}^{k} p_i^2\right)
$$

$$
- \frac{1}{72}\left(p^3 - \sum_{i=1}^{k} p_i^3\right)^2 \bigg/ \left(p^2 - \sum_{i=1}^{k} p_i^2\right)
$$

Now let z be the computed value of $-a \cdot \log(V)$. Then the observed significance level (p value) is approximately

$$\hat{\alpha} = P[\chi_f^2 < z] + (\gamma_2/a^2)\{P[\chi_{f+4}^2 < z] - P[\chi_f^2 < z]\}$$

where $P[\chi_f^2 < z]$ is the probability that a chi-square random variable distributed with f degrees of freedom is less than z.

A Test for the Equality of Several Variance–Covariance Matrices

When considering discriminant analysis in Section 7.2, different discriminant rules were given depending on whether the populations under consideration had equal variance–covariance matrices or not. In this section, the test of equal variance–covariance matrices used there is given. That is, a test of $H_0: \Sigma_1 = \Sigma_2 = \cdots = \Sigma_m$ is considered. This is a multivariate generalization of Bartlett's test for homogeneity of variances in the univariate case.

Let $\mathbf{x}_{rt}$, for $r = 1, 2, \ldots, N_t$, be a random sample of size N_t from the multivariate normal distribution $N_p(\boldsymbol{\mu}_t, \Sigma_t)$ for $t = 1, 2, \ldots, m$. Also suppose that the m samples are collected independently of one another. To test H_0, compute

$$\Lambda = \frac{\prod_{t=1}^{m} |\hat{\Sigma}_t|^{(N_t-1)/2}}{|\hat{\Sigma}|^{(N-m)/2}}, \tag{10.4}$$

where $\hat{\Sigma}_t$ is the sample variance–covariance matrix from the tth sample and

$$\hat{\Sigma} = \left[\sum_{t=1}^{t} (N_t - 1) \Sigma_{\hat{t}}\right] \bigg/ \left[\sum_{t=1}^{t} (N_t - 1)\right]$$

An approximation to the p value for this test statistic is given in Srivastava and Khatri (1979, p. 229).

Computing software does exist for this test statistic and its corresponding p value. This hypothesis can be tested by using the POOL=TEST option in the SAS DISCRIM procedure. (Ignore the rest of the discriminant analysis results if you are only interested in this test.)

10.2
Inference Procedures for a Mean Vector

This section considers methods for making inferences about a vector of population mean parameters. Methods for testing hypotheses and constructing confidence regions are given.

Suppose $x_1, x_2, \ldots, x_N$ is a random sample from a univariate normal distribution, $N(\mu, \sigma^2)$. Let μ_0 be a user-specified positive constant. In a first course in statistics, students usually learn how to test $H_0: \mu = \mu_0$ versus $H_0: \mu \neq \mu_0$ using Student's t test. The test requires one to compute $t = (\bar{x} - \mu_0)/(s/\sqrt{N})$ where $\bar{x}$ is the sample mean and s is the sample standard deviation. Then H_0 is rejected if $|t| > t_{\alpha/2, N-1}$ where $t_{\gamma, v}$ is the upper $\gamma \cdot 100\%$ critical point of Student's t distribution with v degrees of freedom.

Hotelling's T^2 Statistic

To generalize Student's t test to the multivariate case, suppose $\mathbf{x}_1, \mathbf{x}_2, \ldots, \mathbf{x}_N$ is a random sample from a multivariate normal distribution $N_p(\boldsymbol{\mu}, \boldsymbol{\Sigma})$, where both $\boldsymbol{\mu}$ and $\boldsymbol{\Sigma}$ are unknown. Consider testing $H_0: \boldsymbol{\mu} = \boldsymbol{\mu}_0$ versus $H_a: \boldsymbol{\mu} \neq \boldsymbol{\mu}_0$ where $\boldsymbol{\mu}_0$ is a known (i.e., user-specified) $p \times 1$ vector.

Let $T^2 = N(\hat{\boldsymbol{\mu}} - \boldsymbol{\mu})' \hat{\boldsymbol{\Sigma}}^{-1} (\hat{\boldsymbol{\mu}} - \boldsymbol{\mu})$. Then T^2 is called *Hotelling's T^2 statistic*. Note that T^2 is proportional to the Mahalanobis distance between $\hat{\boldsymbol{\mu}}$ and $\boldsymbol{\mu}$. When $p = 1$, Hotelling's T^2 statistic is equal to the square of Student's t statistic; that is, $T^2 = t^2$. We also know that

$$\frac{N-p}{p(N-1)} T^2$$

is distributed as an F distribution with numerator degrees of freedom equal to p and denominator degrees of freedom equal to $N - p$. These results can be used to test hypotheses about $\boldsymbol{\mu}$ and construct confidence regions about $\boldsymbol{\mu}$.

Hypothesis Test for $\boldsymbol{\mu}$

To test $H_0: \boldsymbol{\mu} = \boldsymbol{\mu}_0$ versus $H: \boldsymbol{\mu} \neq \boldsymbol{\mu}_0$, compute

$$T^2 = N(\hat{\boldsymbol{\mu}} - \boldsymbol{\mu}_0)' \hat{\boldsymbol{\Sigma}}^{-1}(\hat{\boldsymbol{\mu}} - \boldsymbol{\mu}_0)$$

and reject H_0 if

$$\frac{N-p}{p(N-1)} T^2 > F_{\alpha, p, N-p} \tag{10.5}$$

where F_{α, v_1, v_2} is the upper $\alpha \cdot 100\%$ critical point of the F distribution with v_1 and v_2 degrees of freedom.

Confidence Region for $\boldsymbol{\mu}$

To find a confidence region for $\boldsymbol{\mu}$, first let the eigenvalues and corresponding eigenvectors of $\hat{\boldsymbol{\Sigma}}$ be denoted by $\hat{\lambda}_1, \hat{\lambda}_2, \ldots, \hat{\lambda}_p$ and $\hat{\mathbf{c}}_1, \hat{\mathbf{c}}_2, \ldots, \hat{\mathbf{c}}_p$, respectively. A $(1 - \alpha) \cdot 100\%$ confidence region for $\boldsymbol{\mu}$ is an ellipsoid centered at $\hat{\boldsymbol{\mu}}$. Its longest axis is of length

$$2\sqrt{\frac{p(N-1)\hat{\lambda}_1 F_{\alpha,p,N-p}}{N(N-p)}}$$

and is in the direction of $\hat{\mathbf{c}}_1$. Its second longest axis is of length

$$2\sqrt{\frac{p(N-1)\hat{\lambda}_2 F_{\alpha,p,N-p}}{N(N-p)}}$$

and is in the direction of $\hat{\mathbf{c}}_2$. The same kind of statement is true for the third, fourth, ..., pth axes of the p-dimensional ellipsoid.

E X A M P L E **10.4**

Suppose a sample of size 25 produces

$$\bar{\mathbf{x}} = \begin{bmatrix} 6 \\ 4 \end{bmatrix} \quad \text{and} \quad \hat{\boldsymbol{\Sigma}} = \begin{bmatrix} 1.5 & 1.1 \\ 1.1 & 2.7 \end{bmatrix}$$

The eigenvalues of $\hat{\boldsymbol{\Sigma}}$ and their corresponding eigenvectors are

$$\hat{\lambda}_1 = 3.353, \hat{\mathbf{c}}_1 = \begin{bmatrix} 0.5105 \\ 0.8599 \end{bmatrix} \quad \text{and} \quad \hat{\lambda}_2 = 0.847, \hat{\mathbf{c}}_2 = \begin{bmatrix} 0.8599 \\ -0.5105 \end{bmatrix}$$

To construct a 95% confidence ellipsoid for

$$\boldsymbol{\mu} = \begin{bmatrix} \mu_1 \\ \mu_2 \end{bmatrix}$$

first note that $F_{0.05,2,23} = 3.422$, that

$$2\sqrt{\frac{p(N-1)\hat{\lambda}_1 F_{\alpha,p,N-p}}{N(N-p)}} = 2\sqrt{\frac{2 \cdot 24 \cdot 3.353 \cdot 3.422}{25 \cdot 23}} = 1.957$$

and that

$$2\sqrt{\frac{p(N-1)\hat{\lambda}_2 F_{\alpha,p,N-p}}{N(N-p)}} = 2\sqrt{\frac{2 \cdot 24 \cdot 0.847 \cdot 3.422}{25 \cdot 23}} = 0.984$$

Thus, a 95% confidence region for $\boldsymbol{\mu}$ is an ellipse centered at

$$\bar{\mathbf{x}} = \begin{bmatrix} 6 \\ 4 \end{bmatrix}$$

with its longest axis in the direction of

$$\hat{\mathbf{c}}_1 = \begin{bmatrix} .510 \\ .860 \end{bmatrix}$$

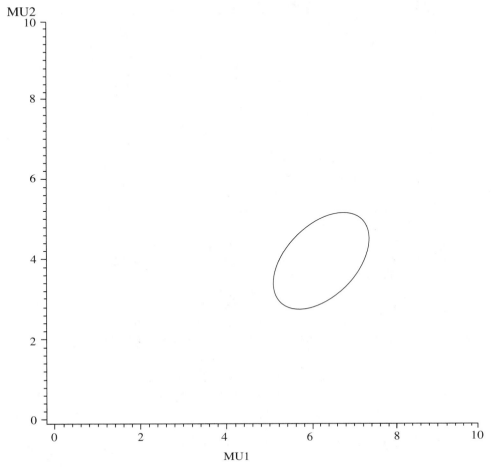

MU2

MU1

F I G. **10.4**
Ninety-five percent confidence ellipse

and of length equal to 1.957. Its second axis is perpendicular to its first and is of length equal to 0.984. See Figure 10.4 for a plot of the 95% confidence ellipse for $\boldsymbol{\mu}$. We can be 95% confident that the true value of $\boldsymbol{\mu}$ falls within this region.

A More General Result

Hotelling's T^2 procedure can be generalized for use in testing other useful hypotheses.

Let **H** be any $q \times p$ matrix of rank q. A test of $H_0: \mathbf{H}\boldsymbol{\mu} = \mathbf{h}$ versus $H_1:$ $\mathbf{H}\boldsymbol{\mu} \neq \mathbf{h}$ can be obtained by computing

$$T^2 = N\,(\mathbf{H}\hat{\boldsymbol{\mu}} - \mathbf{h})'(\mathbf{H}\hat{\boldsymbol{\Sigma}}\mathbf{H}')^{-1}(\mathbf{H}\hat{\boldsymbol{\mu}} - \mathbf{h}) \qquad (10.6)$$

If

$$\frac{(N-q)}{q(N-1)}\,T^2 > F_{\alpha,q,N-q}$$

then H_0 is rejected.

Special Case—A Test of Symmetry

Suppose p responses over time are measured. Let μ_i represent the response expected at time i on a randomly selected experimental unit for $i = 1, 2, \ldots,$ p. A researcher would, no doubt, be interested in knowing whether the μ_i's change over time. Thus, in terms of the μ_i's, a researcher might like to test $H_0: \mu_1 = \mu_2 = \cdots = \mu_p$ versus an alternative that at least two of the μ_i's differ. As an example consider a very simple repeated measures experiment where a random sample of subjects is selected from a single population, and suppose the subjects are evaluated or measured at several different points in time. For example, suppose we take a random sample of 40 3-year-old girls and measure their heights at yearly intervals for the next 8 years. We might be interested in determining if there has been a significant change in heights over time.

Caution   Many introductory-level books suggest a test for the above situation that is often not statistically valid. Many books suggest that you can arrange repeated measures data into a two-way table with SUBJECTS as rows and TIMES as columns, and then test $H_0: \mu_1 = \mu_2 = \cdots = \mu_p$ by performing a randomized complete block analysis of variance with SUBJECTS as blocks. The subjects in this case are the experimental units that are being measured, and the measurements across time are very likely correlated with one another because they are measurements taken on the same experimental unit. It has been shown that if these repeated measures satisfy the Huynh–Feldt conditions described in Section 10.1, then the randomized block analysis correctly analyzes the data. But if the H-F conditions are not satisfied, then the randomized block analysis is not correct.

A test for $H_0: \mu_1 = \mu_2 = \cdots \mu_p$ that is always statistically valid, regardless of whether the H-F conditions are satisfied or not, can be obtained by using Hotelling's T^2 statistic in Eq. (10.6). Let $\boldsymbol{\mu}$ be the $p \times 1$ vector of the μ_is, and then note that H_0 is true if and only if $\mathbf{H}\boldsymbol{\mu} = \mathbf{0}$ where

$$\mathbf{H} = \begin{bmatrix} 1 & -1 & 0 & \cdots & 0 \\ 1 & 0 & -1 & \cdots & 0 \\ \vdots & \vdots & \vdots & \ddots & \vdots \\ 1 & 0 & 0 & \cdots & -1 \end{bmatrix}$$

Remark Other **H** matrices exist so that $\mathbf{H}\boldsymbol{\mu} = \mathbf{0}$ if and only if $\mu_1 = \mu_2 = \cdots = \mu_p$. Fortunately, it does not matter which **H** matrix you use, because all such matrices will produce the same value for Hotelling's T^2.

The next example illustrates how you can use the IML procedure in SAS to perform the test. In addition to performing Hotelling's T^2 test, the program also produces a test of whether the H-F conditions are satisfied.

E X A M P L E 10.5

This example uses the data in the AX23 drug group of Table 26.2 in Milliken and Johnson (1992) to illustrate the multivariate procedures discussed in this section. In this example, eight humans were given a drug after which their heart rate was measured every 5 minutes for 20 minutes. The researcher is interested in knowing whether heart rates change over time.

The following commands were used to input the data into a SAS data file:

```
DATA AX23;
 INPUT PERSON T1-T4 @@;
CARDS;
1 72 86 81 77   2 78 83 88 81
3 71 82 81 75   4 72 83 83 69
5 66 79 77 66   6 74 83 84 77
7 62 73 78 70   8 69 75 76 70
DATA X; SET AX23; DROP PERSON;
```

The following commands invoke IML, read the data into a data matrix, and compute $\overline{\mathbf{x}}$, the sample mean vector, and $\hat{\boldsymbol{\Sigma}}$, the sample variance–covariance matrix:

```
PROC IML; RESET NOLOG;
 TITLE 'Ex. 10.5 - Testing for the H-F Conditions.';
  USE X; READ ALL INTO X;
  N = NROW(X); P = NCOL(X);
  PRINT, "X = " X;
  XBAR = (X[+,])`/N;
  SUMSQ = X`*X - N#XBAR*XBAR`;
  SIGHAT = SUMSQ/(N-1);
  PRINT, "XBAR = "XBAR, "SIGHAT =" SIGHAT;
```

The following commands are used to test whether the H-F conditions are satisfied for these data:

```
 QQ=ORPOL(1:P, P-1);
 PP=QQ(|1 : P,2 : P|);
```

```
SIGSTAR=PP`*SIGHAT*PP;
PRINT, "SIGSTAR = " SIGSTAR;
P=P-1;
LAMDA = DET(SIGSTAR)/(((1/P)#TRACE(SIGSTAR))**P);
M=N-1-(2#P#P+P+2)/(6#P);
A=(P+1)#(P-1)#(P+2)#(2#P#P#P+6#P#P+3#P+2)/(288#P#P);
F=P#(P+1)/2-1;
Z=-M#LOG(LAMDA);
PRINT, "Z = " Z;
ALPHA=(1-PROBCHI(Z,F))
     +(A/(M#M))#(PROBCHI(Z,F)-PROBCHI(Z,F+4));
PRINT, "LAMDA =" LAMDA, "ALPHA FOR H-F CONDITIONS = "
ALPHA;
```

The following commands are used to perform a multivariate test of no time effect using Eq. (10.6).

```
TITLE 'Ex. 10.5 - Hotelling's Test of no time effect.';
 P=P+1;
  H = {1 -1  0  0,
       1  0 -1  0,
       1  0  0 -1};
   SMALLH = {0 0 0}`;
Q = NROW(H);
PRINT/ "H = " H, "SMALLH = " SMALLH;
V = INV(H*SIGHAT*H`);
T2 = N#(H*XBAR - SMALLH)`*V*(H*XBAR - SMALLH);
F = (N - Q)#T2/(Q#(N - 1));
DFN = Q;
DFD = N - Q;
ALPHA = 1 - PROBF (F,DFN,DFD);
PRINT, "T2 = " T2, "F = " F,
     "DEGREES OF FREEDOM ARE" DFN DFD;
PRINT, "ALPHA = " ALPHA;
QUIT;
```

The results from the preceding sets of commands can be found on the pages of computer printout shown in Figure 10.5. Page 1 shows the data matrix, $\bar{x}$, and $\hat{\Sigma}$, followed by a test for the H-F conditions. The H-F conditions cannot be rejected by these data because we would reject for small values of ALPHA. The commands can be found in the file named EX10-5.IML on the enclosed disk.

Because the H-F conditions are not violated by these data, we could test for time effects by completing an analysis of variance table for these data treating PERSONS as blocks. However, for illustration purposes, the multivariate test is shown here. The results of the multivariate (Hotelling's)

```
Ex. 10.5 - Testing for the H-F Conditions.                                    1

                               X
           X =        72        86        81        77
                      78        83        88        81
                      71        82        81        75
                      72        83        83        69
                      66        79        77        66
                      74        83        84        77
                      62        73        78        70
                      69        75        76        70


                              XBAR
           XBAR =      70.5
                       80.5
                         81
                      73.125


                  SIGHAT
   SIGHAT =           24            17 16.714286        19.5
                      17            20 12.285714        13.5
               16.714286 12.285714        16          16
                    19.5      13.5        16      26.125


                    SIGSTAR
   SIGSTAR = 6.4133929 2.8210393 -1.159821
             2.8210393 5.8169643 -1.375581
            -1.159821 -1.375581 4.8633929


                              Z
           Z =       1.9411002

                          LAMDA
          LAMDA = 0.7123247
                                          ALPHA
      ALPHA FOR H-F CONDITIONS =   0.8648123
```

F I G. 10.5
Pages 1, 2, 6, and 7 of computer printout, Example 10.5

test for equal time effects is shown on page 2 of the computer printout. This test produces an observed significance level of 0.00059, so we would conclude that differences do occur with respect to time.

The preceding tests can also be obtained through the use of SAS-GLM with a REPEATED option as illustrated by the following SAS commands. The following commands produce some of the same analyses that were produced by the preceding IML commands. I make no attempt to explain why these commands work because that is beyond the scope of this book.

```
TITLE 'Ex. 10.5 - Using SAS-GLM with its REPEATED option';
PROC GLM;
 MODEL T1-T4 =/NOUNI;
 REPEATED TIME 4/PRINTE;
RUN;
```

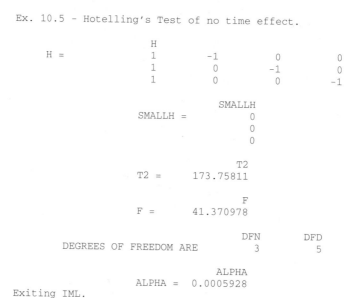

F I G. 10.5
Pages 1, 2, 6, and
7 of computer
printout, Example 10.5

Only the relevant pages of the output are shown in Figure 10.5. The interested reader can see all of the output by executing the file labeled `EX10_5.IML` on the enclosed disk. GLM's test of the H-F conditions is shown on page 6 of the computer printout under the label of `Applied to Orthogonal Components: Test for Sphericity`. Readers should note that the value of the chi-square approximation of 1.9411002 agrees with the value of Z on page 1. There is a slight difference between the two p values shown: 0.8648 from the IML commands and 0.8572 from the GLM commands. This slight difference results because a better (two-stage) approximation of the p value is given by my IML commands.

The multivariate test for equal `TIME` effects on the bottom of page 6 is equivalent to Hotelling's T^2 test shown on page 2. Note that both tests produce the same F value of $F = 41.371$ with 3 and 5 degrees of freedom, and the same p value of 0.006. The test for equal time effects using analysis of variance methods (treating `PERSONS` as blocks) is shown on page 7 where the F value is $F = 39.05$ with 3 and 21 degrees of freedom. The adjusted p values on the far right of page 7 are adjustments that are sometimes made when the H-F conditions are not satisfied. These adjustments have been proposed as an alternative to a multivariate approach when the H-F conditions are not satisfied. See Milliken and Johnson (1992) for additional information about analyses of repeated measures experiments.

General Linear Models Procedure
Repeated Measures Analysis of Variance

Partial Correlation Coefficients from the Error SS&CP Matrix
of the Variables Defined by the Specified Transformation / Prob > |r|

DF = 6	TIME.1	TIME.2	TIME.3
TIME.1	1.000000 0.0	0.694136 0.0561	0.691522 0.0575
TIME.2	0.694136 0.0561	1.000000 0.0	0.640345 0.0872
TIME.3	0.691522 0.0575	0.640345 0.0872	1.000000 0.0

Applied to Orthogonal Components:
Test for Sphericity: Mauchly's Criterion = 0.7123247
Chisquare Approximation = 1.9411002 with 5 df   Prob > Chisquare = 0.8572

Manova Test Criteria and Exact F Statistics for
the Hypothesis of no TIME Effect
H = Type III SS&CP Matrix for TIME   E = Error SS&CP Matrix

S=1   M=0.5   N=1.5

Statistic	Value	F	Num DF	Den DF	Pr > F
Wilks' Lambda	0.03872579	41.3710	3	5	0.0006
Pillai's Trace	0.96127421	41.3710	3	5	0.0006
Hotelling-Lawley Trace	24.82258667	41.3710	3	5	0.0006
Roy's Greatest Root	24.82258667	41.3710	3	5	0.0006

F I G. 10.5
Pages 1, 2, 6, and 7 of computer printout, Example 10.5

General Linear Models Procedure
Repeated Measures Analysis of Variance
Univariate Tests of Hypotheses for Within Subject Effects

Source: TIME

DF	Type III SS	Mean Square	F Value	Pr > F	Adj G - G	Pr > F H - F
3	667.59375000	222.53125000	39.05	0.0001	0.0001	0.0001

Source: Error(TIME)

DF	Type III SS	Mean Square
21	119.65625000	5.69791667

Greenhouse-Geisser Epsilon = 0.8049
Huynh-Feldt Epsilon = 1.2588

F I G. 10.5
Pages 1, 2, 6, and 7 of computer printout, Example 10.5

A Test for Linear Trend

Suppose the p response variables are measured across equally spaced time periods. Also suppose $H_0: \mu_1 = \mu_2 = \cdots = \mu_p$ is rejected. This implies the means are not all equal. In this case, a question of considerable interest is do the means fall onto a straight line? Such a hypothesis is tested in this section.

Consider $H_0: \mu_1, \mu_2, \ldots, \mu_p$ lie on a straight line. This is equivalent to $H_0:$ $\mu_2 - \mu_1 = \mu_3 - \mu_2$, $\mu_3 - \mu_2 = \mu_4 - \mu_3$, $\ldots$, $\mu_{p-1} - \mu_{p-2} = \mu_p - \mu_{p-1}$ for equally spaced time periods since the μ_i's can all fall on a straight line if and only if the slopes between adjacent time points are all equal. H_0 is also equivalent to

$$H_0: \mu_3 - 2\mu_2 + \mu_1 = 0, \qquad \mu_4 - 2\mu_3 + \mu_2 = 0, \ldots, \mu_p - 2\mu_{p-1} + \mu_{p-2} = 0$$

Thus H_0 can be tested by testing $\mathbf{H}\boldsymbol{\mu} = \mathbf{0}$ where

$$\mathbf{H} = \begin{bmatrix} 1 & -2 & 1 & 0 & \cdots & 0 & 0 & 0 \\ 0 & 1 & -2 & 1 & \cdots & 0 & 0 & 0 \\ \vdots & \vdots & \vdots & \vdots & \ddots & \vdots & \vdots & \vdots \\ 0 & 0 & 0 & 0 & \cdots & 1 & -2 & 1 \end{bmatrix}$$

Remark   Tests can also be constructed for quadratic trend, cubic trend, etc. See Beyer (1968, pp. 504–517) for a table of orthogonal polynomials that provides the appropriate contrasts to place into the rows of $\mathbf{H}$.

Fitting a Line to Repeated Measures

As a final look at repeated measures in this section, suppose we have determined that $\mu_1, \mu_2, \ldots, \mu_p$ lie on a straight line. The next important question is "What is the equation of that line?" Many analysts might be tempted to find the equation of this line by using simple linear regression. One of the important assumptions for simple linear regression methods is that the observations being analyzed have independent probability distributions, but this is not true for repeated measures data. To find a more appropriate line for repeated measures data, assume $\mu_i = \alpha_0 + \alpha_1 T_i$, for $i = 1, 2, \ldots, p$. This means that

$$\boldsymbol{\mu} = \begin{bmatrix} 1 & T_1 \\ 1 & T_2 \\ \vdots & \vdots \\ 1 & T_p \end{bmatrix} \begin{bmatrix} \alpha_0 \\ \alpha_1 \end{bmatrix} = \mathbf{B}\boldsymbol{\alpha} \text{ (say)}$$

Then the multivariate estimates of α_0 and α_1 are given by

$$\hat{\boldsymbol{\alpha}} = \begin{bmatrix} \hat{\alpha}_0 \\ \hat{\alpha}_1 \end{bmatrix} = (\mathbf{B}'\hat{\boldsymbol{\Sigma}}^{-1}\mathbf{B})^{-1}\mathbf{B}'\hat{\boldsymbol{\Sigma}}^{-1}\hat{\boldsymbol{\mu}}$$

Simultaneous $(1 - \alpha) \cdot 100\%$ confidence intervals for $\alpha_0 + \alpha_1 T$, $-\infty < T < \infty$, are given by

$$\hat{\alpha}_0 + \hat{\alpha}_1 T \mp \left\{ \frac{T_\alpha^2[1 + U^2/(N-1)]}{N} \right\}^{1/2} \left([1\,T](\mathbf{B}'\hat{\boldsymbol{\Sigma}}^{-1}\mathbf{B})^{-1} \begin{bmatrix} 1 \\ T \end{bmatrix} \right)^{1/2} \quad \text{for } -\infty < T < \infty$$

where

$$T_\alpha^2 = \frac{2(N-1)}{(N-p)} F_{\alpha,2,N-p}$$

and $U^2 = N\hat{\boldsymbol{\mu}}'\mathbf{C}'(\mathbf{C}\hat{\boldsymbol{\Sigma}}\mathbf{C}')^{-1}\mathbf{C}\hat{\boldsymbol{\mu}}$ where

$$\mathbf{C} = \mathbf{D}'(\mathbf{I} - \mathbf{B}(\mathbf{B}'\mathbf{B})^{-1}\mathbf{B}') \quad \text{and} \quad \mathbf{D} = \begin{bmatrix} T_1^2 & T_1^3 & \cdots & T_1^{p-1} \\ T_2^2 & T_2^3 & \cdots & T_2^{p-1} \\ \vdots & \vdots & \ddots & \vdots \\ T_p^2 & T_p^3 & \vdots & T_p^{p-1} \end{bmatrix}$$

Multivariate Quality Control

Hotelling's T^2 statistic can also be used to generalize univariate quality control techniques to multivariate responses. To see how this is accomplished, let $\boldsymbol{\mu}_T$ represent a vector of the desired target values for a process. It is recommended that you estimate $\boldsymbol{\Sigma}$, the variance–covariance matrix of the multivariate response data vectors, by using process data collected during short time intervals over a long period of time. Let $\mathbf{x}_{sr}$ be the vector of measured responses of the rth sample during the sth sampling period. Then

$$\hat{\boldsymbol{\Sigma}} = \frac{\sum\limits_{s=1}^{k} \sum\limits_{r=1}^{n} (\mathbf{x}_{sr} - \bar{\mathbf{x}}_s.)(\mathbf{x}_{sr} - \bar{\mathbf{x}}_s.)'}{k(n-1)}$$

where k equals the number of sampling periods and n equals the number of samples taken per period.

For multivariate process control, you should take a sample of $t \geq 1$ observations and compute

$$F = \frac{t[k(n-1) + 1 - p]}{k(n-1)p} \cdot (\bar{\mathbf{x}} - \boldsymbol{\mu}_T)'\hat{\boldsymbol{\Sigma}}^{-1}(\bar{\mathbf{x}} - \boldsymbol{\mu}_T)$$

where $\bar{\mathbf{x}}$ is the mean of the t observation(s). If $F > F_{\alpha,p,k(n-1)+1-p}$, then we conclude that the process is out of control.

For additional information about multivariate quality control methods and how PCA can also be used, see Jackson (1991).

10.3

Two Sample Procedures

Suppose $\mathbf{x}_{i1}, \mathbf{x}_{i2}, \ldots, \mathbf{x}_{iN_i}$ are random samples from two multivariate normal populations, $N_p(\boldsymbol{\mu}_i, \boldsymbol{\Sigma}_i)$ for $i = 1, 2$. Further, suppose the two random samples are independent (not paired) of each other. Consider testing $H_0: \boldsymbol{\mu}_1 = \boldsymbol{\mu}_2$ versus $H_a: \boldsymbol{\mu}_1 \neq \boldsymbol{\mu}_2$. If $\boldsymbol{\Sigma}_1 = \boldsymbol{\Sigma}_2$, an exact test of H_0 exists. If the two variance–covariance matrices are not equal, only an approximate test of H_0 is available. Recall that a test for $\boldsymbol{\Sigma}_1 = \boldsymbol{\Sigma}_2$ was given in Section 10.1. When $\boldsymbol{\Sigma}_1 = \boldsymbol{\Sigma}_2$, let their common value be denoted by $\boldsymbol{\Sigma}$. In this case, the best estimate of $\boldsymbol{\Sigma}$ is

$$\hat{\boldsymbol{\Sigma}} = \frac{(N_1 - 1)\hat{\boldsymbol{\Sigma}}_1 + (N_2 - 1)\hat{\boldsymbol{\Sigma}}_2}{N_1 + N_2 - 2}$$

To test $H_0: \boldsymbol{\mu}_1 = \boldsymbol{\mu}_2$ versus $H_a: \boldsymbol{\mu}_1 \neq \boldsymbol{\mu}_2$ when $\boldsymbol{\Sigma}_1 = \boldsymbol{\Sigma}_2$, compute

$$T^2 = \frac{N_1 N_2}{N_1 + N_2}(\hat{\boldsymbol{\mu}}_1 - \hat{\boldsymbol{\mu}}_2)'\hat{\boldsymbol{\Sigma}}^{-1}(\hat{\boldsymbol{\mu}}_1 - \hat{\boldsymbol{\mu}}_2)$$

If

$$\frac{N_1 + N_2 - p - 1}{p(N_1 + N_2 - 2)}T^2 > F_{\alpha, p, N_1 + N_2 - p - 1}$$

then reject H_0.

The preceding test can be generalized to obtain a test for $\mathbf{C}\boldsymbol{\mu}_1 = \mathbf{C}\boldsymbol{\mu}_2$ when $\boldsymbol{\Sigma}_1 = \boldsymbol{\Sigma}_2$ and $\mathbf{C}$ is any user-specified $q \times p$ matrix of rank q. In this case compute

$$T^2 = \frac{N_1 N_2}{N_1 + N_2}(\hat{\boldsymbol{\mu}}_1 - \hat{\boldsymbol{\mu}}_2)'\mathbf{C}'(\mathbf{C}\hat{\boldsymbol{\Sigma}}\mathbf{C}')^{-1}\mathbf{C}(\hat{\boldsymbol{\mu}}_1 - \hat{\boldsymbol{\mu}}_2)$$

If

$$\frac{N_1 + N_2 - q - 1}{q(N_1 + N_2 - 2)}T^2 > F_{\alpha, q, N_1 + N_2 - q - 1}$$

then we should reject H_0 at the $\alpha \cdot 100\%$ significance level.

Repeated Measures Experiments

Suppose we have two groups of individuals in which the individuals in one group are assigned one treatment while those in the other group are assigned a different treatment. Then suppose that all individuals are measured repeatedly over time. The resulting experiment is usually classified as a repeated measures experiment. These experiments can be analyzed using the techniques discussed in this section. The measurements across time result in multivariate response vectors where the first element in each vector is the measurement taken at time 1, the second element is the measurement taken at time 2, etc.

Caution Many introductory-level books suggest tests for repeated measures experiments that are often not statistically valid. These books suggest that an analysis of variance table can be partitioned into a BETWEEN PERSON analysis and a WITHIN PERSON analysis in which each partition has its own experimental error term. In these analyses the treatment main effect mean square is compared to the BETWEEN PERSON error mean square (this test is always correct), while the TIME main effect and the TIME × TREATMENT interaction effect are compared to the WITHIN PERSON error mean square. These latter two tests are correct if the repeated measures satisfy the H-F conditions described in Section 10.1, but they are not correct when the H-F conditions are not satisfied. Multivariate analyses of repeated measures are almost always appropriate.

To illustrate the procedures introduced in this section, the data from a repeated measures experiment are used.

E X A M P L E **10.6**

This example uses the data in both the AX23 drug group and the BWW9 drug group of Table 26.2 in Milliken and Johnson (1992) to illustrate the multivariate procedures discussed in this section. In this example, eight humans were given the drug AX23 and another eight humans were given the drug BWW9. Each individual's heart rate was measured every 5 minutes for 20 minutes. The researcher is interested in knowing whether heart rates change over time, whether there is a difference in the two drugs, and whether there is a TIME × DRUG interaction.

The following commands are used to input the data. A list of the data can be found on page 1 of the computer printout shown in Figure 10.6.

```
DATA AX23;
  INPUT PERSON T1-T4 @@;
  DRUG='AX23';
CARDS;
1 72 86 81 77    2 78 83 88 81
3 71 82 81 75    4 72 83 83 69
5 66 79 77 66    6 74 83 84 77
7 62 73 78 70    8 69 75 76 70
RUN;

DATA BWW9;
  INPUT PERSON T1-T4 @@;
  DRUG='BWW9';
```

```
                    Ex. 10.6 - Drug Data                              1
            OBS    PERSON   T1    T2    T3    T4   DRUG

             1       1      72    86    81    77   AX23
             2       2      78    83    88    81   AX23
             3       3      71    82    81    75   AX23
             4       4      72    83    83    69   AX23
             5       5      66    79    77    66   AX23
             6       6      74    83    84    77   AX23
             7       7      62    73    78    70   AX23
             8       8      69    75    76    70   AX23
             9       1      85    86    83    80   BWW9
            10       2      82    86    80    84   BWW9
            11       3      71    78    70    75   BWW9
            12       4      83    88    79    81   BWW9
            13       5      86    85    76    76   BWW9
            14       6      85    82    83    80   BWW9
            15       7      79    83    80    81   BWW9
            16       8      83    84    78    81   BWW9
```

F I G. **10.6**

Pages 1–5, 8, and 15–18 of computer printout, Example 10.6

```
CARDS;
1 85 86 83 80   2 82 86 80 84
3 71 78 70 75   4 83 88 79 81
5 86 85 76 76   6 85 82 83 80
7 79 83 80 81   8 83 84 78 81
RUN;

DATA COMB; SET AX23 BWW9;
PROC PRINT DATA=COMB;
 TITLE 'Ex. 10.6 - Drug Data';
```

The following commands are used to transport the SAS data sets into data matrices in IML, compute the sample means, sample variance–covariance matrices, and compute a pooled estimate of a common variance–covariance matrix:

```
DATA X1; SET AX23; DROP PERSON DRUG;
DATA X2; SET BWW9; DROP PERSON DRUG;

 TITLE 'Ex. 10.6 - IML Analyses';

PROC IML; RESET NOLOG;
USE X1; READ ALL INTO X1;
USE X2; READ ALL INTO X2 ;
N1= NROW(X1); N2= NROW(X2); P= NCOL(X1);

XBAR1=( X1[+,])`/N1;
XBAR2=( X2[+,])`/N2;
PRINT, "XBAR1 = " XBAR1, "XBAR2 = " XBAR2;

SUMSQ1=X1`*X1-XBAR1*XBAR1`#N1;
SUMSQ2=X2`*X2-XBAR2*XBAR2`#N2;
```

Ex. 10.6 - IML Analyses   2

```
                    XBAR1
        XBAR1 =      70.5
                     80.5
                       81
                   73.125

                    XBAR2
        XBAR2 =     81.75
                       84
                   78.625
                    79.75

     SIGHAT1
          24            17 16.714286        19.5
          17            20 12.285714        13.5
  16.714286     12.285714            16          16
       19.5          13.5            16      26.125


     SIGHAT2
  23.642857            11 15.178571   5.3571429
         11     9.4285714  7.4285714   5.2857143
  15.178571     7.4285714  17.696429   8.6071429
  5.3571429     5.2857143  8.6071429         8.5


     SIGHAT
  23.821429            14 15.946429   12.428571
         14     14.714286  9.8571429   9.3928571
  15.946429     9.8571429  16.848214   12.303571
  12/428571     9.3928571  12.303571     17.3125

                        LAMDA
            LAMDA = 0.0094304

                          Z
        Z =       6.4627244

                        F
        DF =       10

                     ALPHA
        ALPHA =  0.7869061
```

FIG. 10.6

Pages 1–5, 8, and 15–18 of computer printout, Example 10.6

```
SIGHAT1 = SUMSQ1/(N1-1);
SIGHAT2 = SUMSQ2/(N2-1);

SIGHAT=(SUMSQ1+SUMSQ2)/(N1+N2-2);
PRINT, SIGHAT1,, SIGHAT2,, SIGHAT;
```

The following commands perform a test of $H_0 \colon \Sigma_1 = \Sigma_2$ using the results in Section 10.1:

```
NU1=N1-1; NU2=N2-1; NU=NU1+NU2;
```

```
C
1              0          0          0
0              1          0          0
0              0          1          0
0              0          0          1
```

```
        TSQ
    119.45432
```

```
         F
    23.464242
```

```
        DF1
         4
```

```
        DF2
         11
```

```
       ALPHA
    0.0000244
```

F I G. 10.6

Pages 1–5, 8, and 15–18 of computer printout, Example 10.6

```
C
1            -1          0          0
1             0         -1          0
1             0          0         -1
```

```
        TSQ
    119.38384
```

```
         F
    34.109669
```

```
        DF1
         3
```

```
        DF2
         12
```

```
       ALPHA
    3.7404E-6
```

F I G. 10.6

Pages 1–5, 8, and 15–18 of computer printout, Example 10.6

```
LAMDA =
  (DET(SUMSQ1)##(NU1/2))#(DET(SUMSQ2)##(NU2/2))#NU##(P#NU/2)/
  (DET(SUMSQ1+SUMSQ2)##(NU/2))/NU1##(P#NU1/2)/NU2##(P#NU2/2);

  T1=(2#P#P+3#P-1)/6/(P+1);
  T2=1/NU1+1/NU2-1/NU;
  T3=1/NU1##2+1/NU2##2-1/NU##2;
  T4=P#(P+1)/48#((P-1)#(P+2)#T3-6#T1##2#T2##2);
```

```
Ex. 10.6 - IML Analyses                              5

        C
        1           1           1           1

                        TSQ
                   6.5471622

                         F
                   6.5471622

                        DF1
                         1

                        DF2
                         14

                       ALPHA
                    0.0227286
```

F I G. **10.6**

Pages 1–5, 8, and 15–18 of computer printout, Example 10.6

```
Exiting IML.
```

```
Z=-2#(1-T1#T2)#LOG(LAMDA);
F=P#(P+1)/2;
B=T4/(1-T1#T2)##2;
ALPHA=1-(PROBCHI(Z,F)+B#(PROBCHI(Z,F+4)-PROBCHI(Z,F)));
PRINT, "LAMDA =" LAMDA, "Z = " Z, "DF = " F, "ALPHA = "
ALPHA;
```

The following commands compute Hotelling's T^2 test statistic for testing $H_0: C\mu_1 = C\mu_2$ for any matrix C. Since the calculations are the same once C is given, the commands are imbedded in a DO loop that allows one to specify several C matrices in the same program. The C matrices selected here are listed next:

$$(1) \qquad C = \begin{bmatrix} 1 & 0 & 0 & 0 \\ 0 & 1 & 0 & 0 \\ 0 & 0 & 1 & 0 \\ 0 & 0 & 0 & 1 \end{bmatrix}$$

an identity matrix that provides a test of $\mu_1 = \mu_2$,

$$(2) \qquad C = \begin{bmatrix} 1 & -1 & 0 & 0 \\ 1 & 0 & -1 & 0 \\ 1 & 0 & 0 & -1 \end{bmatrix}$$

a matrix of contrasts that tests whether the differences between successive time periods are the same for both drug groups. This is equivalent to testing whether the differences between drugs are the same at all time periods. In analysis of variance terminology, this is a test for DRUG*TIME interac-

```
          Ex. 10.6 - Using DISCRIM to Test for Equal Covariance Matrices        8

DISCRIMINANT ANALYSIS     TEST OF HOMOGENEITY OF WITHIN COVARIANCE MATRICES

       Notation: K    = Number of Groups

                 P    = Number of Variables

                 N    = Total Number of Observations - Number of Groups

                 N(i) = Number of Observations in the i'th Group - 1


                         __                             N(i)/2
                         ||  |Within SS Matrix(i)|
             V      = ----------------------------------
                                                N/2
                         |Pooled SS Matrix|


                      _                  _      2
                     |      1        1   |  2P + 3P - 1
             RHO  = 1.0 - | SUM -----  -  ---  | -------------
                     |_      N(i)       N _|  6(P+1)(K-1)


             DF     = .5(K-1)P(P+1)


                                 _                    _
                                |      PN/2            |
                                |     N         V      |
    Under null hypothesis:  -2 RHO ln |  ----------------  |
                                |      __      PN(i)/2  |
                                |_     ||  N(i)        _|

    is distributed approximately as chi-square(DF)


    Test Chi-Square Value =   6.462724
    with      10 DF      Prob > Chi-Sq = 0.7750


    Since the chi-square value is significant at the  0.1000 level,
    a pooled covariance matrix will be used in the discriminant function.

    Reference: Morrison, D.F. (1976)    Multivariate Statistical Methods p252.
```

F I G. **10.6**
Pages 1–5, 8, and 15–18 of computer printout, Example 10.6

tion. This test is also sometimes called a test of parallel profiles for the two drugs.

$$(3) \quad \mathbf{C} = [1 \quad 1 \quad 1 \quad 1]$$

a row matrix of ones that tests whether the means of the two drug groups are equal when summed across the four time measurements. This is equivalent to testing whether the averages of the time measurements are equal in

Ex. 10.6 - Using GLM for Profile Analysis and H-F Conditions   15

General Linear Models Procedure
Repeated Measures Analysis of Variance

Partial Correlation Coefficients from the Error SS&CP Matrix
of the Variables Defined by the Specified Transformation / Prob > |r|

DF = 13	TIME.1	TIME.2	TIME.3
TIME.1	1.000000	0.646501	0.683783
	0.0	0.0092	0.0049
TIME.2	0.646501	1.000000	0.486629
	0.0092	0.0	0.0658
TIME.3	0.683783	0.486629	1.000000
	0.0049	0.0658	0.0

Applied to Orthogonal Components:
Test for Sphericity: Mauchly's Criterion = 0.7906116
Chisquare Approximation = 2.9890663 with 5 df   Prob > Chisquare = 0.7017

Manova Test Criteria and Exact F Statistics for
the Hypothesis of no TIME Effect
H = Type III SS&CP Matrix for TIME   E = Error SS&CP Matrix

S=1   M=0.5   N=5

Statistic	Value	F	Num DF	Den DF	Pr > F
Wilks' Lambda	0.11657647	30.3122	3	12	0.0001
Pillai's Trace	0.88342353	30.3122	3	12	0.0001
Hotelling-Lawley Trace	7.57806038	30.3122	3	12	0.0001
Roy's Greatest Root	7.57806038	30.3122	3	12	0.0001

F I G. **10.6**
Pages 1–5, 8, and 15–18 of computer printout, Example 10.6

the two drug groups. In analysis of variance terminology, this is a test for
DRUG **main effect.**

```
CON=(N1#N2)/(N1+N2);

DO II=1 TO 3;
  IF II=1 THEN C  = I(4);
  IF II=2 THEN C={1 -1 0 0, 1 0 -1 0, 1 0 0 -1};
  IF II=3 THEN C={1 1 1 1};

Q= NROW(C);
PRINT /, C;
V = INV(C*SIGHAT*C`);
```

```
          Ex. 10.6 - Using GLM for Profile Analysis and H-F Conditions        16

                         General Linear Models Procedure
                       Repeated Measures Analysis of Variance

                   Manova Test Criteria and Exact F Statistics for
                        the Hypothesis of no TIME*DRUG Effect
          H = Type III SS&CP Matrix for TIME*DRUG    E = Error SS&CP Matrix

                         S=1     M=0.5     N=5

Statistic                    Value           F     Num DF    Den DF   Pr > F

Wilks' Lambda            0.10496024     34.1097        3        12    0.0001
Pillai's Trace           0.89503976     34.1097        3        12    0.0001
Hotelling-Lawley Trace   8.52741724     34.1097        3        12    0.0001
Roy's Greatest Root      8.52741724     34.1097        3        12    0.0001
```

```
          Ex. 10.6 - Using GLM for Profile Analysis and H-F Conditions        17

                         General Linear Models Procedure
                       Repeated Measures Analysis of Variance
                    Tests of Hypotheses for Between Subjects Effects
```

Source	DF	Type III SS	Mean Square	F Value	Pr > F
DRUG	1	361.0000	361.0000	6.55	0.0227
Error	14	771.9375	55.1384		

```
          Ex. 10.6 - Using GLM for Profile Analysis and H-F Conditions        18

                         General Linear Models Procedure
                       Repeated Measures Analysis of Variance
               Univariate Tests of Hypotheses for Within Subject Effects
```

Source: TIME

DF	Type III SS	Mean Square	F Value	Pr > F	Adj G - G	Pr > F H - F
3	409.31250000	136.43750000	23.31	0.0001	0.0001	0.0001

Source: TIME*DRUG

DF	Type III SS	Mean Square	F Value	Pr > F	Adj G - G	Pr > F H - F
3	392.37500000	130.79166667	22.35	0.0001	0.0001	0.0001

Source: Error(TIME)

DF	Type III SS	Mean Square
42	245.81250000	5.85267857

```
                   Greenhouse-Geisser Epsilon = 0.8811
                     Huynh-Feldt Epsilon = 1.1827
```

F I G. **10.6**

Pages 1–5, 8, and 15–18 of computer printout, Example 10.6

```
U=C*(XBAR1-XBAR2);
TSQ = CON#U`*V*U;
F=(N1+N2-Q-1)#TSQ/(Q#(N1+N2-2));
DF1=Q;
DF2=N1+N2-Q-1;
ALPHA = 1-PROBF(F,DF1,DF2);
PRINT, TSQ, F, DF1, DF2, ALPHA;
END;
QUIT;
```

Some of the results produced by the preceding commands can be found on the pages of the computer printout shown in Figure 10.6. The whole printout can be seen by executing the file named EX10_6.IML on the enclosed disk.

Page 1 shows a print of the data being analyzed; page 2 shows the two sample means, labeled XBAR1 and XBAR2, the two sample variance–covariance matrices, labeled SIGHAT1 and SIGHAT2, and the pooled variance–covariance matrix, labeled SIGHAT.

The last three statistics on the bottom of page 2 provide a test of the hypothesis that $\Sigma_1 = \Sigma_2$. The statistic labeled LAMDA is the computed value of Eq. (10.4), $Z = 6.4627$ is the observed value of a statistic that is approximately distributed as a chi-square random variable with 10 degrees of freedom. An approximate significance level can be computed by computing the probability that $U > 6.4627$ where U has a chi-square distribution with 10 degrees of freedom. The approximate significance level is equal to 0.7750 for these data. ALPHA on page 2 of the printout is an improved calculation of the significance probability of Z using the approximation given in Srivastava and Khatri (1979, p. 229). For these data, we accept the hypothesis that $\Sigma_1 = \Sigma_2$.

The results shown on pages 3–5 of Figure 10.6 correspond to the three different **C** matrices used. On each of these pages TSQ is the value of Hotelling's T^2 statistic, F is the value of an F statistic with q degrees of freedom for the numerator and $N_1 + N_2 - q - 1$ degrees of freedom for the denominator. ALPHA is the observed significance level of the F statistic.

From page 3, we can see that the hypothesis of equal mean vectors for the two drug groups should be rejected. From page 4, we can see that there is significant DRUG*TIME interaction, indicating that the difference between drugs do change with respect to time. Finally, from page 5, we can see that there is a significant main effect due to drugs. We might have to interpret this result cautiously since there is significant DRUG*TIME interaction.

Next we illustrate some alternative ways of producing the results just presented. These alternatives may be preferable to many researchers because they do not require use of the IML procedure.

Recall that one of the options in SAS's DISCRIM procedure allows us to test the equality of sample variance–covariance matrices. The following commands illustrate the use of the DISCRIM procedure in SAS to test H_0: $\Sigma_1 = \Sigma_2$:

```
PROC DISCRIM DATA=COMB POOL=TEST;
 CLASS DRUG;
 VAR T1-T4;
 TITLE 'Ex. 10.6 - Using DISCRIM to Test for Equal
Covariance    Matrices';
RUN;
```

The following GLM commands produce some of the same results that were obtained previously by using IML commands:

```
PROC GLM DATA=COMB;
 CLASS DRUG;
 MODEL T1-T4 = DRUG/NOUNI;
 REPEATED TIME 4/PRINTE;
 TITLE 'Ex. 10.6 - Using GLM for Profile Analysis and H-F
    Conditions';
RUN;
```

Page 8 of the Figure 10.6 computer printout shows the test of $\Sigma_1 = \Sigma_2$ from the DISCRIM procedure. Readers might note that the chi-square value of 6.462724 is equal to Z on page 2, and that the observed significance level agrees with that given by my approximation.

Page 15 shows the test of the H-F conditions; this test produces a p value of 0.7017, and so the repeated measures seem to satisfy the H-F conditions. On the bottom of page 15 are multivariate tests of the TIME main effect, and on the top of page 16 are multivariate tests of TIME*DRUG interaction. These tests for TIME main effect and TIME*DRUG interaction are often used when the H-F conditions are not satisfied. The TIME*DRUG test agrees with the test obtained from the IML analysis. The TIME main effect test was not given in the IML analysis.

On page 17, you can see the test for DRUG main effect. Readers should note that this is the same test as that shown on page 5. The Error Mean Square = 55.1384 on page 17 is the BETWEEN PERSON error mean square.

Page 18 shows tests for the TIME main effect and TIME*DRUG interaction using the WITHIN PERSON error mean square, which has a value of 5.8527 with 42 degrees of freedom. These tests should be all right for this example since the H-F conditions seem to be satisfied. The adjusted p values on the far right of page 18 are adjustments that are sometimes made when the H-F conditions are not satisfied. These adjustments have been proposed as an alternative to the multivariate approach. See Milliken and Johnson (1992) for additional information on the analysis of repeated measures experiments.

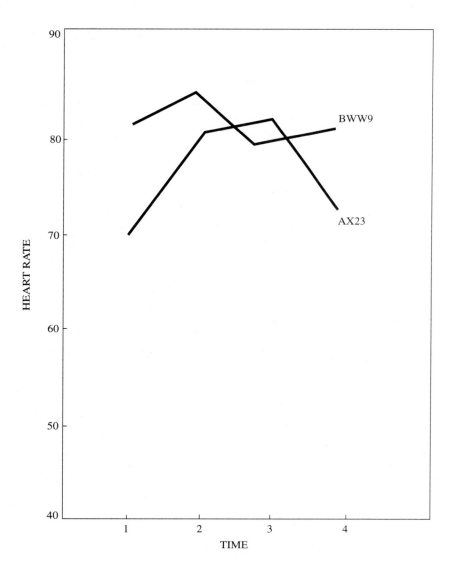

F I G. 10.7
Profile plot of
two drugs

10.4
Profile Analyses

The analyses given in the previous example are often presented as profile
analyses. The values of a set of treatment means over time determine the
profile for the treatment. If we have two treatment groups, then the two sets
of time means for each treatment group define the profiles for each treatment
group. Figure 10.7 shows a profile plot for the two drug groups in Example
10.6. The questions that are of interest in profile analyses are these:

1   Are the profiles identical? This question is answered by testing $\boldsymbol{\mu}_1 = \boldsymbol{\mu}_2$ where $\boldsymbol{\mu}_i$ represents the population mean vector of the time responses for the *i*th treatment group. This question is often broken down into several other questions as follows.

2   Are the profiles parallel? This question is answered by testing for treatment by time interaction.

3   If the profiles are parallel, do they coincide? For parallel profiles, this question can be answered by testing the treatment main effect.

4   If the profiles coincide, do they show any trends? This can be answered by looking for a linear trend, quadratic trend, etc., similar to analyses discussed in Section 10.2.

5   If the profiles are parallel but do not coincide, do the parallel profiles show any trends? This can be answered by looking for a linear trend, quadratic trend, etc., similar to analyses discussed in Section 10.2.

10.5
Additional Two-Group Analyses

In this section two additional situations are described. The first is for the case when the multivariate response vectors in one group are naturally paired with corresponding multivariate response vectors in a second group. The second is for the case when the variance–covariance matrices of independent samples cannot be assumed to be equal.

Paired Samples

Suppose multivariate measurements are taken on a sample of experimental units prior to a treatment being given, then similar multivariate measurements are taken on these same experimental units following the treatment. This obviously creates a natural pairing between the "before" measurements and the "after" measurements.

More formally, suppose $\mathbf{x}_{11}, \mathbf{x}_{12}, \ldots, \mathbf{x}_{1N}$ are measurements taken on N experimental units from a population that has mean $\boldsymbol{\mu}_1$, and suppose that $\mathbf{x}_{21}, \mathbf{x}_{22}, \ldots, \mathbf{x}_{2N}$ make up a second set of measurements that are paired with the corresponding measurements in the first set and that these come from a population that has mean $\boldsymbol{\mu}_2$. To test $H_0: \boldsymbol{\mu}_1 = \boldsymbol{\mu}_2$, we should compute the differences between all pairs of vectors. That is, let $\mathbf{d}_i = \mathbf{x}_{1i} - \mathbf{x}_{2i}$ for $i = 1, 2, \ldots, N$. It can be assumed that the $\mathbf{d}_i$'s are a random sample from a population with mean $\boldsymbol{\mu} = \boldsymbol{\mu}_1 - \boldsymbol{\mu}_2$. A test of H_0 can be made by testing $\boldsymbol{\mu} = \mathbf{0}$ using the methods in Section 10.2. That is, compute

$$T^2 = N\overline{\mathbf{d}}'\hat{\boldsymbol{\Sigma}}_d^{-1}\overline{\mathbf{d}} \qquad \text{where } \hat{\boldsymbol{\Sigma}}_d = \frac{1}{N-1}\sum_{i=1}^{N}(\mathbf{d}_i - \overline{\mathbf{d}})(\mathbf{d}_i - \overline{\mathbf{d}})'$$

and reject H$_0$ if

$$\frac{N-p}{p(N-1)} T^2 > F_{\alpha,p,N-p}$$

where F_{α,v_1,v_2} is the upper $\alpha \cdot 100\%$ critical point of the F distribution with v_1 and v_2 degrees of freedom.

Unequal Variance–Covariance Matrices

In Section 10.3, a test of $\mu_1 = \mu_2$ was given for the case when $\Sigma_1 = \Sigma_2$. How does one compare the means of two groups when the variance–covariance matrices are not equal. The answer to this question depends on whether you have large sample sizes (both sample sizes greater than 25) or small sample sizes.

Large Sample Sizes

For the large sample size case and normally distributed data, we assume that

$$U = (\bar{\mathbf{x}}_1 - \bar{\mathbf{x}}_2)' \left[\frac{1}{N_1} \hat{\Sigma}_1 + \frac{1}{N_2} \hat{\Sigma}_2 \right]^{-1} (\bar{\mathbf{x}}_1 - \bar{\mathbf{x}}_2)$$

is approximately distributed as a chi-square random variable with p degrees of freedom when H$_0$: $\mu_1 = \mu_2$ is true. If a computed value of U is greater than the upper $\alpha \cdot 100\%$ critical point of the chi-square distribution with p degrees of freedom, then H$_0$ would be rejected.

Small Sample Sizes

For the case when the sample sizes are small and when $N_1 = N_2$, we can randomly pair the observation vectors in one group with observation vectors in the other group, and then proceed as if the data were paired observations using the techniques in the paired sample section. The biggest disadvantage to this suggestion is that it is not user invariant. Every data analyst is likely to form different pairs since all analysts should be pairing by some random procedure. Thus all analysts would have different values for their test statistics. We expect, although it cannot be guaranteed, that all analysts would come to the same conclusion about the equality of the two population means.

If the sample sizes are not equal, we can randomly throw some of the observation vectors in the larger sample away so that each sample has equal numbers and then proceed as before. Certainly this has its disadvantages, but should allow us to make the right decision about the equality of the two population means most of the time.

Summary

This chapter introduces multivariate generalizations of most of the statistical inference procedures that are taught in a first course in statistics. In particular, there were multivariate generalizations of t tests and tests for homogeneity of variances.

Exercises

1   The following artificial data refer to a sample of approximately 2-year-old boys from a high-altitude region in Asia. CC is chest circumference and MUAC is mid-upper-arm circumference. All measurements are in centimeters.

Individual	Height	CC	MUAC
1	78	60.6	16.5
2	76	58.1	12.5
3	92	63.2	14.5
4	81	59.0	14.0
5	81	60.8	15.5
6	84	59.5	14.0

a   For lowland children of the same age in the same country, the height, CC, and MUAC means are considered to be 90, 58, and 16 centimeters, respectively. Test the hypothesis that the highland children have the same height, CC, and MUAC means as lowland children.

b   Test the hypothesis that

$$\Sigma = \begin{bmatrix} 30 & 6 & 3 \\ 6 & 2 & 1 \\ 3 & 1 & 2 \end{bmatrix}$$

for these data.

c Test the hypothesis that the height, CC, and MUAC means are in the ratio 6:4:1, respectively.

d Construct a 95% confidence ellipsoid for the mean vector of high-land children.

2 Consider the data in the following table:

Time after treatment

Patient	5	10	15	20	25
1	11	18	15	18	15
2	33	27	31	21	17
3	20	28	27	23	29
4	28	26	18	18	18
5	22	23	22	16	10
6	20	22	16	15	12
7	24	27	22	21	24
8	30	26	30	24	20

a Test the hypothesis that the mean responses at all time periods are equal using Hotelling's T^2 statistic.

b It is desirable to determine if the response varies linearly with time. This may be stated as $H_0: \mu_t = \alpha + \beta t$. Test H_0.

c Find the best estimates of α and β.

d Test whether the Hyuhn–Feldt conditions are satisfied for these data.

e Assume that the Hyuhn–Feldt conditions are satisfied for these repeated measures. Construct an analysis of variance table for these data that tests the time means all equal. How does this test compare to the one in part a?

f Find the usual regression analysis estimates of α and β in the model $\mu_t = \alpha + \beta t$. How do these compare to the estimates in part c?

g Assume that the first three individuals in this data set were controls and that the last five individuals were treated with some treatment. Test whether these two groups have equal means assuming they have equal covariance matrices.

h In part g, determine whether the two groups have parallel profiles. Show a profile plot for these two groups.

3 The following table gives data for 2-year-old girls corresponding to that in Exercise 1 for boys.

Individual	Height	CC	MUAC
1	80	58.4	14.0
2	75	59.2	15.0
3	78	60.3	15.0
4	75	57.4	13.0
5	79	59.5	14.0
6	78	58.1	14.5
7	75	58.0	12.5
8	64	55.5	11.0
9	80	59.2	12.5

a Test the hypothesis that the boys and girls have equal covariance matrices.

b Assuming equal covariance matrices, test the hypothesis that the boys and girls have equal mean vectors.

c Assuming equal covariance matrices, test simultaneously the hypothesis that the boys and girls have equal means on CC and MUAC.

4 Consider the data on girls in Exercise 3. Test the hypothesis that the covariance matrix for these data is equal to $\sigma^2 \mathbf{I}$ for some σ^2.

5 Consider the data in Exercise 3. Test the hypothesis that the correlation matrix for these data is equal to the identity matrix.

6 Consider the data in Example 2.1 and the five groups of variables that were formed in this example. Test the hypothesis that these five groups of variables are independent of one another.

7 Consider the five new variables that were formed, namely, G_1–G_5, from the five groups in Exercise 6. Determine if the covariance matrix of these variables satisfies the sphericity condition.

8 Consider the hotel data described in Exercise 8 of Chapter 1. Use Hotelling's T^2 test to determine whether minimum and maximum room rates for hotels with pools are equal to those of hotels without pools. Discuss your results.

9 Consider the turkey data described in Example 7.1. Consider only the variables HUM, RAD, ULN, and FEMUR.

a Test whether wild turkeys and domestic turkeys have equal variance–covariance matrices. Discuss your results.

b Use Hotelling's T^2 test to determine whether wild turkeys and domestic turkeys have equal mean vectors assuming the two groups of turkeys have equal covariance matrices. Discuss your results.

10 Consider the organizational commitment variables (questions 1–15) in the Pizzazz survey data described in Appendix B.

a Restricting yourself to those employees who have no supervisory responsibilities, determine if the males and females have equal variance–covariance matrices. Assume multivarite normal distributions for this problem.

b Determine whether there is a difference between the mean vectors of the males and females on the organizational commitment variables using Hotelling's T^2 test. Assume that males and females have equal variance–covariance matrices.

c Repeat parts a and b for those employees who have supervisory responsibilities.

11 Repeat Exercise 10 using the job satisfaction variables (questions 61–80).

12 Consider the survey of 304 married adults described in Appendix C and suppose, for this problem, that all 304 individuals come from independent multivariate normal distributions.

a Test whether males and females have equal variance–covariance matrices for their responses to the three-item Kansas Marital Satisfaction Scale.

b Test whether males and females have equal variance–covariance matrices for their responses to the four-item Kansas Family Life Satisfaction Scale.

c Test whether males and females have equal variance–covariance matrices for their responses to the 30-item FACES questionnaire.

d Test whether males and females have equal variance–covariance matrices for their responses to the 60-item FAD questionnaire.

e Consider the four subsets of variables defined by (1) the 60-item FAD questionnaire, (2) the 3-item KMS questionnaire, (3) the 4-item KFLS questionnaire, and (4) the 30-item FACES questionnaire. Perform a statistical test that tests the hypothesis that these four subsets of variables are independent. Assume the data follow a normal distribution.

13 Consider the survey data described in Appendix C. Consider only the data from couples, that is, those individuals where both spouses provided data.

a Test the hypothesis that husbands and wives have equal mean vectors with respect to the variables defined by the Kansas Marital Satisfaction questionnaire.

    b   Test the hypothesis that husbands and wives have equal mean vectors with respect to the variables defined by the Kansas Family Life Satisfaction questionnaire.

    c   Test the hypothesis that husbands and wives have equal mean vectors with respect to the variables defined by the FACES questionnaire.

14   Consider the survey data described in Appendix C. Consider only the data from females.

    a   Test the hypothesis that females whose husbands provided data and females whose husbands did not provide data have equal mean vectors with respect to the variables defined by the Kansas Marital Satisfaction questionnaire.

    b   Test the hypothesis that females whose husbands provided data and females whose husbands did not provide data have equal mean vectors with respect to the variables defined by the Kansas Family Life Satisfaction questionnaire.

    c   Test the hypothesis that females whose husbands provided data and females whose husbands did not provide data have equal mean vectors with respect to the variables defined by the FACES questionnaire.

Multivariate Analysis of Variance

11

Consider male responses to the FACES questionnaire in the married adult survey described in Appendix C and suppose these males are partitioned into four groups: (1) those whose income is less than \$34,999, (2) those whose income is between \$35,000 and \$49,999, (3) those whose income is between \$50,000 and \$74,999, and (4) those whose income is at least \$75,000. A researcher is likely to be interested in how the individuals in these four income categories compare to one another. If there were only two groups to compare, the researcher could use the procedures described in Chapter 10. When there are more than two groups to compare, however, he or she must use multivariate analysis of variance methods. Multivariate analysis of variance methods can also be used to compare two groups of individuals.

In Chapter 10, inference procedures were introduced for comparing the means of two populations. These procedures can be extended to methods for comparing the means of many populations similar to the way that unpaired t tests are generalized to analysis of variance procedures in a first course in statistics. Hotelling's T^2 test for comparing two vectors of means generalizes to multivariate analysis of variance (MANOVA) for comparing more than two vectors of means. This chapter assumes the reader has considerable experience with analysis of variance methods for one-way, two-way, etc., types of experiments. Readers who need a review should examine Milliken and Johnson (1992).

11.1
MANOVA

Suppose an experiment involving m different populations or treatment groups is conducted. Also suppose there are p response variables measured on each experimental unit. Let μ_{ij} represent the expected response for variable j in treatment group i, for $i = 1, 2, \ldots, m, j = 1, 2, \ldots, p$. Let x_{irj} be the observed value of the jth response variable on the rth experimental unit from the ith population. A multivariate means model might be written as

$$x_{irj} = \mu_{ij} + \varepsilon_{irj}, \quad \text{for } i = 1, 2, \ldots, m, \quad j = 1, 2, \ldots, p, \quad r = 1, 2, \ldots, N_i$$

In matrix form this model can be written as

$$\mathbf{x}_{ir} = \boldsymbol{\mu}_i + \boldsymbol{\varepsilon}_{ir}, \quad \text{for } i = 1, 2, \ldots, m, \quad j = 1, 2, \ldots, N_i$$

where

$$\mathbf{x}_{ir} = \begin{bmatrix} x_{ir1} \\ x_{ir2} \\ \vdots \\ x_{irp} \end{bmatrix}, \quad \boldsymbol{\mu}_i = \begin{bmatrix} \mu_{i1} \\ \mu_{i2} \\ \vdots \\ \mu_{ip} \end{bmatrix}, \quad \text{and} \quad \boldsymbol{\varepsilon}_{ir} = \begin{bmatrix} \varepsilon_{ir1} \\ \varepsilon_{ir2} \\ \vdots \\ \varepsilon_{irp} \end{bmatrix}$$

MANOVA Assumptions

Ideally, MANOVA requires that experimental error vectors be independently and identically distributed as multivariate normal distributions having a zero mean and a common but unknown variance–covariance matrix. That is, under ideal conditions, the ε_{ir}'s are distributed independently and identically N_p $(\mathbf{0}, \mathbf{\Sigma})$ where $\mathbf{\Sigma}$ is an unknown positive definite matrix.

In other words, MANOVA requires that the error vectors corresponding to different experimental units be uncorrelated, but allows for elements in the error vector for the same experimental unit to be correlated with one another.

Test Statistics

Recall that in (univariate) analysis of variance (ANOVA), there are sums of squares that correspond to each of the effects in a model. In multivariate analysis of variance (MANOVA), there are sums of squares and cross-products matrices corresponding to each of the model effects.

The error sum of squares and cross-products matrix will be denoted by $\mathbf{E}$ and, for now, any particular hypothesis sum of squares and cross-products matrix will be denoted by $\mathbf{H}$. In many experiments where the analysis requires MANOVA, several hypothesis matrices may need to be considered. For now, $\mathbf{H}$ is used to represent any one of these.

All MANOVA testing procedures are based on functions of the elements of $\mathbf{H}$ and $\mathbf{E}$, and all of the "good" testing procedures are actually functions of the nonzero eigenvalues of $\mathbf{HE}^{-1}$ and/or $\mathbf{H}(\mathbf{H} + \mathbf{E})^{-1}$.

The number of nonzero eigenvalues of $\mathbf{HE}^{-1}$ and/or $\mathbf{H}(\mathbf{H} + \mathbf{E})^{-1}$ is denoted by s and we can show that $s = \min(h, p)$ where h is the degrees of freedom corresponding to the hypothesis matrix and p, as always, is the number of response variables being analyzed. Also let ν equal the degrees of freedom of the error sums of squares and cross-products matrix.

Some of the MANOVA tests that have been proposed are given next. The most popular MANOVA testing procedures are listed here:

1 *Roy's test:* Roy's test is based on the largest eigenvalue of $\mathbf{HE}^{-1}$.

2 *Lawley and Hotelling's test:* Lawley and Hotelling's test statistic is $T = \text{tr}(\mathbf{HE}^{-1})$.

3 *Pillai's test:* Pillai's test statistic is a function of $V = \text{tr}[\mathbf{H}(\mathbf{H} + \mathbf{E})^{-1}]$.

4 *Roy's second test:* Another test statistic by Roy depends on $U = |\mathbf{H}(\mathbf{H} + \mathbf{E})^{-1}|$.

5 *Wilks' likelihood ratio test:* The likelihood ratio test first derived by Wilks depends on $\Lambda = |\mathbf{E}|/|\mathbf{H} + \mathbf{E}|$.

Many of these multivariate analysis of variance test statistics and their corresponding significance probabilities are available in popular computing packages. There is little to be gained by studying methods proposed for computing significance probabilities for these MANOVA test statistics, because I assume that most readers of this book are using statistical computing packages to analyze data. Readers who are interested in the formulas for estimating p values should reference the text by Srivastava and Carter (1983).

Test Comparisons

One question that certainly arises is "Why are there so many different MANOVA tests?" The basic reason is that none of the MANOVA test procedures can be shown to be best for all possible situations. For univariate cases, when experimental errors are independent and normally distributed with equal variances, it has been proven that the F test used in ANOVA is a uniformly most powerful unbiased testing procedure. This basically means that, given that the experimental errors satisfy ideal conditions, there is no better method than ANOVA for testing the hypothesis of equal means. Hence, one and only one method is usually taught in introductory statistics courses. When only one variable is measured, that is, when $p = 1$, all the preceding multivariate tests are equivalent to one another. Furthermore, when $p = 1$, they are all equivalent to the F test in ANOVA.

Power comparisons involving these multivariate analysis of variance tests have led to inconclusive results. The test that is most powerful depends on the structure of the alternative hypothesis. In particular, Roy's test is the best when the treatment means are nearly collinear under the alternative hypothesis. This would be the case if all of the μ_is fell along a straight line within the p-dimensional sample space. The other multivariate test statistics are asymptotically equivalent to one another and differ very little in power for small samples. This book recommends Wilks' likelihood ratio test because likelihood ratio tests generally have very good properties.

Why Do We Use MANOVAs?

The basic hypothesis testing results of a multivariate analysis of variance are not very exciting. If the multivariate tests are significant, then we know the groups being compared are different, but we do not know which ones are different or why they are different. Are all groups different? Are they different on all variables? Some variables? Which variables?

If the multivariate tests are not significant, then does this mean that the groups being compared do not differ on any of the variables? Or does it mean that they do not differ on very many variables, so that the overall test has low power and, as a result, the MANOVA test statistic is not significant?

Many unanswered questions remain regardless of whether a MANOVA test statistic is statistically significant or not. A primary reason for conducting a MANOVA is to gain some guidance as to how you should appropriately proceed to answer the important questions raised in the preceding two paragraphs.

If MANOVA shows there are significant differences between the groups being compared, then we can recommend looking at the measured variables one at a time to assess where the differences among populations really occur. Using MANOVA in this way is somewhat analogous to conducting an *F* test before looking at multiple comparisons among a set of means—a common and wise practice for many researchers when doing univariate analyses.

However, if MANOVA does not show any significant differences, then performing statistical analyses on the individual variables one at a time is extremely dangerous, because differences that appear to occur may not be real; that is, these differences could be showing up by chance alone. The definition of a real difference is one that would be confirmed if another set of data were collected and statistical analyses on these new data were conducted. Differences that are not likely to be verified by new experiments are often called *false positives*. Data analysts should be wary of statistical methods that have a tendency to produce false positives.

Some books recommend that analyses on individual variables should not be undertaken unless the MANOVA tests show significant differences among the group means being compared. I believe that it is appropriate to consider individual variables in the absence of significant MANOVA results as long as you take a very conservative approach when examining the individual variables. That is, if you are going to consider one-variable-at-a-time analyses, then you should use a significance level much smaller than 0.05.

A Conservative Approach to Multiple Comparisons

If the MANOVA test is not significant and if a researcher is going to look at *p* variables one at a time then the researcher should not claim that there is a real difference between the groups being compared with respect to a particular variable unless that variable's significance level is less than α/p where α is the significance level selected initially for the MANOVA tests. This type of approach is called a *Bonferroni approach*. It guarantees that the proportion of experiments analyzed that declare false positives to be real differences is less than α.

E X A M P L E   11.1

A food scientist was studying the effectiveness of phosphate salts in conjunction with vacuum packaging on the preservation of precooked ground turkey meat during long storage periods. In particular, the researcher wanted to compare five phosphate salt treatments to determine the particular salt treatment that would be most effective. The five phosphate salt treatments consisted of (1) a control (no phosphate salt), (2) sodium tripoly-phosphate (STP) at a 0.3% level, (3) STP at a 0.5% level, (4) sodium ascorbate monophosphate (SAsMP) at a 0.3% level, and (5) SAsMP at a 0.5% level. Samples of cooked ground turkey meat were vacuum packaged using one of the five salt treatments, frozen, and stored for 150 days at approximately −14°C. One complete replication of each of the five salt treatments was started on each of five different days, creating five blocked replicates for each set of treatments.

Some of the variables measured included cooking loss (CKG_LOSS), pH (PH), moistness after cooking (MOIST), fat content (FAT), hexanal content (HEX), bathophenathroline-chelateable iron contents (NONHEM), and the cooking time it took to reach an optimum cooking temperature (CKG_TIME). The data are available in the file named EX11_1.SAS on the enclosed floppy disk. They can be found on page 1 of the computer printout shown in Figure 11.1. These data were analyzed initially by using the following SAS commands:

```
PROC PRINT;
 TITLE 'Ex. 11.1 - MANOVA''S on Cooked Turkey Data';
PROC GLM;
 CLASSES REP TRT;
 MODEL CKG_LOSS--CKG_TIME = REP TRT/SS3;
 MEANS TRT/LSD;
 MANOVA H=TRT/PRINTE CANONICAL;
 RUN;
```

The first four commands in the GLM procedure produce univariate ANOVAs and multiple comparisons on each of the measured response variables using the LSD multiple comparison technique. The MANOVA option produces multivariate analysis of variance test statistics. A portion of the results from these commands is discussed next. The output created by the CANONICAL option is discussed in Section 11.2. A complete printout of the computer output can be obtained by executing the file labeled EX11_1.SAS on the enclosed disk.

Pages 3 through 9 of the Figure 11.1 computer output give univariate analyses of variance tables for each of the measured variables: CKG_LOSS, PH, MOIST, FAT, HEX, NONHEM, and CKG_TIME. Since this book assumes its readers are familiar with one-variable-at-a-time analyses, very little discussion of these analyses is given except to say that the phosphate salt treatments differed significantly for the percentage cooking loss ($P < 0.0001$), moisture content ($P = 0.0052$), hexanal content ($P < 0.0001$), and bathophenathroline-chelateable (BPC) iron contents ($P < 0.0001$). The one-at-a-time analyses that were not statistically significant were those for pH level ($P = 0.6049$), fat content ($P = 0.2223$), and cooking time ($P = 0.6542$).

Pages 10 through 16 give multiple comparisons among the five salt treatments using the least significant difference (LSD) method. For information

Ex. 11.1 - MANOVA'S on Cooked Turkey Data   1

OBS	REP	TRT	CKG_LOSS	PH	MOIST	FAT	HEX	NONHEM	CKG_TIME
1	1	1	36.2	6.20	57.01	11.72	1.32	13.1	47
2	1	2	36.2	6.34	59.95	10.00	1.18	9.2	47
3	1	3	31.5	6.52	61.93	9.85	0.50	5.3	46
4	1	4	33.0	6.49	59.95	10.32	1.55	7.8	48
5	1	5	32.5	6.50	60.98	9.88	0.88	8.8	44
6	2	1	34.8	6.49	61.40	9.35	1.00	8.6	48
7	2	2	33.0	6.62	59.96	9.32	0.40	6.8	52
8	2	3	26.8	6.73	63.37	10.08	0.30	4.8	54
9	2	4	34.0	6.67	59.81	9.20	0.75	6.4	50
10	2	5	32.8	6.86	60.37	9.18	0.32	7.2	48
11	3	1	37.5	6.20	57.01	9.92	0.58	9.8	50
12	3	2	33.2	6.67	60.86	10.18	0.48	8.8	47
13	3	3	27.8	6.78	61.92	9.38	0.20	5.8	47
14	3	4	34.2	6.64	59.34	11.32	0.73	8.0	49
15	3	5	32.0	6.78	58.50	10.48	0.35	7.2	48
16	4	1	38.5	7.34	59.25	10.58	1.48	8.2	48
17	4	2	35.0	6.61	61.12	10.05	0.90	7.0	48
18	4	3	33.8	6.65	60.40	9.52	0.32	6.6	50
19	4	4	35.8	6.47	61.08	10.52	1.58	6.6	49
20	4	5	36.5	6.72	61.61	9.70	0.55	7.4	48
21	5	1	38.5	6.40	56.25	10.18	0.90	9.6	47
22	5	2	34.2	6.67	61.37	9.48	0.65	6.8	47
23	5	3	33.5	6.74	61.60	9.60	0.22	5.2	45
24	5	4	37.5	6.47	60.78	10.18	1.30	7.2	48
25	5	5	36.2	6.70	59.57	10.12	0.88	7.0	48

F I G. 11.1

Pages 1, 3–10, 12, and 14–20 of computer printout, Example 11.1

about this method, see Milliken and Johnson (1992). An examination of those pages corresponding to the response variables that showed significant differences (these are the only pages shown in Figure 11.1) in the ANOVAs reveals that the control treatment (TRT = 1) and STP at 0.5% (TRT = 3) are always significantly different. STP at 0.3% (TRT = 2), SAsMP at 0.3% (TRT = 4), and SAsMP at 0.5% (TRT = 5) were not different for cooking loss, moisture, and BPC iron. For hexanal content, there was no significant difference between the control and SAsMP at 0.3%, and there was no significant difference between STP at 0.3% and SAsMP at 0.5%. However, each of the treatments in the first pair of these was significantly different from each of those in the second pair. Finally, there were no significant differences between any of the noncontrol treatments with respect to moisture. A large number of statements about many variables and about which treatments are different and which ones are not have just been made. Are these real differences or are these differences false positives? To help answer this question, consider the output generated by the MANOVA option.

The matrix printed on the bottom of page 16 (see Figure 11.1) is **E**, the error sum of squares and cross-products matrix. If we were to divide each of the elements in this matrix by the error degrees of freedom ($\nu = 16$, for

Ex. 11.1 - MANOVA'S on Cooked Turkey Data   3

General Linear Models Procedure

Dependent Variable: CKG_LOSS

Source	DF	Sum of Squares	Mean Square	F Value	Pr > F
Model	8	164.24000	20.53000	10.06	0.0001
Error	16	32.64000	2.04000		
Corrected Total	24	196.88000			

R-Square	C.V.	Root MSE	CKG_LOSS Mean
0.834214	4.176274	1.4283	34.20000

Source	DF	Type III SS	Mean Square	F Value	Pr > F
REP	4	57.51600	14.37900	7.05	0.0018
TRT	4	106.72400	26.68100	13.08	0.0001

F I G. 11.1
Pages 1, 3–10, 12, and 14–20 of computer printout, Example 11.1

this example), we would have an unbiased estimate of Σ. That is, $\hat{\Sigma} = \frac{1}{\nu}\mathbf{E}$.

The matrix printed on page 17 is $\mathbf{R}$, the estimated correlation matrix computed from $\hat{\Sigma}$. The top number of the pair of numbers in each cell is the within treatment group correlation between the corresponding pair of response variables, and the lower number is the p value corresponding to a test that the true correlation between this pair of variables is zero.

Remark: If we were going to perform a factor analysis on these response variables to look for relationships among the variables, the factor analysis (FA) would need to be performed on either $\mathbf{R}$ or $\hat{\Sigma}$. That is, we would not dump all of the data into a single FA program and try to interpret the results. Factor analysis assumes that the data come from a single population with a common mean vector and a common variance–covariance matrix. Here, we are sampling from five different populations, one corresponding to each of the five treatments, and so we should expect that these populations could have different mean vectors.

The multivariate analysis of variance test statistics for comparing the five treatments on all seven of the measured variables simultaneously are shown on the bottom of page 20. We can see that Wilks' likelihood ratio test is significant at the 0.0011 significance level. This tells us that there are differences between the treatments on some of the measured variables, which reassures us that the differences seen in the univariate analyses on

Ex. 11.1 - MANOVA'S on Cooked Turkey Data   4

General Linear Models Procedure

Dependent Variable: PH

Source	DF	Sum of Squares	Mean Square	F Value	Pr > F
Model	8	0.4688720	0.0586090	1.19	0.3655
Error	16	0.7906240	0.0494140		
Corrected Total	24	1.2594960			

R-Square	C.V.	Root MSE	PH Mean
0.372270	3.362771	0.2223	6.610400

Source	DF	Type III SS	Mean Square	F Value	Pr > F
REP	4	0.3310560	0.0827640	1.67	0.2048
TRT	4	0.1378160	0.0344540	0.70	0.6049

Ex. 11.1 - MANOVA'S on Cooked Turkey Data   5

General Linear Models Procedure

Dependent Variable: MOIST

Source	DF	Sum of Squares	Mean Square	F Value	Pr > F
Model	8	42.071592	5.258949	3.37	0.0185
Error	16	24.968024	1.560502		
Corrected Total	24	67.039616			

R-Square	C.V.	Root MSE	MOIST Mean
0.627563	2.074546	1.2492	60.21560

Source	DF	Type III SS	Mean Square	F Value	Pr > F
REP	4	7.220696	1.805174	1.16	0.3662
TRT	4	34.850896	8.712724	5.58	0.0052

F I G. 11.1

Pages 1, 3–10, 12, and 14–20 of computer printout, Example 11.1

Ex. 11.1 - MANOVA'S on Cooked Turkey Data 6

General Linear Models Procedure

Dependent Variable: FAT

Source	DF	Sum of Squares	Mean Square	F Value	Pr > F
Model	8	4.5166720	0.5645840	1.96	0.1207
Error	16	4.6195440	0.2887215		
Corrected Total	24	9.1362160			

R-Square	C.V.	Root MSE	FAT Mean
0.494370	5.370918	0.5373	10.00440

Source	DF	Type III SS	Mean Square	F Value	Pr > F
REP	4	2.6672560	0.6668140	2.31	0.1025
TRT	4	1.8494160	0.4623540	1.60	0.2223

Ex. 11.1 - MANOVA'S on Cooked Turkey Data 7

General Linear Models Procedure

Dependent Variable: HEX

Source	DF	Sum of Squares	Mean Square	F Value	Pr > F
Model	8	3.8700880	0.4837610	13.23	0.0001
Error	16	0.5852160	0.0365760		
Corrected Total	24	4.4553040			

R-Square	C.V.	Root MSE	HEX Mean
0.868647	24.74748	0.1912	.7728000

Source	DF	Type III SS	Mean Square	F Value	Pr > F
REP	4	1.3824640	0.3456160	9.45	0.0004
TRT	4	2.4876240	0.6219060	17.00	0.0001

F I G. 11.1

Pages 1, 3–10, 12, and 14–20 of computer printout, Example 11.1

Ex. 11.1 - MANOVA'S on Cooked Turkey Data   8

General Linear Models Procedure

Dependent Variable: NONHEM

Source	DF	Sum of Squares	Mean Square	F Value	Pr > F
Model	8	61.272800	7.659100	9.51	0.0001
Error	16	12.881600	0.805100		
Corrected Total	24	74.154400			

R-Square	C.V.	Root MSE	NONHEM Mean
0.826287	11.85615	0.8973	7.568000

Source	DF	Type III SS	Mean Square	F Value	Pr > F
REP	4	13.638400	3.409600	4.24	0.0158
TRT	4	47.634400	11.908600	14.79	0.0001

Ex. 11.1 - MANOVA'S on Cooked Turkey Data   9

General Linear Models Procedure

Dependent Variable: CKG_TIME

Source	DF	Sum of Squares	Mean Square	F Value	Pr > F
Model	8	55.280000	6.910000	2.44	0.0616
Error	16	45.360000	2.835000		
Corrected Total	24	100.640000			

R-Square	C.V.	Root MSE	CKG_TIME Mean
0.549285	3.499056	1.6837	48.12000

Source	DF	Type III SS	Mean Square	F Value	Pr > F
REP	4	48.240000	12.060000	4.25	0.0156
TRT	4	7.040000	1.760000	0.62	0.6542

F I G. 11.1

Pages 1, 3–10, 12, and 14–20 of computer printout, Example 11.1

Ex. 11.1 - MANOVA'S on Cooked Turkey Data   10

T tests (LSD) for variable: CKG_LOSS
NOTE: This test controls the type I comparisonwise error rate not
      the experimentwise error rate.
            Alpha= 0.05 df= 16 MSE= 2.04
               Critical Value of T= 2.12
            Least Significant Difference= 1.915

Means with the same letter are not significantly different.

T Grouping	Mean	N	TRT
A	37.100	5	1
B			
B	34.900	5	4
B	34.320	5	2
B			
B	34.000	5	5
C	30.680	5	3

Ex. 11.1 - MANOVA'S on Cooked Turkey Data   12

T tests (LSD) for variable: MOIST

NOTE: This test controls the type I comparisonwise error rate not
      the experimentwise error rate.

            Alpha= 0.05 df= 16 MSE= 1.560502
               Critical Value of T= 2.12
            Least Significant Difference= 1.6749

Means with the same letter are not significantly different.

T Grouping	Mean	N	TRT
A	61.844	5	3
A			
A	60.652	5	2
A			
A	60.206	5	5
A			
A	60.192	5	4
B	58.184	5	1

F I G. 11.1
Pages 1, 3–10, 12, and 14–20 of computer printout, Example 11.1

Ex. 11.1 - MANOVA'S on Cooked Turkey Data   14

T tests (LSD) for variable: HEX

NOTE: This test controls the type I comparisonwise error rate not
      the experimentwise error rate.

Alpha= 0.05 df= 16 MSE= 0.036576
Critical Value of T= 2.12
Least Significant Difference= 0.2564

Means with the same letter are not significantly different.

T Grouping		Mean	N	TRT
A				
A		1.182	5	4
A				
A		1.056	5	1
B		0.722	5	2
B				
B		0.596	5	5
C		0.308	5	3

Ex. 11.1 - MANOVA'S on Cooked Turkey Data   15

T tests (LSD) for variable: NONHEM

NOTE: This test controls the type I comparisonwise error rate not
      the experimentwise error rate.

Alpha= 0.05 df= 16 MSE= 0.8051
Critical Value of T= 2.12
Least Significant Difference= 1.203

Means with the same letter are not significantly different.

T Grouping		Mean	N	TRT
A		9.860	5	1
B		7.720	5	2
B				
B		7.520	5	5
B				
B		7.200	5	4
C		5.540	5	3

F I G. 11.1

Pages 1, 3–10, 12, and 14–20 of computer printout, Example 11.1

Ex. 11.1 - MANOVA'S on Cooked Turkey Data   16

T tests (LSD) for variable: CKG_TIME

NOTE: This test controls the type I comparisonwise error rate not
the experimentwise error rate.

Alpha= 0.05   df= 16   MSE= 2.835
Critical Value of T= 2.12
Least Significant Difference= 2.2575

Means with the same letter are not significantly different.

T Grouping	Mean	N	TRT
A	48.800	5	4
A			
A	48.400	5	3
A			
A	48.200	5	2
A			
A	48.000	5	1
A			
A	47.200	5	5

E = Error SS&CP Matrix   Note: $\hat{\Sigma} = \frac{1}{16} E$

	CKG_LOSS	PH	MOIST	FAT
CKG_LOSS	32.64	-1.0914	-10.6042	-2.8444
PH	-1.0914	0.790624	0.778336	0.201764
MOIST	-10.6042	0.778336	24.968024	-1.388804
FAT	-2.8444	0.201764	-1.388804	4.619544
HEX	-1.7046	0.212748	1.685432	0.395968
NONHEM	3.728	-1.28412	-5.94608	2.54748
CKG_TIME	-9.01279E-13	-1.4888	-12.7612	5.8132

	HEX	NONHEM	CKG_TIME
CKG_LOSS	-1.7046	3.728	-9.01279E-13
PH	0.212748	-1.28412	-1.4888
MOIST	1.685432	-5.94608	-12.7612
FAT	0.395968	2.54748	5.8132
HEX	0.585216	-0.63944	0.3604
NONHEM	-0.63944	12.8816	2.684
CKG_TIME	0.3604	2.684	45.36

F I G. 11.1

Pages 1, 3–10, 12, and 14–20 of computer printout, Example 11.1

General Linear Models Procedure
Multivariate Analysis of Variance

Partial Correlation Coefficients from the Error SS&CP Matrix / Prob > |r|

DF = 15	CKG_LOSS	PH	MOIST	FAT	HEX	NONHEM
CKG_LOSS	1.000000	-0.214844	-0.371459	-0.231641	-0.390022	0.181809
	0.0	0.4076	0.1421	0.3710	0.1217	0.4849
PH	-0.214844	1.000000	0.175182	0.105575	0.312768	-0.402379
	0.4076	0.0	0.5013	0.6868	0.2216	0.1093
MOIST	-0.371459	0.175182	1.000000	-0.129315	0.440921	-0.331554
	0.1421	0.5013	0.0	0.6208	0.0765	0.1936
FAT	-0.231641	0.105575	-0.129315	1.000000	0.240825	0.330237
	0.3710	0.6868	0.6208	0.0	0.3518	0.1955
HEX	-0.390022	0.312768	0.440921	0.240825	1.000000	-0.232893
	0.1217	0.2216	0.0765	0.3518	0.0	0.3684
NONHEM	0.181809	-0.402379	-0.331554	0.330237	-0.232893	1.000000
	0.4849	0.1093	0.1936	0.1955	0.3684	0.0
CKG_TIME	0.000000	-0.248608	-0.379195	0.401587	0.069950	0.111035
	1.0000	0.3360	0.1333	0.1101	0.7896	0.6714

DF = 15	CKG_TIME
CKG_LOSS	0.000000
	1.0000
PH	-0.248608
	0.3360
MOIST	-0.379195
	0.1333
FAT	0.401587
	0.1101
HEX	0.069950
	0.7896
NONHEM	0.111035
	0.6714
CKG_TIME	1.000000
	0.0

F I G. 11.1

Pages 1, 3–10, 12, and 14–20 of computer printout, Example 11.1

Ex. 11.1 - MANOVA'S on Cooked Turkey Data   18

```
                  General Linear Models Procedure
                  Multivariate Analysis of Variance
                         Canonical Analysis

H = Type III SS&CP Matrix for TRT    E = Error SS&CP Matrix
```

	Canonical Correlation	Adjusted Canonical Correlation	Approx Standard Error	Squared Canonical Correlation
1	0.972689	0.961184	0.012047	0.946124
2	0.796585	0.721566	0.081718	0.634548
3	0.428091	0.159732	0.182628	0.183262
4	0.273911	0.131628	0.206830	0.075027

```
                  Eigenvalues of INV(E)*H
                   = CanRsq/(1-CanRsq)
```

	Eigenvalue	Difference	Proportion	Cumulative
1	17.5610	15.8247	0.8958	0.8958
2	1.7363	1.5120	0.0886	0.9844
3	0.2244	0.1433	0.0114	0.9959
4	0.0811	.	0.0041	1.0000

```
    Test of H0: The canonical correlations in the current row
                 and all that follow are zero
```

	Likelihood Ratio	Approx F	Num DF	Den DF	Pr > F
1	0.01487446	2.9617	28	37.47772	0.0011
2	0.27608492	1.0115	18	31.59798	0.4738
3	0.75546072	0.3612	10	24	0.9518
4	0.92497262	0.2636	4	13	0.8961

Identifies the dimensionality of the alternative hypothesis.

F I G. 11.1

Pages 1, 3–10, 12, and 14–20 of computer printout, Example 11.1

General Linear Models Procedure
Multivariate Analysis of Variance
Canonical Analysis

Standardized Canonical Coefficients

	CAN1	CAN2	CAN3	CAN4
CKG_LOSS	1.0872	0.1188	0.4830	1.2121
PH	-0.1264	-0.6941	0.0423	0.4149
MOIST	-0.6621	-0.1327	1.2669	0.5932
FAT	-0.0483	0.4983	-0.7841	-0.0208
HEX	2.3618	1.1717	0.0954	-0.0471
NONHEM	0.7622	-1.9341	1.0527	-0.4579
CKG_TIME	-0.3828	-0.0849	1.0779	-0.4543
	$\hat{a}_1^*$	$\hat{a}_2^*$	$\hat{a}_3^*$	$\hat{a}_4^*$

can be ignored

FIG. 11.1

Pages 1, 3–10, 12, and 14–20 of computer printout, Example 11.1

General Linear Models Procedure
Multivariate Analysis of Variance
Canonical Analysis

can be ignored

Raw Canonical Coefficients

	$\hat{a}_1$ CAN1	$\hat{a}_2$ CAN2	$\hat{a}_3$ CAN3	$\hat{a}_4$ CAN4
CKG_LOSS	0.3795941525	0.041486777	0.1686239393	0.4232013394
PH	-0.551914399	-3.030095511	0.1848635339	1.8112490348
MOIST	-0.396167034	-0.079424406	0.7580468256	0.3549481952
FAT	-0.078241806	0.8076902258	-1.270774901	-0.03371606
HEX	5.4816353042	2.7195127915	0.2214546202	-0.109420301
NONHEM	0.4336033043	-1.10028629	0.5988938591	-0.260510396
CKG_TIME	-0.186950577	-0.041440283	0.5263832192	-0.221849694

Manova Test Criteria and F Approximations for
the Hypothesis of no Overall TRT Effect
H = Type III SS&CP Matrix for TRT   E = Error SS&CP Matrix

S=4   M=1   N=4

Statistic	Value	F	Num DF	Den DF	Pr > F
Wilks' Lambda	0.01487446	2.96171	28	37.4777	0.0011
Pillai's Trace	1.838960138	1.58036	28	52	0.0762
Hotelling-Lawley Trace	19.60283423	5.95086	28	34	0.0001
Roy's Greatest Root	17.56100437	32.6133	7	13	0.0001

NOTE: F Statistic for Roy's Greatest Root is an upper bound.

FIG. 11.1

Pages 1, 3–10, 12, and 14–20 of computer printout, Example 11.1

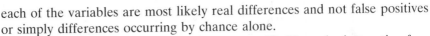

each of the variables are most likely real differences and not false positives or simply differences occurring by chance alone.

Some of the other output shown on pages 18–20 can be interesting for many experiments. These portions of the output are discussed in Sections 11.2 and 11.3.

11.2
Dimensionality of the Alternative Hypothesis

Consider an experiment involving m treatment groups and suppose the mean vectors are given by $\mu_1, \mu_2, \ldots, \mu_m$. These m mean vectors must lie in an $(m - 1)$-dimensional subspace of the p-dimensional sample space. If all of the μ_is are equal to one another, they would lie in a zero-dimensional subspace. This is the case when the MANOVA hypothesis $H_0: \mu_1 = \mu_2 = \cdots = \mu_m$ is true.

If the means were unequal, but fell onto a straight line, then they would lie in a one-dimensional subspace; if the means were unequal and, hence, did not fall onto a line, but fell onto a plane, then they would lie in a two-dimensional subspace. It would, no doubt, be interesting to determine which of these possible alternatives might be the case whenever $H_0: \mu_1 = \mu_2 = \cdots = \mu_m$ is rejected.

In general, the maximum dimensionality of the alternative hypothesis is s, which is the minimum of h and p, and the actual dimensionality is estimated by the number of eigenvalues $\mathbf{HE}^{-1}$ that are significantly greater than zero. To test H_0: the dimensionality of the space in which the μ_is fall is equal to k (for some $k < s$), we compute

$$d_k = \frac{1}{2}(r - p + h) \sum_{i=k+1}^{s} [\log(1 + \hat{\theta}_i)]$$

where $\hat{\theta}_1 \geq \hat{\theta}_2 \geq \cdots \geq \hat{\theta}_s$ are the nonzero eigenvalues $\mathbf{HE}^{-1}$, and reject H_0 if $d_k > \chi^2_{\alpha,(p-k)(h-k)}$.

Remark: When $H_0: \mu_1 = \mu_2 = \cdots = \mu_m$ is rejected, you should determine the dimensionality of the alternative hypothesis by considering $k = 1$, $2, \cdots, s$ until an acceptance is obtained. Such tests are conducted automatically by many MANOVA computing packages.

E X A M P L E **11.1** (continued)

The middle of page 18 of the computer output of Figure 11.1, shows information that can be used to determine the dimensionality of the space in which the sample mean vectors lie. The relevant portion of page 18 is reproduced here:

```
Test of H0: The canonical correlations in the current row
                and all that follow are zero


            Likelihood
               Ratio        Approx F      Num DF       Den DF       Pr > F

      1     0.01487446        2.9617        28        37.47772      0.0011
      2     0.27608492        1.0115        18        31.59798      0.4738
      3     0.75546072        0.3612        10            24        0.9518
      4     0.92497262        0.2636         4            13        0.8961
```

In the likelihood ratio test statistics shown here only the first is statistically significant. This implies that the sample means tend to fall onto a straight line within the seven-dimensional sample space. Thus the dimensionality of the space in which the sample means lie is equal to one. In the next section, a method for comparing the treatments with each other along this line is introduced.

11.3
Canonical Variates Analysis

Suppose it has been determined that the m population means lie in a subspace of dimension k. Canonical variates analysis (CVA) is a method that allows us to compare the m sample means to one another in this subspace. This is particularly interesting and useful when $k \leq 3$ because, in these cases, the m sample means can be plotted in these subspaces. Such plots provide visual information as to which means are quite different from one another and which ones are close to one another using all of the measured variables simultaneously. This greatly simplifies the analysis and, even more important, the discussion of the important results.

The First Canonical Variate

Let $U_{ir} = \mathbf{a}'\mathbf{X}_{ir}$ for $i = 1, 2, \ldots, m$ and $r = 1, 2, \ldots, N_i$. An ANOVA on the U's can be obtained from the MANOVA matrices. In particular, the F value for testing H_0: $\mathbf{a}'\boldsymbol{\mu}_1 = \mathbf{a}'\boldsymbol{\mu}_2 = \cdots = \mathbf{a}'\boldsymbol{\mu}_m$ is given by

$$F = [\mathbf{a}'\mathbf{Ha}/h]/[\mathbf{a}'\mathbf{Ea}/v]$$

The first canonical variate V_1 is defined by the vector $\mathbf{a}$ for which the above F is maximized. Without any loss of generality, a vector $\mathbf{a}$ can be chosen to satisfy $\mathbf{a}'\hat{\boldsymbol{\Sigma}}\mathbf{a} = 1$ where $\hat{\boldsymbol{\Sigma}}$ is the pooled estimate of the variance–covariance matrix, $\boldsymbol{\Sigma}$. That is $\hat{\boldsymbol{\Sigma}} = (1/v)\mathbf{E}$.

We can show that the maximum value of $\mathbf{a}'\mathbf{Ha}/\mathbf{a}'\mathbf{Ea}$ is $\hat{\theta}_1$ where $\hat{\theta}_1$ is the largest eigenvalue of $\mathbf{E}^{-1}\mathbf{H}$; furthermore, this maximum occurs when $\mathbf{a} = \mathbf{a}_1$

where $\mathbf{a}_1$ is an eigenvector of $\mathbf{E}^{-1}\mathbf{H}$ corresponding to the eigenvalue $\hat{\theta}_1$ and satisfying $\mathbf{a}_1'\hat{\mathbf{\Sigma}}\mathbf{a}_1 = 1$.

The Second Canonical Variate

The second canonical variate V_2 is defined by the vector $\mathbf{a}$ that maximizes $F = [\mathbf{a}'\mathbf{Ha}/h]/[\mathbf{a}'\mathbf{Ea}/\nu]$ subject to the constraint that the estimated correlation between the first and second canonical variates is equal to zero; that is, that $\mathbf{a}_1'\hat{\mathbf{\Sigma}}\mathbf{a} = 0$.

The maximum value of $\mathbf{a}'\mathbf{Ha}/\mathbf{a}'\mathbf{Ea}$ subject to the above constraint is $\hat{\theta}_2$, the second largest eigenvalue of $\mathbf{E}^{-1}\mathbf{H}$, and this maximum occurs when $\mathbf{a} = \mathbf{a}_2$ where $\mathbf{a}_2$ is an eigenvector of $\mathbf{E}^{-1}\mathbf{H}$ corresponding to the eigenvalue $\hat{\theta}_2$ and satisfying $\mathbf{a}_2'\hat{\mathbf{\Sigma}}\mathbf{a}_2 = 1$.

Other Canonical Variates

Similarly we can define V_3, V_4, etc., the 3rd, 4th, etc., canonical variates. In each case and without loss of generality, the eigenvectors that satisfy $\mathbf{a}_i'\hat{\mathbf{\Sigma}}\mathbf{a}_i = 1$ for i = 1, 2, $\cdot \cdot \cdot$, s are selected.

E X A M P L E 11.1 (continued)

In Section 11.2, we determined that the space in which the sample means of Example 11.1 fell is one dimensional. That is, the sample means tend to fall along a straight line. The first canonical variate is defined by the first Raw Canonical Coefficients vector given at the top of page 20 of the computer output shown in Figure 11.1. That is, the first canonical variate is defined by

```
CAN1=0.37959*CKG_LOSS-0.55191*PH-0.39617*MOIST-0.078242*FAT
      +5.4816*HEX+0.43360*NONHEM-0.18695*CKG_TIME.
```

This is the linear combination of the response variables that will produce the largest possible F value in a univariate analysis of variance.

The first canonical variate is also given in standardized units under the label Standardized Canonical Coefficients on page 19 of the computer printout (see Figure 11.1). The standardized units may be useful when you are trying to interpret the canonical variates because the standardized canonical coefficients show the relative contributions of each variable to the first canonical variate. You can also compute the correlations between the original variables and a canonical variate; these correlations are also useful when trying to interpret a canonical variate. They could be interpreted similar to the way factor loadings are interpreted.

11.4
Confidence Regions for Canonical Variates

Suppose for now that the experiment being analyzed is balanced with respect to sample size for each treatment group. Let the sample size on which each of the sample mean vectors is computed be denoted by n. We can show that

$$\text{var}[\mathbf{a}_j'(\hat{\boldsymbol{\mu}}_i - \boldsymbol{\mu}_i)] = \mathbf{a}_j'[(1/n)\hat{\boldsymbol{\Sigma}}]\mathbf{a}_j$$
$$= (1/n)\mathbf{a}_j'\hat{\boldsymbol{\Sigma}}\mathbf{a}_j$$
$$= 1/n$$

since $\mathbf{a}_j'\hat{\boldsymbol{\Sigma}}\mathbf{a}_j = 1$. Let k equal the dimension of the space in which the sample means fall. Then $\mathbf{a}_j'(\hat{\boldsymbol{\mu}}_i - \boldsymbol{\mu}_i)$ is approximately $N(0, 1/n)$ for $j = 1, 2, \ldots, k$, and since the canonical variates are independent, it follows that

$$n\left(\sum_{j=1}^{k} [\mathbf{a}_j'(\hat{\boldsymbol{\mu}}_i - \boldsymbol{\mu}_i)]^2\right)$$

is approximately distributed as a chi-square probability distribution with k degrees of freedom for $i = 1, 2, \ldots, m$.

Thus an approximate $(1 - \alpha) \cdot 100\%$ confidence region for $\mathbf{a}_j'\boldsymbol{\mu}_i$, for $j = 1, 2, \ldots, k$ is a spheroid with its center at $[\mathbf{a}_1'\hat{\boldsymbol{\mu}}_i, \mathbf{a}_2'\hat{\boldsymbol{\mu}}_i, \ldots, \mathbf{a}_k'\hat{\boldsymbol{\mu}}_i]'$ and radius equal to $\sqrt{\chi_{\alpha,k}^2/n}$, for $i = 1, 2, \ldots m$. If $k = 1$, then confidence intervals for the first canonical variate can be obtained as shown in the following example.

E X A M P L E 11.1 (continued)

The following SAS commands compute the value of the first canonical variate for each observation vector in the ground turkey study, perform an ANOVA on this newly created variable, compute the locations of the five treatment means in the one-dimensional canonical space, perform approximate multiple comparisons on the first canonical variate, and construct approximate 95% confidence intervals on the first canonical variate for each of the treatments.

```
DATA CVA; SET ;

 CAN1=0.37959*CKG_LOSS-0.55191*PH-0.39617*MOIST-0.078242*FAT
     +5.4816*HEX+0.43360*NONHEM-0.18695*CKG_TIME;

PROC GLM DATA=CVA;
 CLASSES REP TRT;
 MODEL CAN1=REP TRT/SS3;
 MEANS TRT/LSD;
 MEANS TRT/LSD CLM;
```

The results of the preceding commands can be found on pages 22–24 of the computer output shown in Figure 11.2.

Page 22 gives the ANOVA results on the first canonical variate. Note that the largest F value obtainable by considering linear combinations of the seven response variables is shown to be F = 70.24. Also note that its error mean square is equal to 1 (actually 0.99999) on the computer printout. This verifies that the vector $\mathbf{a}_1$, which defines the first canonical variate, was, in fact, selected so that $\mathbf{a}_1'\hat{\mathbf{\Sigma}}\mathbf{a}_1 = 1$.

Note that each of the canonical variate treatment means shown on page 23 is negative. Readers should note that if $\mathbf{a}_i$ defines a canonical variate then $-\mathbf{a}_i$ would also define the same canonical variate. Thus, all of the signs on the canonical variate vector could be changed. This would create a set of canonical variate means that are all positive. In the discussion that follows, no signs have been changed.

On page 23, multiple comparisons among the first canonical variate treatment group means are given. These comparisons suggest that there is little difference between treatments 2 and 5, and that treatments 1, 4, 3, and the pair containing 2 and 5 are all significantly different. Note how the first canonical variate summarizes all of the information in the measured variables in a much simpler, cleaner, and clearer way than did the combination of all of the one-variable-at-a-time analyses. For the interested reader, approximate 95% confidence intervals on the canonical variate means are shown on page 24 (see Figure 11.2).

Figure 11.3 shows a plot of the relative locations of the treatments with respect to one another in the one-dimensional canonical space. Figure 11.4 shows a plot of the first canonical variate means with error bars that have lengths equal to one-half the LSD for comparing a pair of means. In this plot, treatments whose error bars overlap are not significantly different, whereas those whose error bars do not overlap are significantly different.

The ideal turkey product should be low in hexanal content, low in BPC iron content, high in moisture, and low in cooking loss. Each of these responses has positive coefficients in the computation of the first canonical variate. Thus, the better treatments are those that have the smallest values for its first canonical variate mean. From Figure 11.4, treatment 3, which has STP at 0.5%, is significantly better than all of the other treatments. Treatments 2 and 5 are significantly better than treatments 1 and 4, and are not significantly different from one another.

The following commands produce some additional pages of computer output that may help clarify some of the things stated earlier. The MEANS procedure is used to create a data set that has the treatment means for each of the original response variables and for the first canonical variate. Next scatter plots were created by the PLOT procedure to show the relationships between each of the original response variables and the first canonical variate. Finally, the CORR procedure was used to compute the correlations

Ex. 11.1 - MANOVA'S on Cooked Turkey Data   22

Dependent Variable: CAN1

Source	DF	Sum of Squares	Mean Square	F Value	Pr > F
Model	8	387.78657	48.47332	48.47	0.0001
Error	16	15.99981	0.99999		
Corrected Total	24	403.78638			

R-Square	C.V.	Root MSE	CAN1 Mean
0.960376	-5.958334	1.0000	-16.7831

Source	DF	Type III SS	Mean Square	F Value	Pr > F
REP	4	106.81384	26.70346	26.70	0.0001
TRT	4	280.97273	70.24318	70.24	0.0001

Ex. 11.1 - MANOVA'S on Cooked Turkey Data   23

General Linear Models Procedure

T tests (LSD) for variable: CAN1

NOTE: This test controls the type I comparisonwise error rate not
the experimentwise error rate.

Alpha= 0.05 df= 16 MSE= 0.999988
Critical Value of T= 2.12
Least Significant Difference= 1.3407

Means with the same letter are not significantly different.

T Grouping		Mean	N	TRT
A		-12.289	5	1
B		-14.541	5	4
C		-17.107	5	2
C				
C		-17.719	5	5
D		-22.260	5	3

F I G. 11.2
Pages 22–32 of computer printout, Example 11.1

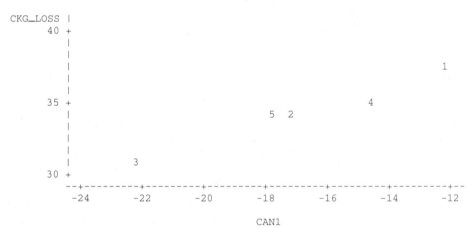

```
       Ex. 11.1 - MANOVA'S on Cooked Turkey Data                    24

              General Linear Models Procedure

           T Confidence Intervals for variable: CAN1

     Alpha= 0.05  Confidence= 0.95  df= 16  MSE= 0.999988
                    Critical Value of T= 2.12
              Half Width of Confidence Interval= 0.948045

                             Lower                      Upper
             TRT       N   Confidence      Mean      Confidence
                             Limit                      Limit

              1        5     -13.237      -12.289      -11.341
              4        5     -15.489      -14.541      -13.593
              2        5     -18.055      -17.107      -16.159
              5        5     -18.667      -17.719      -16.771
              3        5     -23.208      -22.260      -21.312


       Ex. 11.1 - MANOVA'S on Cooked Turkey Data                    25

          Plot of CKG_LOSS*CAN1.  Symbol is value of TRT.

CKG_LOSS |
      40 +
         |
         |
         |                                                    1
         |
         |
      35 +                                          4
         |                      5   2
         |
         |
         |
         |            3
      30 +
         --+---------+---------+---------+---------+---------+---------+--
          -24       -22       -20       -18       -16       -14       -12

                                 CAN1
```

F I G. 11.2
Pages 22–32 of computer printout, Example 11.1

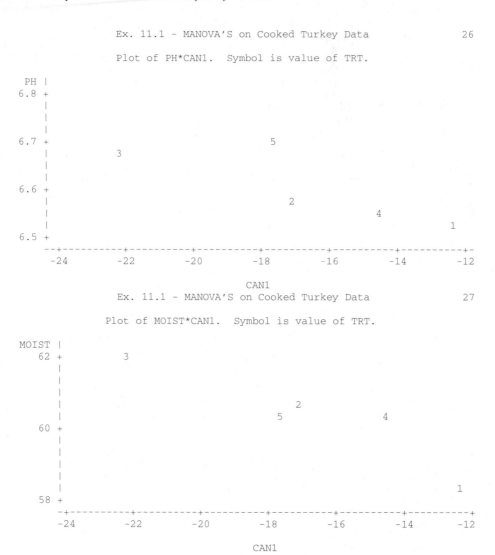

Plot of PH*CAN1.   Symbol is value of TRT.

Plot of MOIST*CAN1.   Symbol is value of TRT.

F I G. 11.2
Pages 22–32 of computer printout, Example 11.1

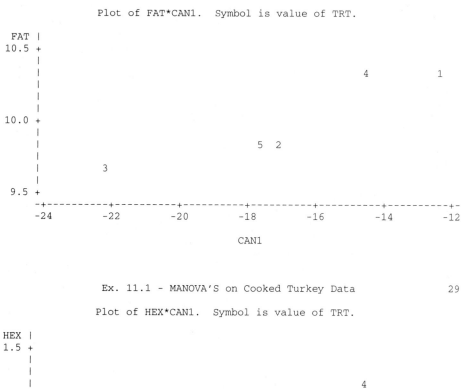

FIG. 11.2
Pages 22–32 of computer printout, Example 11.1

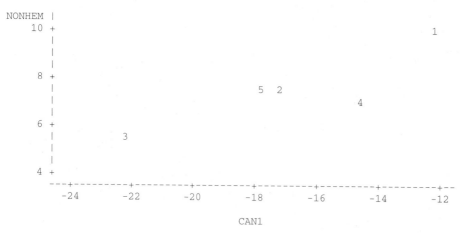

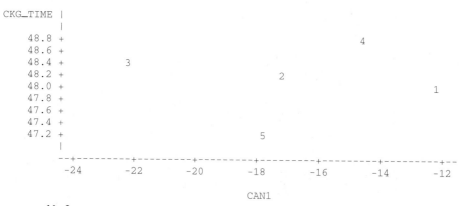

F I G. 11.2

Pages 22–32 of computer printout, Example 11.1

Ex. 11.1 - MANOVA'S on Cooked Turkey Data   32

                          CORRELATION ANALYSIS

1 'WITH' Variables: CAN1
7 'VAR' Variables: CKG_LOSS PH   MOIST   FAT   HEX
                     NONHEM   CKG_TIME

                          Simple Statistics

Variable	N	Mean	Std Dev	Sum	Minimum	Maximum
CAN1	5	-16.7831	3.7482	-83.9156	-22.2596	-12.2893
CKG_LOSS	5	34.2000	2.3100	171.0000	30.6800	37.1000
PH	5	6.6104	0.0830	33.0520	6.5260	6.7120
MOIST	5	60.2156	1.3201	301.0780	58.1840	61.8440
FAT	5	10.0044	0.3041	50.0220	9.6860	10.3500
HEX	5	0.7728	0.3527	3.8640	0.3080	1.1820
NONHEM	5	7.5680	1.5433	37.8400	5.5400	9.8600
CKG_TIME	5	48.1200	0.5933	240.6000	47.2000	48.8000

Pearson Correlation Coefficients / Prob > |R| under Ho: Rho=0 / N = 5

	CKG_LOSS	PH	MOIST	FAT	HEX	NONHEM	CKG_TIME
CAN1	0.98250	-0.81005	-0.92127	0.91369	0.93002	0.88932	0.03226
	0.0028	0.0965	0.0262	0.0300	0.0220	0.0435	0.9589

FIG. 11.2
Pages 22–32 of computer printout, Example 11.1

between the treatment means on each of the original response variables
and the first canonical variate.

```
PROC SORT; BY TRT;

PROC MEANS NOPRINT; BY TRT;
 VAR CAN1 CKG_LOSS PH MOIST FAX HEX NONHEM CKG_TIME;
 OUTPUT OUT=MEANS MEAN=CAN1 CKG_LOSS PH MOIST FAT HEX NONHEM
   CKG_TIME;
 RUN;

OPTIONS PAGESIZE=24;
PROC PLOT;
 PLOT (CKG_LOSS PH MOIST FAT HEX NONHEM CKG_TIME)*CAN1 =
     TRT;
 RUN;

OPTIONS PAGESIZE=54;
PROC CORR;
 VAR CKG_LOSS PH MOIST FAX HEX NONHEM CKG_TIME; WITH CAN1;
 RUN;
```

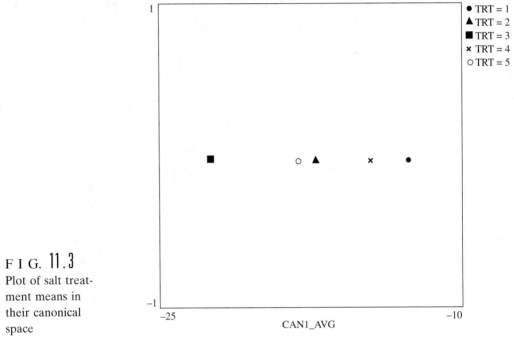

F I G. 11.3
Plot of salt treatment means in their canonical space

F I G. 11.4
Plot of treatment means on first canonical variate ±
($\frac{1}{2}$)LSD

Results from the preceding commands are shown on pages 25–32 of the computer printout of Figure 11.2. The correlations between the treatment means for the original variables and the first canonical variate are shown on page 32. For many experiments, such correlations may be useful in determining how to interpret the canonical variates. It is interesting to note that all of the response variables except for CKG_TIME are quite highly correlated with the first canonical variate with absolute correlations ranging from 0.81 to 0.98. This includes both FAT and PH and suggests that trends exist in these response variables consistent with that seen in the other response variables even though the one-variable-at-a-time analyses on these two variables were not statistically significant.

As you examine each of the plots on pages 25–31, notice the linear trends between the measured variable and the first canonical variate on each plot except for cooking time. Also note that the linear trends appear stronger for those original response variables whose means were significantly different than for those that were not significantly different. Also note how treatments 2 and 5 seem to hang together on each of the plots and that treatment 1 always falls to the far right on each plot. The next example illustrates a case where two canonical functions are required.

E X A M P L E 11.2

A researcher wanted to study the influence of soluble dietary fiber and cholesterol intake on rats. Diets containing 10% dietary fiber as cellulose (control), pectin, psyllium, or oat bran with or without 0.3% added cholesterol were fed to rats for 3 weeks. Hence, eight diets were studied: four dietary fiber diets crossed with each of two cholesterol levels. Seventy-two male rats were grouped into nine blocks according to weight, and each one of the eight FIBER × CHOLESTEROL diet combinations was randomly assigned to one of the 8 rats in each block. The measured responses included total liver lipids (TLIP), total liver cholesterol (TCHOL), final wet liver weight (FWW), final dry liver weight (FWD), total cholesterol in the blood plasma (TPCHOL), and serum high-density lipoprotein cholesterol (HDL). In addition, the percent liver dry matter (PCTDRYMT) was computed from FWW and FWD by the formula PCTDRYMT = (FWD/FWW)*100. The data are in the file labeled EX11_2.SAS on enclosed floppy disk. A portion of the data (that coming from the first three blocks) is shown on page 1 of the computer printout of Figure 11.5.

Initially the data were analyzed using the following SAS commands:

```
PROC PRINT DATA=RATDAT;
  VAR BLK DIET CHOL FIBER TLIP TCHOL TPCHOL HDL PCTDRYMT;
  TITLE 'Ex. 11.2 - Analyses of Rat Data';
```

Ex. 11.2 - Analyses of Rat Data   1

OBS	BLK	DIET	CHOL	FIBER	TLIP	TCHOL	TPCHOL	HDL	PCTDRYMT
1	1	A	1	1	53.54	1.16	56.49	48.41	79.6782
2	2	A	1	1	47.01	2.12	66.06	28.52	74.6070
3	3	A	1	1	48.35	2.52	62.92	24.91	73.6120
10	1	B	2	1	59.52	5.10	79.04	24.99	67.8646
11	2	B	2	1	61.96	8.63	92.72	29.92	85.1869
12	3	B	2	1	53.26	5.75	65.72	35.02	86.6071
19	1	C	1	2	44.57	1.28	73.54	50.37	83.3915
20	2	C	1	2	44.84	2.03	39.43	16.02	80.5000
21	3	C	1	2	45.11	1.02	59.51	32.94	85.7453
28	1	D	2	2	41.31	2.47	80.64	35.74	73.8142
29	2	D	2	2	39.40	0.95	47.07	14.28	87.7888
30	3	D	2	2	38.32	1.90	63.77	33.37	76.5250
37	1	E	1	3	48.92	1.44	79.40	42.63	49.8678
38	2	E	1	3	40.76	1.91	50.80	31.54	58.4488
39	3	E	1	3	45.93	1.25	61.28	52.87	59.0308
46	1	F	2	3	49.45	4.09	95.74	17.85	59.0332
47	2	F	2	3	51.09	2.04	52.22	38.21	74.5079
48	3	F	2	3	48.10	2.88	70.88	44.75	72.3275
55	1	G	1	4	57.33	1.92	78.16	51.25	68.3817
56	2	G	1	4	41.28	1.75	52.04	49.52	81.6667
57	3	G	1	4	44.84	1.33	84.19	74.43	77.0245
64	1	H	2	4	48.10	4.44	93.61	47.30	81.0921
65	2	H	2	4	64.13	5.38	52.58	25.08	77.0755
66	3	H	2	4	40.22	2.55	60.75	20.62	76.8467

F I G. 11.5

Pages 1, 4–11, 13–19, 21, and 23–26 of computer printout, Example 11.2

```
PROC GLM;
  CLASSES CHOL FIBER BLK;
  MODEL TLIP TCHOL TPCHOL HDL PCTDRYMT = BLK CHOL|FIBER/SS3;
  MANOVA H=CHOL|FIBER;
  TITLE2 'ANOVA''s and MANOVA''s on Rat Data';
RUN;
```

Some of the output from these commands is shown in Figure 11.5. The total output can be seen by executing the file named EX11_2.SAS on the enclosed disk. Analysis of variance tables for each of the measured variables are shown on pages 4–8. These results are not discussed here except to say that the diets seemed to have some effects on each of the measured variables.

A portion of the results from the MANOVA option is shown on pages 9–11 of the computer printout. Note that there are MANOVA tests for FIBER main effect, CHOLESTEROL main effect, and FIBER × CHOLESTEROL interaction. Also note that each of these multivariate analysis of variance tests is statistically significant. As a result, we are reassured that the significant differences seen in the one-variable-at-a-time analyses are real and not false positives.

Ex. 11.2 - Analyses of Rat Data   4
ANOVA's and MANOVA's on Rat Data
General Linear Models Procedure

Dependent Variable: TLIP

Source	DF	Sum of Squares	Mean Square	F Value	Pr > F
Model	15	3639.37578	242.62505	6.23	0.0001
Error	56	2179.16741	38.91370		
Corrected Total	71	5818.54319			

R-Square	C.V.	Root MSE	TLIP Mean
0.625479	12.82904	6.23808	48.624722

Source	DF	Type III SS	Mean Square	F Value	Pr > F
BLK	8	446.01934	55.75242	1.43	0.2034
CHOL	1	806.27894	806.27894	20.72	0.0001
FIBER	3	1468.33905	489.44635	12.58	0.0001
CHOL*FIBER	3	918.73845	306.24615	7.87	0.0002

Ex. 11.2 - Analyses of Rat Data   5
ANOVA's and MANOVA's on Rat Data
General Linear Models Procedure

Dependent Variable: TCHOL

Source	DF	Sum of Squares	Mean Square	F Value	Pr > F
Model	15	163.953865	10.930258	14.03	0.0001
Error	56	43.614167	0.778824		
Corrected Total	71	207.568032			

R-Square	C.V.	Root MSE	TCHOL Mean
0.789880	33.72830	0.88251	2.6165278

Source	DF	Type III SS	Mean Square	F Value	Pr > F
BLK	8	13.0517444	1.6314681	2.09	0.0515
CHOL	1	66.8746125	66.8746125	85.87	0.0001
FIBER	3	42.1660597	14.0553532	18.05	0.0001
CHOL*FIBER	3	41.8614486	13.9538162	17.92	0.0001

F I G. 11.5

Pages 1, 4–11, 13–19, 21, and 23–26 of computer printout, Example 11.2

Ex. 11.2 - Analyses of Rat Data

ANOVA's and MANOVA's on Rat Data

General Linear Models Procedure

6

Dependent Variable: TPCHOL

Source	DF	Sum of Squares	Mean Square	F Value	Pr > F
Model	15	8761.15370	584.07691	2.54	0.0060
Error	56	12865.81150	229.74663		
Corrected Total	71	21626.96520			

R-Square	C.V.	Root MSE	TPCHOL Mean
0.405103	20.12444	15.1574	75.318333

Source	DF	Type III SS	Mean Square	F Value	Pr > F
BLK	8	5016.26527	627.03316	2.73	0.0129
CHOL	1	323.93609	323.93609	1.41	0.2401
FIBER	3	3221.22334	1073.74111	4.67	0.0055
CHOL*FIBER	3	199.72899	66.57633	0.29	0.8326

Ex. 11.2 - Analyses of Rat Data

ANOVA's and MANOVA's on Rat Data

General Linear Models Procedure

7

Dependent Variable: HDL

Source	DF	Sum of Squares	Mean Square	F Value	Pr > F
Model	15	11397.7981	759.8532	5.72	0.0001
Error	56	7433.9672	132.7494		
Corrected Total	71	18831.7652			

R-Square	C.V.	Root MSE	HDL Mean
0.605243	31.24762	11.5217	36.872222

Source	DF	Type III SS	Mean Square	F Value	Pr > F
BLK	8	9886.28762	1235.78595	9.31	0.0001
CHOL	1	66.24005	66.24005	0.50	0.4829
FIBER	3	1117.47453	372.49151	2.81	0.0479
CHOL*FIBER	3	327.79588	109.26529	0.82	0.4866

F I G. 11.5

Pages 1, 4–11, 13–19, 21, and 23–26 of computer printout, Example 11.2

```
                  Ex. 11.2 - Analyses of Rat Data                 8
               ANOVA's and MANOVA's on Rat Data
                General Linear Models Procedure
```

Dependent Variable: PCTDRYMT

Source	DF	Sum of Squares	Mean Square	F Value	Pr > F
Model	15	6532.41997	435.49466	7.01	0.0001
Error	56	3476.84320	62.08649		
Corrected Total	71	10009.26317			

R-Square	C.V.	Root MSE	PCTDRYMT Mean
0.652637	10.66479	7.87950	73.883284

Source	DF	Type III SS	Mean Square	F Value	Pr > F
BLK	8	376.21394	47.02674	0.76	0.6411
CHOL	1	45.61661	45.61661	0.73	0.3950
FIBER	3	6091.02268	2030.34089	32.70	0.0001
CHOL*FIBER	3	19.56673	6.52224	0.11	0.9568

```
                  Ex. 11.2 - Analyses of Rat Data                 9
               ANOVA's and MANOVA's on Rat Data

                General Linear Models Procedure
                Multivariate Analysis of Variance


            Manova Test Criteria and Exact F Statistics for
               the Hypothesis of no Overall CHOL Effect
       H = Type III SS&CP Matrix for CHOL   E = Error SS&CP Matrix

                     S=1    M=1.5    N=25
```

Statistic	Value	F	Num DF	Den DF	Pr > F
Wilks' Lambda	0.37991027	16.9749	5	52	0.0001
Pillai's Trace	0.62008973	16.9749	5	52	0.0001
Hotelling-Lawley Trace	1.63220051	16.9749	5	52	0.0001
Roy's Greatest Root	1.63220051	16.9749	5	52	0.0001

F I G. 11.5

Pages 1, 4–11, 13–19, 21, and 23–26 of computer printout, Example 11.2

```
                     Ex. 11.2 - Analyses of Rat Data              10
                     ANOVA's and MANOVA's on Rat Data

                     General Linear Models Procedure
                     Multivariate Analysis of Variance


                 Manova Test Criteria and F Approximations for
                    the Hypothesis of no Overall FIBER Effect
          H = Type III SS&CP Matrix for FIBER    E = Error SS&CP Matrix


                      S=3     M=0.5     N=25

Statistic                    Value         F      Num DF    Den DF   Pr > F

Wilks' Lambda             0.13321277   10.3215       15   143.9505   0.0001
Pillai's Trace            1.31555535    8.4348       15        162   0.0001
Hotelling-Lawley Trace    3.52638652   11.9114       15        152   0.0001
Roy's Greatest Root       2.58013412   27.8654        5         54   0.0001

        NOTE: F Statistic for Roy's Greatest Root is an upper bound.



                     Ex. 11.2 - Analyses of Rat Data              11
                     ANOVA's and MANOVA's on Rat Data

                     General Linear Models Procedure
                     Multivariate Analysis of Variance


           Characteristic Roots and Vectors of: E Inverse * H, where
         H = Type III SS&CP Matrix for CHOL*FIBER    E = Error SS&CP Matrix


                 Manova Test Criteria and F Approximations for
                  the Hypothesis of no Overall CHOL*FIBER Effect
         H = Type III SS&CP Matrix for CHOL*FIBER    E = Error SS&CP Matrix


                      S=3     M=0.5     N=25

Statistic                    Value         F      Num DF    Den DF   Pr > F

Wilks' Lambda             0.45761874    3.1413       15   143.9505   0.0002
Pillai's Trace            0.56634990    2.5133       15        162   0.0023
Hotelling-Lawley Trace    1.13317341    3.8276       15        152   0.0001
Roy's Greatest Root       1.08581040   11.7268        5         54   0.0001

        NOTE: F Statistic for Roy's Greatest Root is an upper bound.
```

F I G. 11.5

Pages 1, 4–11, 13–19, 21, and 23–26 of computer printout, Example 11.2

```
           Ex. 11.2 - Analyses of Rat Data                   13
  Determining the Dimensionality of the Alternative Hypothesis

               General Linear Models Procedure

              T tests (LSD) for variable: TLIP

NOTE: This test controls the type I comparisonwise error rate not
      the experimentwise error rate.

            Alpha= 0.05  df= 56  MSE= 38.9137
                  Critical Value of T= 2.00
            Least Significant Difference= 5.8909

Means with the same letter are not significantly different.

          T Grouping          Mean     N  DIET

                  A          63.743     9  B

                  B          54.651     9  H

                  C          47.102     9  F
                  C
                  C          46.698     9  A
                  C
                  C          45.699     9  G
                  C
                  C          44.486     9  C
                  C
                  C          44.231     9  E
                  C
                  C          42.388     9  D
```

F I G. **11.5**

Pages 1, 4–11, 13–19, 21, and 23–26 of computer printout, Example 11.2

The results from the MEANS option from another GLM analysis (commands given later) are given on pages 13–17 of the computer printout. These pages give multiple comparisons on the diets by using each response variable one at a time. It is difficult to make general conclusions from these comparisons because of the dissimilarity of the diet comparisons across the variables being analyzed. To make general conclusions, we need to determine the dimensionality of the alternative hypothesis and examine the treatment group means in a reduced dimensional canonical space.

Thus, the data are reanalyzed to determine the dimensionality of the alternative hypothesis and to develop canonical functions of the diet means in the canonical space.

The SAS commands used follow:

```
PROC GLM;
 CLASSES BLK DIET;
 MODEL TLIP TCHOL TPCHOL HDL PCTDRYMT = BLK DIET/NOUNI;
```

```
                      Ex. 11.2 - Analyses of Rat Data                    14
       Determining the Dimensionality of the Alternative Hypothesis

                      General Linear Models Procedure

                      T tests (LSD) for variable: TCHOL

      NOTE: This test controls the type I comparisonwise error rate not
            the experimentwise error rate.

                 Alpha= 0.05  df= 56  MSE= 0.778824
                      Critical Value of T= 2.00
                 Least Significant Difference= 0.8334

      Means with the same letter are not significantly different.

              T Grouping              Mean      N  DIET

                           A          5.966      9  B

                           B          3.804      9  H

                           C          2.581      9  F
                           C
                  D        C          1.970      9  D
                  D
                  D                   1.709      9  E
                  D
                  D                   1.659      9  A
                  D
                  D                   1.630      9  G
                  D
                  D                   1.613      9  C
```

FIG. 11.5

Pages 1, 4–11, 13–19, 21, and 23–26 of computer printout, Example 11.2

```
      MEANS DIET/LSD;
      MANOVA H=DIET/CANONICAL;
      TITLE2 'Determining the Dimensionality of the Alternative
          Hypothesis';
      RUN;
```

A portion of the results from the CANONICAL option on the MANOVA option in the preceding SAS commands is shown on pages 18 and 19 of the Figure 11.5 computer printout. From the likelihood ratio tests just below the middle of page 18, we can see that the first two canonical functions are statistically significant, while the others are not. This suggests that the dimensionality of the alternative space is two. We can also see that the first two canonical variates account for 93.17% of the total variability among the diet means.

The bottom of page 19 lists the Standardized Canonical Coefficients that can be used to compute canonical variates on standardized

```
          Ex. 11.2 - Analyses of Rat Data                    15
   Determining the Dimensionality of the Alternative Hypothesis

               General Linear Models Procedure

            T tests (LSD) for variable: TPCHOL

 NOTE: This test controls the type I comparisonwise error rate not
       the experimentwise error rate.

            Alpha= 0.05  df= 56  MSE= 229.7466
                  Critical Value of T= 2.00
            Least Significant Difference= 14.314

 Means with the same letter are not significantly different.

        T Grouping              Mean      N  DIET

                    A           85.161     9  H
                    A
                    A           80.346     9  B
                    A
                    A           80.344     9  F
                    A
            B       A           78.708     9  G
            B       A
            B       A   C       77.130     9  E
            B       A   C
            B       A   C       72.351     9  A
            B           C
            B           C       64.600     9  C
                        C
                        C       63.907     9  D
```

F I G. 11.5

Pages 1, 4–11, 13–19, 21, and 23–26 of computer printout, Example 11.2

data and the Raw Canonical Coefficients that can be used to compute canonical variates on the unstandardized data.

The following commands were used to create values for each of the experimental units in the canonical variates space, to compute the means of the diet treatment groups in this canonical space, and to plot these means in the canonical space. A scatter plot of these means is shown on page 21 of the computer printout.

```
DATA CVSCRS;
 SET RATDAT;

   CV1 = 0.0542*TLIP + 0.8674*TCHOL - 0.01178*TPCHOL
             - 0.00023*HDL + 0.0602*PCTDRYMT;

   CV2 = -0.03414*TLIP + 0.6164*TCHOL + 0.00537*TPCHOL
             + 0.00571*HDL - 0.1125*PCTDRYMT;
```

```
              Ex. 11.2 - Analyses of Rat Data                    16
    Determining the Dimensionality of the Alternative Hypothesis

                   General Linear Models Procedure

                   T tests (LSD) for variable: HDL

    NOTE: This test controls the type I comparisonwise error rate not
          the experimentwise error rate.

                   Alpha= 0.05  df= 56  MSE= 132.7494
                       Critical Value of T= 2.00
                   Least Significant Difference= 10.88

    Means with the same letter are not significantly different.

            T Grouping               Mean       N  DIET

                          A          45.487      9  G
                          A
                 B        A          40.198      9  F
                 B        A
                 B        A          38.238      9  H
                 B        A
                 B        A          37.951      9  E
                 B        A
                 B        A          37.559      9  A
                 B
                 B                   32.627      9  D
                 B
                 B                   32.590      9  B
                 B
                 B                   30.329      9  C
```

F I G. 11.5
Pages 1, 4–11, 13–19, 21, and 23–26 of computer printout, Example 11.2

```
PROC SORT; BY DIET;

PROC MEANS NOPRINT; BY DIET;
 VAR CV1 CV2;
 OUTPUT OUT=MEANS MEAN=CV1 CV2;

PROC PLOT;
 PLOT CV2*CV1=DIET;
 TITLE2 'Plot of Sample Means Using the First Two Canonical
Variates';
RUN;
```

An examination of the plot on page 21 seems to suggest that diets A, C, D, and G are quite similar to one another and that diets E and F are similar to each other. Unfortunately, these assessments are being made subjectively rather than statistically. To quantify these assessments, we can con-

```
            Ex. 11.2 - Analyses of Rat Data                    17
     Determining the Dimensionality of the Alternative Hypothesis

                   General Linear Models Procedure

              T tests (LSD) for variable: PCTDRYMT

     NOTE: This test controls the type I comparisonwise error rate not
           the experimentwise error rate.

                   Alpha= 0.05  df= 56  MSE= 62.08649
                       Critical Value of T= 2.00
                   Least Significant Difference= 7.4409

     Means with the same letter are not significantly different.

            T Grouping           Mean      N  DIET

                   A            82.500      9  B
                   A
                   A            80.605      9  A
                   A
                   A            79.127      9  H
                   A
                   A            78.579      9  D
                   A
                   A            77.959      9  C
                   A
                   A            75.959      9  G

                   B            58.511      9  F
                   B
                   B            57.827      9  E
```

FIG. 11.5
Pages 1, 4–11, 13–19, 21, and 23–26 of computer printout, Example 11.2

struct approximate confidence regions (circles) about each of the diet means in the canonical space. To get the information necessary for this, we can use the following SAS commands:

```
PROC GLM DATA=CVSCRS;
 CLASSES BLK DIET;
 MODEL CV1 CV2 = BLK DIET/SS3;
 MEANS DIET/LSD;
 TITLE2 'Univariate Analyses of the First Two Canonical
Variates';
 RUN;
```

A portion of the results from the preceding commands is shown on pages 23–26 of the computer printout of Figure 11.5.

Pages 23 and 24 give ANOVA tables for each of the canonical variates. Note that the F for comparing diets with respect to CV1 is 34.79. This is the

Ex. 11.2 - Analyses of Rat Data   18
Determining the Dimensionality of the Alternative Hypothesis

General Linear Models Procedure
Multivariate Analysis of Variance
Canonical Analysis

H = Type III SS&CP Matrix for DIET   E = Error SS&CP Matrix

	Canonical Correlation	Adjusted Canonical Correlation	Approx Standard Error	Squared Canonical Correlation
1	0.901688	0.884464	0.023555	0.813041
2	0.775945	0.747286	0.050132	0.602091
3	0.508696	0.444725	0.093386	0.258772
4	0.255636	0.149115	0.117755	0.065350
5	0.103531	-.013186	0.124638	0.010719

Eigenvalues of INV(E)*H
= CanRsq/(1-CanRsq)

	Eigenvalue	Difference	Proportion	Cumulative
1	4.3488	2.8356	0.6912	0.6912
2	1.5131	1.1640	0.2405	0.9317
3	0.3491	0.2792	0.0555	0.9872
4	0.0699	0.0591	0.0111	0.9983
5	0.0108	.	0.0017	1.0000

Test of H0: The canonical correlations in the current row
and all that follow are zero

	Likelihood Ratio	Approx F	Num DF	Den DF	Pr > F
1	0.05098609	6.5020	35	221.174	0.0001
2	0.27271219	3.4995	24	186.105	0.0001
3	0.68536330	1.4615	15	149.4716	0.1267
4	0.92463203	0.5494	8	110	0.8168
5	0.98928143	0.2022	3	56	0.8944

Dimensionality of the alternative hypothesis is 2.

Standardized Canonical Coefficients

	CAN1	CAN2	CAN3	CAN4	CAN5
TLIP	0.4907	-0.3091	0.6040	-1.3608	-0.5657
TCHOL	1.4831	1.0539	-0.7377	1.2394	0.1530
TPCHOL	-0.2057	0.0937	0.8498	-0.2221	0.8293
HDL	-0.0037	0.0929	0.7839	0.9948	-0.7213
PCTDRYMT	0.7146	-1.3353	0.2483	0.1794	0.1616
	$\hat{a}_1^*$	$\hat{a}_2^*$			

Can be ignored

F I G. 11.5

Pages 1, 4–11, 13–19, 21, and 23–26 of computer printout, Example 11.2

```
              Ex. 11.2 - Analyses of Rat Data            19
      Determining the Dimensionality of the Alternative Hypothesis

                    General Linear Models Procedure
                    Multivariate Analysis of Variance
                          Canonical Analysis

                      Raw Canonical Coefficients

              CAN1         CAN2         CAN3         CAN4         CAN5

   TLIP     0.05420033  -0.03414083  0.066722296  -0.15032421  -0.06249383
   TCHOL    0.867411007  0.616354307  -0.43143685  0.724887478  0.089491747
   TPCHOL  -0.01178451   0.005370411  0.04869251   -0.01272416  0.047516026
   HDL     -0.00022842   0.005706718  0.048135885  0.06108523   -0.04429163
   PCTDRYMT 0.060183124  -0.1124594   0.020916629  0.015107172  0.013606352
```

$\hat{a}_1$ $\hat{a}_2$

can be ignored

F I G. 11.5

Pages 1, 4–11, 13–19, 21, and 23–26 of computer printout, Example 11.2

maximum F attainable by considering linear combinations of the measured response variables. Pages 25 and 26 give the values of the canonical variate means and multiple comparisons on the diets using these canonical variate means. The critical value used here for these multiple comparisons is not actually a correct one, since the data are being used to define the canonical variates. Still the comparisons are useful, particularly if we are able to interpret the canonical variates.

Earlier in this section we showed how confidence spheroids can be constructed for each of the diet means. Because the canonical space is two dimensional, this means the confidence spheroids will actually be circles. Because $k = 2$, the radius of each circle will be equal to $\sqrt{\chi^2_{\alpha,k}/n} = \sqrt{5.99/9} = 0.816$ for approximately 95% confidence regions. Figure 11.6 shows a plot of the diet means in the two-dimensional canonical space with 95% confidence circles about each mean. Circles that overlap indicate diets that are probably not significantly different, while diets whose circles do not overlap are those that are significantly different. An examination of Figure 11.6 reveals that diets A, C, D, and G do not differ from one another significantly, and that diets E and F do not differ. Diets B, H, {A, C, D, G} and {E, F} do differ significantly from one another.

Also note that with respect to the first canonical variate, B≫H≫{A, C, D, G} ≈ {E, F} where "≫" means "significantly larger than" and "≈" means "not significantly different than." This is consistent with the LSD comparisons shown on page 25 of the computer printout (see Figure 11.5). Next consider the standardized canonical coefficients printed on page 18 of

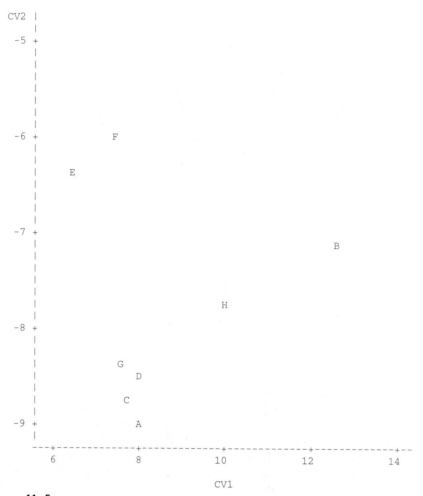

F I G. 11.5

Pages 1, 4–11, 13–19, 21, and 23–26 of computer printout, Example 11.2

the computer printout. From these it appears that TLIP, TCHOL, and PCTDRYMT are positively correlated with the first canonical variate. It also appears that TPCHOL and HDL do not seem to have any significant relationship with the first canonical variate. To summarize what can be deduced from the preceding observations, we can say that for the response variables TLIP, TCHOL, and PCTDRYMT, B≫H≫{A, C, D, G} ≈ {E, F}.

Ex. 11.2 - Analyses of Rat Data   23
Univariate Analyses of the First Two Canonical Variates

General Linear Models Procedure

Dependent Variable: CV1

Source	DF	Sum of Squares	Mean Square	F Value	Pr > F
Model	15	260.964441	17.397629	17.40	0.0001
Error	56	56.005626	1.000100		
Corrected Total	71	316.970067			

	R-Square	C.V.	Root MSE	CV1 Mean
	0.823309	11.82501	1.00005	8.4570793

Source	DF	Type III SS	Mean Square	F Value	Pr > F
BLK	8	17.409607	2.176201	2.18	0.0432
DIET	7	243.554835	34.793548	34.79	0.0001

Ex. 11.2 - Analyses of Rat Data   24
Univariate Analyses of the First Two Canonical Variates

General Linear Models Procedure

Dependent Variable: CV2

Source	DF	Sum of Squares	Mean Square	F Value	Pr > F
Model	15	94.2935975	6.2862398	6.28	0.0001
Error	56	56.0331432	1.0005918		
Corrected Total	71	150.3267407			

	R-Square	C.V.	Root MSE	CV2 Mean
	0.627258	-12.91689	1.00030	-7.744090

Source	DF	Type III SS	Mean Square	F Value	Pr > F
BLK	8	9.5077343	1.1884668	1.19	0.3229
DIET	7	84.7858632	12.1122662	12.11	0.0001

F I G. 11.5

Pages 1, 4–11, 13–19, 21, and 23–26 of computer printout, Example 11.2

```
                Ex. 11.2 - Analyses of Rat Data                    25
      Univariate Analyses of the First Two Canonical Variates

                  General Linear Models Procedure

                  T tests (LSD) for variable: CV1

 NOTE: This test controls the type I comparisonwise error rate not
       the experimentwise error rate.

              Alpha= 0.05  df= 56  MSE= 1.0001
                   Critical Value of T= 2.00
            Least Significant Difference= 0.9444

Means with the same letter are not significantly different.

         T Grouping             Mean      N  DIET

                  A            12.642      9  B

                  B            10.014      9  H

                  C             7.976      9  D
                  C
                  C             7.961      9  A
                  C
                  C             7.736      9  C
                  C
                  C             7.526      9  G
                  C
              D   C             7.358      9  F
              D
              D                 6.443      9  E
```

F I G. 11.5

Pages 1, 4–11, 13–19, 21, and 23–26 of computer printout, Example 11.2

Differences between diet groups on the second canonical variate are not quite as obvious as they were with the first canonical variate. From Figure 11.6 and page 26 of the computer printout, it appears that diets F ≈ E, E ≈ B, B ≈ H, H ≈ G, and G ≈ D ≈ C ≈ A. It also appears that F≫{E, B, H, G, D, C, A}, E≫{H, G, D, C, A}, and B≫{G, D, C, A}. From the standardized canonical coefficients for the second canonical variate printed on page 18 of the computer printout, it appears that TCHOL is positively correlated with the second canonical variate while PCTDRYMT is negatively correlated with the second canonical variate. None of the other variables seems to be related to the second canonical variate.

```
            Ex. 11.2 - Analyses of Rat Data                    26
      Univariate Analyses of the First Two Canonical Variates

                 General Linear Models Procedure

                 T tests (LSD) for variable: CV2

NOTE: This test controls the type I comparisonwise error rate not
      the experimentwise error rate.

           Alpha= 0.05  df= 56  MSE= 1.000592
                  Critical Value of T= 2.00
           Least Significant Difference= 0.9446

Means with the same letter are not significantly different.

           T Grouping          Mean      N  DIET

                        A      -5.939     9  F
                        A
                 B      A      -6.331     9  E
                 B
                 B      C      -7.163     9  B
                        C
                 D      C      -7.747     9  H
                 D
                 D      E      -8.418     9  G
                 D      E
                 D      E      -8.543     9  D
                        E
                        E      -8.775     9  C
                        E
                        E      -9.037     9  A
```

F I G. 11.5
Pages 1, 4–11, 13–19, 21, and 23–26 of computer printout, Example 11.2

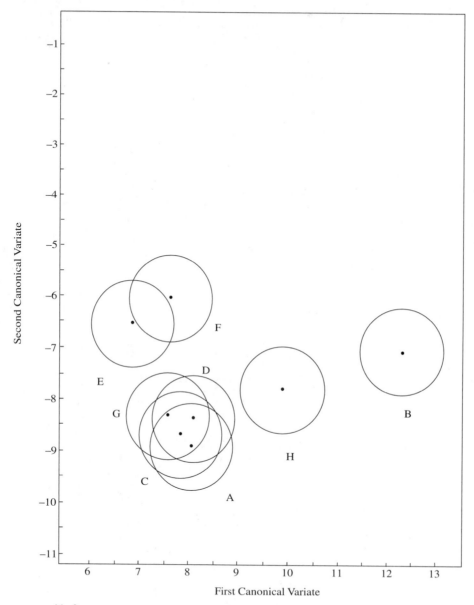

F I G. 11.6

Canonical variates plot for the rat experiment

Summary

This chapter introduces multivariate analysis of variance. This procedure is useful as a preliminary test prior to considering one-variable-at-a-time analyses to protect yourself from claiming false positives to be real. Canonical variates analysis was introduced as a simple way of comparing all treatment groups to one another using all measured variables simultaneously. This is accomplished by comparing the group means to one another in the canonical space in which the treatment means lie.

Exercises

1 Consider the data in the file labeled BEEF.DAT on the enclosed disk. This file contains data from a beef study that was conducted to evaluate the effects that different washing treatments might have on the quality of beef after storage. The experiment was conducted in a randomized block design with three blocks or replicates. Some of the treatments were not evaluated in the first replicate. There were three wash treatments: normal (N–N), chlorine (C–C), and lactic acid (L–L). Storage times were for 10, 40, 80, and 120 days of storage.

Six panelists evaluated meat samples for each wash TREATMENT*STORAGE TIME*REP combination. The samples were assessed for beefy aroma (BEEFY), bloody/serumy aroma (BLOODY), metallic aroma (METAL), grassy/barnyard aroma (GRASSY), sour aroma (SOUR), and spoiled aroma (SPOILED). Each of these was evaluated on a scale ranging from 0 to 15 where 0 indicated no such aroma and 15 indicated a very intense aroma. The numbers in the data set consist of the means of the six panelists' scores.

Using multivariate methods rather than analysis of variance methods, answer each of the following questions:

a Use a MANOVA to determine if there is interaction between the washing treatments and storage time. Explain your answer.

b Are there significant differences due to the washing treatments? Explain your answer.

c Are there significant differences due to the storage times? Explain your answer.

d What is the dimensionality of the space in which the 12 treatment group means lie? Explain your answer.

e What linear combination of the data vectors would produce the largest F value in an ANOVA for comparing the 12 treatment groups? What is the value of this largest F value? Can you interpret this linear combination? What is the value of the largest F value for comparing the 12 treatment groups among the individual response variables?

f   Evaluate each of the 12 treatment group means on the first canonical function. (*Hint:* The computer can be used to help find these means.)

g   Construct 95% confidence intervals about each of the 12 treatment group means using the first canonical function. Which treatment groups appear to be significantly different?

2   There is a file on the enclosed disk labeled NCAA.DAT that contains information for the NCAA graduation rates report appearing in *USA Today* on August 13, 1992. The information is based on the 297 schools that play Division I basketball. Excluded are the 8 Ivy League schools and 3 military academies that do not give athletic scholarships and therefore had no data to report. The data are based on the combined freshman classes of 1983–1984 and 1984–1985, the most current data reported to the NCAA.

The numbers in the data set are the percentage of students, by category, from the above combined classes who graduated within 6 years of enrollment. For most schools, the report covers 30 or more scholarship athletes in football, 5 to 15 in baseball, and fewer than 10 in men's and women's basketball. In determining graduation rates, students who transferred out of the school in good standing were counted as nongraduates, and students who transferred into the school and graduated were not counted. The conference to which each school belongs is identified in the data set. The first two digits in each school's code represent the conference to which the school belongs. The last two digits give a code unique to each school.

a   Do you believe that there are differences in graduation rates between the conferences using all six of the measured graduation rates? Explain your answer.

b   What is the dimensionality of the space in which the 33 conference mean vectors lie? Explain your answer. How would you interpret the first two canonical functions?

c   What linear combination of the data vectors would produce the largest F value in an ANOVA for comparing the 33 conference mean vectors? What is the value of this largest F value? What is the value of the largest F value for comparing the 33 conferences among the individual response variables? What linear combination that is uncorrelated with the first produces the second largest F value? What is the value of this second largest F value? (*Hint:* You will need to use the computer to help you get this F value.)

d   Evaluate each of the 33 conferences on the first two canonical functions and plot these on a graph. (*Hint:* Do not do this one by hand; use the computer.)

e   Construct 95% confidence circles about each of the 33 conference means plotted in part d. Which conferences appear to be significantly different?

Using multivariate methods rather than analysis of variance methods answer each of the following questions.

3 Consider the organizational commitment variables in the Pizzazz survey data described in Appendix B.

    a Use a MANOVA to determine if there is interaction between the gender of the employee and the position filled by the employee. Explain your answer.

    b Are there significant differences due to gender? Explain your answer.

    c Are there significant differences due to the type of position the employee fills? Explain your answer.

    d What is the dimensionality of the space in which these 12 group means lie? Explain your answer.

    e What linear combination of the data vectors would produce the largest F value in an ANOVA for comparing the 12 group means? What is the value of this largest F value? Can you interpret this linear combination? What is the value of the largest F value for comparing the 12 group means among the individual response variables?

    f Evaluate each of the 12 group means on the first canonical function. (*Hint:* The computer can be used to help find these means.)

    g Construct 95% confidence intervals about each of the 12 group means using the first canonical function. Which groups appear to be significantly different?

    h Construct 95% confidence circles about each of the 12 group means using the first two canonical functions. Which groups appear to be significantly different?

4 Repeat Exercise 3 using the job satisfaction variables.

5 Consider the males in the married adult survey described in Appendix C and partition these males into four groups: (1) those whose income is less than \$34,999, (2) those whose income is between \$35,000 and \$49,999, (3) those whose income is between \$50,000 and \$74,999, and (4) those whose income is at least \$75,000. Consider their responses to the FACES questionnaire.

    a Use a MANOVA to determine if there is a difference in mean vectors for these four income groups on the FACES variables.

    b What is the dimensionality of the space in which these four group means lie? Explain your answer.

    c What linear combination of the data vectors would produce the largest F value in an ANOVA for comparing the four income group means? What is the value of this largest F value? Can you interpret this linear combination? What is the value of the largest F value for comparing the four treatment groups among the individual response variables?

d Evaluate each of the four income group means on the first canonical function.

e Construct 95% confidence intervals about each of the four income group means using the first canonical function. Which groups appear to be significantly different?

f Construct 95% confidence circles about each of the four income group means using the first two canonical functions. Which groups appear to be significantly different?

6 Repeat Exercise 5 using the Kansas Marital Satisfaction Scale variables and then the Kansas Family Life Satisfaction Scale variables.

7 Consider the females in the married adult survey described in Appendix C and partition these females into groups according to education level (EDU). Consider their responses to the FACES questionnaire.

a Use a MANOVA to determine if there are differences in mean vectors for the FACES variables in the different educational levels.

b What is the dimensionality of the space in which these group means lie? Explain your answer.

c What linear combination of the data vectors would produce the largest *F* value in an ANOVA for comparing educational levels? What is the value of this largest *F* value? Can you interpret this linear combination? What is the value of the largest *F* value for comparing educational levels among the individual response variables?

d Evaluate each of the educational level means on the first canonical function.

e Construct 95% confidence intervals about each of the educational level means using the first canonical function. Which groups appear to be significantly different?

f Construct 95% confidence circles about each of the educational level means using the first two canonical functions. Which groups appear to be significantly different?

8 Repeat Exercise 7 using the Kansas Marital Satisfaction Scale variables and then the Kansas Family Life Satisfaction Scale variables.

Prediction Models and Multivariate Regression 12

Consider those married females in the married adults survey described in Appendix C whose spouses also filled out a questionnaire. Consider their responses two sets of measured variables—one set containing responses to the Kansas Marital Satisfaction Scale (KMS1–KMS3) and a second set containing their response to the Family Adaptability Cohesion Evaluation Scales (FACES). A researcher might be interested in how these two sets of variables relate to each other for this group of females. In particular, the researcher might be interested in determining if the FACES variables can be used to predict marital satisfaction.

12.1
Multiple Regression

Multiple regression can be used to create models that allow us to model, one by one, each of the marital satisfaction variables as a linear function of the Family Adaptability Cohesion Evaluation Scales as illustrated in the following example.

E X A M P L E 12.1

SAS-REG was used to create models that can be used to predict each of the marital satisfaction responses from the FACES variables. Stepwise regression utilizing the backward elimination procedure was used to simplify each of the models as much as might be possible. The SAS commands, given in the file labeled EX12_1.SAS on the enclosed disk, are repeated here:

```
OPTIONS PAGESIZE=54 NODATE PAGENO=1;
TITLE 'Ex. 12.1 Stepwise regression using the backward
    elimination procedure.';
PROC REG;
 MODEL KMS1-KMS3 = FACES1-FACES30/SELECTION=BACKWARD SLS=.05;
 RUN;
```

More than 50 pages of output are created from these commands, but only a portion of this output is discussed in this section and shown in Figure 12.1. Anyone who wants to see all of the output can execute the file labeled EX12_1.SAS on the enclosed disk. The preceding commands create a separate analysis for each of the dependent variables, KMS1, KMS2, and KMS3. Summaries of the results of the backward elimination procedure are given on pages 19, 36, and 55 of the computer output shown in Figure 12.1,

Ex. 12.1 Stepwise regression using the backward elimination procedure.   18

	DF	Sum of Squares	Mean Square	F	Prob>F
Regression	5	45.95279044	9.19055809	21.22	0.0001
Error	108	46.78405167	0.43318566		
Total	113	92.73684211			

Variable	Parameter Estimate	Standard Error	Type II Sum of Squares	F	Prob>F
INTERCEP	1.76644577	0.56031092	4.30542981	9.94	0.0021
FACES1	0.37679840	0.09785925	6.42226793	14.83	0.0002
FACES6	0.21220732	0.06245630	5.00083145	11.54	0.0010
FACES7	0.20660832	0.08503206	2.55743289	5.90	0.0168
FACES12	-0.24790513	0.07299813	4.99599095	11.53	0.0010
FACES15	-0.14562115	0.06818722	1.97568194	4.56	0.0350

Ex. 12.1 Stepwise regression using the backward elimination procedure.   19
Summary of Backward Elimination Procedure for Dependent Variable KMS1

Step	Variable Removed	Number In	Partial R**2	Model R**2	C(p)	F	Prob>F
1	FACES2	29	0.0000	0.6496	29.0005	0.0005	0.9822
2	FACES3	28	0.0000	0.6496	27.0023	0.0018	0.9660
3	FACES25	27	0.0000	0.6496	25.0055	0.0033	0.9546
4	FACES8	26	0.0000	0.6496	23.0085	0.0031	0.9557
5	FACES23	25	0.0002	0.6494	21.0564	0.0502	0.8232
6	FACES10	24	0.0004	0.6490	19.1506	0.0998	0.7528
7	FACES28	23	0.0006	0.6484	17.2886	0.1477	0.7017
8	FACES14	22	0.0017	0.6467	15.6844	0.4277	0.5148
9	FACES4	21	0.0026	0.6441	14.3056	0.6755	0.4133
10	FACES11	20	0.0059	0.6382	13.6986	1.5201	0.2207
11	FACES24	19	0.0056	0.6326	13.0353	1.4506	0.2315
12	FACES30	18	0.0063	0.6263	12.5192	1.6026	0.2087
13	FACES29	17	0.0079	0.6184	12.3952	2.0133	0.1592
14	FACES21	16	0.0096	0.6088	12.6637	2.4091	0.1239
15	FACES13	15	0.0080	0.6009	12.5563	1.9812	0.1625
16	FACES5	14	0.0083	0.5926	12.5156	2.0307	0.1573
17	FACES20	13	0.0071	0.5855	12.2028	1.7306	0.1914
18	FACES9	12	0.0054	0.5800	11.4855	1.3062	0.2558
19	FACES22	11	0.0095	0.5706	11.7290	2.2776	0.1344
20	FACES19	10	0.0096	0.5610	12.0038	2.2808	0.1341
21	FACES17	9	0.0132	0.5477	13.1388	3.1047	0.0810
22	FACES26	8	0.0148	0.5329	14.6539	3.4122	0.0676
23	FACES16	7	0.0116	0.5213	15.3964	2.6024	0.1097
24	FACES27	6	0.0151	0.5062	16.9802	3.3500	0.0700
25	FACES18	5	0.0107	0.4955	17.5100	2.3140	0.1312

F I G. 12.1

Pages 18, 19, 35, 36, 54, and 55 of computer printout, Example 12.1

Ex. 12.1 Stepwise regression using the backward elimination procedure.   35

	DF	Sum of Squares	Mean Square	F	Prob>F
Regression	7	50.49170045	7.21310006	14.03	0.0001
Error	106	54.49952762	0.51414649		
Total	113	104.99122807			

Variable	Parameter Estimate	Standard Error	Type II Sum of Squares	F	Prob>F
INTERCEP	1.67324258	0.69956508	2.94136180	5.72	0.0185
FACES2	0.26579636	0.08660753	4.84254421	9.42	0.0027
FACES6	0.18167820	0.06944474	3.51895560	6.84	0.0102
FACES7	0.28572493	0.09125843	5.04007652	9.80	0.0023
FACES12	-0.25002911	0.08321362	4.64172230	9.03	0.0033
FACES15	-0.16609476	0.07689922	2.39858329	4.67	0.0330
FACES22	-0.14039240	0.07020233	2.05622667	4.00	0.0481
FACES27	0.20833750	0.10037023	2.21519495	4.31	0.0403

Ex. 12.1 Stepwise regression using the backward elimination procedure.   36

Summary of Backward Elimination Procedure for Dependent Variable KMS2

Step	Variable Removed	Number In	Partial R**2	Model R**2	C(p)	F	Prob>F
1	FACES3	29	0.0001	0.6012	29.0112	0.0112	0.9160
2	FACES28	28	0.0001	0.6011	27.0269	0.0159	0.8999
3	FACES8	27	0.0002	0.6009	25.0675	0.0416	0.8389
4	FACES14	26	0.0003	0.6006	23.1272	0.0618	0.8043
5	FACES23	25	0.0004	0.6002	21.2032	0.0795	0.7787
6	FACES25	24	0.0006	0.5997	19.3181	0.1215	0.7283
7	FACES4	23	0.0009	0.5988	17.5015	0.1959	0.6591
8	FACES16	22	0.0010	0.5978	15.7032	0.2175	0.6421
9	FACES5	21	0.0037	0.5941	14.4780	0.8423	0.3612
10	FACES29	20	0.0033	0.5908	13.1709	0.7545	0.3873
11	FACES21	19	0.0051	0.5857	12.2245	1.1505	0.2862
12	FACES30	18	0.0046	0.5811	11.1751	1.0364	0.3113
13	FACES13	17	0.0047	0.5765	10.1451	1.0571	0.3065
14	FACES17	16	0.0057	0.5708	9.3300	1.2904	0.2588
15	FACES18	15	0.0076	0.5632	8.9165	1.7227	0.1924
16	FACES24	14	0.0076	0.5555	8.5033	1.7105	0.1940
17	FACES11	13	0.0070	0.5486	7.9540	1.5525	0.2157
18	FACES20	12	0.0101	0.5385	8.0605	2.2421	0.1374
19	FACES9	11	0.0102	0.5283	8.1788	2.2273	0.1387
20	FACES19	10	0.0119	0.5164	8.6566	2.5742	0.1117
21	FACES10	9	0.0134	0.5030	9.4417	2.8500	0.0944
22	FACES26	8	0.0141	0.4889	10.3696	2.9437	0.0892
23	FACES1	7	0.0080	0.4809	10.0373	1.6462	0.2023

F I G. 12.1

Pages 18, 19, 35, 36, 54, and 55 of computer printout, Example 12.1

Ex. 12.1 Stepwise regression using the backward elimination procedure.   54

	DF	Sum of Squares	Mean Square	F	Prob>F
Regression	4	55.94503832	13.98625958	23.92	0.0001
Error	109	63.73917221	0.58476305		
Total	113	119.68421053			

Variable	Parameter Estimate	Standard Error	Type II Sum of Squares	F	Prob>F
INTERCEP	2.29841120	0.70228429	6.26339020	10.71	0.0014
FACES1	0.32703279	0.11790198	4.49904472	7.69	0.0065
FACES3	-0.19347446	0.07493341	3.89830823	6.67	0.0112
FACES7	0.34315329	0.09516556	7.60320346	13.00	0.0005
FACES24	-0.24607636	0.08679827	4.69999687	8.04	0.0055

Ex. 12.1 Stepwise regression using the backward elimination procedure.   55

Summary of Backward Elimination Procedure for Dependent Variable KMS3

Step	Variable Removed	Number In	Partial R**2	Model R**2	C(p)	F	Prob>F
1	FACES2	29	0.0000	0.6080	29.0041	0.0041	0.9494
2	FACES23	28	0.0000	0.6080	27.0121	0.0081	0.9284
3	FACES28	27	0.0002	0.6078	25.0547	0.0436	0.8351
4	FACES30	26	0.0002	0.6076	23.0981	0.0450	0.8325
5	FACES11	25	0.0004	0.6072	21.1744	0.0798	0.7782
6	FACES16	24	0.0004	0.6068	19.2560	0.0863	0.7696
7	FACES18	23	0.0007	0.6061	17.4045	0.1588	0.6912
8	FACES8	22	0.0011	0.6050	15.6355	0.2492	0.6189
9	FACES4	21	0.0017	0.6033	13.9999	0.3966	0.5304
10	FACES25	20	0.0014	0.6020	12.2876	0.3150	0.5760
11	FACES29	19	0.0021	0.5998	10.7395	0.4987	0.4819
12	FACES27	18	0.0047	0.5951	9.7430	1.1132	0.2941
13	FACES14	17	0.0049	0.5902	8.7780	1.1467	0.2870
14	FACES21	16	0.0055	0.5847	7.9444	1.2904	0.2588
15	FACES15	15	0.0053	0.5793	7.0747	1.2466	0.2670
16	FACES17	14	0.0073	0.5720	6.6238	1.7044	0.1948
17	FACES19	13	0.0043	0.5677	5.5338	0.9941	0.3212
18	FACES10	12	0.0076	0.5602	5.1398	1.7545	0.1883
19	FACES22	11	0.0050	0.5552	4.1940	1.1432	0.2875
20	FACES26	10	0.0089	0.5462	4.0854	2.0481	0.1555
21	FACES13	9	0.0081	0.5382	3.7913	1.8287	0.1792
22	FACES20	8	0.0104	0.5278	3.9881	2.3362	0.1294
23	FACES5	7	0.0130	0.5148	4.7490	2.8993	0.0916
24	FACES9	6	0.0116	0.5032	5.2038	2.5325	0.1145
25	FACES12	5	0.0173	0.4859	6.8571	3.7156	0.0566
26	FACES6	4	0.0185	0.4674	8.7720	3.8842	0.0513

F I G. 12.1

Pages 18, 19, 35, 36, 54, and 55 of computer printout, Example 12.1

and the final models selected by the procedure are shown on pages 18, 35, and 54. The summaries show the order in which variables are removed from the initial model that includes all of the predictor variables. The fitted model for KMS1 is

```
KMS1= 1.766+0.377·FACES1+0.212·FACES6+0.207·FACES7-
0.248·FACES12-0.146·FACES15
```

This model has $R^2 = 0.4955$. The fitted model for KMS2 is

```
KMS2=1.673+0.266·FACES2+0.182·FACES6+0.286·FACES7-
0.250·FACES12-0.166·FACES15 −0.140·FACES22+0.208·FACES27
```

with $R^2 = 0.4809$. The fitted model for KMS3 is

```
KMS3=2.298+0.327·FACES1-0.193·FACES3+0.343·FACES7-
0.246·FACES24
```

with $R^2 = 0.4674$.

How does one interpret these fitted models? Interpretation is not simple because we do not know for sure why certain predictor variables were eliminated by the backward elimination procedure. Obviously, variables were removed because they were not statistically significant, but why were they not statistically significant? These are two possible explanations:

1 A variable is removed because it has nothing to do with marital satisfaction.

2 A variable is related to marital satisfaction, but it was removed from the model because it contributed nothing over and above what is being contributed by other variables that remain in the model. That is, a variable can be removed if it depends on other variables that remain in the model even though the variable being removed is related to marital satisfaction.

Obviously, it would be nice to know which of these two plausible explanations applies to each of the variables in the predictor set. Unfortunately, this is not always easy to determine. If the dimension of the space in which the predictor variables lie is equal to the number of predictor variables, then we can be assured that when a variable is removed from a predictor set, it is because it has nothing to do with the dependent variable.

However, if the dimension of the space in which the predictor variables lie is less than the number of predictor variables, a variable can be removed for either of the preceding two reasons. In this case, we say that the predictor variables are multicollinear with one another or that there are multicollinearities in the predictor set. Principal components analysis can be used to determine if there are multicollinearities in a predictor set of variables. If any of the eigenvalues of the variance–covariance matrix and/or the correlation matrix of the predictor variables is close to zero, then multicollinearity exists.

In the preceding example, multicollinearity is very likely to exist since all of the predictor variables are trying to get at various aspects of family adaptability cohesion and we really expect them to be correlated with one another.

In the preceding example, in addition to having multicollinearity in the predictor variables, there is likely to be multicollinearity in the dependent variables. This is because each of these variables is trying to get at various aspects of marital satisfaction. How do we handle cases in which there are many predictor variables and many variables to predict and lots of relationships among the variables in each set?

Canonical correlation analysis has been proposed for comparing two sets of variables. This chapter introduces the ideas behind canonical correlation analysis and then introduces an alternative approach involving both factor analysis and regression analysis for comparing two sets of variables.

12.2

Canonical Correlation Analysis

Canonical correlation analysis (CCA) is a generalization of multiple correlation used in multiple regression problems.

Recall that R^2, the *coefficient of determination*, in regression problems is the proportion of the variability in a dependent variable that is accounted for by a set of predictor variables, and $R = \sqrt{R^2}$ is called the *multiple correlation coefficient*. The multiple correlation coefficient can also be interpreted as a measure of the maximum correlation that is attainable between the dependent variable and any linear combination of the predictor variables.

Two Sets of Variables

Suppose that a response vector $\mathbf{x}$ can be partitioned into two parts so that

$$\mathbf{x} = \begin{bmatrix} \mathbf{x}_1 \\ \mathbf{x}_2 \end{bmatrix} \sim N\left(\begin{bmatrix} \boldsymbol{\mu}_1 \\ \boldsymbol{\mu}_2 \end{bmatrix}, \begin{bmatrix} \boldsymbol{\Sigma}_{11} & \boldsymbol{\Sigma}_{12} \\ \boldsymbol{\Sigma}_{21} & \boldsymbol{\Sigma}_{22} \end{bmatrix} \right)$$

where $\mathbf{x}_1$ is a $q \times 1$ $(q < p)$ vector and $\mathbf{x}_2$ is a $(p - q) \times 1$ vector. One question that is of interest is whether the q responses in $\mathbf{x}_1$ are independent of the $p - q$ responses in $\mathbf{x}_2$. If they are not independent, then one might consider using $\mathbf{x}_2$ to predict $\mathbf{x}_1$ or vice versa.

As an example, suppose the q variables in $\mathbf{x}_1$ correspond to measurements that are hard to obtain and/or expensive to measure and that the $p - q$ variables in $\mathbf{x}_2$ correspond to measurements that are easy to obtain and inexpensive to measure. We hope that the measurements that are easy to obtain can be used to predict those that are hard to obtain, thus negating a need to continue measuring variables that are hard to obtain.

If an investigator is interested in relating these two sets of variables when they are not independent, the interrelationships that exist between the two sets may be almost completely described by the correlation between a few linear combinations of the responses within each set of variables.

Caution When one separates a response vector $\mathbf{x}$ into two parts, $\mathbf{x}_1$ and $\mathbf{x}_2$, the separation must always be motivated by the nature of the responses or by some other external means. The responses should never be partitioned into two groups as the result of an inspection of the data.

If $q = 1$, that is, there is only one variable in the first set, then the multiple correlation coefficient is the largest correlation attainable between x_1 and a linear combination of the variables in $\mathbf{x}_2$. Next this is generalized to the case where there is more than one variable in the first set.

The First Canonical Correlation

When $q > 1$, a natural generalization of the multiple correlation concept is to find the largest correlation attainable between a linear combination of the variables in $\mathbf{x}_1$ and a linear combination of the variables in $\mathbf{x}_2$. That is, can we find $\mathbf{a}_1$ and $\mathbf{b}_1$ so that $\text{corr}(\mathbf{a}_1'\mathbf{x}_1, \mathbf{b}_1'\mathbf{x}_2)$ is a maximum?

Assuming that the answer to the preceding question is yes, let ρ_1 represent this maximum correlation. That is,

$$\rho_1 = \max_{\mathbf{a} \neq 0, \mathbf{b} \neq 0} \left[\text{corr}(\mathbf{a}'\mathbf{x}_1, \mathbf{b}'\mathbf{x}_2) \right]$$

DEFINITION 12.1

The first canonical correlation between $\mathbf{x}_1$ and $\mathbf{x}_2$ is defined by ρ_1. In addition,

$$U_1 = \mathbf{a}_1'\mathbf{x}_1 \qquad \text{and} \qquad V_1 = \mathbf{b}_1'\mathbf{x}_2$$

where $\mathbf{a}_1$ and $\mathbf{b}_1$ are the values of $\mathbf{a}$ and $\mathbf{b}$ that produce this maximum correlation, are called the first canonical variates.

Without any loss of generality, we can choose $\mathbf{a}_1$ and $\mathbf{b}_1$ so that

$$\text{var}(U_1) = \text{var}(V_1) = 1$$

This simply requires choosing $\mathbf{a}_1$ so that $\mathbf{a}_1'\Sigma_{11}\mathbf{a}_1 = 1$ and choosing $\mathbf{b}_1$ so that $\mathbf{b}_1'\Sigma_{22}\mathbf{b}_1 = 1$.

The Second Canonical Correlation

Next let $U_2 = \mathbf{a}_2'\mathbf{x}_1$ and $V_2 = \mathbf{b}_2'\mathbf{x}_2$ where $\mathbf{a}_2$ and $\mathbf{b}_2$ are chosen so that

1 U_2 and V_2 are uncorrelated with U_1 and V_1,

2 $\text{var}(U_2) = \text{var}(V_2) = 1$, and

3   the correlation between $\mathbf{a}_2'\mathbf{x}_1$ and $\mathbf{b}_2'\mathbf{x}_2$, denoted by ρ_2, is a maximum over all $\mathbf{a}_2$ and $\mathbf{b}_2$ such that $\mathbf{a}_2'\mathbf{x}_1$ and $\mathbf{b}_2'\mathbf{x}_2$ are uncorrelated with U_1 and V_1.

Then ρ_2 is called the second canonical correlation, and $U_2 = \mathbf{a}_2'\mathbf{x}_1$ and $V_2 = \mathbf{b}_2'\mathbf{x}_2$ are called the second canonical variates.

Similarly one can define 3rd, 4th, etc., canonical correlations and their corresponding canonical variates.

It has been shown that the first canonical correlation coefficient squared, namely, ρ_1^2, is the largest eigenvalue of $\Sigma_{11}^{-1}\Sigma_{12}\Sigma_{22}^{-1}\Sigma_{21}$, and that $\mathbf{a}_1$ is an eigenvector of $\Sigma_{11}^{-1}\Sigma_{12}\Sigma_{22}^{-1}\Sigma_{21}$ corresponding to its largest eigenvalue, ρ_1^2. In addition, $\mathbf{b}_1$ is an eigenvector of $\Sigma_{22}^{-1}\Sigma_{21}\Sigma_{11}^{-1}\Sigma_{12}$ corresponding to this matrix's largest eigenvalue, which is also equal to ρ_1^2. In both cases the eigenvectors $\mathbf{a}_1$ and $\mathbf{b}_1$ are chosen so that $\mathbf{a}_1'\Sigma_{11}\mathbf{a}_1 = 1$ and $\mathbf{b}_1'\Sigma_{22}\mathbf{b}_1 = 1$.

We can also show that ρ_2^2 is the second largest eigenvalue of $\Sigma_{11}^{-1}\Sigma_{12}\Sigma_{22}^{-1}\Sigma_{21}$, $\mathbf{a}_2$ is an eigenvector of $\Sigma_{11}^{-1}\Sigma_{12}\Sigma_{22}^{-1}\Sigma_{21}$ corresponding to ρ_2^2, and $\mathbf{b}_2$ is an eigenvector of $\Sigma_{22}^{-1}\Sigma_{21}\Sigma_{11}^{-1}\Sigma_{12}$ corresponding to its second largest eigenvalue, which is also ρ_2^2. Again, $\mathbf{a}_2$ and $\mathbf{b}_2$ are chosen so that $\mathbf{a}_2'\Sigma_{11}\mathbf{a}_2 = 1$ and $\mathbf{b}_2'\Sigma_{22}\mathbf{b}_2 = 1$.

In a similar manner, additional canonical correlations and their corresponding canonical variates can be defined simply by making sure that each successive pair of canonical variates is uncorrelated with all that preceded them. The results are given by the eigenvalues and eigenvectors of $\Sigma_{11}^{-1}\Sigma_{12}\Sigma_{22}^{-1}\Sigma_{21}$ and/or $\Sigma_{22}^{-1}\Sigma_{21}\Sigma_{11}^{-1}\Sigma_{12}$.

Number of Canonical Correlations

The actual number of canonical correlations possible is equal to the minimum of q and $p - q$. The number of nonzero canonical correlations is equal to rank of the matrix Σ_{12}.

Estimates

The preceding definitions were given in terms of population parameters; in practice, Σ will need to be estimated by $\hat{\Sigma}$, after which we can estimate the canonical correlations and their respective pairs of canonical variates by taking corresponding functions of $\hat{\Sigma}$ that correspond to those functions of Σ.

For example, ρ_1 is estimated by $\hat{\rho}_1$ where $\hat{\rho}_1$ is the square root of the largest eigenvalue of $\hat{\Sigma}_{11}^{-1}\hat{\Sigma}_{12}\hat{\Sigma}_{22}^{-1}\hat{\Sigma}_{21}$, and $\mathbf{a}_1$ is estimated by $\hat{\mathbf{a}}_1$, the eigenvector of $\hat{\Sigma}_{11}^{-1}\hat{\Sigma}_{12}\hat{\Sigma}_{22}^{-1}\hat{\Sigma}_{21}$ corresponding to its largest eigenvalue, $\hat{\rho}_1^2$, and normalized so that $\hat{\mathbf{a}}_1'\hat{\Sigma}_{11}\hat{\mathbf{a}}_1 = 1$. Likewise, $\mathbf{b}_1$ is estimated by $\hat{\mathbf{b}}_1$, the eigenvector of $\hat{\Sigma}_{22}^{-1}\hat{\Sigma}_{21}\hat{\Sigma}_{11}^{-1}\hat{\Sigma}_{12}$ corresponding to its largest eigenvalue, $\hat{\rho}_1^2$, and normalized so that $\hat{\mathbf{b}}_1'\hat{\Sigma}_{22}\hat{\mathbf{b}}_1 = 1$.

Hypothesis Tests on the Canonical Correlations

Before performing a canonical correlation analysis, we should test $H_{01}: \rho_1 = 0$ versus $H_a: \rho_1 > 0$. It is interesting to note that this hypothesis is equivalent to testing $H_{01}: \Sigma_{12} = 0$ versus $H_{a1}: \Sigma_{12} \neq 0$. This test is a special case of the test for independence of sets of variables that was introduced in Chapter 10. The test statistic from Chapter 10 for the case when there are two groups of variables is

$$\Lambda_1 = \frac{|\hat{\Sigma}|}{|\hat{\Sigma}_{11}| \cdot |\hat{\Sigma}_{22}|}$$

$$= \frac{|\hat{R}|}{|\hat{R}_{11}| \cdot |\hat{R}_{22}|}$$

Remark   It can be shown that $\Lambda_1 = \prod_{i=1}^{k}(1 - \hat{\rho}_i^2)$ where $k = \min(q, p - q)$.

The test that ρ_1 is equal to zero and tests that the remaining canonical correlations are zero are provided by many statistical computing packages. The procedure used is the following: To test $H_{0r}: \rho_r = 0$, calculate

$$\Lambda_r = \prod_{i=r}^{k} (1 - \hat{\rho}_i^2)$$

and reject H_{0r}, if

$$-a \log(\Lambda_r) > \chi^2_{\alpha,(q-r+1)(p-q-r+1)}$$

where $a = N - (p + 3)/2$.

If H_{01} is rejected and H_{02} is not, then ρ_1 is the only significant canonical correlation. If H_{02} is also rejected, then there are at least two nonzero canonical correlations and you should test $H_{03}: \rho_3 = 0$, etc. To illustrate a canonical correlation analysis consider the following example.

E X A M P L E   **12.2**

A researcher was conducting an experiment to find an ideal recipe for making great chocolate chip cookies. Fifty-four batches of cookies were made using several different recipe formulations at several different points in time. The resulting cookies were evaluated according to many different criteria. The responses measured on the cookies fell into five distinct groups of variables. Three of the groups contained characteristics that had been evaluated by sensory experts, whereas the other two groups contained responses measured by mechanical instruments. These latter two groups of variables did not require the subjective opinions of sensory experts. Each of the sensory responses was measured on a continuous scale that ran from 0 to 15 with 15 representing a very good response and 0 representing a very poor response.

The variables in the first group of sensory measurements were called "appearance" variables. This group included evaluations for optical surface cracking (CKG), frequency of open holes on the top of the cookie (HOLES), the degree to which raised protrusions occurred on the top of the cookie (PROTUS), the amount of curvature present in the top of the cookie when the cookie is placed on a flat surface (CONTOUR), the size of the majority of the cells on the area of a cut surface (CELLSIZE), and cell uniformity (CELLUNIF).

The variables in the second group required the sensory experts to evaluate each cookie's texture characteristics, and the variables in this group were called "texture" variables. The texture group included sensory evaluations of surface roughness (SURF_RO), which was a measure of the degree to which the top surface of the cookie contained bumps or particles when the sensory expert's lips were passed over the surface of the cookie; the abundance of loose particles on the lips of the expert (LOOSE); firmness (FIRM), which was a measure of the force required to bite completely through the cookie; fracturability (FRACT), which was a measure of the force with which a cookie shatters; the cohesiveness of a chewed sample (COH), which was the degree to which the sample holds together in a mass after chewing the cookie 7 or 8 chews; the number of chews required to hydrate the sample of cookie and bring it to a state where it is ready to swallow (CHEW), the amount of material left in and around the molar teeth of the sensory expert (MOLAR), and the degree of material left in the mouth (MOUTH).

The third group of sensory variables required the experts to evaluate the flavor of each cookie, and the variables in this group are called "flavor" variables. The flavor variables included a measure of caramel flavor (CARMEL), sweet aromatics (AROMA), sweetness (SWEET), saltiness (SALTY), vanilla-like flavor (VAN), sourness (SOUR), and bitterness (BITTER).

The fourth and fifth groups of variables were measured by instruments and did not require the subjective opinions of the panelists. The variables in the fourth set were called "size" variables. They included the specific gravity of the cookies (SPG), the weight of five cookies (W), the thickness of a stack of five cookies (T), the ratio of weight to thickness (RATIO), and cookie softness, which was measured by an instron (INSTRON).

The fifth set of variables was made up of three measures of color. These variables are denoted by L, A, and B.

Five sensory experts evaluated each of the cookies for their sensory characteristics. The data used in this analysis are the averages of the five experts' scores. Portions of the resulting data can be found on pages 1–10 of the computer printout shown in Figure 12.2. The total data set can be seen by executing the file named EX12_2.SAS on the enclosed disk.

Researchers believed that the five groups of variables would be independent of one another. That is, while researchers expected there might be correlation among the variables within a group, they expected there would be

```
        Ex. 12.2 - Canonical Correlation Analysis of Cookie Data       1 & 2
                          APPEARANCE VARIABLES
```

OBS	CKG	HOLES	PROTUS	CONTOUR	CELLSIZE	CELLUNIF
1	13.8	2.2	3.4	5.0	5.0	12.8
2	13.6	2.2	3.2	5.0	4.8	13.1
3	13.8	2.0	2.5	5.0	4.9	12.6
4	13.9	2.0	3.2	5.2	5.2	12.9
5	13.3	2.1	3.0	4.9	5.1	12.9
.	.	.	.	.	.	.
.	.	.	.	.	.	.
.	.	.	.	.	.	.
52	0.2	6.0	2.2	7.5	10.4	6.4
53	0.5	6.9	0.8	7.5	7.1	11.4
54	0.7	6.4	2.2	6.8	10.4	8.1

```
        Ex. 12.2 - Canonical Correlation Analysis of Cookie Data       3 & 4
                            TEXTURE VARIABLES
```

OBS	SURF_RO	LOOSE	FIRM	FRACT	COH	CHEW	MOLAR	MOUTH
1	12.5	10.8	7.0	1.4	7.1	12.2	6.1	1.2
2	12.2	10.7	7.0	1.7	7.4	12.3	6.0	1.0
3	12.6	11.2	7.3	1.7	7.1	12.3	6.1	1.3
4	12.4	10.8	7.1	1.5	7.0	11.7	6.0	1.3
5	12.3	10.8	7.4	1.9	7.1	12.3	6.2	1.1
.	.	.	.	.	.	.	.	.
.	.	.	.	.	.	.	.	.
.	.	.	.	.	.	.	.	.
52	2.7	0.2	4.9	0.0	8.8	13.0	5.6	3.0
53	2.8	0.1	5.3	0.0	8.4	13.2	6.1	1.8
54	1.4	0.2	4.8	0.1	8.7	14.0	6.5	2.3

F I G. **12.2**

Pages 1–14 and 16–19 of computer printout, Example 12.2

no correlation between variables in different groups. In this example, a canonical correlation analysis is performed on two of the groups of measured variables—the flavor set and the texture set. One consequence of this analysis is that it will enable the researchers to determine whether the flavor variables are independent of the texture variables as expected.

The canonical correlation analysis was performed in SAS by using the following commands:

```
PROC CANCORR OUT=SCORES SIMPLE CORR NCAN=2
  VPREFIX=FLAVOR WPREFIX=TEXTURE;
  VAR CARMEL--BITTER;
  WITH SURF_RO--MOUTH;
  TITLE2;
  RUN;
```

Ex. 12.2 - Canonical Correlation Analysis of Cookie Data   5 & 6
FLAVOR VARIABLES

OBS	CARMEL	AROMA	SWEET	SALTY	VAN	SOUR	BITTER
1	6.5	10.5	8.5	3.5	2.0	0.5	1.8
2	6.5	10.5	8.2	3.5	2.0	0.5	1.8
3	6.4	10.3	8.0	3.7	2.0	0.5	2.0
4	6.5	10.5	8.0	3.6	2.0	0.5	1.8
5	6.6	10.5	8.2	3.7	2.0	0.5	1.8
.	.	.	.	.	.	.	.
.	.	.	.	.	.	.	.
52	4.4	9.3	7.1	3.0	1.4	0.8	1.4
53	5.5	9.9	7.6	3.5	2.0	0.6	1.8
54	5.1	9.4	7.4	3.6	1.5	0.7	1.8

Ex. 12.2 - Canonical Correlation Analysis of Cookie Data   7 & 8
SIZE VARIABLES

OBS	INSTRON	SPG	W	T	RATIO
1	9.0	1.0	429.8	69.2	6.2
2	15.1	0.9	441.4	70.0	6.3
3	.	1.1	439.6	62.8	7.0
4	.	1.0	450.7	68.7	6.6
5	11.5	1.0	438.6	78.3	5.6
.	.	.	.	.	.
.	.	.	.	.	.
.	.	.	.	.	.
52	.	1.3	443.3	73.3	6.0
53	2.6	1.2	423.9	75.3	5.6
54	3.0	1.1	407.0	80.2	5.1

Ex. 12.2 - Canonical Correlation Analysis of Cookie Data   9 & 10
COLOR VARIABLES

OBS	L	A	B
1	43.16	5.35	11.91
2	44.17	6.59	13.08
3	46.42	6.22	13.04
4	44.19	6.72	13.16
5	42.53	6.75	12.45
.	.	.	.
.	.	.	.
.	.	.	.
52	44.90	6.76	13.88
53	46.52	6.90	15.71
54	43.74	7.87	13.18

F I G. **12.2**
Pages 1–14 and 16–19 of computer printout, Example 12.2

Ex. 12.2 - Canonical Correlation Analysis of Cookie Data   11

Means and Standard Deviations

7 'VAR' Variables
8 'WITH' Variables
54 Observations

Variable	Mean	Std Dev
CARMEL	5.757407	0.633231
AROMA	10.009259	0.375308
SWEET	7.896296	0.353400
SALTY	3.464815	0.238040
VAN	1.835185	0.198254
SOUR	0.620370	0.127944
BITTER	1.872222	0.207789
SURF_RO	6.225926	4.492643
LOOSE	3.722222	5.011521
FIRM	5.912963	1.013316
FRACT	0.718519	0.991677
COH	7.766667	0.654159
CHEW	12.566667	0.511749
MOLAR	6.005556	0.350427
MOUTH	1.759259	0.409103

F I G. **12.2**

Pages 1–14 and 16–19 of computer printout, Example 12.2

```
PROC PLOT;
 PLOT FLAVOR1*TEXTURE1 FLAVOR2*TEXTURE2;
 TITLE2 'SCATTER PLOT OF CANONICAL VARIABLES';
 RUN;

PROC PRINT;
 VAR TEXTURE1 TEXTURE2 FLAVOR1 FLAVOR2;
 TITLE2 'PRINT OF STANDARDIZED CANONICAL VARIABLES';
 RUN;
```

The first group of commands performs the canonical correlation analysis. The VAR statement is used to identify the variables falling into the first group and the WITH statement is used to identify the variables that are in the second set. The OUT=SCORES option creates an output data set named SCORES that contains canonical scores for each of the cookies in the data set being analyzed.

The SIMPLE option produces the means and standard deviations for each of the variables being analyzed while the CORR option produces correlations between all of the measured variables being analyzed.

The NCAN=2 option limits the number of canonical functions computed for each of the two groups of variables to two. Generally, we would not know how many canonical functions are required without doing a prelimi-

Ex. 12.2 - Canonical Correlation Analysis of Cookie Data   12

Correlations Among the Original Variables

Correlations Among the 'VAR' Variables R_{11}

	CARMEL	AROMA	SWEET	SALTY
CARMEL	1.0000	0.8917	0.7075	0.4705
AROMA	0.8917	1.0000	0.8069	0.2677
SWEET	0.7075	0.8069	1.0000	0.2788
SALTY	0.4705	0.2677	0.2788	1.0000
VAN	0.7306	0.7005	0.5378	0.4465
SOUR	-0.4362	-0.4873	-0.5366	0.0178
BITTER	0.2031	0.1292	0.0551	0.3613

	VAN	SOUR	BITTER
CARMEL	0.7306	-0.4362	0.2031
AROMA	0.7005	-0.4873	0.1292
SWEET	0.5378	-0.5366	0.0551
SALTY	0.4465	0.0178	0.3613
VAN	1.0000	-0.4007	0.4226
SOUR	-0.4007	1.0000	0.3127
BITTER	0.4226	0.3127	1.0000

Correlations Among the 'WITH' Variables R_{22}

	SURF_RO	LOOSE	FIRM	FRACT
SURF_RO	1.0000	0.9912	0.8927	0.7604
LOOSE	0.9912	1.0000	0.8737	0.7385
FIRM	0.8927	0.8737	1.0000	0.8377
FRACT	0.7604	0.7385	0.8377	1.0000
COH	-0.7969	-0.7842	-0.7801	-0.7992
CHEW	-0.3949	-0.3710	-0.3212	-0.3698
MOLAR	0.1869	0.1768	0.2724	0.2196
MOUTH	-0.7412	-0.7508	-0.6418	-0.4767

	COH	CHEW	MOLAR	MOUTH
SURF_RO	-0.7969	-0.3949	0.1869	-0.7412
LOOSE	-0.7842	-0.3710	0.1768	-0.7508
FIRM	-0.7801	-0.3212	0.2724	-0.6418
FRACT	-0.7992	-0.3698	0.2196	-0.4767
COH	1.0000	0.6313	-0.0823	0.6907
CHEW	0.6313	1.0000	0.4187	0.4359
MOLAR	-0.0823	0.4187	1.0000	-0.1932
MOUTH	0.6907	0.4359	-0.1932	1.0000

F I G. 12.2

Pages 1–14 and 16–19 of computer printout, Example 12.2

Ex. 12.2 - Canonical Correlation Analysis of Cookie Data   13

Correlations Among the Original Variables

Correlations Between the 'VAR' Variables and the 'WITH' Variables R_{12}

	SURF_RO	LOOSE	FIRM	FRACT
CARMEL	0.8695	0.8480	0.8195	0.6755
AROMA	0.8770	0.8494	0.7756	0.6799
SWEET	0.7215	0.6980	0.6946	0.6199
SALTY	0.3102	0.3001	0.3641	0.2890
VAN	0.7065	0.6908	0.6335	0.5417
SOUR	-0.5071	-0.4945	-0.4969	-0.4551
BITTER	0.1180	0.1271	0.1478	0.1893

	COH	CHEW	MOLAR	MOUTH
CARMEL	-0.7423	-0.3474	0.2842	-0.7330
AROMA	-0.7872	-0.5043	0.2162	-0.7324
SWEET	-0.6535	-0.4983	0.1266	-0.5453
SALTY	-0.1834	0.1962	0.5566	-0.2824
VAN	-0.6527	-0.2858	0.3230	-0.6938
SOUR	0.4569	0.3938	0.0690	0.3045
BITTER	-0.2707	0.0657	0.4297	-0.3398

F I G. 12.2
Pages 1–14 and 16–19 of computer printout, Example 12.2

nary canonical correlation analysis on the data; the program's default is to choose the number of canonical functions to be equal to the minimum of q and $p - q$. For these data $q = 7$ and $p - q = 8$. As we will soon see, there are only two canonical correlations that are significantly different from zero for these data. This implies that the number of significant canonical functions is equal to two.

The VPREFIX=FLAVOR option causes the two canonical functions in the VAR group to be identified by FLAVOR1 and FLAVOR2 rather than by V1 and V2, which are the default choices. Similarly, the WPREFIX=TEXTURE option causes the two canonical functions in the WITH group to be identified by TEXTURE1 and TEXTURE2 rather than by W1 and W2, which are the default choices.

The second group of commands creates scatter plots of each of the two pairs of canonical scores for the 54 cookies, and the third group of commands prints the canonical scores.

Page 11 of the computer printout (see Figure 11.2), shows the means and standard deviations of the measured variables. The top of page 12 shows correlations among the variables in the VAR group (this matrix is R_{11}) and the bottom of page 12 shows correlations among the variables in the WITH group (this matrix is R_{22}). On page 13 we find the correlations among the variables in the VAR group with those variables in the WITH

Ex. 12.2 - Canonical Correlation Analysis of Cookie Data   14

Canonical Correlation Analysis

	Canonical Correlation		Adjusted Canonical Correlation	Approx Standard Error	Squared Canonical Correlation
1	0.940142	$\hat{\rho}_1$	0.925516	0.015952	0.883867
2	0.674919	$\hat{\rho}_2$	0.579042	0.074791	0.455516
3	0.471888	$\hat{\rho}_3$	0.213761	0.106773	0.222678
4	0.413970	.	.	0.113821	0.171371
5	0.395341	:	.	0.115892	0.156295
6	0.113435		.	0.135593	0.012868
7	0.025059		.	0.137274	0.000628

Eigenvalues of INV(E)*H
= CanRsq/(1-CanRsq)

	Eigenvalue	Difference	Proportion	Cumulative
1	7.6108	6.7742	0.8327	0.8327
2	0.8366	0.5501	0.0915	0.9243
3	0.2865	0.0797	0.0313	0.9556
4	0.2068	0.0216	0.0226	0.9782
5	0.1852	0.1722	0.0203	0.9985
6	0.0130	0.0124	0.0014	0.9999
7	0.0006	.	0.0001	1.0000

Test of H0: The canonical correlations in the current row
and all that follow are zero

	Likelihood Ratio	Approx F	Num DF	Den DF	Pr > F
1	0.03389959	3.3635	56	215.3324	0.0001
2	0.29190346	1.3657	42	191.0687	0.0832
3	0.53611049	0.9332	30	166	0.5708
4	0.68968934	0.8312	20	140.2481	0.6726
5	0.83232584	0.6827	12	114.0588	0.7650
6	0.98651259	0.0999	6	88	0.9962
7	0.99937203	0.0141	2	45	0.9860

Identifies the number of nonzero canonical correlations.

FIG. **12.2**

Pages 1–14 and 16–19 of computer printout, Example 12.2

group (this matrix is $\mathbf{R}_{12}$). Obviously, if we combine these three matrices, we would have the full correlation matrix $\mathbf{R}$.

The primary results of the canonical correlation analysis are shown on page 14. From the first column on the left near the top, we see that the first canonical correlation is $\hat{\rho}_1 = 0.940142$, the second is $\hat{\rho}_2 = 0.674919$, the third is $\hat{\rho}_3 = 0.471888$, etc.

Ex. 12.2 - Canonical Correlation Analysis of Cookie Data   16

Canonical Correlation Analysis

Raw Canonical Coefficients for the 'VAR' Variables

	FLAVOR1	FLAVOR2
CARMEL	0.4295626638	-0.618816176
AROMA	1.5535311708	1.2760037807
SWEET	0.0050929629	-1.685695557
SALTY	-0.047131202	2.3317312878
VAN	0.7743914334	-0.441157066
SOUR	-0.488164274	-0.658995544
BITTER	0.4075555266	3.6242266083
	$\hat{a}_1$	$\hat{a}_2$

Raw Canonical Coefficients for the 'WITH' Variables

	TEXTURE1	TEXTURE2
SURF_RO	0.3708608868	-0.153743428
LOOSE	-0.19909586	-0.020377481
FIRM	-0.072234277	-0.500578321
FRACT	-0.04863786	0.2303303057
COH	-0.306017005	-1.104304011
CHEW	-0.198102188	1.2122335922
MOLAR	0.5156979493	1.5844652351
MOUTH	-0.476214465	-1.568968337
	$\hat{b}_1$	$\hat{b}_2$

F I G. **12.2**

Pages 1–14 and 16–19 of computer printout, Example 12.2

Likelihood ratio tests for the statistical significance of the canonical correlations are shown on page 14 under the label Test of H0: The canonical correlations in the current row and all that follow are zero. An examination of the p values on the far right of these tests reveals that the first canonical correlation is statistically significant at the 0.0001 level and that the second is statistically significant at the 0.0832 level. None of the other canonical correlations approaches statistical significance. Thus we now see why the NCAN=2 option was selected in the SAS command statements given earlier.

Note that the column on top of page 14 labeled Adjusted Canonical Correlation contains estimators of the population canonical correlations that are approximately unbiased. Occasionally, when computing unbiased estimators of the canonical correlations, it is possible for an estimator of a smaller canonical correlation coefficient to actually be larger than the estimator of a larger one. Because this does not make sense, the adjusted estimate is not printed in this case. For these data, the adjusted estimate of ρ_4 was larger than the adjusted estimate of ρ_3. Because ρ_3 is known to be larger than ρ_4, the estimator of ρ_4 is not printed.

Obviously, this analysis indicates that the flavor variables are not completely independent of the texture variables as the original researchers ex-

Canonical Correlation Analysis

Standardized Canonical Coefficients for the 'VAR' Variables

	FLAVOR1	FLAVOR2
CARMEL	0.2720	-0.3919
AROMA	0.5831	0.4789
SWEET	0.0018	-0.5957
SALTY	-0.0112	0.5550
VAN	0.1535	-0.0875
SOUR	-0.0625	-0.0843
BITTER	0.0847	0.7531
	$\hat{a}_1^*$	$\hat{a}_2^*$

Standardized Canonical Coefficients for the 'WITH' Variables

	TEXTURE1	TEXTURE2
SURF_RO	1.6661	-0.6907
LOOSE	-0.9978	-0.1021
FIRM	-0.0732	-0.5072
FRACT	-0.0482	0.2284
COH	-0.2002	-0.7224
CHEW	-0.1014	0.6204
MOLAR	0.1807	0.5552
MOUTH	-0.1948	-0.6419
	$\hat{b}_1^*$	$\hat{b}_2^*$

FIG. 12.2

Pages 1–14 and 16–19 of computer printout, Example 12.2

pected. Since the two sets of variables are not independent, you might ask how the two sets of variables are related to each other.

The top of page 16 shows the values of $\hat{a}_1$ and $\hat{a}_2$. These vectors define the canonical variates in terms of the raw data values of the variables in the VAR group, that is, the flavor variables. The bottom of page 16 lists the values of $\hat{b}_1$ and $\hat{b}_2$. These vectors define the canonical variates in terms of the raw data values of the variables in the WITH group, that is, the texture variables.

On page 17, we find vectors that can be used on standardized data to compute scores for the canonical variates.

To determine whether we can interpret the newly formed canonical variables, we study the correlations that the original variables have with each of the sets of canonical variables. These correlations are similar to factor loadings in factor analysis. Four sets of such correlations are shown on pages 18 and 19. The output contains correlations between each of the two sets of original variables and with their own set of canonical variables and also with the other set of canonical variables.

On the top of page 18 of the computer printout, we see the correlations between the flavor variables and the two canonical functions created from this set of variables. The correlations between the texture variables and the two canonical variables created from the flavor variables are shown on the

Correlations Between the 'VAR' Variables and Their Canonical Variables

	FLAVOR1	FLAVOR2
CARMEL	0.9445	0.0007
AROMA	0.9730	-0.1255
SWEET	0.7823	-0.2921
SALTY	0.3714	0.5643
VAN	0.8174	0.2413
SOUR	-0.5014	0.4533
BITTER	0.2566	0.8398

Correlations Between the 'WITH' Variables and Their Canonical Variables

	TEXTURE1	TEXTURE2
SURF_RO	0.9529	-0.1608
LOOSE	0.9270	-0.1449
FIRM	0.8649	-0.0942
FRACT	0.7506	-0.0214
COH	-0.8634	0.0239
CHEW	-0.4835	0.5062
MOLAR	0.2968	0.7633
MOUTH	-0.8281	-0.1723

Correlations Between the 'VAR' Variables and the
Canonical Variables of the 'WITH' Variables

	TEXTURE1	TEXTURE2
CARMEL	0.8880	0.0004
AROMA	0.9147	-0.0847
SWEET	0.7355	-0.1972
SALTY	0.3492	0.3809
VAN	0.7685	0.1628
SOUR	-0.4714	0.3059
BITTER	0.2413	0.5668

Canonical Structure

Correlations Between the 'WITH' Variables and
the Canonical Variables of the 'VAR' Variables

	FLAVOR1	FLAVOR2
SURF_RO	0.8958	-0.1085
LOOSE	0.8715	-0.0978
FIRM	0.8131	-0.0636
FRACT	0.7056	-0.0144
COH	-0.8117	0.0161
CHEW	-0.4545	0.3416
MOLAR	0.2790	0.5152
MOUTH	-0.7785	-0.1163

FIG. 12.2
Pages 1–14 and 16–19 of computer printout, Example 12.2

bottom of page 18. On the top of page 19, we find correlations among the flavor variables and the canonical variables created from the texture variables. Finally, on the bottom of page 19, we see the correlations among the texture variables and the canonical variables created from the flavor variables.

Interpreting Canonical Functions

Can the canonical functions be interpreted? Interpretations of the canonical functions is usually difficult. Recall from factor analysis that factor loading matrices were usually easier to interpret after performing a rotation on the factor loading matrix. The same is likely to be true here. That is, interpretations of the canonical functions might be made easier if we were to perform some sort of a rotation on the canonical functions. However, if we rotate either of the two sets of canonical functions, we will destroy the properties that were used to define the canonical functions. That is, they will no longer be the linear combinations of variables that are most highly correlated with each other. Furthermore, if we were to perform a rotation on the canonical functions, each of the rotated canonical functions in one set would likely be correlated with each of the rotated canonical functions in the other set, rather than having only one canonical function from the first set of variables correlated with only one canonical function from the second set of variables.

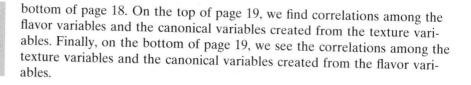

E X A M P L E **12 . 2 (continued)**

Nevertheless, an attempt might be made to interpret the canonical functions for the flavor and texture variables in our example even though such an attempt is often ill advised. The variables in the flavor group that are highly correlated with FLAVOR1 are CARMEL, AROMA, SWEET, and VAN while those highly correlated with FLAVOR2 are BITTER and SALTY. SOUR has some correlation with both FLAVOR1 and FLAVOR2. SOUR is correlated with FLAVOR1 in a negative direction and with FLAVOR2 in a positive direction. How would you interpret FLAVOR1 and FLAVOR2? Keep in mind that the two characteristics measured by these variables are not correlated with each other.

The variables in the texture group that are highly correlated with TEXTURE1 are SURF_RO, LOOSE, FIRM, and FRACT in a positive direction and COH and MOUTH in a negative direction. The variables in the texture group that are correlated with TEXTURE2 are MOLAR and CHEW. How would you interpret TEXTURE1 and TEXTURE2? Again, keep in mind that TEXTURE1 and TEXTURE2 are measuring characteristics that are not uncorrelated with each other.

Now are your interpretations of FLAVOR1 and TEXTURE1 consistent with characteristics that are expected to be highly correlated? Also note that the flavor variables that are highly correlated with FLAVOR1 are also highly correlated with TEXTURE1 and the texture variables that are highly

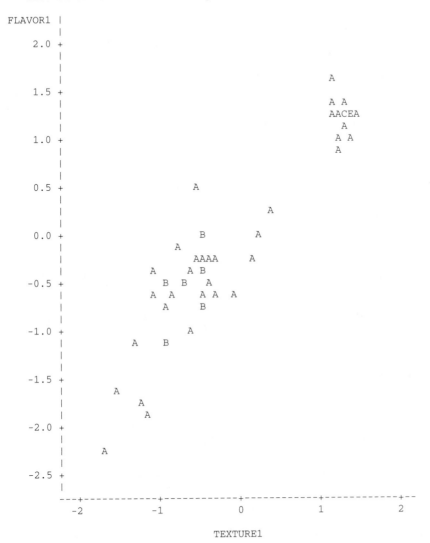

Ex. 12.2 - Canonical Correlation Analysis of Cookie Data   20
              SCATTER PLOT OF CANONICAL VARIABLES

Plot of FLAVOR1*TEXTURE1.   Legend: A = 1 obs, B = 2 obs, etc.

FIG. 12.3
Pages 20–22 of computer printout, Example 12.2

correlated with TEXTURE1 are also highly correlated with FLAVOR1. To a lesser extent, the flavor variables that are correlated with FLAVOR2 are also correlated with TEXTURE2 and the texture variables that are correlated with TEXTURE2 are also correlated with FLAVOR2. This is as it should be—at least mathematically.

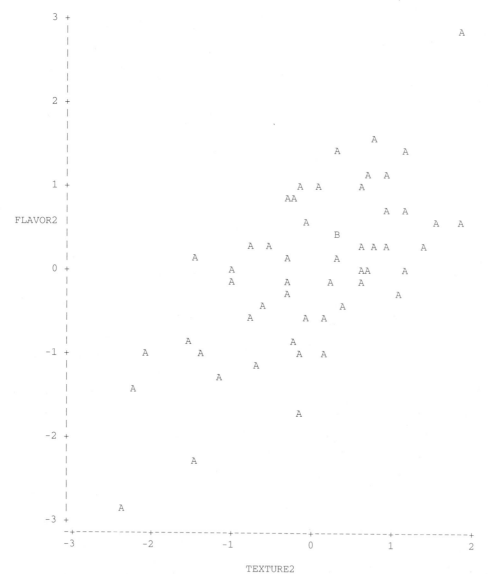

FIG. 12.3

Pages 20–22 of computer printout, Example 12.2

Ex. 12.2 - Canonical Correlation Analysis of Cookie Data   22

PRINT OF STANDARDIZED CANONICAL VARIABLES

OBS	TEXTURE1	TEXTURE2	FLAVOR1	FLAVOR2
1	1.39767	-0.17719	1.23974	-1.02409
2	1.22378	-0.11465	1.23822	-0.51839
3	1.25142	-0.31746	0.95562	0.81663
4	1.37895	-0.99987	1.23248	0.05193
5	1.34966	0.20506	1.27175	-0.11392
.	.	.	.	.
.	.	.	.	.
.	.	.	.	.
52	-1.70032	-2.24965	-2.28425	-1.44437
53	-0.76012	0.89594	-0.17534	0.28035
54	-1.55010	1.87027	-1.56567	0.61486

F I G. 12.3
Pages 20–22 of computer printout, Example 12.2

Scatter plots between the two pairs of canonical variables from each set are shown on pages 20 and 21 of the computer printout of Figure 12.3. Note that the linear relationship between FLAVOR1 and TEXTURE1 on page 20 appears much stronger than the linear relationship between FLAVOR2 and TEXTURE2 on page 21 as would be expected. A portion of a printout of the canonical scores for each of the 54 cookie recipes is shown on page 22.

Canonical Correlation Analysis with SPSS

E X A M P L E 12.3

The commands necessary to perform a canonical correlation analysis on the data in Example 12.2 with SPSS are given in the file labeled EX12_3.SSX on the enclosed disk and repeated here:

```
SET WIDTH = 80
TITLE "Ex. 12.3 - Canonical Correlation Analysis with SPSS"
DATA LIST FREE
  /CKG HOLES PROTUS CONTOUR CELLSIZE CELLUNIF
    SURF_RO LOOSE FIRM FRACT COH CHEW MOLAR MOUTH
  CARMEL AROMA SWEET SALTY VAN SOUR BITTER
      SPG W T RATIO INSTRON
        L A B
```

```
VARIABLE LABELS CKG 'Surface Cracking'/
                HOLES 'Open Holes on Surface'/
                PROTUS 'Protrusions'/
                CONTOUR 'Curvature Contour'/
                CELLSIZE 'Cell Size'/
                CELLUNIF 'Cell Uniformity'/
                SURF_RO 'Surface Roughness'/
                LOOSE 'Loose Particles on Surface'/
                FIRM 'Firmness'/
                FRACT 'Fracturability'/
                COH 'Cohesiveness'/
                CHEW 'Chew Count'/
                MOLAR 'Molar Packing'/
                MOUTH 'Residual Mouth coating'/
                CARMEL 'Caramel'/
                AROMA 'Sweet Aromatics'/
                SWEET 'Sweetness'/
                SALTY 'Saltiness'/
                VAN 'Vanilla'/
                SOUR 'Sourness'/
                BITTER 'Bitterness'/
                SPG 'Specific Gravity'/
                W 'Weight of 5 Cookies'/
                T 'Thickness of 5 Cookies'/
                RATIO 'Ratio of Weight to Thickness'/
                INSTRON 'Machine Measure of Firmness'/
                L 'Color L'/
                A 'Color A'/
PRINT / CKG HOLES PROTUS CONTOUR CELLSIZE CELLUNIF
        SURF_RO LOOSE FIRM FRACT COH CHEW MOLAR MOUTH
        CARMEL AROMA SWEET SALTY VAN SOUR BITTER
        SPG W T RATIO INSTRON
        L A B
BEGIN DATA
.
.   {data were placed here}
.
END DATA
MANOVA CARMEL AROMA SWEET SALTY VAN SOUR BITTER
    WITH SURF_RO LOOSE FIRM FRACT COH CHEW MOLAR MOUTH
    /PRINT SIGNIF (MULTIV EIGEN DIMENR)
        DISCRIM(STAN COR)
    /DESIGN = CONSTANT
```

Ex. 12.3 - Canonical Correlation Analysis with SPSS   Page 12
- -
Eigenvalues and Canonical Correlations

Root No.	Eigenvalue	Pct.	Cum. Pct.	Canon Cor.		Sq. Cor
1	7.611	83.273	83.273	.940	$\hat{\rho}_1$	.884
2	.837	9.154	92.426	.675	$\hat{\rho}_2$	.456
3	.286	3.134	95.561	.472	$\hat{\rho}_3$	.223
4	.207	2.263	97.824	.414	.	.171
5	.185	2.027	99.851	.395	:	.156
6	.013	.143	99.993	.113		.013
7	.001	.007	100.000	.025		.001

- -
Dimension Reduction Analysis

Roots	Wilks L.	F	Hypoth. DF	Error DF	Sig. of F
1 TO 7	.03390	3.36349	56.00	215.33	.000
2 TO 7	.29190	1.36570	42.00	191.07	.083
3 TO 7	.53611	.93323	30.00	166.00	.571
4 TO 7	.68969	.83118	20.00	140.25	.673
5 TO 7	.83233	.68275	12.00	114.06	.765
6 TO 7	.98651	.09992	6.00	88.00	.996
7 TO 7	.99937	.01414	2.00	45.00	.986

Determines the number of nonzero canonical correlations.

F I G. 12.4
Pages 12–14 of computer printout, Example 12.3

As seen from the preceding commands, SPSS requires that the MANOVA procedure be used to perform a canonical correlation analysis. The MANOVA procedure is very general in that it allows us to do many kinds of statistical analyses. Not only can it be used to perform multivariate analyses of variance as its name implies, it can be used to perform regression analyses, complicated analyses of variance, analyses of covariance, and canonical correlation analyses. To use the MANOVA procedure for canonical correlation analysis, the variables belonging to the first set are placed after the MANOVA statement and those belonging to the second set are placed after a WITH statement. It is important that the DESIGN=CONSTANT option be selected when doing a canonical correlation analysis.

Much of the resulting output is similar to that obtained from regression programs; the portion of the output that pertains to canonical correlation analysis is shown in Figure 12.4. All of the output can be seen by executing the file labeled EX12_3.SSX on the enclosed disk.

The values of the canonical correlations are shown in the middle of page 12 (see Figure 12.4) under the column labeled Canon Cor. The likeli-

Ex. 12.3 - Canonical Correlation Analysis with SPSS   Page   13

Standardized canonical coefficients for DEPENDENT variables

Function No.

Variable	1	2
CARMEL	-.272	.392
AROMA	-.583	-.479
SWEET	-.002	.596
SALTY	.011	-.555
VAN	-.154	.087
SOUR	.062	.084
BITTER	-.085	-.753
	$\hat{a}_1^*$	$\hat{a}_2^*$

- -

Correlations between DEPENDENT and canonical variables

Function No.

Variable	1	2
CARMEL	-.945	-.001
AROMA	-.973	.125
SWEET	-.782	.292
SALTY	-.371	-.564
VAN	-.817	-.241
SOUR	.501	-.453
BITTER	-.257	-.840

F I G. 12.4

Pages 12–14 of computer print-out, Example 12.3

hood ratio test statistics that can be used to determine the number of ca-nonical correlations that are significantly different than zero are shown on page 12 in the portion of the output labeled Dimension Reduction Analysis. Two of the canonical correlations are significantly greater than zero just as in the SAS analysis. Because two canonical correlations are sig-nificantly greater than zero, SPSS gives additional details for the first two canonical functions.

On the top of page 13, we see standardized canonical coefficients for the flavor variables, and at the bottom we see the correlations between each of the flavor variables and the first two canonical variates from the flavor variables.

The top of page 14 shows the standardized canonical coefficients for the texture variables, and the bottom shows the correlations between the tex-ture variables and the first two canonical variates from the texture group.

It is interesting to note that the standardized canonical coefficients and their corresponding correlations have signs that are reversed from what they were in the SAS output. This is true for the canonical functions from both sets of variables. This is nothing to worry about, but would have an ef-fect on how the canonical functions are interpreted.

Ex. 12.3 - Canonical Correlation Analysis with SPSS   Page   14

Standardized canonical coefficients for COVARIATES

                      CAN. VAR.

COVARIATE   1   2

SURF_RO   -1.666   .691
LOOSE   .998   .102
FIRM   .073   .507
FRACT   .048   -.228
COH   .200   .722
CHEW   .101   -.620
MOLAR   -.181   -.555
MOUTH   .195   .642
                 $\hat{b}_1^*$ $\hat{b}_2^*$
- -
Correlations between COVARIATES and canonical variables

                      CAN. VAR.
Covariate   1   2

SURF_RO   -.953   .161
LOOSE   -.927   .145
FIRM   -.865   .094
FRACT   -.751   .021
COH   .863   -.024
CHEW   .483   -.506
MOLAR   -.297   -.763
MOUTH   .828   .172

F I G. 12.4
Pages 12–14 of computer printout, Example 12.3

12.3
Factor Analysis and Regression

In canonical correlation analysis, we are basically trying to relate two groups of variables to each other through the use of a fewer number of canonical variables. This is a worthwhile objective, and while canonical correlation analysis achieves this objective mathematically, it rarely produces new variables that can be easily and accurately interpreted. That is, canonical correlation analysis rarely helps with interpretations of the data being analyzed.

An alternative to canonical correlation analysis that can be recommended is to perform factor analyses on each of the two sets of variables, rotate the factors in each of the two sets to improve their interpretability, create factor scores through the use of one of the scoring procedures discussed in Chapter 6, and, finally, relate these two newly created sets of factor scores to each other through the use of multiple regression. The following example illustrates how this can be accomplished with SAS.

E X A M P L E **12.4**

Consider the experiment described in Example 12.2. Factor analysis was performed on the set of flavor variables and another factor analysis was performed on the set of texture variables. The results of these initial factor analyses are summarized here. For each of the two sets of variables, the data were analyzed using the principal factor method with iteration and choosing prior communality estimates by the squared multiple correlation method.

Factor analyses on the flavor variables identified three factors, but one of the three was a trivial factor with only sourness loading on it. So a second factor analysis was performed by removing SOUR from the analysis and requesting two factors. These two factors seem to summarize the remaining information in the flavor data.

Factor analyses on the texture variables identified three factors, and these three factors seem to summarize the information in the texture data.

Thus the final factor analysis (FA) results for the flavor variables are obtained by using the following two sets of SAS commands. The OUT= options in these two sets create new data sets that contain factor scores by using the regression method to compute factor scores as discussed in Chapter 6. These will be used later to compare the set of flavor variables to the set of texture variables.

```
PROC FACTOR DATA=COOKIE METHOD=PRINIT ROTATE=VARIMAX NFACT=2
 PRIORS=SMC OUT=FLAV_SCR;
 VAR CARMEL AROMA SWEET SALTY VAN BITTER;
 TITLE2 'The Final Factor Analysis on the Flavor Variables';
RUN;


PROC FACTOR DATA=COOKIE METHOD=PRINIT ROTATE=VARIMAX NFACT=3
 PRIORS=SMC OUT=TEXT_SCR HEYWOOD;
 VAR SURF_RO LOOSE FIRM FRACT COH CHEW MOLAR MOUTH;
 TITLE2 'The Final Factor Analysis on the Texture Variables';
RUN;
```

All of the output produced from these commands can be seen by executing the file labeled EX12_4.SAS on the enclosed disk. Only the rotated factor loading matrices are shown in Figure 12.5. The rotated factor loading pattern for the flavor variables is shown on page 3 (recall that SOUR was not used in the FA), and the rotated factor loading pattern for the texture variables is shown on page 7.

For the flavor data, the variables loading on the first factor along with their factor loadings are CARMEL (0.86), AROMA (0.98), SWEET (0.80), and VAN (0.61). The variables loading on the second factor along with their factor loadings are SALTY (0.54), VAN (0.59), and BITTER (0.68). Recall also that SOUR was a variable that was uncorrelated with these two factors.

```
        Ex. 12.4 - An Alternative to Canonical Correlation Analysis          3
            The Final Factor Analysis on the Flavor Variables

Rotation Method: Varimax


                        Rotated Factor Pattern

                            FACTOR1    FACTOR2

                CARMEL     0.85857    0.35668
                AROMA      0.98318    0.13946
                SWEET      0.79976    0.09220
                SALTY      0.25497    0.53526
                VAN        0.61182    0.59146
                BITTER     0.00376    0.67523



        Ex. 12.4 - An Alternative to Canonical Correlation Analysis          7
            The Final Factor Analysis on the Texture Variables

Rotation Method: Varimax

                        Rotated Factor Pattern

                        FACTOR1    FACTOR2    FACTOR3

            SURF_RO     0.88335    0.41822    0.03293
            LOOSE       0.89175    0.38541    0.04229
            FIRM        0.71465    0.58914    0.11481
            FRACT       0.42341    0.90934    0.02843
            COH        -0.67277   -0.56453    0.24848
            CHEW       -0.37191   -0.25764    0.90096
            MOLAR       0.17016    0.13705    0.55588
            MOUTH      -0.78176   -0.18365    0.07418


        Scoring Coefficients Estimated by Regression
```

F I G. **12.5**
Pages 3, 7, 9, 11, and 13 of computer printout, Example 12.4

How can these three characteristics of the flavor variables be intepreted? In consultation with a food sensory expert, we determined that factor 1 is a measure of overall sweetness of the cookie, factor 2 is a measure of bitterness/saltiness. The food sensory expert assured us that sweetness and bitterness/saltiness are not opposites and that they are expected to be un-correlated. The third characteristic existing in the flavor data is a measure of sourness. Again, according to the sensory expert, these three characteris-tics—sweetness, bitterness/saltiness, and sourness—are not expected to be related to one another. To those untrained in sensory evaluations, this may not be so obvious.

For the texture data, the variables loading on the first factor along with their factor loadings are SURF_RO (0.88), LOOSE (0.89), FIRM (0.71), COH

Backward Elimination Procedure for Dependent Variable SOUR

Step 0   All Variables Entered   R-square = 0.30107627   C(p) = 4.00000000

	DF	Sum of Squares	Mean Square	F	Prob>F
Regression	3	0.26121154	0.08707051	7.18	0.0004
Error	50	0.60638105	0.01212762		
Total	53	0.86759259			

Variable	Parameter Estimate	Standard Error	Type II Sum of Squares	F	Prob>F
INTERCEP	0.62037037	0.01498618	20.78240741	1713.64	0.0001
TEXTURE1	-0.05261831	0.01540083	0.14156677	11.67	0.0013
TEXTURE2	-0.03992775	0.01497510	0.08621559	7.11	0.0103
TEXTURE3	0.02386654	0.01483586	0.03138551	2.59	0.1140

Bounds on condition number: 1.002088,   9.01253

Step 1   Variable TEXTURE3 Removed R-square = 0.26490087   C(p) = 4.58793604

	DF	Sum of Squares	Mean Square	F	Prob>F
Regression	2	0.22982604	0.11491302	9.19	0.0004
Error	51	0.63776656	0.01250523		
Total	53	0.86759259			

Variable	Parameter Estimate	Standard Error	Type II Sum of Squares	F	Prob>F
INTERCEP	0.62037037	0.01521770	20.78240741	1661.90	0.0001
TEXTURE1	-0.05348590	0.01562916	0.14645329	11.71	0.0012
TEXTURE2	-0.03922128	0.01519990	0.08326325	6.66	0.0128

F I G. **12.5**
Pages 3, 7, 9, 11, and 13 of computer printout, Example 12.4

(−0.67), and MOUTH (−0.78). The variables loading on the second factor along with their factor loadings are FIRM (0.59), FRACT (0.91), and COH (−0.56). The variables loading on the third factor along with their factor loadings are CHEW (0.90) and MOLAR (0.56). How can these three characteristics of the texture variables be intepreted? Again with help from a sensory expert, it was determined that factor 1 is a measure of particle adhesiveness, factor 2 is a measure of cookie crispness and/or hardness, and factor 3 is a measure of masticatability of the cookie or the ease with which a cookie can be consumed. Again, the sensory expert reasoned that these characteristics would be uncorrelated with one another.

The next set of SAS commands was used to rename the variables containing the factor scores for each of the data sets created by the factor anal-

```
       Ex. 12.4 - An Alternative to Canonical Correlation Analysis      11
                   Regression Analyses on Variables Found in FA

     Backward Elimination Procedure for Dependent Variable FLAVOR1

Step 0  All Variables Entered     R-square = 0.75760893   C(p) =    4.00000000

                    DF        Sum of Squares    Mean Square        F  Prob>F
Regression           3          39.70968540    13.23656180    52.09  0.0001
Error               50          12.70480430     0.25409609
Total               53          52.41448970

                Parameter        Standard         Type II
Variable         Estimate           Error  Sum of Squares        F  Prob>F

INTERCEP       -0.00000000      0.06859652      0.00000000     0.00  1.0000
TEXTURE1        0.78828923      0.07049449     31.77304142   125.04  0.0001
TEXTURE2        0.35723432      0.06854581      6.90148482    27.16  0.0001
TEXTURE3       -0.11770413      0.06790846      0.76336742     3.00  0.0892

Bounds on condition number:      1.002088,       9.01253
-------------------------------------------------------------------------
Step 1   Variable TEXTURE3 Removed  R-square = 0.74304488   C(p) =   5.00424706

                    DF        Sum of Squares    Mean Square        F  Prob>F

Regression           2          38.94631798    19.47315899    73.74  0.0001
Error               51          13.46817172     0.26408180
Total               53          52.41448970

                Parameter        Standard         Type II
Variable         Estimate           Error  Sum of Squares        F  Prob>F

INTERCEP       -0.00000000      0.06993142      0.00000000     0.00  1.0000
TEXTURE1        0.79256800      0.07182225     32.15833611   121.77  0.0001
TEXTURE2        0.35375018      0.06984966      6.77334471    25.65  0.0001
```

F I G. **12.5**
Pages 3, 7, 9, 11, and 13 of computer printout, Example 12.4

yses and then to merge these into a single combined data set. The variables
containing the factor scores in the flavor data were named FLAVOR1 and
FLAVOR2, and the variables containing the factor scores in the texture data
were named TEXTURE1–TEXTURE3.

```
DATA FLAV_SCR; SET FLAV_SCR;
  DROP FACTOR1-FACTOR2;
  FLAVOR1=FACTOR1; FLAVOR2=FACTOR2;
RUN;

DATA TEXT_SCR; SET TEXT_SCR;
  DROP FACTOR1-FACTOR3;
  TEXTURE1=FACTOR1; TEXTURE2=FACTOR2; TEXTURE3=FACTOR3;
RUN;
```

Ex. 12.4 - An Alternative to Canonical Correlation Analysis   13
Regression Analyses on Variables Found in FA

Backward Elimination Procedure for Dependent Variable FLAVOR2

Step 0   All Variables Entered   R-square = 0.20512604   C(p) =   4.00000000

	DF	Sum of Squares	Mean Square	F	Prob>F
Regression	3	7.96371189	2.65457063	4.30	0.0089
Error	50	30.85979412	0.61719588		
Total	53	38.82350601			

Variable	Parameter Estimate	Standard Error	Type II Sum of Squares	F	Prob>F
INTERCEP	-0.00000000	0.10690909	0.00000000	0.00	1.0000
TEXTURE1	0.24334125	0.10986711	3.02774338	4.91	0.0314
TEXTURE2	0.13044481	0.10683005	0.92021609	1.49	0.2278
TEXTURE3	0.27392870	0.10583673	4.13452183	6.70	0.0126

Step 1   Variable TEXTURE2 Removed R-square = 0.18142349   C(p) =   3.49096278

	DF	Sum of Squares	Mean Square	F	Prob>F
Regression	2	7.04349580	3.52174790	5.65	0.0061
Error	51	31.78001020	0.62313745		
Total	53	38.82350601			

Variable	Parameter Estimate	Standard Error	Type II Sum of Squares	F	Prob>F
INTERCEP	-0.00000000	0.10742245	0.00000000	0.00	1.0000
TEXTURE1	0.24354542	0.11039455	3.03283345	4.87	0.0319
TEXTURE3	0.27771850	0.10629920	4.25337323	6.83	0.0118

F I G. 12.5

Pages 3, 7, 9, 11, and 13 of computer printout, Example 12.4

```
DATA COMBINED;
  MERGE COOKIE FLAV_SCR TEXT_SCR;
  RUN;
```

   Now that the underlying characteristics of each of the two sets of variables have been identified, we would like to know how the two sets of characteristics relate to each other. To determine how they relate to each other, we can build regression models using one set of characteristics to predict the other set of characteristics.

   The next set of commands was used to regress the three underlying flavor characteristics on the three underlying texture characteristics. To eliminate variables that do not relate, a stepwise regression was performed using the backward selection procedure. Since the predictor variables in this analysis are not correlated with one another, we can be assured that when a variable is removed from a predictor set, that it is removed because it has no effect on the dependent variable being analyzed. Variables are allowed to remain in the model only if they are significant at a 5% significance level.

```
PROC REG DATA=COMBINED;
 MODEL SOUR FLAVOR1 FLAVOR2 = TEXTURE1 TEXTURE2 TEXTURE3
   /SELECTION=BACKWARD SLS=.05;
 TITLE2 'Regression Analyses on Variables Found in FA';
 RUN;
```

The results from these commands are shown on pages 9, 11, and 13 of the computer printout shown in Figure 12.5.

An examination of the results of the first regression analysis shows that both TEXTURE1 (particle adhesiveness) and TEXTURE2 (crispiness) are statistically significant predictors of SOUR (sourness), although the R^2 value is only 0.265. So while the regression is statistically significant, the relationship may not be of much practical importance.

An examination of the results of the second regression analysis shows that TEXTURE (particle adhesiveness) and TEXTURE2 (crispiness) are statistically significant predictors of FLAVOR1 (sweetness). The value of R^2 for this regression is 0.743. This R^2 is large enough that it would appear that the relationship between sweetness and particle adhesiveness and crispiness is likely to be of some practical importance. It might be important to see why such a relationship occurs since it was conjectured that none should occur. Perhaps we need to do some additional training of the sensory panelists to eliminate this apparent relationship in the characteristics being measured.

An examination of the results of the third regression analysis shows that TEXTURE1 (particle adhesiveness) and TEXTURE3 (masticatability) are statistically significant predictors of FLAVOR2 (bitterness/saltiness), although the R^2 value is only 0.181. Once again, even though the regression is statistically significant, the relationship is likely to be of little practical importance.

As a final comment, in this example relationships were discovered that were not expected. More often than not, relationships between two sets of variables will be expected, and we often want to study and explore these relationships.

E X A M P L E 12.5

A researcher wanted to explore relationships between "father's" attitudes and their "daughter's" dating behaviors. Questionnaires were given to a sample of daughters and their fathers. The questionnaires given to daughters were different than the questionnaires given to fathers. There were approximately 200 items in the questionnaire each daughter received and approximately 50 items in the questionnaire each father received. A canonical correlation analysis was performed on these two groups of items, but nothing interpretable resulted from the analysis.

Next a factor analysis on the daughter data revealed approximately seven factors that were clearly interpretable. Another factor analysis was performed on the father data revealing approximately four factors that were interpretable. Next scores were computed for each of the father factors and scores were computed for each of the daughter factors, and these two sets of underlying characteristics were related to each other through the use of stepwise regression methods. Unfortunately, the data for this study no longer exist, and it is not possible to work through this example.

Summary

This chapter discusses procedures for exploring relationships that may exist between two sets of variables. Canonical correlation analysis is a mathematical procedure that directly addresses this problem. Unfortunately, due to an inability to interpret the canonical functions in many cases, canonical correlation analysis cannot always be expected to produce meaningful summaries of the relationships between two sets of variables. An alternative approach involving factor analysis and regression was described and illustrated with an example. This approach is likely to produce much better results than canonical correlation analysis.

Exercises

1   Consider the data in Example 12.2

   a   Suppose two groups of variables are formed—one from the APPEAR-ANCE variables and one by combining the SIZE variables and the COLOR variables. Perform a canonical correlation analysis on these two sets of variables.

   b   In part a, what is the largest correlation that can be attained between a linear combination of the APPEARANCE variables and a linear combination of the SIZE and COLOR variables? What linear combinations of these two groups of variables produce this largest correlation?

   c   In part a, what is the largest correlation that can be attained between linear combinations of the variables in these two groups that are uncorrelated with the linear combinations obtained in part b? What linear combinations of these two groups of variables produce this second largest correlation?

d How many canonical correlations are significantly greater than zero? Explain your answer.

e Perform a factor analysis on each of these two groups of variables and relate their factors to each other via regression.

f Write a short report summarizing what you have learned about these data from your analyses.

2 Consider the employees without supervisory responsibilities in the Pizzazz survey data described in Appendix B.

  a Suppose two groups of variables are formed—one from the organizational commitment variables and a second group from the job satisfaction variables. Perform a canonical correlation analysis on these two sets of variables.

  b In part a, what is the largest correlation that can be attained between a linear combination of the organizational commitment variables and a linear combination of the job satisfaction variables? What linear combinations of these two groups of variables produce this largest correlation?

  c In part a, what is the largest correlation that can be attained between linear combinations of the variables in these two groups that are uncorrelated with the linear combinations obtained in part b? What linear combinations of these two groups of variables produce this second largest correlation?

  d How many canonical correlations are significantly greater than zero? Explain your answer.

  e Perform a factor analysis on each of these two groups of variables and relate their factors to each other via regression.

  f Write a short report summarizing what you have learned about these data from your analyses.

3 Consider all of the females in the married adults survey data described in Appendix C.

  a Suppose two groups of variables are formed—one from the 30-item FACES questionnaire and a second group from the Kansas Marital Satisfaction Scale (KMS) variables. Perform a canonical correlation analysis on these two sets of variables.

  b In part a, what is the largest correlation that can be attained between a linear combination of the FACES variables and a linear combination of the KMS variables? What linear combinations of these two groups of variables produce this largest correlation?

  c In part a, what is the largest correlation that can be attained between linear combinations of the variables in these two groups that are uncorrelated with the linear combinations obtained in part b? What linear combinations of these two groups of variables produce this second largest correlation?

d How many canonical correlations are significantly greater than zero? Explain your answer.

e Perform a factor analysis on the FACES variables and interpret, if possible, rotated factors.

f Perform a factor analysis on the KMS variables and interpret, if possible, rotated factors.

g Relate the two sets of factors in parts e and f to each other using regression methods.

h Write a short report describing what you have learned about the data from the analyses performed in this problem.

4 Consider the males in the married adults survey data described in Appendix C.

a Suppose two groups of variables are formed—one from the 30-item FACES questionnaire and a second group from the 60-item FAD questionnaire. Perform a canonical correlation analysis on these two sets of variables.

b In part a, what is the largest correlation that can be attained between a linear combination of the FACES variables and a linear combination of the FAD variables? What linear combinations of these two groups of variables produce this largest correlation?

c In part a, what is the largest correlation that can be attained between linear combinations of the variables in these two groups that are uncorrelated with the linear combinations obtained in part b? What linear combinations of these two groups of variables produce this second largest correlation?

d How many canonical correlations are significantly greater than zero? Explain your answer.

e Provide interpretations, if possible, of the canonical variates that correspond to statistically significant canonical correlations.

f Perform a factor analysis on the FACES variables and interpret, if possible, rotated factors.

g Perform a factor analysis on the FAD variables and interpret, if possible, rotated factors.

h Relate the two sets of factors in parts f and g to each other using regression methods.

i Write a short report describing those things you have learned about the data from the analyses performed in this problem.

Matrix Results

Appendix A

This book requires very few calculations to be done by hand particularly when computing with matrices and vectors. When such computations are necessary, several very good software packages are available that can complete the required computations. Such packages include SAS-IML, GAUSS, S, MINITAB, and APL. Those who want to write their own routines for doing multivariate analyses should learn to use one of these packages.

The following sections are intended to provide a review of some of the important results that are usually taught in a course dealing with matrix algebra. Hadi (1996) recently wrote a nice little book that provides a good introduction to matrix algebra for those who want to use it as a tool. The portions of Hadi's book that are useful for this text are Chapters 1, 2, Sections 3.1–3.4, 4.1–4.3, Chapter 5, Sections 6.1, 6.2.1, 6.2.2, 9.1, 9.2, 10.1, and 10.2.

In the next section, some basic matrix results are reviewed.

A.1
Basic Definitions and Rules of Matrix Algebra

1　To multiply two matrices, the number of columns in the first must equal the number of rows in the second. Thus matrix multiplication is defined for

$$\underset{m \times n}{\mathbf{A}} \; \underset{n \times p}{\mathbf{B}} = \underset{m \times p}{\mathbf{C}}$$

2　The associative law holds for matrix multiplication when all matrix products are defined. Thus, $\mathbf{A}(\mathbf{BC}) = (\mathbf{AB})\mathbf{C}$.

3　Matrix multiplication is not always commutative. That is, $\mathbf{AB}$ is not always equal to $\mathbf{BA}$. In fact, the two matrices $\mathbf{AB}$ and $\mathbf{BA}$ are not even required to have the same dimensions.

4　If

$$\mathbf{A} = \begin{bmatrix} a_{11} & a_{12} & \cdots & a_{1p} \\ a_{21} & a_{22} & \cdots & a_{2p} \\ \vdots & \vdots & & \vdots \\ a_{n1} & a_{n2} & \cdots & a_{np} \end{bmatrix}$$

then $\mathbf{A}'$ is called the transpose of the matrix $\mathbf{A}$. It is defined by

$$\mathbf{A}' = \begin{bmatrix} a_{11} & a_{21} & \cdots & a_{n1} \\ a_{12} & a_{22} & \cdots & a_{n2} \\ \vdots & & & \\ a_{1p} & a_{2p} & \cdots & a_{np} \end{bmatrix}$$

5　$(\mathbf{AB})' = \mathbf{B}'\mathbf{A}'$.

6   A matrix is called a symmetric matrix if $\mathbf{A} = \mathbf{A}'$. An example of a symmetric matrix is a correlation matrix.

7   The trace of a $p \times p$ square matrix is defined to be the sum of its diagonal elements. We write

$$\text{tr}(\mathbf{A}) = \sum_{i=1}^{p} a_{ii}$$

8   If the matrix products $\mathbf{AB}$ and $\mathbf{BA}$ are both defined, then $\text{tr}(\mathbf{AB}) = \text{tr}(\mathbf{BA})$. For any square matrix $\mathbf{C}$, $\text{tr}(\mathbf{C}') = \text{tr}(\mathbf{C})$.

9   A square matrix is called a diagonal matrix if all of its off-diagonal elements are zero.

10   A special diagonal matrix is the $p \times p$ identity matrix defined by

$$\mathbf{I}_p = \begin{bmatrix} 1 & 0 & 0 & \cdots & 0 \\ 0 & 1 & 0 & \cdots & 0 \\ 0 & 0 & 1 & \cdots & 0 \\ \vdots & \vdots & \vdots & \ddots & \vdots \\ 0 & 0 & 0 & \cdots & 1 \end{bmatrix}_{p \times p}$$

It is called an identity matrix because $\mathbf{AI} = \mathbf{IA} = \mathbf{A}$ for any matrix $\mathbf{A}$.

11   The determinant of a square matrix, $\mathbf{A}$, is defined by

$$|\mathbf{A}| = \sum_{j=1}^{p} a_{ij} A_{ij}$$

where $A_{ij} = (-1)^{i+j}$ times the determinant of the matrix obtained from $\mathbf{A}$ by deleting its ith row and its jth column. Note that the definition usually takes $i = 1$, but the result is the same for any $i = 1, 2, \ldots, p$.

12   If $|A| \neq 0$, then $\mathbf{A}$ is said to be a nonsingular matrix. In this case, the inverse of $\mathbf{A}$ exists and is denoted by $\mathbf{A}^{-1}$. The inverse satisfies $\mathbf{AA}^{-1} = \mathbf{A}^{-1}\mathbf{A} = \mathbf{I}$.

13 $|\mathbf{AB}| = |\mathbf{A}| \cdot |\mathbf{B}|$ when both $\mathbf{A}$ and $\mathbf{B}$ are square matrices. Also $|\mathbf{A}'| = |\mathbf{A}|$.

14   Suppose

$$\mathbf{A} = \begin{bmatrix} \mathbf{A}_{11} & \mathbf{A}_{12} \\ \mathbf{A}_{21} & \mathbf{A}_{22} \end{bmatrix}$$

then

$$|\mathbf{A}| = |\mathbf{A}_{11}| \cdot |\mathbf{A}_{22} - \mathbf{A}_{21}\mathbf{A}_{11}^{-1}\mathbf{A}_{12}| \qquad \text{if } |\mathbf{A}_{11}| \neq 0$$
$$= |\mathbf{A}_{22}| \cdot |\mathbf{A}_{11} - \mathbf{A}_{12}\mathbf{A}_{22}^{-1}\mathbf{A}_{21}| \qquad \text{if } |\mathbf{A}_{22}| \neq 0$$

15   If $\mathbf{A}$ is symmetric, then $\mathbf{A}^{-1}$ is also symmetric.

16   If $\mathbf{A}$ and $\mathbf{B}$ are nonsingular matrices, then $(\mathbf{A}')^{-1} = (\mathbf{A}^{-1})'$, and $(\mathbf{AB})^{-1} = \mathbf{B}^{-1}\mathbf{A}^{-1}$.

17 A set of vectors $\mathbf{x}_1, \mathbf{x}_2, \ldots, \mathbf{x}_p$ is said to be linearly dependent if there exist constants $c_1, c_2, \ldots, c_p$, which are not all equal to zero, such that $\sum_{i=1}^{p} c_i \mathbf{x}_i = \mathbf{0}$. Otherwise the vectors are said to be *linearly independent.*

A set of vectors is linearly dependent if and only if at least one of the vectors can be written as a linear combination of the remaining vectors.

18 The rank of a matrix is defined as the maximum number of rows (columns) in $\mathbf{A}$ which are linearly independent. Equivalently, the rank of $\mathbf{A}$ is the dimension of the vector subspace spanned by the rows (columns) of the matrix $\mathbf{A}$.

19 Each of the following statements about ranks of matrices is true:

a $\quad \text{rank}\left(\underset{m \times n}{\mathbf{A}}\right) \leq \min(m,n)$.

b $\quad \text{rank}(\mathbf{A}) = \text{rank}(\mathbf{A}')$.

c $\quad \text{rank}(\mathbf{A}) = \text{rank}(\mathbf{BA}) = \text{rank}(\mathbf{AC})$ for nonsingular matrices $\mathbf{B}$ and $\mathbf{C}$.

d $\quad \text{rank}\left(\underset{p \times p}{\mathbf{A}}\right) = p$ if and only if $|\mathbf{A}| \neq 0$.

20 Two vectors, $\mathbf{x}$ and $\mathbf{y}$, are orthogonal if $\mathbf{x}'\mathbf{y} = 0$. In addition, two such orthogonal vectors are said to be orthonormal if $\mathbf{x}'\mathbf{x} = 1$ and $\mathbf{y}'\mathbf{y} = 1$.

21 A vector $\mathbf{x}$ is said to be normalized if $\mathbf{x}'\mathbf{x} = 1$.

22 A square matrix $\mathbf{P}$ is said to be orthogonal if $\mathbf{PP}' = \mathbf{I}$ or $\mathbf{P}'\mathbf{P} = \mathbf{I}$. In this case $\mathbf{P}^{-1} = \mathbf{P}'$.

23 If $\mathbf{P}$ is an orthogonal matrix, then $|\mathbf{P}|$ is equal to either 1 or -1.

A.2
Quadratic Forms

The following results are often not taught in elementary courses in matrix algebra, but they are important for statistical analyses.

1 A quadratic form in p variables $x_1, x_2, \ldots, x_p$ is a function of the form $\sum_{i=1}^{p} \sum_{j=1}^{p} a_{ij} x_i x_j$. Such a quadratic form can always be written as $\mathbf{x}'\mathbf{Ax}$ for some

symmetric matrix $\mathbf{A}$ where $\mathbf{x} = \begin{bmatrix} x_1 \\ x_2 \\ \vdots \\ x_p \end{bmatrix}$.

2 A symmetric matrix $\mathbf{A}$ is called

a positive definite (p.d.) if $\mathbf{x}'\mathbf{Ax} > 0$ for every $\mathbf{x} \neq \mathbf{0}$.

b positive semidefinite (p.s.d.) if $\mathbf{x}'\mathbf{Ax} \geq 0$ for every $\mathbf{x}$ and if $\mathbf{x}'\mathbf{Ax} = 0$ for some nonzero $\mathbf{x}$.

c nonnegative (n.n.) if $\mathbf{A}$ is either positive semidefinite or positive definite.

3   A matrix $\mathbf{A}$ is positive definite if and only if there exists a nonsingular matrix $\mathbf{Q}$ such that $\mathbf{A} = \mathbf{Q}\mathbf{Q}'$. *Note:* This implies that $\mathbf{A}$ is symmetric.

4   A matrix $\underset{p \times p}{\mathbf{A}}$ is a positive semidefinite matrix of rank $m < p$ if and only if there exists a $p \times m$ matrix $\mathbf{Q}$ of rank m such that

$$\underset{p \times p}{\mathbf{A}} = \underset{p \times m}{\mathbf{Q}} \ \underset{m \times p}{\mathbf{Q}'}$$

A.3
Eigenvalues and Eigenvectors

The eigenvalues and eigenvectors of a matrix are special functions of the elements of a matrix that play an extremely important role in many multivariate analysis techniques. See Chapter 4 for additional motivation for studying eigenvalues and eigenvectors.

In each of the following, let $\boldsymbol{\Sigma}$ be a $p \times p$ symmetric matrix.

1   The eigenvalues (also called characteristic roots or latent roots by some authors) of $\boldsymbol{\Sigma}$ are the roots of the polynomial equation given by

$$|\boldsymbol{\Sigma} - \lambda \mathbf{I}| = 0$$

Note that this determinantal equation is a pth degree polynomial equation in λ.

2   To each eigenvalue of $\boldsymbol{\Sigma}$ there corresponds a nonzero vector called an *eigenvector* (also called a *characteristic vector* or *latent vector*) that satisfies $\boldsymbol{\Sigma}\mathbf{c}_i = \lambda_i \mathbf{c}_i$, for $i = 1, 2, \ldots, p$.

3   If $\boldsymbol{\Sigma}$ is a symmetric matrix of real numbers, then its eigenvalues and eigenvectors will always consist of real numbers.

4   The eigenvalues of $\boldsymbol{\Sigma}$ are denoted by $\lambda_1 \geq \lambda_2 \geq \cdots \geq \lambda_p$.

5   Eigenvectors are not unique so they are often normalized such that $\mathbf{c}_i' \mathbf{c}_i = 1$.

6   When two eigenvalues are not equal, their corresponding eigenvectors will be orthogonal. When two or more eigenvalues of $\boldsymbol{\Sigma}$ are equal, the corresponding eigenvectors can and always will be chosen to be orthogonal to one another.

7   The trace of a symmetric matrix is equal to the sum of its eigenvalues, that is, $\mathrm{tr}(\boldsymbol{\Sigma}) = \sum_{i=1}^{p} \lambda_i$.

8   The determinant of a symmetric matrix is always equal to the product of its eigenvalues, that is, $|\boldsymbol{\Sigma}| = \prod_{i=1}^{p} \lambda_i$.

9   A symmetric matrix $\boldsymbol{\Sigma}$ is positive definite if and only if $\lambda_i > 0$ for each i.

10   A symmetric matrix $\boldsymbol{\Sigma}$ is positive semidefinite if and only if $\lambda_i \geq 0$ for each i and at least one eigenvalue is equal to zero.

11 If Σ is a nonnegative matrix of rank m, then there will be exactly m nonzero eigenvalues.

12 If Σ is a symmetric matrix, there exists an orthogonal matrix, $\mathbf{C}$, such that $\mathbf{C}'\Sigma\mathbf{C} = \Lambda$ where Λ is a diagonal matrix. Furthermore, the diagonal elements of Λ are the eigenvalues of Σ and the columns of $\mathbf{C}$ are their corresponding eigenvectors. Thus,

$$\Lambda = \begin{bmatrix} \lambda_1 & & & 0 \\ & \lambda_2 & & \\ & 0 & \ddots & \\ & & & \lambda_p \end{bmatrix}$$

and $\mathbf{C} = [\mathbf{c}_1\ \mathbf{c}_2 \cdots \mathbf{c}_p]$.

13 The previous result implies that

$$\Sigma = \mathbf{C}\Lambda\mathbf{C}' = \sum_{i=1}^{p} \lambda_i \mathbf{c}_i \mathbf{c}_i'$$

This is called the spectral decomposition of Σ.

A.4
Distances and Angles

1 The norm of a vector $\mathbf{x}$ is defined by $\|\mathbf{x}\| = \sqrt{\mathbf{x}'\mathbf{x}}$. This measures the distance from $\mathbf{x}$ to the origin in E_p.

2 The distance between two vectors $\mathbf{x}$ and $\mathbf{y}$ is $\|\mathbf{x} - \mathbf{y}\|$.

3 The angle θ between $\mathbf{x}$ and $\mathbf{y}$ is given by

$$\cos \theta = \frac{\mathbf{x}'\mathbf{y}}{\|\mathbf{x}\| \cdot \|\mathbf{y}\|}$$

4 The projection of $\mathbf{x}$ onto $\mathbf{y}$ is

$$\frac{\mathbf{x}'\mathbf{y}}{\|\mathbf{y}\|} \cdot \mathbf{y} = (\|\mathbf{x}\| \cos \theta)\mathbf{y}$$

A.5
Miscellaneous Results

1 If $\mathbf{A}$ is positive definite, then $\mathbf{A}^{-1}$ is positive definite.

2 The nonzero eigenvalues of $\mathbf{AB}$ and $\mathbf{BA}$ are the same. In this result $\mathbf{A}$ and $\mathbf{B}$ are not required to be square matrices, but $\mathbf{AB}$ and $\mathbf{BA}$ are required to be square matrices.

3   If **A** is positive definite and **B** is positive semidefinite, then there exists a nonsingular matrix **C** such that $\mathbf{C'AC} = \mathbf{I}$ and $\mathbf{C'BC} = \mathbf{D}$ where **D** is a diagonal matrix whose diagonal elements are the roots of $|\mathbf{B} - \lambda\mathbf{A}| = 0$ or $|\mathbf{BA}^{-1} - \lambda\mathbf{I}| = 0$.

4   Let **A** be a nonnegative matrix with eigenvalues $\lambda_1 \geq \lambda_2 \geq \cdots \geq \lambda_p$.

   a $\max\limits_{x \neq 0} \dfrac{\mathbf{x'Ax}}{\mathbf{x'x}} = \lambda_1$

   b $\min\limits_{x \neq 0} \dfrac{\mathbf{x'Ax}}{\mathbf{x'x}} = \lambda_p$

   c   These values are attained when **x** is an eigenvector of **A** corresponding to the roots λ_1 or λ_p, respectively.

5   Let **A** be a nonnegative matrix and let **B** be a positive definite matrix.

   a $\max\limits_{x \neq 0} \dfrac{\mathbf{x'Ax}}{\mathbf{x'Bx}} = \mathrm{CH}_{\max}\mathbf{B}^{-1}\mathbf{A}$

   b $\min\limits_{x \neq 0} \dfrac{\mathbf{x'Ax}}{\mathbf{x'Bx}} = \mathrm{CH}_{\min}\mathbf{B}^{-1}\mathbf{A}$

   c   These values are attained when **x** is an eigenvector of $\mathbf{B}^{-1}\mathbf{A}$ corresponding to the root selected.

Work Attitudes Survey

Appendix B

A survey of 975 employees of a national restaurant chain was undertaken by Kansas State University researcher Dr. Ronald Downey. The data set includes 144 variables. The data are real, but for obvious reasons, the true name of the restaurant chain is not being disclosed. In this book the restaurant chain is called Pizzazz. The survey used to collect data from Pizzazz employees is reprinted here.

WORK ATTITUDES SURVEY*

THE FOLLOWING QUESTIONS DEAL WITH A WIDE VARIETY OF EXPERIENCES YOU ENCOUNTER WHILE ON THE JOB. THEY ARE PART OF A STUDY WE ARE CONDUCTING TO BETTER UNDERSTAND OUR PIZZAZZ CREW MEMBERS, SHIFT SUPERVISORS, HOSTS/HOSTESSES, DRIVERS, ASSISTANT MANAGERS AND MANAGERS. <u>YOUR RESPONSES ARE CONFIDENTIAL AND NONE OF YOUR INDIVIDUAL ANSWERS WILL BE SEEN BY ANYONE AT PIZZAZZ</u>. PLEASE TAKE A FEW MOMENTS OF YOUR TIME AND COMPLETE THE SURVEY. FOLLOW THE DIRECTIONS GIVEN AT THE TOP OF EACH SECTION.

USE THE FOLLOWING SCALE TO ANSWER QUESTIONS 1–18: (CIRCLE A NUMBER)						
1	2	3	4	5	6	7
STRONGLY DISAGREE	MODERATELY DISAGREE	SLIGHTLY DISAGREE	NEITHER DISAGREE NOR AGREE	SLIGHTLY AGREE	MODERATELY AGREE	STRONGLY AGREE

1. I AM WILLING TO WORK HARDER THAN MOST PEOPLE AT PIZZAZZ.   1 2 3 4 5 6 7
2. I TALK UP PIZZAZZ TO MY FRIENDS AS A GREAT COMPANY TO WORK FOR.   1 2 3 4 5 6 7
3. I FEEL VERY LITTLE LOYALTY TO PIZZAZZ.   1 2 3 4 5 6 7
4. I WOULD ACCEPT ALMOST ANY TYPE OF JOB IN ORDER TO STAY WITH PIZZAZZ.   1 2 3 4 5 6 7
5. MY VALUES AND PIZZAZZ'S VALUES ARE SIMILAR.   1 2 3 4 5 6 7
6. I AM PROUD TO TELL OTHERS THAT I WORK FOR PIZZAZZ.   1 2 3 4 5 6 7
7. I COULD JUST AS WELL BE WORKING FOR ANOTHER COMPANY AS LONG AS THE WORK WAS SIMILAR.   1 2 3 4 5 6 7
8. PIZZAZZ REALLY INSPIRES ME TO DO MY BEST.   1 2 3 4 5 6 7
9. IT WOULD TAKE VERY LITTLE CHANGE IN MY WORK TO CAUSE ME TO LEAVE PIZZAZZ.   1 2 3 4 5 6 7
10. I AM GLAD THAT I CHOSE PIZZAZZ TO WORK FOR.   1 2 3 4 5 6 7
11. THERE'S NOT MUCH TO BE GAINED BY STAYING WITH PIZZAZZ A LONG TIME.   1 2 3 4 5 6 7
12. I OFTEN FIND IT DIFFICULT TO AGREE WITH PIZZAZZ'S ATTITUDES TOWARDS ITS EMPLOYEES.   1 2 3 4 5 6 7
13. I REALLY CARE ABOUT WHAT HAPPENS TO PIZZAZZ.   1 2 3 4 5 6 7
14. FOR ME THIS IS THE BEST COMPANY TO WORK FOR.   1 2 3 4 5 6 7
15. DECIDING TO WORK FOR PIZZAZZ WAS A MISTAKE.   1 2 3 4 5 6 7
16. I AM VERY MUCH PERSONALLY INVOLVED IN MY WORK.   1 2 3 4 5 6 7
17. THE MAJOR SATISFACTION IN MY LIFE COMES FROM MY JOB.   1 2 3 4 5 6 7
18. THE MOST IMPORTANT THINGS WHICH HAPPEN TO ME INVOLVE MY JOB.   1 2 3 4 5 6 7

* Reprinted by permission of Dr. Ronald Downey.

USE THE FOLLOWING SCALE TO ANSWER QUESTIONS 19–60: (CIRCLE A NUMBER)

1 STRONGLY DISAGREE	2 DISAGREE	3 NEITHER AGREE NOR DISAGREE	4 AGREE	5 STRONGLY AGREE

19. INTERACTING WITH CUSTOMERS IS ENJOYABLE   1 2 3 4 5
20. IT IS IMPORTANT TO ME THAT THE CUSTOMER IS SATISFIED.   1 2 3 4 5
21. THE EMPLOYEES AT MY RESTAURANT/UNIT PROVIDE EXCELLENT CUSTOMER SERVICE.   1 2 3 4 5
22. MY MANAGER ENCOURAGES ME TO PROVIDE EXCELLENT CUSTOMER SERVICE.   1 2 3 4 5
23. THE TRAINING I RECEIVED AT PIZZAZZ PREPARED ME TO PROVIDE EXCELLENT
    CUSTOMER SERVICE.   1 2 3 4 5
24. CUSTOMERS TREAT ME WITH RESPECT.   1 2 3 4 5
25. AT PEAK HOURS IT'S SO BUSY (E.G., FRIDAY NIGHT) WE CAN'T PROVIDE EXCELLENT
    CUSTOMER SERVICE.   1 2 3 4 5
26. OUR SERVICE PROCEDURES MAKES IT EASY FOR ME TO GIVE EXCELLENT CUSTOMER
    SERVICE.   1 2 3 4 5
27. MY MANAGER EXPECTS US TO ALWAYS FOLLOW PROCEDURES, EVEN IF IT MEANS GIVING
    LESS THAN EXCELLENT CUSTOMER SERVICE.   1 2 3 4 5
28. MY JOB LEAVES ME ENOUGH TIME TO SPEND WITH MY FAMILY AND FRIENDS.   1 2 3 4 5
29. WHEN I AM AT WORK, I OFTEN WORRY ABOUT MY HOME, SCHOOL AND/OR FAMILY.   1 2 3 4 5
30. I GET SO INVOLVED IN MY JOB THAT IT CONFLICTS WITH MY WORK AND HOME OR
    SCHOOL RESPONSIBILITIES.   1 2 3 4 5
31. MY JOB GIVES ME A WELCOME BREAK FROM OTHER RESPONSIBILITIES.   1 2 3 4 5
32. MY FAMILY THINKS IT'S A GOOD IDEA FOR ME TO WORK.   1 2 3 4 5
33. MY FRIENDS APPROVE OF MY JOB AT PIZZAZZ.   1 2 3 4 5
34. PIZZAZZ REALLY MAKES ME FEEL WANTED.   1 2 3 4 5
35. I HAVE CONFIDENCE IN THE HONESTY AND FAIRNESS OF MANAGEMENT.   1 2 3 4 5
36. PIZZAZZ HAS THE WELFARE OF THE EMPLOYEES AT HEART.   1 2 3 4 5
37. CHANGES ARE MADE HERE WITH LITTLE REGARD FOR WELFARE OF THE EMPLOYEES.   1 2 3 4 5
38. THE EMPLOYEES HAVE AN ACTIVE PART IN RUNNING PIZZAZZ.   1 2 3 4 5
39. I CANNOT TRUST PIZZAZZ TO STAND BEHIND THEIR PROMISES.   1 2 3 4 5
40. PEOPLE WORK TOGETHER AROUND HERE.   1 2 3 4 5
41. I ALWAYS KNOW WHERE I STAND WITH PIZZAZZ.   1 2 3 4 5
42. I AM NOT REALLY SURE OF HOW LONG MY JOB WILL LAST.   1 2 3 4 5
43. I AM AFRAID OF LOSING MY JOB.   1 2 3 4 5
44. IT IS VERY UNLIKELY THAT MY JOB WILL BE TERMINATED.   1 2 3 4 5
45. I CAN KEEP MY JOB HERE FOR AS LONG AS I DO GOOD WORK.   1 2 3 4 5
46. I CAN BE SURE OF MY JOB AS LONG AS I DO GOOD WORK.   1 2 3 4 5
47. LAYOFFS ARE A TYPICAL OCCURRENCE AROUND HERE.   1 2 3 4 5
48. I WORK HARD AND DO MY JOB TO THE BEST OF MY ABILITIES.   1 2 3 4 5
49. I THINK I AM DOING A GOOD JOB.   1 2 3 4 5
50. MY WORK MEETS OR EXCEEDS PIZZAZZ'S STANDARDS.   1 2 3 4 5
51. I GET ALONG WELL WITH MY FELLOW EMPLOYEES.   1 2 3 4 5
52. MY SUPERVISOR KNOWS I AM DOING A GOOD JOB.   1 2 3 4 5
53. MY WORK IS AT LEAST AS GOOD AS THAT OF OTHERS WITH WHOM I WORK.   1 2 3 4 5
54. I OFTEN THINK ABOUT QUITTING MY JOB.   1 2 3 4 5
55. I AM ABSENT MORE OFTEN THAN OTHER EMPLOYEES.   1 2 3 4 5
56. ATTENDANCE IS IMPORTANT IN MY RESTAURANT/UNIT.   1 2 3 4 5
57. IF I PERFORM WELL, I KNOW MY EFFORTS WILL BE REWARDED THROUGH MY SALARY.   1 2 3 4 5

58. CONSIDERING THE WORK I DO, AND WHAT I MIGHT EARN WITH ANOTHER COMPANY, I AM PAID FAIRLY.   1 2 3 4 5
59. THIS JOB MEASURES UP TO WHAT I EXPECTED WHEN I TOOK IT.   1 2 3 4 5
60. I AM AWARE OF THE CAREER OPPORTUNITIES AT PIZZAZZ.   1 2 3 4 5

PLEASE INDICATE <u>HOW YOU FEEL ABOUT YOUR JOB AT PIZZAZZ.</u> USING THE SCALE BELOW, CIRCLE THE NUMBER THAT DESCRIBES YOUR VIEW.

1	2	3	4	5
VERY DISSATISFIED	DISSATISFIED	I CAN'T DECIDE	SATISFIED	VERY SATISFIED

---

61. BEING ABLE TO KEEP BUSY ALL THE TIME.   1 2 3 4 5
62. THE CHANCE TO WORK ALONE ON THE JOB.   1 2 3 4 5
63. THE CHANCE TO DO DIFFERENT JOBS FROM TIME TO TIME.   1 2 3 4 5
64. THE CHANCE TO BE "SOMEBODY" IN THE COMMUNITY.   1 2 3 4 5
65. THE WAY MY MANAGER HANDLES PEOPLE.   1 2 3 4 5
66. THE ABILITY OF MY MANAGER TO MAKE GOOD DECISIONS.   1 2 3 4 5
67. BEING ABLE TO DO THINGS THAT DON'T GO AGAINST BY CONSCIENCE.   1 2 3 4 5
68. THE WAY MY JOB PROVIDES FOR STEADY EMPLOYMENT.   1 2 3 4 5
69. THE CHANCE TO DO THINGS FOR OTHER PEOPLE.   1 2 3 4 5
70. THE CHANCE TO TELL OTHER PEOPLE WHAT TO DO.   1 2 3 4 5
71. THE CHANCE TO DO SOMETHING THAT MAKES USE OF MY ABILITIES.   1 2 3 4 5
72. THE WAY PIZZAZZ'S POLICIES ARE PUT INTO PRACTICE.   1 2 3 4 5
73. THE PAY AND THE AMOUNT OF WORK I DO.   1 2 3 4 5
74. THE CHANCES FOR ADVANCEMENT ON THE JOB.   1 2 3 4 5
75. THE FREEDOM TO USE MY OWN JUDGMENT.   1 2 3 4 5
76. THE CHANCE TO TRY MY OWN METHODS OF DOING THE JOB.   1 2 3 4 5
77. THE PHYSICAL WORKING CONDITIONS AT PIZZAZZ.   1 2 3 4 5
78. THE WAY MY CO-WORKERS GET ALONG WITH EACH OTHER.   1 2 3 4 5
79. THE PRAISE I GET FOR DOING A GOOD JOB.   1 2 3 4 5
80. THE FEELING OF ACCOMPLISHMENT I GET FROM THE JOB.   1 2 3 4 5
81. THE FORMAL PERFORMANCE REVIEWS GIVEN BY MY MANAGER.   1 2 3 4 5
82. THE TEAM SPIRIT IN MY RESTAURANT/UNIT.   1 2 3 4 5

USE THE FOLLOWING SCALE TO ANSWER QUESTIONS 83–85

1	2	3	4	5
VERY LIKELY	UNLIKELY	NEUTRAL	LIKELY	VERY LIKELY

---

83. HOW LIKELY IS IT THAT YOU COULD FIND A GOOD JOB IF YOU WERE TO LEAVE PIZZAZZ?   1 2 3 4 5
84. HOW LIKELY IS IT THAT YOU WILL LOOK FOR ANOTHER JOB DURING THE NEXT THREE MONTHS?   1 2 3 4 5
85. HOW LIKELY IS IT THAT YOU WILL QUIT YOUR JOB AT PIZZAZZ DURING THE NEXT THREE MONTHS?   1 2 3 4 5
86. HOW MANY <u>TOTAL DAYS</u> HAVE YOU BEEN ABSENT DURING THE PAST TWO MONTHS WHEN SCHEDULED TO WORK? _____

DIRECTIONS: CIRCLE ONE OF THE THREE LETTERS THAT BEST DESCRIBES HOW YOU FEEL ABOUT THE FOLLOWING STATEMENTS

	AGREE	?	DIS-AGREE
87. I GET RECOGNITION WHEN I DO GOOD WORK.	A	?	D
88. PAY INCREASES HAVE BEEN ABOUT WHAT I EXPECTED.	A	?	D
89. MY SUPERVISOR IS FAIR IN DEALINGS WITH ME.	A	?	D
90. OUR COMPANY HAS GOOD WORKING CONDITIONS.	A	?	D
91. I AM SATISFIED WITH MY JOB.	A	?	D
92. I AM PROUD TO WORK FOR PIZZAZZ.	A	?	D
93. OUR FRINGE BENEFITS ARE AS GOOD AS OTHER FOOD SERVICE COMPANIES IN OUR AREA.	A	?	D
94. I KNOW WHAT COMPANY BENEFITS ARE AVAILABLE TO ME.	A	?	D
95. WITHIN THE LAST FEW MONTHS I HAVE NOT LOOKED FOR A JOB WITH ANOTHER COMPANY.	A	?	D
96. I HAVE BEEN PROPERLY TRAINED TO DO MY JOB.	A	?	D
97. I FEEL FREE TO TAKE MY COMPLAINTS TO MY SUPERVISOR.	A	?	D
98. I AGREE WITH MY LAST PERFORMANCE REVIEW.	A	?	D
99. I AM SATISFIED WITH THE WAY THE WORK SCHEDULE IS BEING HANDLED.	A	?	D
100. I GET PAID FOR ALL HOURS WORKED.	A	?	D

HOW DOES YOUR WORK GROUP COMPARE WITH OTHER WORK GROUPS THAT YOU KNOW ON EACH OF THE FOLLOWING QUESTIONS?

	BETTER THAN MOST	ABOUT THE SAME	NOT AS GOOD
101. THE WAY WORKERS GET ALONG TOGETHER.	1	2	3
102. THE WAY THE WORKERS STICK TOGETHER.	1	2	3
103. THE WAY THE WORKERS HELP EACH OTHER ON THE JOB.	1	2	3

104. HOW DID YOU HEAR ABOUT YOUR JOB OPENING AT PIZZAZZ? CHECK ALL THAT APPLY.
   • SOMEONE WHO WORKED HERE _____
   • AS A CUSTOMER _____
   • NEWSPAPER AD _____
   • RADIO/TV AD _____
   • SIGN ON WINDOW _____
   • JOB SERVICE _____
   • FRIEND _____

105. WHAT WERE THE MOST IMPORTANT THINGS THAT MADE YOU WANT TO APPLY FOR A JOB AT PIZZAZZ (CHECK 3)
   • FLEXIBLE SCHEDULE _____
   • CHANCE TO MEET NEW PEOPLE _____
   • FUN PLACE _____
   • CLOSE TO HOME _____
   • LIKE PIZZA _____
   • JOB EASY TO GET _____
   • CHANCE FOR ADVANCEMENT _____
   • CHANCE TO LEARN NEW SKILLS _____
   • TYPE OF WORK (MAKING PIZZA, SERVING CUSTOMERS) _____
   • GOOD MONEY/GOOD TIPS _____
   • FRIENDS WORK HERE _____
   • REPUTATION OF THE COMPANY _____

BACKGROUND INFORMATION

PLEASE ANSWER EACH OF THE FOLLOWING QUESTIONS TO THE BEST OF YOUR KNOWLEDGE. WE WILL USE THIS FOR RESEARCH PURPOSES ONLY. YOU WILL NOT BE IDENTIFIED FROM THIS INFORMATION.

---

1. YOUR AGE (IN YEARS)? _____
2. SEX: ( ) MALE   ( ) FEMALE
3. RACE: ( ) CAUCASIAN   ( ) BLACK   ( ) HISPANIC   ( ) OTHER
4. MARITAL STATUS: ( ) SINGLE   ( ) MARRIED   ( ) OTHER
5. IF MARRIED, HOW MANY CHILDREN DO YOU HAVE? _____
6. WHAT IS THE HIGHEST LEVEL OF FORMAL EDUCATION YOU HAVE COMPLETED?
    ( ) GRADE SCHOOL   ( ) HIGH SCHOOL   ( ) COLLEGE
7. WHAT IS YOUR POSITION?
    ( ) KITCHEN CREW   ( ) DELIVERY DRIVER   ( ) WAITER/WAITRESS
    ( ) SHIFT SUPERVISOR   ( ) ASS'T MANAGER   ( ) MANAGER
8. ARE YOU SATISFIED WITH YOUR CURRENT WORK SCHEDULE?
    ( ) YES   ( ) NO, I WANT TO WORK LESS   ( ) NO, I WANT TO WORK MORE
9. IN GENERAL IS YOUR WORK SCHEDULE:
    ( ) THE SAME FROM WEEK TO WEEK   ( ) DIFFERENT FROM WEEK TO WEEK
10. HOW MUCH SAY DO YOU HAVE IN YOUR WORK SCHEDULE?
    ( ) I HAVE NO SAY   ( ) I HAVE SOME SAY   ( ) I HAVE A LOT OF SAY
11. WHERE IS YOUR RESTAURANT LOCATED?
    ( ) VERY LARGE CITY   ( ) AVERAGE SIZE CITY   ( ) SMALL TOWN
12. HOW MANY HOURS ARE YOU <u>SCHEDULED</u> TO WORK EACH WEEK? _____
13. DO YOU ATTEND SCHOOL (CHECK ONE)?
    ( ) NO   ( ) HIGH SCHOOL   ( ) COLLEGE   ( ) OTHER
14. ARE YOU EMPLOYED IN ANOTHER JOB? (CHECK ONE)
    ( ) NO   ( ) YES, PART-TIME   ( ) YES, FULL-TIME
15. DO YOU WORK FOR (CHECK ONE)?
    ( ) RESTAURANT   ( ) DELIVERY   ( ) RESTAURANT BASED DELIVERY (RBD)
16. HOW LONG HAVE YOU WORKED FOR PIZZAZZ?
    ( ) LESS THAN 3 MONTHS
    ( ) 3 TO 6 MONTHS
    ( ) 6 TO 12 MONTHS
    ( ) MORE THAN 2 YEARS
17. HOW MUCH LONGER DO YOU PLAN TO WORK AT PIZZAZZ?
    ( ) 1 TO 3 MONTHS
    ( ) 3 TO 6 MONTHS
    ( ) 6 TO 12 MONTHS
    ( ) MORE THAN 2 YEARS
18. IF YOU PLAN TO LEAVE PIZZAZZ DURING THE NEXT 6 MONTHS, WHAT WILL BE THE MAJOR REASON?
    ( ) SCHOOL
    ( ) PERSONAL REASONS
    ( ) DISLIKE PIZZAZZ AS A PLACE TO WORK
    ( ) MORE ATTRACTIVE JOB

THE FOLLOWING THREE QUESTIONS DO NOT APPLY TO MANAGERS AND ASSISTANT MANAGERS

---

19. WHAT IS YOUR MAJOR REASON FOR WORKING (CHECK ONE)?
    ( ) NEED MONEY   ( ) EXTRA MONEY   ( ) ENJOYMENT
20. IF YOU WORK LESS THAN 35 HOURS A WEEK AT PIZZAZZ, IS IT BECAUSE
    ( ) YOU DON'T WANT TO WORK MORE THAN THAT.
    ( ) YOU COULD NOT FIND FULL-TIME WORK.
    ( ) YOU HAVE ANOTHER JOB.
21. IF YOU WANT TO WORK LESS THAN 35 HOURS A WEEK, HOW MANY HOURS DO YOU WANT TO WORK?
    _____

B.1

Data File Structure

A portion (10 records) of the data from the above survey is given below. The full data set can be found in the file labeled PIZZAZZ.DAT on the enclosed disk. SAS commands that can be used to read these data are given in the file labeled PIZZAZZ.SAS on the enclosed disk and SPSS commands that can be used to read this data are given in the file labeled PIZZAZZ.SSX on the enclosed disk. Each complete record of data from an employee requires two lines.

```
651567162566561644444444424233235441245453224441555442315344444433444432243452
444542121211121133131111112121222112111221222421112232223611243111 514625712
561647243544441564344543141341423343333432153424544442244434443344443434444443
4443223121311223112112222211111112111111122113342124111122371225 1 512607417
643444244435424333 4    2 335334234333224441554435315234344234444333332334242
3323531321322231122212222111212211221111111317111 112223421225112 999000001
4263114141774174424454232222422411115234343333144542441513245251134343311131142
11334303333331111331232211111121111121111113261112212123711253412 999000002
571567271712771654453545343455353554242543244515555441253445445454555354355444
4335213121112111121111211121111121121221111111113191432312323711225 999000003
771177713733311711445534151141544353112541153315555551154555551155552454444454
4452110111111111111111111121111211111111122113311323231122633254423265058808641
724717741471425744554554444344255111515413344415555551352235353232555452211334 1
43245573232313131133123311212111211221211123311124233111351125221 510649227
64353324545543544444344424323342231424333333334444445131132333333333333333333333
333441121122322111112222111111111111121111113331113313232351123 4   999000004
66213726665526116664334433323244433242423224442444444324232344443334333432442233
333222013232221111121111112111122121111111340112311223511255 12 509523481
671567362626651632455555444242245543543534443515555443154244454555454454334434342
4343222312112111122111111111121211111122111326112023122237112554411 511646147
```

Columns 1–79 of the first line contain the circled survey responses for questions 1–79 with one space for each circled response. Columns 1–6 of the second line contain the circled survey responses for questions 80–85, column 7 contains an entry for the total number of days absent from work during the last two months (if the number of days absent is greater than 9, this entry is coded as 'A' for 10 days absent, 'B' for 11 days absent, etc.). Columns 8–21 contain entries for questions 87–100, respectively, ('1'=AGREE, '2'=?, and '3'=DISAGREE) and columns 22–24 contain entries for questions 101–103. Columns 25–31 contain 7 entries corresponding to each of the 7 parts of question 104 ('1'=not checked, '2'=checked) and columns 32–43 contain 12 entries corresponding to each of the 12 parts of question 105 ('1'=not checked, '2'=checked). Column 44 contains an entry indicating the division worked in ('1'=LA/ORANGE CO., '2'=SAN DIEGO DEL., '3'=KC DELIVERY, '4'=DALLAS, '5'=CHICAGO CENTRAL, '6'=THREE RIVERS, '7'=METRO NY, '8'=DELAWARE VALLEY, '9'=MARYLAND, '0'=ATLANTIC DELIVERY).

The remaining entries on the second line are concerned with background information about the employee. Columns 45–46 contain the employee's age, column 47 contains the employee's gender ('1'=MALE, '2'=FEMALE), column 48 identifies the employee's race ('1'=CAUCASIAN, '2'=BLACK, '3'=HISPANIC, '4'=OTHER), column 49 gives the employee's marital status ('1'=SINGLE, '2'=MARRIED, '3'=OTHER), column 50 gives the number of children the employee has, column 51 gives the employee's education level ('1'=GRADE SCHOOL, '2'=HIGH SCHOOL, '3'=COLLEGE), column 52 provides the employee's position at Pizzazz ('1'=KITCHEN CREW, '2'=SHIFT SUPERVISOR, '3'=DELIVERY DRIVER, '5'=ASST MANAGER, '4'=WAITER/WAITRESS, '6'=MANAGER). Column 53 provides the employee's satisfaction with current work schedule ('1'=YES, '2'=NO WANT TO WORK LESS, '3'=NO WANT TO WORK MORE). Column 54 gives the type of work schedule ('1'=THE SAME EACH WEEK, '2'=DIFFERENT EACH WEEK). Column 55 shows how much an employee has to say about the work schedule ('1'=I HAVE NO SAY, '2'=I HAVE SOME SAY, '3'=I HAVE A LOT TO SAY). Column 56 identifies the type of setting Pizzazz is located in ('1'=VERY LARGE CITY, '2'=AVERAGE SIZE CITY, '3'=SMALL TOWN). Columns 57 and 58 give the number of hours the employee is scheduled to work each week. Column 59 tells the type of school, if any, the employee attends ('1'=NO, '2'=HIGH SCHOOL, '3'=COLLEGE, '4'=OTHER). Column 60 indicates whether the employee is employed in another job ('1'=NO, '2'=YES, PART TIME, '3'=YES FULL TIME). Column 61 contains the type of job ('1'=RESTAURANT, '2'=DELIVERY, '3'=RESTAURANT BASED DELIVERY). Column 62 gives the length of time employee has worked at Pizzazz ('1'=LESS THAN 3 MONTHS, '2'=3 TO 6 MONTHS, '3'=6 TO 12 MONTHS, '4'=1 TO 2 YEARS, '5'=MORE THAN 2 YEARS). Column 63 gives an estimate of time the employee expects to continue working for Pizzazz ('1'=1 TO 3 MONTHS, '2'=3 TO 12 MONTHS, '3'=1 TO 2 YEARS, '4'=2 TO 5 YEARS, '5'=I HAVE NO PLANS TO LEAVE). Column 64 identifies a major reason employee will leave Pizzazz ('1'=SCHOOL, '2'=PERSONAL REASONS, '3'=DISLIKE PIZZAZZ AS A PLACE TO WORK, '4'=MORE ATTRACTIVE JOB). Column 65 gives the major reason for working ('1'=NEED MONEY, '2'=EXTRA MONEY, '3'=ENJOYMENT). Column 66 gives the reason for working less than 35 hours per week ('1'=DO NOT WANT TO WORK MORE THAN THAT, '2'=COULD NOT FIND FULL TIME WORK, '3'=HAVE ANOTHER JOB). Columns 67 and 68 give the number hours per week the employee would like to work. Columns 69–78 should be ignored.

B.2
SPSS Data Entry Commands

The SPSS commands that can be used to read these data are shown here; they can also be found in the file labeled `PIZZAZZ.SSX` on the enclosed disk.

```
UNNUMBERED
DATA LIST FILE=INLINE RECORDS=2
 /1 COMMIT1 TO COMMIT15 1-15 JOBINV1 TO JOBINV3 16-18
   SERVA1 TO SERVA2 19-20 SERVE1 TO SERVE7 21-27 INST1 TO INST5 28-32
   PHI 33 JOBSEC1 TO JOBSEC20 34-53
   TQUIT 54
   ABSENT1 TO ABSENT2 55-56 PHIA1 TO PHIA4 57-60
   JOBSAT1 TO JOBSAT19 61-79
 /2 JOBSAT20 1 PHIB1 TO PHIB2 2-3 QUIT1 TO QUIT3 4-6#ABSENCE 7(A)
   PHIC1 TO PHIC14 8-21
   TEAM1 TO TEAM3 22-24
   PHID1 TO PHID7 25-31 PHIE1 TO PHIE12 32-43 DIVISION 44
   AGE 45-46 SEX 47 RACE 48
   MSTATUS 49 NCHILD 50 EDUC 51 POSITION 52 SATSCHED 53
   TYPESCH 54 FLEX 55
   URBRUR 56 HOURS 57-58 SCHOOL 59 ANJOB 60 TYPERES 61
   LENGTHE 62 HOWLONG 63 REASONQ 64 REASONW 65 WHY 66 NHOURS 67-68
   SOCSEC 69-77
VARIABLE LABELS
   COMMIT1  '#1:   ORGANIZATIONAL COMMITMENT 1'
   COMMIT2  '#2:   ORGANIZATIONAL COMMITMENT 2'
   COMMIT3  '#3: (R)ORGANIZATIONAL COMMITMENT 3'
   COMMIT4  '#4:   ORGANIZATIONAL COMMITMENT 4'
   COMMIT5  '#5:   ORGANIZATIONAL COMMITMENT 5'
   COMMIT6  '#6:   ORGANIZATIONAL COMMITMENT 6'
   COMMIT7  '#7: (R)ORGANIZATIONAL COMMITMENT 7'
   COMMIT8  '#8:   ORGANIZATIONAL COMMITMENT 8'
   COMMIT9  '#9: (R)ORGANIZATIONAL COMMITMENT 9'
   COMMIT10 '#10:   ORGANIZATIONAL COMMITMENT 10'
   COMMIT11 '#11: (R)ORGANIZATIONAL COMMITMENT 11'
   COMMIT12 '#12: (R)ORGANIZATIONAL COMMITMENT 12'
   COMMIT13 '#13:   ORGANIZATIONAL COMMITMENT 13'
   COMMIT14 '#14:   ORGANIZATIONAL COMMITMENT 14'
   COMMIT15 '#15: (R)ORGANIZATIONAL COMMITMENT 15'
   JOBINV1 '#16: JOB INVOLVEMENT 1'
   JOBINV2 '#17: JOB INVOLVEMENT 2'
   JOBINV3 '#18: JOB INVOLVEMENT 3'
   SERVA1  '#19: SERVICE ATTITUDES 1'
   SERVA2  '#20: SERVICE ATTITUDES 2'
```

```
SERVE1   '#21: SERVICE EVALUATIONS 1'
SERVE2   '#22: SERVICE EVALUATIONS 2'
SERVE3   '#23: SERVICE EVALUATIONS 3'
SERVE4   '#24: SERVICE EVALUATIONS 4'
SERVE5   '#25: (R)SERVICE EVALUATIONS 5'
SERVE6   '#26: SERVICE EVALUATIONS 6'
SERVE7   '#27: (R)SERVICE EVALUATIONS 7'
INST1    '#28: (R)INTERACTION STRAIN 1'
INST2    '#29: INTERACTION STRAIN 2'
INST3    '#30: INTERACTION STRAIN 3'
INST4    '#31: (R)INTERACTION STRAIN 4'
INST5    '#32: (R)INTERACTION STRAIN 5'
PHI   '#33: JOB APPROVAL BY FRIENDS/FAMILY'
JOBSEC1  '#34: JOB SECURITY 1-CO. CONCERN FOR IND'
JOBSEC2  '#35: JOB SECURITY 2-CO. CONCERN FOR IND'
JOBSEC3  '#36: JOB SECURITY 3-CO. CONCERN FOR IND'
JOBSEC4  '#37: (R)JOB SECUR 4-CO. CONCERN FOR IND'
JOBSEC5  '#38: JOB SECURITY 5-CO. CONCERN FOR IND'
JOBSEC6  '#39: (R)JOB SECURITY 6-CO. CONC. FO IND'
JOBSEC7  '#40: JOB SECURITY 7-CO. CONCERN FOR IND'
JOBSEC8  '#41: JOB SECURITY 8-CO. CONCERN FOR IND'
JOBSEC9  '#42: (R)JOB SECURITY 9-JOB PERMANENCE'
JOBSEC10 '#43: (R)JOB SECURITY 10-JOB PERMANANCE'
JOBSEC11 '#44: JOB SECURITY 11-JOB PERMANANCE'
JOBSEC12 '#45: JOB SECURITY 12-JOB PERMANANCE'
JOBSEC13 '#46: JOB SECURITY 13-JOB PERMANANCE'
JOBSEC14 '#47: (R)JOB SECURITY 14-JOB PERMANANCE'
JOBSEC15 '#48: JOB SECURITY 15-SELF RATED JOB PERF.'
JOBSEC16 '#49: JOB SECURITY 16-SELF RATED JOB PERF.'
JOBSEC17 '#50: JOB SECURITY 17-SELF RATED JOB PERF.'
JOBSEC18 '#51: JOB SECURITY 18-SELF RATE JOB PERF.'
JOBSEC19 '#52: JOB SECURITY 19-SELF RATED JOB PERF.'
JOBSEC20 '#53: JOB SECURITY 20-SELF RATED JOB PERF.'
TQUIT '#54: THINKING ABOUT QUITTING'
ABSENT1  '#55: ABSENTEEISM 1'
ABSENT2  '#56: ABSENTEEISM 2'
PHIA1 '#57: PIZZAZZ NORMATIVE QUESTION A 1'
PHIA2 '#58: PIZZAZZ NORMATIVE QUESTION A 2'
PHIA3 '#59: PIZZAZZ NORMATIVE QUESTION A 3'
PHIA4 '#60: PIZZAZZ NORMATIVE QUESTION A 4'
JOBSAT1  '#61: JOB SATISFACTION 1'
JOBSAT2  '#62: JOB SATISFACTION 2'
JOBSAT3  '#63: JOB SATISFACTION 3'
JOBSAT4  '#64: JOB SATISFACTION 4'
JOBSAT5  '#65: JOB SATISFACTION 5'
```

```
JOBSAT6   '#66: JOB SATISFACTION 6'
JOBSAT7   '#67: JOB SATISFACTION 7'
JOBSAT8   '#68: JOB SATISFACTION 8'
JOBSAT9   '#69: JOB SATISFACTION 9'
JOBSAT10  '#70: JOB SATISFACTION 10'
JOBSAT11  '#71: JOB SATISFACTION 11'
JOBSAT12  '#72: JOB SATISFACTION 12'
JOBSAT13  '#73: JOB SATISFACTION 13'
JOBSAT14  '#74: JOB SATISFACTION 14'
JOBSAT15  '#75: JOB SATISFACTION 15'
JOBSAT16  '#76: JOB SATISFACTION 16'
JOBSAT17  '#77: JOB SATISFACTION 17'
JOBSAT18  '#78: JOB SATISFACTION 18'
JOBSAT19  '#79: JOB SATISFACTION 19'
JOBSAT20 '#80: JOB SATISFACTION 20'
PHIB1 '#81: PIZZAZZ QUESTION B 1'
PHIB2 '#82: PIZZAZZ QUESTION B 2'
QUIT1 '#83: INTENTION TO QUIT 1'
QUIT2 '#84: INTENTION TO QUIT 2'
QUIT3 '#85: INTENTION TO QUIT 3'
ABSENCES '#86: DAYS ABSENT DURING PAST 2 MONTHS'
PHIC1 '#87: PIZZAZZ NORMATIVE QUESTION C 1'
PHIC2 '#88: PIZZAZZ NORMATIVE QUESTION C 2'
PHIC3 '#89: PIZZAZZ NORMATIVE QUESTION C 3'
PHIC4 '#90: PIZZAZZ NORMATIVE QUESTION C 4'
PHIC5 '#91: PIZZAZZ NORMATIVE QUESTION C 5'
PHIC6 '#92: PIZZAZZ NORMATIVE QUESTION C 6'
PHIC7 '#93: PIZZAZZ NORMATIVE QUESTION C 7'
PHIC8 '#94: PIZZAZZ NORMATIVE QUESTION C 8'
PHIC9 '#95: PIZZAZZ NORMATIVE QUESTION C 9'
PHIC10 '#96: PIZZAZZ NORMATIVE QUESTION C 10'
PHIC11 '#97: PIZZAZZ NORMATIVE QUESTION C 11'
PHIC12 '#98: PIZZAZZ NORMATIVE QUESTION C 12'
PHIC13 '#99: PIZZAZZ NORMATIVE QUESTION C 13'
PHIC14 '#100: PIZZAZZ NORMATIVE QUESTION C 14'
TEAM1   '#101: (R)TEAM ORIENTATION 1'
TEAM2   '#102: (R)TEAM ORIENTATION 2'
TEAM3   '#103: (R)TEAM ORIENTATION 3'
PHID1  '#104: JOB SEARCH-SOMEONE WHO WORKED HERE'
PHID2  '#104: JOB SEARCH-AS A CUSTOMER'
PHID3  '#104: JOB SEARCH-NEWSPAPER AD'
PHID4  '#104: JOB SEARCH-RADIO/TV AD'
PHID5  '#104: JOB SEARCH-SIGN ON WINDOW'
PHID6  '#104: JOB SEARCH-JOB SERVICE'
```

```
PHID7   '#104: JOB SEARCH-FRIEND'
PHIE1   '#105: WHY APPLIED-FLEXIBLE SCHEDULE'
PHIE2   '#105: WHY APPLIED-MEET NEW PEOPLE'
PHIE3   '#105: WHY APPLIED-FUN PLACE'
PHIE4   '#105: WHY APPLIED-CLOSE TO HOME'
PHIE5   '#105: WHY APPLIED-LIKE PIZZA'
PHIE6   '#105: WHY APPLIED-JOB EASY TO GET'
PHIE7   '#105: WHY APPLIED-CHANCE FOR ADVANCEMENT'
PHIE8   '#105: WHY APPLIED-LEARN NEW SKILLS'
PHIE9   '#105: WHY APPLIED-TYPE OF WORK'
PHIE10  '#105: WHY APPLIED-GOOD MONEY/GOOD TIPS'
PHIE11  '#105: WHY APPLIED-FRIENDS WORK HERE'
PHIE12  '#105: WHY APPLIED-COMPANY REPUTATION'
AGE     'BACKGROUND #1: AGE'
SEX     'BACKGROUND #2: SEX'
RACE    'BACKGROUND #3: RACE'
MSTATUS 'BACKGROUND #4: MARITAL STATUS'
NCHILD  'BACKGROUND #5: NUMBER OF CHILDREN'
EDUC    'BACKGROUND #6: EDUCATION'
POSITION 'BACKGROUND #7: POSITION'
SATSCHED 'BACKGROUND #8: SCHEDULE SATISFACTION'
TYPESCH 'BACKGROUND #9: REG. VS IRREG. SCHEDULE'
FLEX    'BACKGROUND #10: FLEXIBILTY OF SCHEDULE'
URBRUR  'BACKGROUND #11: SIZE OF CITY'
HOURS   'BACKGROUND #12: HOURS SCHEDULED/WEEK'
SCHOOL  'BACKGROUND #13: SCHOOL ATTENDANCE'
ANJOB   'BACKGROUND #14: ANOTHER JOB'
TYPERES 'BACKGROUND #15: TYPE OF RESTAURANT'
LENGTHE 'BACKGROUND #16: LENGTH OF EMPLOYMENT'
HOWLONG 'BACKGROUND #17: HOW MUCH LONGER WORKING'
REASONQ 'BACKGROUND #18: REASON FOR QUITTING'
REASONW 'BACKGROUND #19: REASON FOR WORKING'
WHY     'BACKGROUND #20: WHY LESS THAN 35 HOURS'
NHOURS  'BACKGROUND #21: NO. HOURS WANTED/WEEK'
SOCSEC  'SOCIAL SECURITY NUMBER'
DIVISION 'DIVISION WORKING IN'
VALUE LABELS
COMMIT1 TO JOBINV3 1 'STRONGLY DISAGREE' 2 'MODERATELY DISAGREE'
        3 'SLIGHTLY DISAGREE' 4 'NEITHER DISAGREE NOR AGREE'
        5 'SLIGHTLY AGREE' 6 'MODERATELY AGREE' 7 'STRONGLY
        AGREE'/
SERVA1 TO PHIA4  1 'STRONGLY DISAGREE' 2 'DISAGREE'
          3 'NEITHER AGREE NOR DISAGREE' 4 'AGREE'
          5 'STRONGLY AGREE'/
```

```
JOBSAT1 TO PHIB2 1 'VERY DISSATISFIED' 2 'DISSATISFIED'
        3 'I CANNOT DECIDE' 4 'SATISFIED' 5 'VERY SATISFIED'/
QUIT1 TO QUIT3  1 'VERY UNLIKELY' 2 'UNLIKELY' 3 'NEUTRAL' 4 'LIKELY'
        5 'VERY LIKELY'/
PHIC1 TO PHIC14 1 'AGREE' 2 'DO NOT KNOW' 3 'DISAGREE'/
TEAM1 TO TEAM3 1'BETTER THAN MOST' 2 'ABOUT THE SAME' 3 'NOT AS GOOD'/
PHID1 TO PHIE12 1 'DID NOT CHECK' 2 'CHECKED'/
   SEX 1 'MALE' 2 'FEMALE'/
   RACE 1 'CAUCASIAN' 2 'BLACK' 3 'HISPANIC' 4 'OTHER'/
   MSTATUS 1 'SINGLE' 2 'MARRIED' 3 'OTHER'/
   EDUC 1 'GRADE SCHOOL' 2 'HIGH SCHOOL' 3 'COLLEGE'/
   POSITION 1 'KITCHEN CREW' 2 'SHIFT SUPERVISOR' 3 'DELIVERY DRIVER'
    5 'ASST MANAGER' 4 'WAITER/WAITRESS' 6 'MANAGER'/
   SATSCHED 1 'YES' 2 'NO WANT TO WORK LESS' 3 'NO WANT TO WORK MORE'/
   TYPESCH 1 'THE SAME EACH WEEK' 2 'DIFFERENT EACH WEEK'/
   FLEX 1 'I HAVE NO SAY' 2 'I HAVE SOME SAY' 3 'I HAVE A LOT TO SAY'/
   URBRUR 1 'VERY LARGE CITY' 2 'AVERAGE SIZE CITY' 3 'SMALL TOWN'/
   SCHOOL 1 'NO' 2 'HIGH SCHOOL' 3 'COLLEGE' 4 'OTHER'/
   ANJOB 1 'NO' 2 'YES, PART TIME' 3 'YES FULL TIME'/
   DIVISION 1 'LA/ORANGE CO.' 2 'SAN DIEGO DEL.' 3 'KC DELIVERY'
    4 'DALLAS' 5 'CHICAGO CENTRAL' 6 'THREE RIVERS'
    7 'METRO NY' 8 'DELAWARE VALLEY' 9 'MARYLAND'
    0 'ATLANTIC DELIVERY'/
   TYPERES 1 'RESTAURANT' 2 'DELIVERY'
    3 'RESTAURANT BASED DELIVERY'/
   LENGTHE 1 'LESS THAN 3 MONTHS' 2 '3 TO 6 MONTHS'
    3 '6 TO 12 MONTHS' 4 '1 TO 2 YEARS' 5 'MORE THAN 2 YEARS'/
   HOWLONG 1 '1 TO 3 MONTHS' 2 '3 TO 12 MONTHS' 3 '1 TO 2 YEARS'
    4 '2 TO 5 YEARS' 5 'I HAVE NO PLANS TO LEAVE'/
   REASONQ 1 'SCHOOL' 2 'PERSONAL REASONS'
    3 'DISLIKE PIZZAZZ AS A PLACE TO WORK' 4 'MORE ATTRACTIVE JOB'/
   REASONW 1 'NEED MONEY' 2 'EXTRA MONEY' 3 'ENJOYMENT'/
   WHY 1 'DO NOT WANT TO WORK THAN THAT'
    2 'COULD NOT FIND FULL TIME WORK' 3 'HAVE ANOTHER JOB'/
BEGIN DATA

********************************************************
**   INSERT THE DATA FROM THE FILE LABELED          **
**   PIZZAZZ.DAT HERE                      **
********************************************************
END DATA
```

B.3
SAS Data Entry Commands

The SAS commands that can be used to read this data are shown below; they can also be found in the file labeled PIZZAZZ.SAS on the enclosed disk.

```
DATA PIZZAZZ;
INPUT (COMMIT1-COMMIT15 JOBINV1-JOBINV3 SERVA1 SERVA2 SERVE1-SERVE7
   INST1-INST5 PHI JOBSEC1-JOBSEC20 TQUIT ABSENT1 ABSENT2
   PHIA1-PHIA4 JOBSAT1-JOBSAT19)(79*1.)#2
   (JOBSAT20 PHIB1 PHIB2 QUIT1-QUIT3)(6*1.) (ABSENCES)($1.)
   (PHIC1-PHIC14 TEAM1-TEAM3 PHID1-PHID7 PHIE1-PHIE12 DIVISION)(37*1.)
   (AGE SEX RACE MSTATUS NCHILD EDUC POSITION SATSCHED TYPESCH FLEX
    URBRUR HOURS SCHOOL ANJOB TYPERES LENGTHE HOWLONG REASONQ REASONW
    WHY NHOURS)(2.10*1.2.8*1.2.);
LABEL
   COMMIT1 = '#1:   ORGANIZATIONAL COMMITMENT 1'
   COMMIT2 = '#2:   ORGANIZATIONAL COMMITMENT 2'
   COMMIT3 = '#3: (R)ORGANIZATIONAL COMMITMENT 3'
   COMMIT4 = '#4:   ORGANIZATIONAL COMMITMENT 4'
   COMMIT5 = '#5:   ORGANIZATIONAL COMMITMENT 5'
   COMMIT6 = '#6:   ORGANIZATIONAL COMMITMENT 6'
   COMMIT7 = '#7: (R)ORGANIZATIONAL COMMITMENT 7'
   COMMIT8 = '#8:   ORGANIZATIONAL COMMITMENT 8'
   COMMIT9 = '#9: (R)ORGANIZATIONAL COMMITMENT 9'
   COMMIT10= '#10:   ORGANIZATIONAL COMMITMENT 10'
   COMMIT11= '#11: (R)ORGANIZATIONAL COMMITMENT 11'
   COMMIT12= '#12: (R)ORGANIZATIONAL COMMITMENT 12'
   COMMIT13= '#13:   ORGANIZATIONAL COMMITMENT 13'
   COMMIT14= '#14:   ORGANIZATIONAL COMMITMENT 14'
   COMMIT15= '#15: (R)ORGANIZATIONAL COMMITMENT 15'
   JOBINV1 = '#16:   JOB INVOLVEMENT 1'
   JOBINV2 = '#17:   JOB INVOLVEMENT 2'
   JOBINV3 = '#18:   JOB INVOLVEMENT 3'
   SERVA1  = '#19:   SERVICE ATTITUDES 1'
   SERVA2  = '#20:   SERVICE ATTITUDES 2'
   SERVE1  = '#21:   SERVICE EVALUATIONS 1'
   SERVE2  = '#22:   SERVICE EVALUATIONS 2'
   SERVE3  = '#23:   SERVICE EVALUATIONS 3'
   SERVE4  = '#24:   SERVICE EVALUATIONS 4'
   SERVE5  = '#25: (R)SERVICE EVALUATIONS 5'
   SERVE6  = '#26:   SERVICE EVALUATIONS 6'
   SERVE7  = '#27: (R)SERVICE EVALUATIONS 7'
   INST1 = '#28: (R)INTERACTION STRAIN 1'
   INST2 = '#29:   INTERACTION STRAIN 2'
```

```
INST3  = '#30:   INTERACTION STRAIN 3'
INST4  = '#31: (R)INTERACTION STRAIN 4'
INST5  = '#32: (R)INTERACTION STRAIN 5'
PHI    = '#33: JOB APPROVAL BY FRIENDS/FAMILY'
JOBSEC1  = '#34:   JOB SECURITY 1-CO. CONCERN FOR IND'
JOBSEC2  = '#35:   JOB SECURITY 2-CO. CONCERN FOR IND'
JOBSEC3  = '#36:   JOB SECURITY 3-CO. CONCERN FOR IND'
JOBSEC4  = '#37: (R)JOB SECURITY 4-CO. CONCERN FOR IND'
JOBSEC5  = '#38:   JOB SECURITY 5-CO. CONCERN FOR IND'
JOBSEC6  = '#39: (R)JOB SECURITY 6-CO. CONCERN FOR IND'
JOBSEC7  = '#40:   JOB SECURITY 7-CO. CONCERN FOR IND'
JOBSEC8  = '#41:   JOB SECURITY 8-CO. CONCERN FOR IND'
JOBSEC9  = '#42: (R)JOB SECURITY 9-JOB PERMANANCE'
JOBSEC10= '#42: (R)JOB SECURITY 10-JOB PERMANANCE'
JOBSEC11= '#44:   JOB SECURITY 11-JOB PERMANANCE'
JOBSEC12= '#45:   JOB SECURITY 12-JOB PERMANANCE'
JOBSEC13= '#46:   JOB SECURITY 13-JOB PERMANANCE'
JOBSEC14= '#47: (R)JOB SECURITY 14-JOB PERMANANCE'
JOBSEC15= '#48:   JOB SECURITY 15-SELF RATED JOB PERF.'
JOBSEC16= '#49:   JOB SECURITY 16-SELF RATED JOB PERF.'
JOBSEC17= '#50:   JOB SECURITY 17-SELF RATED JOB PERF.'
JOBSEC18= '#51:   JOB SECURITY 18-SELF RATE JOB PERF.'
JOBSEC19= '#52:   JOB SECURITY 19-SELF RATED JOB PERF.'
JOBSEC20= '#53:   JOB SECURITY 20-SELF RATED JOB PERF.'
TQUIT = '#54: THINKING ABOUT QUITTING'
ABSENT1 = '#55: ABSENTEEISM 1'
ABSENT2 = '#56: ABSENTEEISM 2'
PHIA1 = '#57: PIZZAZZ NORMATIVE QUESTION A 1'
PHIA2 = '#58: PIZZAZZ NORMATIVE QUESTION A 2'
PHIA3 = '#59: PIZZAZZ NORMATIVE QUESTION A 3'
PHIA4 = '#60: PIZZAZZ NORMATIVE QUESTION A 4'
JOBSAT1  = '#61: JOB SATISFACTION 1'
JOBSAT2  = '#62: JOB SATISFACTION 2'
JOBSAT3  = '#63: JOB SATISFACTION 3'
JOBSAT4  = '#64: JOB SATISFACTION 4'
JOBSAT5  = '#65: JOB SATISFACTION 5'
JOBSAT6  = '#66: JOB SATISFACTION 6'
JOBSAT7  = '#67: JOB SATISFACTION 7'
JOBSAT8  = '#68: JOB SATISFACTION 8'
JOBSAT9  = '#69: JOB SATISFACTION 9'
JOBSAT10= '#70: JOB SATISFACTION 10'
JOBSAT11= '#71: JOB SATISFACTION 11'
JOBSAT12= '#72: JOB SATISFACTION 12'
JOBSAT13= '#73: JOB SATISFACTION 13'
JOBSAT14= '#74: JOB SATISFACTION 14'
```

```
JOBSAT15= '#75: JOB SATISFACTION 15'
JOBSAT16= '#76: JOB SATISFACTION 16'
JOBSAT17= '#77: JOB SATISFACTION 17'
JOBSAT18= '#78: JOB SATISFACTION 18'
JOBSAT19= '#79: JOB SATISFACTION 19'
JOBSAT20= '#80: JOB SATISFACTION 20'
PHIB1 = '#81: PIZZAZZ QUESTION B 1'
PHIB2 = '#82: PIZZAZZ QUESTION B 2'
QUIT1 = '#83: INTENTION TO QUIT 1'
QUIT2 = '#84: INTENTION TO QUIT 2'
QUIT3 = '#85: INTENTION TO QUIT 3'
ABSENCES= '#86: DAYS ABSENT DURING PAST 2 MONTHS'
PHIC1 = '#87: PIZZAZZ NORMATIVE QUESTION C 1'
PHIC2 = '#88: PIZZAZZ NORMATIVE QUESTION C 2'
PHIC3 = '#89: PIZZAZZ NORMATIVE QUESTION C 3'
PHIC4 = '#90: PIZZAZZ NORMATIVE QUESTION C 4'
PHIC5 = '#91: PIZZAZZ NORMATIVE QUESTION C 5'
PHIC6 = '#92: PIZZAZZ NORMATIVE QUESTION C 6'
PHIC7 = '#93: PIZZAZZ NORMATIVE QUESTION C 7'
PHIC8 = '#94: PIZZAZZ NORMATIVE QUESTION C 8'
PHIC9 = '#95: PIZZAZZ NORMATIVE QUESTION C 9'
PHIC10 = '#96: PIZZAZZ NORMATIVE QUESTION C 10'
PHIC11 = '#97: PIZZAZZ NORMATIVE QUESTION C 11'
PHIC12 = '#98: PIZZAZZ NORMATIVE QUESTION C 12'
PHIC13 = '#99: PIZZAZZ NORMATIVE QUESTION C 13'
PHIC14 = '#100: PIZZAZZ NORMATIVE QUESTION C 14'
TEAM1 = '#101: (R) TEAM ORIENTATION 1'
TEAM2 = '#102: (R) TEAM ORIENTATION 2'
TEAM3 = '#103: (R) TEAM ORIENTATION 3'
PHID1 = '#104: JOB SEARCH-SOMEONE WHO WORKED HERE'
PHID2 = '#104: JOB SEARCH-AS A CUSTOMER'
PHID3 = '#104: JOB SEARCH-NEWSPAPER AD'
PHID4 = '#104: JOB SEARCH-RADIO/TV AD'
PHID5 = '#104: JOB SEARCH-SIGN ON WINDOW'
PHID6 = '#104: JOB SEARCH-JOB SERVICE'
PHID7 = '#104: JOB SEARCH-FRIEND'
PHIE1 = '#105: WHY APPLIED-FLEXIBLE SCHEDULE'
PHIE2 = '#105: WHY APPLIED-MEET NEW PEOPLE'
PHIE3 = '#105: WHY APPLIED-FUN PLACE'
PHIE4 = '#105: WHY APPLIED-CLOSE TO HOME'
PHIE5 = '#105: WHY APPLIED-LIKE PIZZA'
PHIE6 = '#105: WHY APPLIED-JOB EASY TO GET'
PHIE7 = '#105: WHY APPLIED-CHANCE FOR ADVANCEMENT'
PHIE8 = '#105: WHY APPLIED-LEARN NEW SKILLS'
PHIE9 = '#105: WHY APPLIED-TYPE OF WORK'
```

```
PHIE10 = '#105: WHY APPLIED-GOOD MONEY/GOOD TIPS'
PHIE11 = '#105: WHY APPLIED-FRIENDS WORK HERE'
PHIE12 = '#105: WHY APPLIED-COMPANY REPUTATION'
AGE  = 'BACKGROUND #1: AGE'
SEX  = 'BACKGROUND #2: SEX'
RACE  = 'BACKGROUND #3: RACE'
MSTATUS = 'BACKGROUND #4: MARITAL STATUS'
NCHILD = 'BACKGROUND #5: NUMBER OF CHILDREN'
EDUC  = 'BACKGROUND #6: EDUCATION'
POSITION= 'BACKGROUND #7: POSITION'
SATSCHED= 'BACKGROUND #8: SCHEDULE SATISFACTION'
TYPESCH = 'BACKGROUND #9: REG. VS IRREG. SCHEDULE'
FLEX  = 'BACKGROUND #10: FLEXIBILITY OF SCHEDULE'
URBRUR = 'BACKGROUND #11: SIZE OF CITY'
HOURS  = 'BACKGROUND #12: HOURS SCHEDULED/WEEK'
SCHOOL = 'BACKGROUND #13: SCHOOL ATTENDANCE'
ANJOB  = 'BACKGROUND #14: ANOTHER JOB'
TYPERES = 'BACKGROUND #15: TYPE OF RESTAURANT'
LENGTHE = 'BACKGROUND #16: LENGTH OF EMPLOYMENT'
HOWLONG = 'BACKGROUND #17: HOW MUCH LONGER WORKING'
REASONQ = 'BACKGROUND #18: REASON FOR QUITTING'
REASONW = 'BACKGROUND #19: REASON FOR WORKING'
WHY  = 'BACKGROUND #20: WHY LESS THAN 35 HOURS'
NHOURS = 'BACKGROUND #21: NO. HOURS WANTED/WEEK'
DIVISION= 'DIVISION WORKING IN';


*************************************************
**                                           **
**   INSERT THE DATA FROM THE FILE LABELED    **
**   PIZZAZZ.DAT AFTER                        **
**   THE CARDS SATEMENT.                      **
**                                           **
*************************************************;
CARDS;
```

Family Control Study

Appendix C

Stephen Rathbun (1995) collected data to study the theoretical utility of conceptualizing control as being a multidimensional construct rather than a unidimensional construct. His sample data come from currently married church-attending adults with at least one child less than 19 years old living at home. The data were obtained from surveys of married adults participating in Protestant Sunday School classes in two large Midwestern cities. The data can be found in an ASCII file named FIRO.SAS on the enclosed disk. An SPSS system file is also on the enclosed disk and is named FIRO.SYS.

Data were collected from 304 married adults. Two hundred and forty-two of these adults were married couples who both filled out the survey. Respondents were given questionnaires that included several survey instruments. The questionnaire used is reprinted here.

About Your Family*

DIRECTIONS

The following are a number of statements about families. Please read each statement carefully and decide how well it describes your own family. You should answer according to how YOU see your family. Please <u>CIRCLE</u> your response.

SA = Strongly Agree   A = Agree   D = Disagree   SD = Strongly Disagree

001. Planning family activities is difficult because we misunderstand each other. SA A D SD

002. We resolve most everyday problems around the house. SA A D SD

003. When someone is upset the others know why. SA A D SD

004. When you ask someone to do something, you have to check that they did it. SA A D SD

005. If someone is in trouble, the others become too involved. SA A D SD

006. In times of crisis we can turn to each other for comfort. SA A D SD

007. We don't know what to do when an emergency comes up. SA A D SD

008. We sometimes run out of things that we need. SA A D SD

009. We are reluctant to show our affection for each other. SA A D SD

010. We make sure members meet their family responsibilities. SA A D SD

011. We cannot talk to each other about the sadness we feel. SA A D SD

012. We usually act on our decisions regarding problems. SA A D SD

013. You only get the interest of others when something is important to them. SA A D SD

014. You can't tell how a person is feeling from what they are saying. SA A D SD

015. Family tasks don't get spread around enough. SA A D SD

* Reprinted by permission of Professor Steven Rathbun.

016. Individuals are accepted for what they are.	SA	A	D	SD
017. You can easily get away with breaking the rules.	SA	A	D	SD
018. People come right out and say things instead of hinting at them.	SA	A	D	SD
019. Some of us just don't respond emotionally.	SA	A	D	SD
020. We know what to do in an emergency.	SA	A	D	SD
021. We avoid discussing our fears and concerns.	SA	A	D	SD
022. It is difficult to talk to each other about tender feelings.	SA	A	D	SD
023. We have trouble meeting our bills.	SA	A	D	SD
024. After our family tries to solve a problem, we usually discuss whether it worked or not.	SA	A	D	SD
025. We are too self-centered.	SA	A	D	SD
026. We can express feelings to each other.	SA	A	D	SD
027. We have no clear expectations about toilet habits.	SA	A	D	SD
028. We do not show our love for each other.	SA	A	D	SD
029. We talk to people directly rather than through a go-between.	SA	A	D	SD
030. Each of us has particular duties and responsibilities.	SA	A	D	SD
031. There are lots of bad feelings in the family.	SA	A	D	SD
032. We have rules about hitting people.	SA	A	D	SD
033. We get involved with each other only when something interests us.	SA	A	D	SD
034. There's little time to explore personal interests.	SA	A	D	SD
035. We often don't say what we mean.	SA	A	D	SD
036. We feel accepted for what we are.	SA	A	D	SD
037. We show interest in each other when we can get something out of it personally.	SA	A	D	SD
038. We resolve most emotional upsets that come up.	SA	A	D	SD
039. Tenderness takes second place to other things in our family.	SA	A	D	SD
040. We discuss who is to do household jobs.	SA	A	D	SD
041. Making decisions is a problem for our family.	SA	A	D	SD
042. Our family shows interest in each other only when they can get something out of it.	SA	A	D	SD
043. We are frank with each other.	SA	A	D	SD
044. We don't hold to any rules or standards.	SA	A	D	SD
045. If people are asked to do something, they need reminding.	SA	A	D	SD
046. We are able to make decisions about how to solve problems.	SA	A	D	SD
047. If the rules are broken, we don't know what to expect.	SA	A	D	SD
048. Anything goes in our family.	SA	A	D	SD
049. We express tenderness.	SA	A	D	SD
050. We confront problems involving feelings.	SA	A	D	SD
051. We don't get along well together.	SA	A	D	SD
052. We don't talk to each other when we are angry.	SA	A	D	SD
053. We are generally dissatisfied with the family duties assigned to us.	SA	A	D	SD
054. Even though we mean well, we intrude too much into each other's lives.	SA	A	D	SD
055. There are rules about dangerous situations.	SA	A	D	SD

056. We confide in each other.	SA	A	D	SD
057. We cry openly.	SA	A	D	SD
058. We don't have reasonable transport.	SA	A	D	SD
059. When we don't like what someone has done, we tell them.	SA	A	D	SD
060. We try to think of different ways to solve problems.	SA	A	D	SD

DIRECTIONS: PLEASE RESPOND TO THE FOLLOWING. <u>CIRCLE</u> the appropriate number.

1 = Very Dissatisfied   2 = Dissatisfied   3 = Neutral   4 = Satisfied   5 = Very Satisfied

061. How satisfied are you with your relationship with your children?	1	2	3	4	5
062. How satisfied are you with your children's relationship with each other?	1	2	3	4	5
063. Check here if you have only one child: ____	1	2	3	4	5
064. Overall, how satisfied are you with your current family relationships?	1	2	3	4	5

1 = Very Dissatisfied   2 = Dissatisfied   3 = Neutral   4 = Satisfied   5 = Very Satisfied

<u>CIRCLE THE APPROPRIATE NUMBER</u>

065. How satisfied are you with your marriage?	1	2	3	4	5
066. How satisfied are you with your relationship with your mate?	1	2	3	4	5
067. How satisfied are you with your mate as a spouse or close companion?	1	2	3	4	5

For the following set of statements, you are to decide which statements are true of your marriage and which are false.

068. Every new thing I have learned about my spouse has pleased me.	True	False
069. I have some needs that are not being met by my marriage.	True	False
070. My spouse and I get angry with each other sometimes.	True	False
071. My marriage is a perfect success.	True	False
072. There are times when my spouse and I argue.	True	False
073. I don't think any couple could live together with greater harmony than my spouse and I.	True	False

DIRECTIONS: Below you will find a series of statements about family relationships. Read each statement carefully and decide the extent to which you never or always agree with the item. For each statement <u>CIRCLE THE NUMBER</u> which seems to best describe your opinion. Please do not omit any statements.

1 = Almost Never   2 = Once in a While   3 = Sometimes   4 = Frequently   5 = Almost Always

074. Family members are supportive of each other during difficult times.	1	2	3	4	5
075. In our family, it is easy for everyone to express his/her opinion.	1	2	3	4	5

076.	It is easier to discuss problems with people outside the family than with other family members.	1	2	3	4	5
077.	Each family member has input regarding major family decisions.	1	2	3	4	5
078.	Our family gathers together in the same room.	1	2	3	4	5
079.	Children have a say in their discipline.	1	2	3	4	5
080.	Our family does things together.	1	2	3	4	5
081.	Family members discuss problems and feel good about the solutions.	1	2	3	4	5
082.	In our family, everyone goes his/her own way.	1	2	3	4	5
083.	We shift household responsibilities from person to person.	1	2	3	4	5
084.	Family members know each other's close friends.	1	2	3	4	5
085.	It is hard to know what the rules are in our family.	1	2	3	4	5
086.	Family members consult other family members on personal decisions.	1	2	3	4	5
087.	Family members say what they want.	1	2	3	4	5
088.	We have difficulty thinking of things to do as a family.	1	2	3	4	5
089.	In solving problems, the children's suggestions are followed.	1	2	3	4	5
090.	Family members feel very close to each other.	1	2	3	4	5
091.	Discipline is fair in our family.	1	2	3	4	5
092.	Family members feel closer to people outside the family than to other family members.	1	2	3	4	5
093.	Our family tries new ways of dealing with problems.	1	2	3	4	5
094.	Family members go along with what the family decides to do.	1	2	3	4	5
095.	In our family, everyone shares responsibilities.	1	2	3	4	5
096.	Family members like to spend their free time with each other.	1	2	3	4	5
097.	It is difficult to get a rule changed in our family.	1	2	3	4	5
098.	Family members avoid each other at home.	1	2	3	4	5
099.	When problems arise, we compromise.	1	2	3	4	5
100.	We approve of each other's friends.	1	2	3	4	5
101.	Family members are afraid to say what is on their minds.	1	2	3	4	5
102.	Family members pair up rather than do things as a total family.	1	2	3	4	5
103.	Family members share interests and hobbies with each other.	1	2	3	4	5

PLEASE DESCRIBE YOURSELF BY ANSWERING THE FOLLOWING QUESTIONS

104. Your Age:____

105. Your Sex:____

106. What is your race/ethnicity?

____Caucasian (non-Hispanic)   ____Asian

____African-American   ____American Indian

____Caucasian (Hispanic)   ____Other

107. What is the highest level of education you have achieved? (Please CHECK the one that best applies.)

____(1) Grade School   ____(5) Bachelor's Degree

____(2) High School or G.E.D.   ____(6) Master's Degree

____(3) Vocational/Technical Degree   ____(7) Ph.D. or equivalent

____(4) Associate Degree

108. What is your annual family income?

____Under $15,000	____$50,000–74,999
____$15,000–24,999	____$75,000–99,999
____$25,000–34,999	____$100,000–149,999
____$35,000–49,999	____over $150,000

109. What is your <u>current</u> marital status? (check all that apply)

____Single, never married	____Living together (not married)
____Married (only once)	____Divorced (not remarried)
____Married (two or more times)	____Widowed (not remarried)

110. How many years have you been living with your spouse?____

Information about your children. (Include those children of minor age who may live with you only part of the time because of joint custody arrangements.)

PLEASE FILL OUT ONLY ONE LINE FOR EACH CHILD BY CIRCLING THE CORRECT RESPONSE.

Sex		Age	Natural Child	Step Child	Foster Child
F	M	_____	NC	SC	FC
F	M	_____	NC	SC	FC
F	M	_____	NC	SC	FC
F	M	_____	NC	SC	FC
F	M	_____	NC	SC	FC
F	M	_____	NC	SC	FC
F	M	_____	NC	SC	FC
F	M	_____	NC	SC	FC

A portion of the data in this study is shown below. Each individual's record includes four lines of data.

```
41233142424133313231444131441141344413142
44242134114333111422559555522111115524434
43241552354135353255234312153212511219
9999999999999999999999922

41243143414233314241414241442131334413133
43242134114343211322449444522111115514434
43241342154144441144135301133212511219
9999999999999999999999912

41223143323233424234333232442241333323232
33243234224333112421449555422111115414544
4134144245413444314421434214529161251
11999999999999999999999923
```

```
41234142414143414241443141441141444141414 1
44444144114244111432449555512111155155 15
41151451155135541115114311345291612511
1199999999999999999999913

31233132323234214232332232442241333122 32
44243144224334223422459554521221154354 24
33351443344133432234222345215622321719999 9
999999999999999999999999901

31231244323133223222333222432232333232 32
3324323422333322332244945555221211552554 5
4435141334513553212514546215529921612151 1
14199999999999999999999923

32233233323232232222232222232223233323 232
23232233223233222332254955555121111432354 4
43341343344334443233334421154219216121511
14199999999999999999999913

32133142424233323333323332323411414314322324 2
4424213422444411292143944442112125434434
423423423442344431342143411252172911619
999999999999999999999902
```

The first 60 variables (FAD01–FAD60) in the FIRO data sets are responses to questions on the Family Assessment Device (FAD) developed by Epstein, Baldwin, and Bishop (1982). This instrument utilizes a four-point Likert-type format with endpoints "strongly agree" and "strongly disagree." In the FIRO data sets the responses are coded with numerical values 1–4 with 1 = strongly agree, 2 = agree, 3 = disagree, and 4 = strongly disagree.

Variables 61–64 (KFLS1–KFLS4) are responses to the Kansas Family Life Satisfaction Scale developed by Schumm *et al.* (1986a). Variables 65–67 (KMS1–KMS3) are responses to the Kansas Marital Satisfaction Scale, which was developed by Schumm *et al.* (1986b). These two instruments use a five-point Likert-type format with endpoints "very dissatisfied" and "very satisfied." The coding for these responses is shown on the questionnaire.

Variables 68–73 (EMC1–EMC6) are true/false responses to Edmonds' Marital Conventionalization Scale (Anderson, Russell, and Schumm, 1983). These answers have been coded with 1 = true and 2 = false on the data files.

Variables 74–103 (FACES1–FACES30) are responses to the Family Adaptability Cohesion Evaluation Scales developed by Olson, Portner, and Bell (1982). This instrument utilizes a five-point Likert-type format with endpoints "Almost Never" and "Almost Always." The coding for these responses is shown on the questionnaire.

The remaining variables are demographic variables. The first seven of these give the respondents AGE, SEX (1 = female, 2 = male), RACE [1 = Caucasion (non-Hispanic), 2 = African-American, 3 = Caucasian (Hispanic), 4 = Asian, 5 = American Indian, 6 = other], education level (EDU) (1 = Grade School, 2 = High School or G.E.D., 3 = Vocational/Technical Degree, 4 = Associate Degree, 5 = Bachelor's Degree, 6 = Master's Degree, 7 = Ph.D. or equivalent), annual income level (INCOME) (1 = Under $15,000, 2 = $15,000–24,999, 3 = $25,000–34,999, 4 = $35,000–49,999, 5 = $50,000–74,999, 6 = $75,000–99,999, 7 = $100,000–149,999, 8 = over $150,000), current marital status (MARITAL) [1 = Single, never married, 2 = Married (only once), 3 = Married (two or more times), 4 = Living together (not married), 5 = Divorced (not remarried), 6 = Widowed (not remarried)], and years living with spouse (YRSMAR). The next 24 variables give information about minor children who live at home. For each successive child, the data file includes the child's sex, age, and type (CHLD1SEX, CHLD1AGE, CHLD1TYP, CHLD2SEX, CHLD2AGE, CHLD2TYP, . . . , CHLD8SEX, CHLD8AGE, CHLD8TYP). Missing data are recorded as either a "9" or a "99."

The last two variables are COUPLE and FAMSIZE. COUPLE has a value of 0 if the individual's spouse did not fill out a questionnaire, 1 if the individual is female and the individual's spouse filled out a questionnaire, and 2 if the individual is a male and the individual's spouse filled out a questionnaire. FAMSIZE gives the number of minor children living at home.

The first line of data in each record contains values for FAD01–FAD40. The second line contains values for FAD41–FAD60, KFLS1–KFLS4, KMS1–KMS3, EMC1–EMC6, and FACES1–FACES7. The third line of data in each record contains values for FACES8–FACES30, AGE, SEX, RACE, EDU, INCOME, MARITAL, YRSMAR, CHLD1SEX, CHLD1AGE, CHLD1TYP, CHLD2SEX, CHLD2AGE, CHLD2TYP, and CHLD3SEX. The last line of data in each record contains values for CHLD3AGE, CHLD3TYP, CHLD4SEX, CHLD4AGE, CHLD4TYP, CHLD5SEX, CHLD5AGE, CHLD5TYP, CHLD6SEX, CHLD6AGE, CHLD6TYP, CHLD7SEX, CHLD7AGE, CHLD7TYP, CHLD8SEX, CHLD8AGE, CHLD8TYP, COUPLE, and FAMSIZE.

The SAS statements that can be used to read this data file are given below. These commands, along with the data, can be found in the file labeled FIRO.SAS on the enclosed disk.

```
DATA FIRO;
 INPUT (FAD01-FAD40) (1. 39*2) #2 (FAD41-FAD60 KFLS1-KFLS4
    KMS1-KMS3 EMC1-EMC6 FACES1-FACES7) (1. 39*2.)
    #3 (FACES8-FACES30) (1. 22*2.) (AGE SEX RACE EDU INCOME
    MARITAL YRSMAR CHLD1SEX CHLD1AGE CHLD1TYP CHLD2SEX CHLD2AGE
    CHLD2TYP CHLD3SEX) (3. 5*2. 3. 2. 3. 2. 2. 3. 2. 2.)
    #4 (CHLD3AGE CHLD3TYP CHLD4SEX CHLD4AGE CHLD4TYP CHLD5SEX
    CHLD5AGE CHLD5TYP CHLD6SEX CHLD6AGE CHLD6TYP CHLD7SEX
CHLD7AGE CHLD7TYP CHLD8SEX CHLD8AGE CHLD8TYP COUPLE FAMSIZE)
    (2. 2. 2. 3. 2. 2. 3. 2. 2. 3. 2. 2. 3. 2. 2. 3. 2. 2. 2.);
```

```
CARDS;
412331424241333132314441314411413441314142
442421341143331114225595555221111552 4434
432415523541353532552343121532122511219
99999999999999999999999922
412431434142333142414142414421313341313 3
432421341143432113224494445221111551 4434
432413421541444411441353011332122511219
99999999999999999999999912
                  .
                  .
                  .
```

References

Abramowitz, M., and L. A. Stegun, eds., *Handbook of Mathematical Functions,* U.S. Department of Commerce, National Bureau of Standards Applied Mathematical Series 55, 1964.

Adelman, I., M. Greer, and C. T. Morris, "Instruments and Goals in Economic Development," *American Economic Review,* **59,** no. 2 (1969), 409–426.

Agresti, A., *Categorical Data Analysis,* New York: John Wiley, 1988.

Akaike, H., "Information Theory and the Extension of the Maximum Likelihood Principle," in *2nd International Symposium on Information Theory,* ed. V. N. Petrov and F. Csaki, Budapest: Akailseoniai-Kiudo, 267–281, 1973.

Akaike, H., "A New Look at the Statistical Identification Model," *IEEE Transactions on Automatic Control,* **19** (1974), 716–723.

Anderberg, M. R., *Cluster Analysis for Applications,* New York: Academic Press, 1973.

Anderson, S. A., C. S. Russell, and W. R. Schumm, "Perceived Marital Quality and Family Lifecycle Categories: A Further Analysis," *Journal of Marriage and the Family,* **45** (1983) 127–139.

Anderson, T. W., "Asymptotic Theory for Principal Components Analysis," *Annals of Mathematical Statistics,* **34** (1963), 122–148.

Anderson, T. W., *An Introduction to Multivariate Statistical Analysis* (2nd ed.), New York: John Wiley, 1984.

Andrews, D. F., "Plots of High Dimensional Data," *Biometrics,* **28** (1972), 125–136.

Andrews, D. F., R. Gnanadesikan, and J. L. Warner, "Transformations of Multivariate Data," *Biometrics,* **27,** no. 4 (1971), 825–840.

Atkinson, A. C., *Plots, Transformations and Regression,* Oxford, England: Oxford University Press, 1985.

Bacon-Shone, J., and W. K. Fung, "A New Graphical Method for Detecting Single and Multiple Outliers in Univariate and Multivariate Data," *Applied Statistics,* **36,** no. 2 (1987), 153–162.

Bartholomew, D. J., *Latent Variable Models and Factor Analysis,* London: Griffin, 1987.

Bartlett, M. S., "Properties of Sufficiency and Statistical Tests," *Proceedings of the Royal Society of London (A),* **160** (1937), 268–282.

Bartlett, M. S., "The Statistical Conception of Mental Factors," *British Journal of Psychology,* **28** (1937), 97–104.

Bartlett, M. S., "Further Aspects of the Theory of Multiple Regression," *Proceedings of the Cambridge Philosophical Society,* **34** (1938), 33–40.

Bartlett, M. S., "A Note on Tests of Significance in Multivariate Analysis," *Proceedings of the Cambridge Philosophical Society,* **35** (1939), 180–185.

Bartlett, M. S., "Multivariate Analysis," *Journal of the Royal Statistical Society Supplement (B),* **9** (1947), 176–197.

Bartlett, M. S., "An Inverse Matrix Adjustment Arising in Discriminant Analysis," *Annals of Mathematical Statistics,* **22** (1951), 107–111.

Bartlett, M. S., "A Note on Multiplying Factors for Various Chi-Squared Approximations," *Journal of the Royal Statistical Society (B),* **16** (1954), 296–298.

Bartlett, M. S., "A Note on the Multiplying Factors for Various Z' Approximations," *Journal of the Royal Statistical Society (B),* **16** (1954), 296–298.

Becker, R. A., W. S. Cleveland, and A. R. Wilks, "Dynamic Graphics for Data Analysis," *Statistical Science,* **2,** no. 4 (1987), 355–395.

Begg, C. B., and R. Gray, "Calculation of Polychotomous Logistic Regression Parameters Using Individualized Regression," *Biometrika,* **71** (1984), 1–10.

Bellman, R., *Introduction to Matrix Analysis* (2nd ed.), New York: McGraw-Hill, 1970.

Belsley, D. A., E. Kuh, and R. E. Welsh, *Regression Diagnostics,* New York: John Wiley, 1980.

Bentler, P. M., "Multivariate Analysis with Latent Variables: Causal Models," *Annual Review of Psychology,* **31** (1980), 419–456.

Beyer, W. H., *CRC Handbook of Tables for Probability and Statistics* (2nd ed.), Boca Raton, FL: CRC Press, 1968.

Bhattacharyya, G. K., and R. A. Johnson, *Statistical Concepts and Methods,* New York: John Wiley, 1977.

Bickel, P. J., and K. A. Doksum, *Mathematical Statistics: Basic Ideas and Selected Topics,* San Francisco: Holden-Day, 1977.

Bielby, W. T., and R. M. Hauser, "Structural Equation Models," *Annual Review of Sociology,* **3** (1977), 137–161.

Bishop, Y. M. M., S. E. Feinberg, and P. W. Holland, *Discrete Multivariate Analysis: Theory and Practice,* Cambridge, MA: The MIT Press, 1975.

Bliss, C. I., "Statistics in Biology," *Statistical Methods for Research in the Natural Sciences,* Vol. 2, New York: McGraw-Hill, 1967.

Bollen, K. A., *Structural Equations with Latent Variables,* New York: John Wiley, 1989.

Boomsma, A. "Comparing Approximations of Confidence Intervals for the Product-Moment Correlation Coefficient," *Statistica Neerlandica,* **31** (1977), 179–186.

Bouma, B. N., *et al.,* "Evaluation of the Detection Rate of Hemophilia Carriers," *Statistical Methods for Clinical Decision Making,* **7,** no. 2 (1975), 339–350.

Bowerman, B. L., and R. T. O'Connell, *Linear Statistical Models: An Applied Approach* (2nd ed.), Boston: PWS-Kent, 1990.

Box, G. E. P., "A General Distribution Theory for a Class of Likelihood Criteria," *Biometrika,* **36** (1949), 317–346.

Box, G. E. P., and D. R. Cox, "An Analysis of Transformations (with discussion)," *Journal of the Royal Statistical Society (B),* **26,** no. 2 (1964), 211–252.

Box, G. E. P., and N. R. Draper, *Evolutionary Operation: A Statistical Method for Process Improvement,* New York: John Wiley, 1969.

Box, G. E. P., W. G. Hunter, and J. S. Hunter, *Statistics for Experimenters,* New York: John Wiley, 1978.

Breiman, L., J. Friedman, R. Olshen, and C. Stone, *Classification and Regression Trees,* Belmont, CA: Wadsworth, 1984.

Chambers, J. M., W. S. Cleveland, B. Kleiner, and P. A. Tukey, *Graphical Methods for Data Analysis,* New York: Chapman and Hall, 1983.

Chatfield, C., and A. J. Collins, *Introduction to Multivariate Analysis,* London: Chapman and Hall, 1980.

Chatterjee, S., and B. Price, *Regression Analysis by Example,* New York: John Wiley, 1977.

Chernoff, H., "Using Faces to Represent Points in K-Dimensional Space Graphically," *Journal of the American Statistical Association,* **68,** no. 342 (1973), 361–368.

Cleveland, W. S., *The Elements of Graphing Data,* New York: Chapman and Hall, 1985.

Cleveland, W. S., and M. E. McGill, *Dynamic Graphics for Statistics,* New York: Chapman and Hall, 1988.

Cleveland, W. S., and D. A. Relles, "Clustering by Identification with Special Application to Two-Way Tables of Counts," *Journal of the American Statistical Association,* 70, no. 351 (1975), 626–630.

Cliff, N. *Analyzing Multivariate Data,* Orlando, FL: Harcourt Brace Jovanovich, 1987.

Cochran, W. G., *Sampling Techniques* (3rd ed.), New York: John Wiley, 1977.

Cochran, W. G., and G. M. Cox, *Experimental Designs* (2nd ed.), New York: John Wiley, 1957.

Cook, R. D., and S. Weisberg, *Residuals and Influence in Regression,* London: Chapman and Hall, 1982.

Cormack, R. M., "A Review of Classification (with discussion)," *Journal of the Royal Statistical Society (A),* **134,** no. 3 (1971), 321–367.

Daniel, C., and F. S. Wood, *Fitting Equations to Data* (2nd ed.), New York: John Wiley, 1980.

Darroch, J. N., and J. E. Mosimann, "Canonical and Principal Components of Shape," *Biometrika,* **72,** no. 1 (1985), 241–252.

David, F. N., *Tables of the Correlation Coefficient,* Cambridge: Cambridge University Press, 1938.

David, F. N., *Tables of the Correlation Coefficient,* London: Cambridge University Press, issued by *Biometrika,* 1954.

Davis, J. C., "Information Contained in Sediment Size Analysis," *Mathematical Geology,* **2,** no. 2 (1970), 105–112.

Dawkins, B., "Multivariate Analysis of National Track Records," *The American Statistician,* **43,** no. 2 (1989), 110–115.

Dempster, A. P., N. M. Laird, and D. B. Rubin, "Maximum Likelihood from Incomplete Data via the EM Algorithm (with discussion)," *Journal of the Royal Statistical Society (B),* **39,** no. 1 (1977), 1–38.

Digby, P. G. N., and R. A. Kempton, *Multivariate Analysis of Ecological Communities,* New York: Chapman and Hall, 1987.

Dillon, W. R., and M. Goldstein, *Multivariate Analysis: Methods and Applications,* New York: John Wiley, 1984.

Draper, N. R., and H. Smith, *Applied Regression Analysis* (2nd ed.), New York: John Wiley, 1981.

Duncan, O. D., *Introduction to Structural Equation Models,* New York: Academic Press, 1975.

Dunham, R. B., "Reaction to Job Characteristics: Moderating Effects of the Organization," *Academy of Management Journal,* **20,** no. 1 (1977), 42–65.

Dunham, R. B., and D. J. Kravetz, "Canonical Correlation Analysis in a Predictive System," *Journal of Experimental Education,* **43,** no. 4 (1975), 35–42.

Dunn, G., Everitt, B., and Pickles, A., *Modelling Covariances and Latent Variables using EQS,* New York: Chapman and Hall, 1994.

Dunn, L. C., "The Effect of Inbreeding on the Bones of the Fowl," *Storrs Agricultural Experimental Station Bulletin,* **52** (1928), 1–112.

Durbin, J., and G. S. Watson, "Testing for Serial Correlation in Least Squares Regression II," *Biometrika,* **38** (1951), 159–178.

Eaton, M., *Multivariate Statistics: A Vector Space Approach,* New York: John Wiley, 1983.

Eaton, M., and M. Periman, "The Non-singularity of Generalized Sample Covariance Matrices," *Annals of Statistics,* **1** (1973), 710–717.

Eisenbeis, R. A., "Pitfalls in the Application of Discriminant Analysis in Business, Finance and Economics," *Journal of Finance,* **32,** no. 3 (1977), 875–900.

Epstein, N. B., L. M. Baldwin, and D. S. Bishop, *Family Assessment Device.* Providence, RI: Brown/Butler Hospital Family Research Program, 1982.

Everitt, B., *Cluster Analysis,* London: Heinemann Educational Books, 1974.

Everitt, B., *Graphical Techniques for Multivariate Data,* New York: North-Holland, 1978.

Fienberg, S. E., "Graphical Methods in Statistics," *The American Statistician,* **33,** no. 4 (1979), 165–178.

Filliben, J. J., "The Probability Plot Correlation Coefficient Test for Normality," *Technometrics,* **17,** no. 1 (1975), 111–117.

Fisher, R. A., "The Use of Multiple Measurements in Taxonomic Problems," *Annals of Eugenics,* **7** (1936), 179–188.

Fisher, R. A., "The Statistical Utilization of Multiple Measurements," *Annals of Eugenics,* **8** (1938), 376–386.

Flury, B., *Common Principal Components and Related Multivariate Models,* New York: John Wiley, 1988.

Flury, B., and Riedwyl, H., *Multivariate Statistics, A Practical Approach,* New York: Chapman and Hall, 1988.

Fong, C., "Malaysian Integrated Population Program Performance: Its Relation to Organizational and Integration Factors," *Management Science,* **31,** no. 1 (1985), 50–65.

Fritts, H. C., T. J. Biasing, B. P. Hayden, and J. E. Kutzbach, "Multivariate Techniques for Specifying Tree-Growth and Climate Relationships and Reconstructing Anomalies in Paleoclimate," *Journal of Applied Meteorology,* **10,** no. 5 (1971), 845–864.

Galton, F., "Regression Toward Mediocrity in Heredity Stature," *Journal of the Anthropological Institute,* **15** (1885), 246–263.

Ganesalingam, S., "Classification and Mixture Approaches to Clustering via Maximum Likelihood," *Applied Statistics,* **38,** no. 3 (1989), 455–466.

Geisser, S., "Discrimination, Allocatory and Separatory, Linear Aspects," in *Classification and Clustering* (J. Van Ryzin, ed.), pp. 301–330. New York: Academic Press, 1977.

Gerrild, P. M., and R. J. Lantz, "Chemical Analysis of 75 Crude Oil Samples from Pliocene Sand Units, Elk Hills Oil Field, California," *U.S. Geological Survey Open-File Report,* 1969.

Girschick, M. A., "On the Sampling Theory of Roots of Determinantal Equations," *Annals of Mathematical Statistics,* **10** (1939), 203–224.

Gnanadesikan, R., *Methods for Statistical Data Analysis of Multivariate Observations,* New York: John Wiley, 1977.

Goldberger, A. S., *Econometric Theory,* New York: John Wiley, 1964.

Goldberger, A. S., "Structural Equation Methods in the Social Sciences," *Econometrica,* **40** (1972), 979–1001.

Graybill, F. A., *Introduction to Matrices with Applications in Statistics,* Belmont, CA: Wadsworth, 1969.

Graybill, F. A., *Theory and Application of the Linear Model,* Belmont, CA: Wadsworth, 1976.

Gunst, R., J. D. F. Habbema, J. Hermans, and K. Van Den Brock, "A Stepwise Discriminant Analysis Program Using Density Estimation," in *Compstat 1974 Proceedings Computational Statistics,* pp. 101–110. Vienna: Physics, 1974.

Haagen, K., D. J. Bartholomew, and M. Deistler, *Statistical Modelling and Latent Variables,* New York: Elsevier Science Publishing, 1993.

Hadi, A. S., *Matrix Algebra as a Tool,* Belmont, CA: Duxbury Press, 1996.

Halinar, J. C., "Principal Component Analysis in Plant Breeding," unpublished report based on data collected by Dr. F. A. Bliss, University of Wisconsin, 1979.

Halmos, P. R., *Finite Dimensional Vector Spaces* (2nd ed.), Princeton, NJ: D. Van Nostrand, 1958.

Hand, D. J., *Discrimination and Classification,* New York: John Wiley, 1981.

Hand, D. J., and C. C. Taylor, *Multivariate Analysis of Variance and Repeated Measures,* New York: Chapman and Hall, 1987.

Harmon, H. H., *Modern Factor Analysis,* Chicago, IL: The University of Chicago Press, 1967.

Hartley, H. O., "Maximum Likelihood Estimation from Incomplete Data," *Biometrics,* **14** (1958), 174–194.

Hartley, H. O., and R. R. Hocking, "The Analysis of Incomplete Data," *Biometrics,* **27** (1971), 783–808.

Hayduk, L. A., *Structural Equation Modeling with LISREL,* Baltimore: The Johns Hopkins University Press, 1987.

Heck, D. L., "Charts of Some Upper Percentage Points of the Distribution of the Largest Characteristic Root," *Annals of Mathematical Statistics,* **31** (1960), 625–642.

Heise, D. R., *Causal Analysis,* New York: John Wiley, 1975.

Hernandez, F., and R. A. Johnson, "The Large-Sample Behavior of Transformations to Normality," *Journal of the American Statistical Association,* **75,** no. 372 (1980), 855–861.

Hills, M., "Allocation Rules and Their Error Rates," *Journal of the Royal Statistical Society (B),* **28** (1966), 1–31.

Hills, M., Book Review, *Applied Statistics,* **26** (1977), 339–340.

Hirschey, M., and D. W. Wichern, "Accounting and Market-Value Measures of Profitability: Consistency, Determinants and Uses," *Journal of Business and Economic Statistics,* **2,** no. 4 (1984), 375–383.

Hogg, R. V., and A. T. Craig, *Introduction to Mathematical Statistics* (4th ed.), New York: Macmillan, 1978.

Hosmer, D. W., and S. Lemeshow, *Applied Logistic Regression,* New York: John Wiley & Sons, 1989.

Hotelling, H., "Analysis of a Complex of Statistical Variables into Principal Components," *Journal of Educational Psychology,* **24** (1933), 417–441, 498–520.

Hotelling, H., "The Most Predictable Criterion," *Journal of Educational Psychology,* **26** (1935), 139–142.

Hotelling, H., "Relations Between Two Sets of Variables," *Biometrika,* **28** (1936), 321–377.

Hotelling, H., "Simplified Calculation of Principal Components," *Psychometrika,* **1** (1936), 27–35.

Hudlet, R., and R. A. Johnson, "Linear Discrimination and Some Further Results on Best Lower Dimensional Representations" in *Classification and Clustering* (J. Van Ryzin, ed.), pp. 371–394. New York: Academic Press, 1977.

Jackson, J. Edward, *A User's Guide to Principal Components,* New York: John Wiley, 1991.

Jambu, M., *Exploratory and Multivariate Data Analysis,* San Diego: Academic Press, 1991.

Jobson, J. D., *Applied Multivariate Data Analysis, Vol. I: Regression and Experimental Design,* New York: Springer-Verlag, 1991.

Jobson, J. D., *Applied Multivariate Data Analysis, Vol. II: Categorical and Multivariate Methods,* New York: Springer-Verlag, 1992.

John, P. W. M., *Statistical Design and Analysis of Experiments,* New York: Macmillan, 1971.

Johnson, N. L., and S. Kotz, *Continuous Multivariate Distributions,* New York: John Wiley, 1972.

Johnson, R. A., and T. Wehrly, "Measures and Models for Angular Correlation and Angular-Linear Correlation," *Journal of the Royal Statistical Society (B),* **39** (1977), 222–229.

Johnson, W., "The Detection of Influential Observations for Allocation, Separation, and the Determination of Probabilities in a Bayesian Framework," *Journal of Business and Economic Statistics,* **5,** no. 3 (1987), 369–381.

Jolicoeur, P., "The Multivariate Generalization of the Allometry Equation," *Biometrics,* **19** (1963), 497–499.

Jolicoeur, P., and J. E. Mosimann, "Size and Shape Variation in the Painted Turtle: A Principal Component Analysis," *Growth,* **24** (1960), 339–354.

Joreskog, K. G., "Factor Analysis by Least Squares and Maximum Likelihood" in *Statistical Methods for Digital Computers* (K. Enstein, A. Ralston, and H. S. Wilf, eds.), New York: John Wiley, 1975.

Joreskog, K. G., and D. Sorbom, *Advances in Factor Analysis and Structural Equation Models,* Cambridge, MA: Abt Books, 1979.

Joreskog, K. G., and D. Sorbom, "LISREL VI Analysis of Linear Structural Relations by Maximum Likelihood, Instrumental Variables, and Least Squares Methods," User's Guide, Department of Statistics, University of Uppsala, Sweden, 1984.

Kaiser, H. F., "The Varimax Criterion for Analytic Rotation in Factor Analysis," *Psychometrika,* **23** (1958), 187–200.

Kaufman, L., and P. J. Rousseeuw, *Finding Groups in Data: An Introduction to Cluster Analysis,* New York: John Wiley, 1990.

Kendall, M. G., *Multivariate Analysis,* New York: Hafner Press, 1975.

Kendall, M. G., and A. Stuart, *The Advanced Theory of Statistics, Vol. 2, Inference and Relationship.* New York: Hafner Press, 1961.

Kim, L., and Y. Kim, "Innovation in a Newly Industrializing Country: A Multiple Discriminant Analysis," *Management Science,* **31,** no. 3 (1985), 312–322.

King, B., "Market and Industry Factors in Stock Price Behavior," *Journal of Business,* **39** (1966), 139–190.

Klatzky, S. R., and R. W. Hodge, "A Canonical Correlation Analysis of Occupational Mobility," *Journal of the American Statistical Association,* **66,** no. 333 (1971), 16–22.

Krishnaiah, P. R., and L. N. Kanal, *Classification, Pattern Recognition and Reduction of Dimensionality,* New York: Elsevier Science Publishing, 1982.

Krzanowski, W. J., "The Performance of Fisher's Linear Discriminant Function Under Non-Optimal Conditions," *Technometrics,* **19,** no. 2 (1977), 191–200.

Krzanowski, W. J., *Principles of Multivariate Analysis: A User's Perspective,* New York: Clarendon Press, 1988.

Kshirsagar, A. M., *Multivariate Analysis,* New York: Marcel Dekker, 1972.

Lachenbruch, P. A., *Discriminant Analysis,* New York: Hafner Press, 1975.

Lachenbruch, P. A., and M. R. Mickey, "Estimation of Error Rates in Discriminant Analysis," *Technometrics,* **10,** no. 1 (1968), 1–11.

Lawley, D. N., "Tests of Significance in Canonical Analysis," *Biometrika,* **46** (1959), 59–66.

Lawley, D. N., "On Testing a Set of Correlation Coefficients for Equality," *Annals of Mathematical Statistics,* **34** (1963), 149–151.

Lawley, D. N., and A. E. Maxwell, *Factor Analysis as a Statistical Method* (2nd ed.), New York: American Elsevier Publishing, 1971.

Linden, M., "A Factor Analytic Study of Olympic Decathlon Data," *Research Quarterly,* **48,** no. 3 (1977), 562–568.

Looney, S. W., and T. R. Gulledge, Jr., "Use of the Correlation Coefficient with Normal Probability Plots," *The American Statistician,* **39,** no. 1 (1985), 75–79.

MacCrimmon, K., and D. Wehrung, "Characteristics of Risk Taking Executives," *Management Science,* **36,** no. 4 (1990), 422–435.

Manly, B. F. J., *Multivariate Statistical Methods, A Primer,* New York: Chapman and Hall, 1986.

Marriott, F. H. C., *The Interpretation of Multiple Observations,* London: Academic Press, 1974.

Mather, P. M., "Study of Factors Influencing Variation in Size Characteristics in Fluvioglacial Sediments," *Mathematical Geology,* **4,** no. 3 (1972), 219–234.

Maxwell, A. E., *Multivariate Analysis in Behavioral Research,* London: Chapman and Hall, 1977.

McLachlan, G. J., *Discriminant Analysis and Statistical Pattern Recognition,* New York: John Wiley, 1992.

Milligan, G. W., "An Examination of the Effect of Six Types of Error Perturbation on Fifteen Clustering Algorithms," *Psychometrika,* **45** (1980), 325–342.

Milliken, G. A., and D. E. Johnson, *Analysis of Messy Data, Vol. 1, Designed Experiments,* London: Chapman and Hall, 1992.

Morrison, D. F., *Multivariate Statistical Methods* (2nd ed.), New York: McGraw-Hill, 1976.

Mucciardi, A. N., and E. E. Gose, "A Comparison of Seven Techniques for Choosing Subsets of Pattern Recognition Properties," *IEEE Transactions on Computers,* **C20** (1971), 1023–1031.

Murray, G. D., "A Cautionary Note on Selection of Variables in Discriminant Analysis," *Applied Statistics,* **26,** no. 3 (1977), 246–250.

Myers, Raymond H., *Classical and Modern Regression with Applications,* Boston: PWS-Kent, 1990.

Noble, B., and J. W. Daniel, *Applied Linear Algebra* (3rd ed.), Englewood Cliffs, NJ: Prentice Hall, 1988.

Number Cruncher Statistical System (NCSS), Version 5.1—Graphics, Dr. Jerry Hintze, Kaysville, UT.

Olson, D. H., J. Portner, and R. Bell, *FACES II: Family Adaptability and Cohesion Scales.* St. Paul, MN: Family Social Sciences, University of Minnesota, 1982.

Parker, R. N., and M. D. Smith, "Deterrence, Poverty, and Type of Homicide," *American Journal of Sociology,* **85** (1979), 614–624.

Pearson, E. S., and H. O. Hartley, eds., *Biometrika Tables for Statisticians,* Vol. 11, England: Cambridge University Press, 1972.

Pillai, K. C. S., "Upper Percentage Points of the Largest Root of a Matrix in Multivariate Analysis," *Biometrika,* **54** (1967), 189–193.

Rao, C. R., *Linear Statistical Inference and Its Applications* (2nd ed.), New York: John Wiley, 1973.

Rathbun, S. W., Control structures and control processes in a sample of church attending caucasian families: A contribution to the family interaction literature. Ph.D Dissertation at Kansas State U., 1995.

Ruben, H., "Some New Results on the Distribution of the Sample Correlation Coefficient," *Journal of the Royal Statistical Society (B),* **28** (1966) 513–525.

Sarle, W. S., *The Cubic Clustering Criterion,* SAS Technical Report A-108, Cary, NC: SAS Institute, Inc., 1983.

Schafer, J. L., *Analysis and Simulation of Incomplete Multivariate Data,* New York: Chapman and Hall, 1994.

Scheffe, H., *The Analysis of Variance,* New York: John Wiley, 1959.

Schumm, W. R., E. E. McCollum, M. A. Bugaighis, A. P. Jurich, and S. Bollman, "Characteristics of the Kansas Family Life Satisfaction Scale in a Regional Sample," *Psychological Reports,* **58** (1986), 975–980.

Schumm, W. R., L. A. Paff-Bergen, R. C. Hatch, F. C. Obiorah, J. M. Copeland, L. D. Meens, and M. A. Bugaighis, "Concurrent and Discriminant Validity of the Kansas Marital Satisfaction Scale," *Journal of Marriage and the Family,* **48** (1986), 381–387.

Scott, D. W., *Multivariate Density Estimation,* New York: John Wiley, 1992.

Seber, G. A., *Multivariate Observations,* New York: John Wiley, 1984.

Shapiro, S. S., and M. B. Wilk, "An Analysis of Variance Test for Normality (Complete Samples)," *Biometrika,* **52,** no. 4 (1965), 591–611.

Simon, H. A., "Spurious Correlations: A Causal Interpretation," *Journal of the American Statistical Association,* **49** (1954), 467–479.

Spearman, C. E., "General Intelligence Objectively Determined and Measured," *American Journal of Psychology* **15,** 201–293.

Spenner, K. I., "From Generation to Generation: The Transmission of Occupation," Ph.D. dissertation, University of Wisconsin, 1977.

Srivastava, M. S., and E. M. Carter, *An Introduction to Applied Multivariate Statistics,* New York: Elsevier Science Publishing, 1983.

Srivastava, M. S., and Khatri, C. G., *An Introduction to Multivariate Statistics,* New York: Elsevier Science, 1979.

Stoetzel, J., "A Factor Analysis of Liquor Preference," *Journal of Advertising Research,* **1** (1960), 7–11.

Tabakoff, B., *et al.,* "Differences in Platelet Enzyme Activity Between Alcoholics and Nonalcoholics," *New England Journal of Medicine,* **318,** no. 3 (1988), 134–139.

Tiku, M. L., and N. Balakrishnan, "Testing the Equality of Variance–Covariance Matrices the Robust Way," *Communications in Statistics—Theory, and Methods,* **14,** no. 12 (1985), 3033–3051.

Tiku, M. L., and M. Singh, "Robust Statistics for Testing Mean Vectors of Multivariate Distributions," *Communications in Statistics—Theory and Methods,* **11,** no. 9 (1982), 985–1001.

Timm, N. H., *Multivariate Analysis with Applications in Education and Psychology,* Monterey, CA: Brooks/Cole, 1975.

Trieschmann, J. S., and G. E. Pinches, "A Multivariate Model for Predicting Financially Distressed P-L Insurers," *Journal of Risk and Insurance,* **40,** no. 3 (1973), 327–338.

Tukey, J. W., *Exploratory Data Analysis,* Reading, MA: Addison-Wesley, 1977.

Verrill, S., and R. A. Johnson, "Tables and Large-Sample Distribution Theory for Censored-Data Correlation Statistics for Testing Normality," *Journal of the American Statistical Association,* **83,** no. 404 (1988), 1192–1197.

Wald, A., "On a Statistical Problem Arising in the Classification of an Individual into One of Two Groups," *Annals of Mathematical Statistics,* **15** (1944), 145–162.

Welch, B. L., "Note on Discriminant Functions," *Biometrika,* **31** (1939), 218–220.

Wilks, S. S., "Certain Generalizations in the Analysis of Variance," *Biometrika,* **24** (1932), 471–494.

Wright, S., "Statistical Methods in Biology," *Journal of the American Statistical Association Supplement; Papers and Proceedings of the 92nd Annual Meeting,* **26** (1931), 155–163.

Wright, S., "The Method of Path Coefficients," *Annals of Mathematical Statistics,* **5** (1934), 161–215.

Wright, S., "The Interpretation of Multivariate Systems" in *Statistics and Mathematics in Biology* (O. Kempthorne *et al.,* eds.), pp. 11–33. Ames, IA: Iowa State University Press, 1954.

Wright, S., *Evolution and the Genetics of Population,* Vol. 1, Chicago, IL: University of Chicago Press, 1968.

Young, F. W., and Hamer, R. M., *Multidimensional Scaling: History, Theory, and Applications,* Hillsdale, NJ: Lawrence Erlbaum Associates, 1987.

Zehna, P., "Invariance of Maximum Likelihood Estimators," *Annals of Mathematical Statistics,* **37,** no. 3 (1966), 744.

Index